WITHDRAWN
UTSA LIBRARIES

D1608876

OFDMA Mobile Broadband Communications

A Systems Approach

Written by the pioneers of Flash-OFDM, arguably the first commercially developed OFDMA-based mobile broadband system in the world, this book teaches OFDMA from first principles, enabling readers to understand mobile broadband as a whole.

The book examines the key requirements for data-centric mobile; how OFDMA fits well with data networks; why mobile broadband needs to be IP-based; and how to bridge communications theory to real-world air interface design and make a good system choice between performance and complexity. It also explores the future of wireless technologies beyond conventional cellular architecture.

One of the key challenges faced by newcomers to this field is how to apply the wireless communications theory and principles to the real world and how to understand sophisticated commercial systems such as LTE. The authors use their first-hand experience to help graduate students, researchers, and professionals working on 4G to bridge the gap between theory and practice.

Junyi Li is Vice President of Engineering at Qualcomm, responsible for conceptualizing and developing next-generation wireless networking solutions. He was a key inventor of Flash-OFDM and a founding member of Flarion Technologies. He is a Fellow of the IEEE.

Xinzhou Wu is Principal Engineer/Manager at Qualcomm. He is well known for his research and innovation in the area of wireless communications and networking.

Rajiv Laroia is Senior Vice President of Engineering and CTO at Sonus Networks. He was the founder and CTO of Flarion Technologies and then Senior Vice President of Engineering at Qualcomm. He is widely recognized as a pioneer of OFDMA-based cellular technologies. He is a Fellow of the IEEE.

"Li, Wu and Laroia's book fulfils a major need – an authoritative reference on OFDMA, which is the foundation for 4G cellular technology, and likely the dominant air interface technology for many years to come. The authors are renowned innovators and system engineers who among them pioneered the use of OFDM for cellular systems, and the book contains many crisp insights that no other team of authors could provide. The academic rigor of the book is also remarkable, particularly from practicing system engineers. In fact, they have independently derived several cutting edge research results in order to make various technical points! A must-have book for any wireless system engineer's personal library."

Jeffrey Andrews, The University of Texas at Austin

"This is a must read book for both students and engineers who are interested in learning about how the principles of OFDM can be used to design and control wireless networks. This self-contained book begins with a gentle introduction and basics of OFDMA systems, builds the material to advanced state-of-the-art techniques currently being used, and then looks ahead at the design of future wireless systems. A remarkable book written by the pioneers of Flash OFDM technologies."

Ness B. Shroff, The Ohio State University

"This is a very special book, written by industry pioneers of OFDMA technology, taking the unconventional and very timely system view. It provides a beautiful perspective of how important theoretical ideas and understanding of the needs of real world communications systems were harnessed to develop flash-OFDM from first principles. LTE has since adopted OFDMA technology as the converged global 4G standard. It presents in a unique way the principles, basics and advanced elements of wireless OFDMA technology, and also provides a future centric perspective of practical aspects of wireless communications. This three part book offers insights for scholars, researchers and wireless industrial engineers, leaders and visionaries. In particular, this book is beneficial for students as they are exposed to more than standard theoretical perspectives, unlike of most books and courses in the area. This book is a pleasure to read."

Shlomo Shamai, Technion – Israel Institute of Technology

"This book provides a thoroughly-researched holistic viewpoint of wireless communications using OFDMA technology. The authors' perspective on the relationship between theory and practice illustrated through the use of practical examples makes the book unique. I expect the book to be a valuable resource for researchers and practitioners in the area of wireless networks."

R. Srikant, University of Illinois at Urbana-Champaign

"System design involves taking a holistic view of all the aspects and angles involved – this is a particularly tough challenge in cellular wireless system which involves widely disparate aspects such as unreliability of physical wireless medium, arbitrating the shared wireless medium among many users, infrastructure management (base stations and sectorization) and interface with external networks such as the internet). Flash-OFDM personifies the philosophy of system design and this book, coming from pioneers of the technology is a must read for anyone interested in a system view of the entire cellular system as well as anyone interested in understanding the process of system design."

Pramod Viswanath, University of Illinois at Urbana-Champaign

"Most texts in communications focus either on theory or on detailed description of standards. This unique book takes cutting-edge theory and shows how it can be applied to real-world systems. An invaluable guide on the bridge from theory to practice, written by some of the best system engineers in the field."

David Tse, University of California at Berkeley

OFDMA Mobile Broadband Communications

A Systems Approach

JUNYI LI
Qualcomm

XINZHOU WU
Qualcomm

RAJIV LAROIA
Sonus Networks

CAMBRIDGE UNIVERSITY PRESS
Cambridge, New York, Melbourne, Madrid, Cape Town,
Singapore, São Paulo, Delhi, Mexico City

Cambridge University Press
The Edinburgh Building, Cambridge CB2 8RU, UK

Published in the United States of America by Cambridge University Press, New York

www.cambridge.org
Information on this title: www.cambridge.org/9781107001602

© Cambridge University Press 2013

This publication is in copyright. Subject to statutory exception
and to the provisions of relevant collective licensing agreements,
no reproduction of any part may take place without the written
permission of Cambridge University Press.

First published 2013

Printed and bound in the United Kingdom by the MPG Books Group

A catalogue record for this publication is available from the British Library

Library of Congress Cataloging in Publication data
Li, Junyi, 1969–
OFDMA mobile broadband communications : a systems approach / Junyi Li, Qualcomm,
Bridgewater, New Jersey, Xinzhou Wu, Qualcomm, Bridgewater, New Jersey, Rajiv Laroia,
Sonus Networks.
   pages   cm
Includes bibliographical references and index.
ISBN 978-1-107-00160-2 (hardback)
1. Orthogonal frequency division multiplexing.   I. Wu, Xinzhou.   II. Laroia, Rajiv.   III. Title.
TK5103.484.L578   2013
621.39′8 – dc23   2012042953

ISBN 978-1-107-00160-2 Hardback

Cambridge University Press has no responsibility for the persistence or
accuracy of URLs for external or third-party internet websites referred to
in this publication, and does not guarantee that any content on such
websites is, or will remain, accurate or appropriate.

To my parents, Songnian Li and Xiangzhen Huang
Junyi Li

To my parents, Yaping Wu and Liying Ji
Xinzhou Wu

To my parents, Krishan and Anuradha Laroia
Rajiv Laroia

Contents

	Foreword		*page* xiii
	Preface		xiv
	List of Notation		xvii
	List of Abbreviations		xix
1	**Introduction**		1
	1.1	Evolution towards mobile broadband communications	1
	1.2	System design principles of wireless communications	3
	1.3	Why OFDMA for mobile broadband?	4
	1.4	Systems approach and outline of the book	6
2	**Elements of OFDMA**		9
	2.1	OFDM	9
		2.1.1 Tone signals	9
		2.1.2 Cyclic prefix	10
		2.1.3 Time-frequency resource	13
		2.1.4 Block signal processing	14
		Discussion notes 2.1 FFT/IFFT	15
		Discussion notes 2.2 Filtering	16
		Discussion notes 2.3 Equalization	17
	2.2	From OFDM to OFDMA	18
		2.2.1 Basic principles	18
		2.2.2 Comparison: OFDMA, CDMA, and FDMA	21
		2.2.3 Inter-cell interference averaging: OFDMA versus CDMA	21
		2.2.4 Tone hopping: averaging versus peaking	24
		Practical example 2.1 Physical resource block allocation and hopping in LTE data channels	26
		2.2.5 Time-frequency synchronization and control	30
		2.2.6 Block signal processing	33
		Discussion notes 2.4 Block front-end processing at the base station	34
		Discussion notes 2.5 Wideband processing at the user	34
	2.3	Peak-to-average power ratio and SC-FDMA	34
		2.3.1 PAPR problem	34

		2.3.2	PAPR of OFDMA	35
		2.3.3	SC-FDMA and PAPR reduction	35
		2.3.4	Frequency domain equalization at the SC-FDMA receiver	40
		Discussion notes 2.6 SINR degradation in SC-FDMA		42
		2.3.5	System aspects of SC-FDMA	45
		Practical example 2.2 Uplink data and control channels in LTE		46
	2.4	Real-world impairments		52
		2.4.1	Carrier frequency offset and Doppler effect	52
		2.4.2	Arrival time beyond the cyclic prefix	55
		2.4.3	Sampling rate mismatch	56
		2.4.4	I/Q imbalance	60
		2.4.5	Power amplifier nonlinear distortion	61
		Discussion notes 2.7 Determination of OFDMA parameters		61
	2.5	Cross interference and self-noise models		63
		2.5.1	Cross interference and self-noise due to ICI	63
	2.6	Self-noise due to imperfect channel estimation		64
		2.6.1	Self-noise measurement via null pilot	67
	2.7	Summary of key ideas		68

3 System design principles — 70

	3.1	System benefits of OFDMA		70
	3.2	Fading channel mitigation and exploitation		74
		3.2.1	Fading mitigation	75
		3.2.2	Fading exploitation	75
		3.2.3	Mitigation or exploitation?	77
	3.3	Intra-cell user multiplexing		77
	3.4	Inter-cell interference management		80
		3.4.1	Interference averaging and active control	81
		3.4.2	Universal versus fractional frequency reuse	82
	3.5	Multiple antenna techniques		84
		3.5.1	System benefits	84
		3.5.2	OFDMA advantages	86
	3.6	Scheduling		87
	3.7	Network architecture and airlink support		89
		3.7.1	Unplanned deployment of base stations	90
		3.7.2	Mobile IP-based handoff	91
	3.8	Summary of key ideas: evolution of system design principles		92

4 Mitigation and exploitation of multipath fading — 94

	4.1	Multipath fading channel		97
		4.1.1	Impulse response model	97
		4.1.2	Amplitude statistics	99

	4.1.3	Channel variation in time	100
	4.1.4	Channel variation in frequency	103
	4.1.5	Gaussian-Markov model	105
4.2	Communications over a fading channel: the single-user case		106
	4.2.1	Performance penalty due to multipath fading	106
	4.2.2	Mitigation of fading via channel state feedback	108
	Discussion notes 4.1 Practical consideration of feedback-based approaches		112
	4.2.3	Mitigation of fading via diversity	115
	Discussion notes 4.2 Tradeoff considerations for achieving diversity		122
	4.2.4	Feedback or diversity	123
4.3	Communications over a fading channel: the multiuser case		126
	4.3.1	Fading channel and multiuser diversity	126
	Practical example 4.1 Multiuser diversity in the downlink: EV-DO		130
	Practical example 4.2 Multiuser diversity in the uplink: Flash-OFDM and LTE		133
	4.3.2	Exploring multiuser diversity in frequency and space	135
	4.3.3	Multiuser or single-user diversity	144
4.4	Summary of key ideas		148

5 Intra-cell user multiplexing   150

5.1	Orthogonal multiplexing		151
	5.1.1	Orthogonal multiplexing in the perfect model	151
	Discussion notes 5.1 An analysis of optimal power and bandwidth allocation in a cellular downlink		157
	Practical example 5.1 Downlink user multiplexing: EV-DO, HSDPA, and LTE		160
	5.1.2	Orthogonal multiplexing in the cross interference model	167
	Discussion notes 5.2 An analysis of optimal power and bandwidth allocation for orthogonal uplink multiplexing with cross interference in the power limited regime		169
	5.1.3	Orthogonal multiplexing in the self-noise model	172
5.2	Non-orthogonal multiplexing		174
	5.2.1	Non-orthogonal multiplexing in the perfect model	176
	5.2.2	Non-orthogonal multiplexing in the cross interference and self-noise models	180
	5.2.3	Superposition-by-position coding	183
5.3	Inter-sector interference management		189
	5.3.1	Sectorization	189
	5.3.2	Synchronized sectors	190
	5.3.3	Users at sector edge	192
5.4	Summary of key ideas		195

6 Inter-cell interference management — 196

- 6.1 Analysis of SIR distributions — 198
  - 6.1.1 Downlink SIR — 199
  - Discussion notes 6.1 An analysis of C/I distribution with randomly-placed base stations — 202
  - 6.1.2 Uplink SIR — 205
- 6.2 Uplink power control and SINR assignment in OFDMA — 209
  - 6.2.1 SINR feasibility region — 210
  - 6.2.2 Distributed power control — 211
  - 6.2.3 SINR assignment — 212
  - 6.2.4 Joint bandwidth and SINR assignment — 215
  - 6.2.5 Utility maximization in SINR assignment — 216
  - Practical example 6.1 Uplink power control in LTE — 217
- 6.3 Fractional frequency reuse — 219
  - 6.3.1 A two-cell analysis — 220
  - Discussion notes 6.2 Motivation of fractional frequency reuse from a different angle — 225
  - 6.3.2 Static FFR in a multi-cell scenario — 226
  - 6.3.3 Breathing cells: FFR in the time domain — 230
  - 6.3.4 Adaptive FFR — 233
  - Practical example 6.2 Inter-cell interference coordination in LTE — 236
- 6.4 Summary of key ideas — 237

7 Use of multiple antennas — 239

- 7.1 MIMO channel modeling — 240
  - 7.1.1 Linear antenna arrays — 241
  - 7.1.2 Polarized antennas — 247
- 7.2 SU-MIMO techniques — 251
  - 7.2.1 Channel state information at both transmitter and receiver — 251
  - 7.2.2 Channel state information only at receiver — 252
  - 7.2.3 Multiplexing with polarized antennas — 254
- 7.3 Multiuser MIMO techniques — 254
  - 7.3.1 Uplink SDMA — 256
  - 7.3.2 Downlink beamforming — 261
- 7.4 Multi-cell MIMO techniques — 267
  - 7.4.1 Coordinated beamforming — 268
  - 7.4.2 Inter-sector beamforming — 271
  - 7.4.3 Inter-cell interference avoidance with polarized antennas — 273
  - Practical example 7.1 Multiple antenna techniques in LTE — 273
- 7.5 Summary of key ideas — 280

8 Scheduling — 282

- 8.1 Scheduling for infinitely backlogged traffic — 283

		8.1.1	Fairness based on utility functions	283
		8.1.2	Gradient-based scheduling schemes	286
	8.2	Scheduling for elastic traffic		289
		8.2.1	Congestion control and scheduling	290
		Discussion notes 8.1 TCP performance over wireless		292
	8.3	Scheduling for inelastic traffic		293
		8.3.1	Throughput optimal scheduling	294
		8.3.2	Tradeoff between queue-awareness and channel-awareness	296
		8.3.3	Admission control	299
	8.4	Multi-class scheduling		300
	8.5	Flow level scheduling		301
	8.6	Signaling for scheduling		304
		8.6.1	Dynamic packet scheduling	304
		Practical example 8.1 Signaling for scheduling in LTE		307
		8.6.2	Semi-persistent scheduling	310
		Practical example 8.2 Semi-persistent scheduling in LTE for VoIP		311
		8.6.3	MAC state scheduling	311
		Practical example 8.3 LTE DRX mode and Flash-OFDM HOLD state		312
	8.7	Summary of key ideas		313
9	**Handoff in IP-based network architecture**			315
	9.1	IP-based cellular network architecture		317
		9.1.1	Motivation for IP-based cellular network architecture	317
		9.1.2	Description of IP-based cellular networks	317
	9.2	Soft handoff in CDMA		319
	9.3	Make-before-break handoff in OFDMA		323
		9.3.1	Parallel independent links to multiple base stations	324
		9.3.2	Mobile IP-based MBB handoff procedure	327
		9.3.3	Uplink macro-diversity	328
		9.3.4	Downlink macro-diversity	333
		9.3.5	MBB handoff in an FFR or multi-carrier scenario	335
	9.4	Break-before-make handoff in OFDMA		337
		9.4.1	BBM handoff in an FFR or multi-carrier scenario	338
		9.4.2	Expedited BBM handoff	339
	9.5	Handoff initiation		342
		9.5.1	The universal frequency reuse case	342
		Practical example 9.1 Flash signaling in Flash-OFDM		351
		Practical example 9.2 Handoff in a railway Flash-OFDM network		353
		9.5.2	The non-universal frequency reuse cases	354
	9.6	Mobile-controlled versus network-controlled handoff		356
		Practical example 9.3 Cell search and random access in LTE handoff		357
	9.7	Summary of key ideas		363

10 Beyond conventional cellular frameworks — 365

- 10.1 Heterogeneous topology — 366
  - 10.1.1 Relays — 367
  - 10.1.2 Femtocells — 383
  - 10.1.3 Device-to-device communications — 398
  - Discussion notes 10.1 Gaussian interference channel capacity — 412
- 10.2 Cooperative communication — 415
  - 10.2.1 User cooperation — 417
  - 10.2.2 Network cooperation — 425
- 10.3 Cognitive radio — 431
  - 10.3.1 Spectrum sensing — 433
  - 10.3.2 Spectrum sharing — 438
  - Practical example 10.1 LTE-Advanced — 444
  - Practical example 10.2 Cognitive radio RAN in TV white spaces (IEEE 802.22) — 456
- 10.4 Summary of key ideas — 458

A Overview of system operations — 461

- A.1 Cell search, synchronization, and identification — 461
- A.2 Link establishment — 462
- A.3 Traffic control and transmission — 463
- A.4 Sleep state — 465
- A.5 Handoff — 465

B OFDM point-to-point communications — 467

- B.1 Signal-presence detection — 467
- B.2 Synchronization — 471
- B.3 Channel estimation — 477
- B.4 Error correction — 487

C Brief review of channel capacity — 495

- C.1 AWGN channel — 495
- C.2 Flat fading channel — 496
  - C.2.1 Channel side information only at receiver — 496
  - C.2.2 Channel side information at both receiver and transmitter — 497
- C.3 Frequency selective fading channel — 499
- C.4 Multiuser capacity — 499

References — 503
Index — 514

Foreword

It is not so long ago that links in a wired network were not able to cooperate freely but were subject to central control by ISDN and ATM protocols. We now live in a world where Internet protocols have made it possible for networks to grow like weeds. We take it for granted that links should regulate their own use by generating prices that reflect congestion and that users adjust rates in response to the cost of traversing the network.

The IP revolution that has transformed the wireline world is coming to wireless. Migration from cellphones to smartphones has created demand for capacity that simply cannot be met by circuit switched networks engineered to provide worst case coverage at the cell boundary. This is a monograph written by revolutionaries that maps the new world of what is possible when wireless resources are properly shared.

The monograph is remarkable for starting with services, with medium access, and then asking how to engineer the physical layer that the higher layers want to see. The authors answered this question themselves by making a journey from concept to working system and then staging field trials. This monograph is the result of a virtuous cycle where engineering challenges led to theoretical insights and new theory was proved out in working systems.

The authors know all about the strengths and weaknesses of different wireless systems like CDMA and LTE and are able to motivate the design choices that informed OFDMA. There are some beautiful new ideas, from fractional reuse, to dumb antennas, even to a new self-noise model for the wireless channel. There is also a final chapter that looks out beyond the Flarion OFDMA system to a world in which relays and femtocells work in wireless harmony.

It is an extraordinary book by an extraordinary team, and I recommend it highly.

Robert Calderbank

Dean of Natural Sciences and Professor of Computer Science, Electrical Engineering, and Mathematics at Duke University.

Professor of Electrical Engineering, Mathematics and Applied and Computational Mathematics at Princeton University.

Preface

Why we wrote this book

Back in the late 1990s, when CDMA was widely considered the dominant technology for cellular 3G, two of the authors and a few colleagues in Bell Laboratories designed an alternative technology called *Flash-OFDM* with two basic yet fundamental ideas: *OFDMA-based airlink* and *all IP-based network architecture*. In early 2000, we founded a startup company, Flarion Technologies, to prove Flash-OFDM in the market by building terminals and base stations, and testing and deploying the networks in a wide variety of locations, configurations, and frequency bands. As arguably the first commercially deployed OFDMA/IP-based cellular system, Flash-OFDM helped make those two ideas the key enabling features in 4G mobile broadband LTE.

From the remarkable journey of designing, developing, and deploying Flash-OFDM, we have learned, and in some cases "unlearned," a few important lessons:

- While early cellular wireless communications design focuses predominantly on the physical layer, *mobile broadband requires more system-level thinking across different protocol layers than just the physical layer*. For example, OFDMA, in comparison with CDMA, more readily facilitates a simplified IP-based network architecture design, where air interface specific technology functions and processing are collapsed into a base station and IP layer protocols are used for handoff.
- *Conventional wisdom developed in early cellular wireless communications needs to be reexamined from first principles*. For example, the wireless channel is conventionally modeled with additive noise and multiplicative channel response; we found that self-noise should also be included when multiplexing users with large signal dynamic range in OFDMA. As another example, universal frequency reuse is conventionally considered the most spectrally efficient; we found that for data, fractional frequency reuse improves both cell edge and cell average performance.

The writing of this book was motivated by a desire to share our firsthand experience. The book is somewhat unorthodox in the sense that it does not follow a pedagogic treatment of wireless communications theory or describe the details of a specific standard (e.g., LTE). Rather, we take a *systems approach* to explain the design principles of OFDMA mobile broadband communications. We believe that such an approach will benefit the readers to appreciate design rationales, maximize the performance of present 4G systems, and develop new ideas for future evolution.

Who will benefit from this book?

- Graduate or senior undergraduate students in electrical engineering. It can be used as a text or reference book in an advanced course on wireless communications and systems.
- Professors and researchers interested in advanced research topics, who may wish to broaden their research scope or to gain new insights by understanding the system-level picture.
- Systems engineers of wireless communications equipment and semiconductor vendors, who need to thoroughly understand the system issues and to look for new ideas to improve the performance of their products.
- R&D staff of wireless operators, who need to appreciate design tradeoffs and to optimize the performance of deployed systems.

The book consists of three parts. The first part (Chapters 2 and 3) describes the basics and salient features of OFDMA, and outlines qualitatively the high level system design principles of OFDMA mobile broadband communications. The second part (Chapters 4 to 9) addresses various system issues: wireless channels, power/bandwidth allocation, interference management, spatial signal processing, scheduling, handoff, and interaction between the airlink and the network architecture. The third part (Chapter 10) expands beyond the conventional cellular framework and covers the latest research topics of femtocells, relays, device-to-device communications, cooperative communications, and cognitive radio. To bridge theory and practice, a number of "discussion notes" and "practical examples" are included throughout the book in which we share real world system experience. Most of the practical examples are from the OFDMA-based systems (LTE and Flash-OFDM), although we also consider the CDMA-based systems (e.g., IS-95, EV-DO, and HSPA) for comparison. [32, 58, 138] provide excellent description of the LTE standards and related techniques.

The prerequisites of the book are a solid understanding of signal processing and digital communications, and familiarity with CDMA and FDMA/TDMA cellular technologies; [61, 159] are excellent textbooks that cover the required knowledge. To help the readers to appreciate the context of system-level issues, Appendix A describes typical operations in a mobile broadband airlink system. Appendix B reviews a few basic point-to-point communications techniques as the building blocks of the system operations. Appendix C summarizes a small set of capacity results used in the book; no additional information theory background is required.

Acknowledgments

Our exploration of OFDMA was originally inspired by the late Aaron Wyner's visionary work in the early 1990s at Bell Labs. We own an tremendous debt to our colleagues at Flarion Technologies (which became Qualcomm Flarion Technologies after the acquisition in 2006), for developing, implementing, testing, and relentlessly improving

Flash-OFDM. Collectively they proved to the world that the OFDMA-based airlink and the IP-based network architecture are the right choice for mobile broadband communications. Among the major contributors are Pablo Anigstein, Scott Corson, Arnab Das, Mike DiMare, Hui Jin, Samir Kapoor, Frank Lane, Vladimir Parizhsky, Vince Park, Sundeep Rangan, Tom Richardson, Murari Srinivasan, Chuck Stanski, George Tsirtsis, Sathya Uppala, and Michaela Vanderveen. We would also like to thank the members of Technical Advisory Board of Flarion, including Robert Calderbank, Sharad Malik, Jan Rabaey, Shlomo Shamai, David Tse, Sergio Verdu, Andrew Viterbi, and Jacob Ziv.

For the realization of this book, we are particularly indebted to Pramod Viswanath, who advised us on the structure of the book. Special thanks go to Robert Calderbank for writing the Foreword. We appreciate Jeffrey Andrews, Yingbin Liang, Vince Park, Sanjay Shakkottai, Ness Shroff, R. Srikant, David Tse, George Tsirtsis, and Lei Ying for reviewing the final text and providing constructive comments. We thank Fredric Ridder for careful copy editing. Finally, at Cambridge University Press, Phil Meyler has provided constant encouragement to our book project from the very beginning, and Sarah Matthews and Elizabeth Horne have given us lots of detailed support to bring the project to its completion.

Notation

Variables

C	Capacity
c_i	Index of the serving base station index of user i
$f[k]$	Frequency of tone k
G	Channel power gain
H	Channel complex gain (amplitude and phase)
$H[k]$	Frequency domain baseband channel gain at tone k
$H(f)$	Frequency domain baseband channel gain at frequency f
$H[s,k]$	Frequency domain baseband channel gain at tone k at OFDM symbol s
$H(t,f)$	Frequency domain baseband channel gain at frequency f at time t
$h(t,\tau)$	Time domain impulse channel response at time t τ to an input at time $t-\tau$
j	$\sqrt{-1}$
K	Frequency reuse factor
K_r	Number of receive antennas
K_S	Number of subbands
K_t	Number of transmit antennas
K_u	Number of users multiplexed in multiuser MIMO in a cell
l_i	Load of user i
L_b	Aggregate load of base station b
L_D	Diversity order
M	Number of users in a cell
M_R	Number of relays in a cell
N_c	Number of sub-carriers
P	Transmit power
P_m	Maximum transmit power
R	Data rate
Q_m	Inter-cell interference budget
q_i	Total inter-cell interference plus noise power at base station serving user i
$S_{i,j}$	Channel gain ratio of user j between base stations c_j and c_i
s_i	Spillage of user i

List of Notation

T_{cp}	OFDM symbol cyclic prefix (CP) duration
$T_i[t]$	Empirical throughput of user i at time t
T_p	Sensing period in cognitive spectrum sensing
T_q	Quiet period in cognitive spectrum sensing
T_s	OFDM symbol duration excluding CP
U	Utility function in scheduling
u_i	Receive power of user i in the uplink
$W[k], w[k]$	Frequency domain and time domain noise
$\mathcal{W}$	Frequency bandwidth
$X[k]$	Transmitted complex symbol at tone k
$x[l]$	Transmitted complex symbol at discrete time l
$x(t)$	Transmitted complex signal at time t
$Y[k]$	Received complex symbol at tone k
$y[l]$	Received complex symbol at discrete time l
$y(t)$	Received complex signal at time t
z	Relative channel gain variation due to fading, equal to the ratio of instantaneous and average channel gains
α_i	Fraction of bandwidth assigned to user i
β_i	Fraction of power or inter-cell interference budget assigned to user i
γ	Operating SNR or SINR with given power and bandwidth allocation
$\tilde{\gamma}$	Nominal SNR or SINR when the entire power and bandwidth are assigned to the user

Functions

$\mathbf{1}_{\{x\}}$	Indicator function, equal to 1 if x is true or equal to 0 otherwise
$\mathcal{CN}(0, \sigma^2)$	Circularly symmetric complex Gaussian random variable with zero mean and variance σ^2
$\delta(k)$	Delta function, equal to 1 if $k = 0$, or equal to 0 if $k \neq 0$
$\mathbf{H}^*$	Complex conjugate-transpose of $\mathbf{H}$
$\mathbf{H}^T$	Transpose of $\mathbf{H}$
$\mathbb{E}(x)$	Expectation of a random variable x
$\mathbb{F}_X(x)$	Cumulative probability density function (CDF) of a random variable X, evaluated at x
$\log(x)$	Natural logarithm
$\Re(x)$	The real component of x
$\text{sinc}(x)$	$\frac{\sin(\pi x)}{\pi x}$
$\mathcal{N}(0, \sigma^2)$	Real Gaussian random variable with zero mean and variance σ^2
$\mathbb{P}(X)$	Probability of an event X
$x \mod N$	x modular N
$(x)^+$	Equal to x if $x > 0$, or equal to 0 otherwise

Abbreviations

General

ARQ	Automatic Retransmission Request
BBM	Break-before-Make
CDM	Code Division Multiplexing
CDMA	Code Division Multiple Access
C/I	Carrier to Interference Ratio
D2D	Device-to-Device
DFT/IDFT	Discrete Fourier Transform/Inverse Discrete Fourier Transform
EV-DO	Evolution-Data Optimized or Evolution-Data Only
FDM	Frequency Division Multiplexing
FDMA	Frequency Division Multiple Access
FFR	Fractional Frequency Reuse
FFT/IFFT	Fast Fourier Transform/Inverse Fast Fourier Transform
H-ARQ	Hybrid ARQ
HSDPA	High-Speed Downlink Packet Access
HSUPA	High-Speed Uplink Packet Access
ICI	Inter-Carrier Interference
INR	Interference-to-Noise Ratio
IP	Internet Protocol
ISI	Inter-Symbol Interference
LOS	Line-of-Sight
LTE	Long Term Evolution
MBB	Make-before-Break
MIMO	Multiple Input Multiple Output
MMSE	Minimum Mean-Square Error
MU-MIMO	Multiuser MIMO
OFDM	Orthogonal Frequency Division Multiplexing
OFDMA	Orthogonal Frequency Division Multiple Access
PAPR	Peak-to-Average Power Ratio
QoS	Quality of Service
RF	Radio Frequency
SC-FDMA	Single Carrier FDMA
SDMA	Spatial Division Multiple Access

SINR	Signal-to-Interference plus Noise Ratio
SNR	Signal-to-Noise Ratio
SU-MIMO	Single User MIMO
SVD	Singular Value Decomposition
TCP	Transmission Control Protocol
TDM	Time Division Multiplexing
TDMA	Time Division Multiple Access
UMTS	Universal Mobile Telephone Service
WCDMA	Wideband CDMA
W-WAN	Wireless Wide Area Network
XPD	Cross polarization discrimination

LTE-specific

CCE	Control Channel Elements
CQI	Channel Quality Indicator
CS-RS	Cell Specific Reference Signal
CSI-RS	Channel-State-Information Reference Signal
DM-RS	Demodulation Reference Signal
eNB	eNodeB, base station
H-ARQ-ACK	Hybrid ARQ Acknowledgment
HII	High Interference Indicator
OI	Overload Indicator
PBCH	Physical Broadcast Channel
PCFICH	Physical Control Format Indicator Channel
PDCCH	Physical Downlink Control Channel
PDSCH	Physical Downlink Shared Channel
PMI	Precoding Matrix Indicator
PHICH	Physical Hybrid-ARQ Indicator Channel
PUCCH	Physical Uplink Control Channel
PUSCH	Physical Uplink Shared Channel
PRACH	Physical Random Access Channel
PRB	Physical Resource Block
PSS	Primary Synchronization Signal
RBG	Resource Block Group
RE	Resource Element
RI	Rank Information
RNTP	Relative Narrowband Transmit Power
SR	Scheduling Request
SRS	Sounding Reference Signal
SSS	Secondary Synchronization Signal
UE	User Equipment
VRB	Virtual Resource Block

1 Introduction

1.1 Evolution towards mobile broadband communications

Explosive growth of wireless communications services and products in the past three decades or so has fundamentally changed the way by which the majority of the world's population exchange, distribute and access information. As a strong driver of the growth, cellular telephony has so far been the most successful application of wireless communications. All forms of wireless communications utilize the radio spectrum, a scarce natural resource. Spectrum access is a general term of the technologies by which users utilize the radio spectrum. Cellular telephony uses a *cellular* concept, which provides an effective spectrum access solution to improve the efficiency of radio spectrum utilization.

In a cellular system, many base stations are deployed to cover a large service area. The service area is divided into a number of *cells*, each served by a base station, as shown in Figure 1.1. When a user makes a call, it is connected to the base station with the best RF propagation.[1] The base stations are connected to the operator's core networks via backhaul connections such as T1 or fiber optics. Spectrum is reused among the cells. This is possible because a signal decays fast as it travels through the wireless channel. If a signal utilizing some spectrum in a cell is sufficiently attenuated in another cell, then the same spectrum can be reused.

Two technical issues arise when the cellular concept is applied in practice. The first issue is *interference management*. To maximize the spectrum utilization, spectrum should be reused as tightly as possible, thereby resulting in interference between signals sharing the same spectrum to be managed. Figure 1.1 shows that interference comes from inter-cell or intra-cell. The second issue is *handoff*. When a mobile user moves from one cell to another, it switches to a different base station. Handoff should be seamless to avoid dropping calls.

Cellular systems have evolved through several generations. Figure 1.2 lists major commercially deployed cellular systems and their main services and spectrum access technologies, where the time line roughly represents the time periods during which those systems were introduced to the marketplace. It seems that a major technology shift occurs every ten years or so.

To ensure interoperability among operators and equipment vendors and to build ecosystems to foster technology adoption, those systems are based on technologies

---

[1] Other criteria, such as network loading, can also be used for base station selection.

Introduction

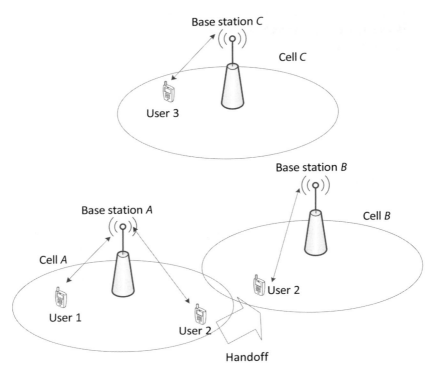

Figure 1.1 Illustration of the cell concept. Base station A serves users 1 and 2 in cell A, and base station C serves user 3 in cell C. Cells A and C reuse the same spectrum and therefore interfere with each other (inter-cell interference). In cell A, users 1 and 2 may interfere with each other (intra-cell interference) if they share the same spectrum. Handoff occurs when user 2 moves from cell A and B.

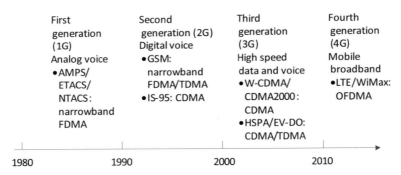

Figure 1.2 Evolution of cellular technologies.

developed in national or international standards organizations. The latest trend is that unlike the previous generations where multiple technologies compete in the marketplace, 4G will likely be dominated by a single standard called LTE (Long-Term Evolution). LTE was developed by 3GPP, a partnership project in which operating partners (regional standards developing organizations) have de facto "surrendered" responsibility to 3GPP for all areas under its scope. On the one hand, a single dominant standard facilitates global adoption of the technology and helps penetrate into new device markets beyond traditional cellular phones, such as consumer electronics, tablet computers, gaming devices, and machine-to-machine communications devices. On the other hand, getting the maximum mileage out of the same standard, as opposed to competing among different technologies, is now of great importance to gain a competitive edge among different operators and vendors.

What are the requirements of 4G cellular mobile broadband? Simply put, the performance should be on par with wired broadband even when a user is mobile. The demand side drivers of mobile broadband come from hardware (smart mobile devices) and software (high-bandwidth applications). A conventional set of exemplary applications includes web access, file transferring, video streaming, gaming, instant messaging, and voice over IP. For those applications throughput, latency, and reliability are the most important performance metrics of interest. However, new requirements will likely arise as new usages emerge. For example, when location-based services and social networking applications grow popular, it is desirable to efficiently transport short, chatty messages for a large user population. Below are some general system design considerations to address ever-evolving requirements of mobile broadband.

- Performance should be consistent throughout a cell in order to deliver predictable services. Not only peak or average but also cell edge performance is important.
- Cost per megabyte should be sufficiently low to stimulate mobile broadband usage. This requires not only high spectral efficiency of the airlink, but also low cost of deploying and operating the networks and utilizing the available spectrum.
- Devices are becoming increasingly intelligent (e.g., multimode and multiband devices) and can play a significant role in system operations. For example, devices can choose between available access interfaces (LTE versus Wi-Fi) or search for available bandwidth depending on policy, channel, interference, and congestion conditions. As another example, because battery technologies have not progressed as rapidly as microelectronics, system solutions of reducing battery power consumption are much desirable.

1.2 System design principles of wireless communications

Mobile broadband exhibits very different requirements, challenges, and characteristics from cellular telephony.

Traditionally, the dominant traffic in cellular telephony has been circuit-switched voice. Voice traffic is transported at a constant and low data rate (on the order of 10 kbps), delay sensitive (a voice frame missing its deadline becomes useless), but error tolerant (our ears are tolerant to a few percent of voice frame errors).

On the other hand, mobile broadband deals with a variety of traffic among which voice is just one type. A significant portion of the traffic is bursty and requires a very high burst data rate (on the order of a few Mbps). Moreover, it can tolerate some delays but is

Table 1.1. A brief comparison of the system design ideas between the cellular telephony and mobile broadband systems.

Cellular telephony [168]	Mobile broadband
Universal frequency reuse	Fractional frequency reuse
Power control	Power control and rate control
Soft handoff	Make-before-break handoff
Forward error correction	Forward and feedback error correction
Voice activity and interference averaging	Packet-switched scheduling

in general sensitive to packet errors. The reason is that upper layer protocols employed in broadband networks, such as Transmission Control Protocol (TCP), are originally designed for the wired environment. In the wired networks, the dominant source of packet loss is congestion; congestion control algorithms in TCP are largely based on packet loss. However, in the wireless networks, packet errors can also be caused by channel fading or interference, and therefore using TCP congestion control can have undesirable consequences at higher layers.

The epilog section of [168] lists a few system design principles of cellular telephony in the CDMA context. We summarize these principles in left column of Table 1.1. In particular, note that with tight delay constraint, forward error correction and power control are required for link reliability against channel fading and interference. Soft handoff introduces macrodiversity to further improve link reliability when a user is between multiple base stations. Interference averaging makes universal frequency reuse possible, thereby increasing the system capacity in terms of the total number of voice users and making CDMA an interference-limited system. Any reduction in interference, such as by exploiting voice activity, directly translates into capacity increase.

As cellular systems migrate to mobile broadband, the system design ideas also evolve. For comparison, we contrast the design principles of mobile broadband versus cellular telephony in Table 1.1; we will expand the list and present details in Chapter 3. Note that because of delay tolerance, feedback error correction using packet retransmission can be additionally used to improve link reliability. Since the data rate does not have to be constant, rate control works in conjunction with power control to take advantage of channel fluctuation rather than just mitigating it. As traffic becomes bursty, simple statistical multiplexing with voice activity no longer suffices; packet-switched scheduling is required to meet rapidly changing traffic needs. Furthermore, make-before-break handoff replaces soft handoff to better support IP-based network architecture, and fractional frequency reuse replaces universal frequency reuse in particular to enhance cell edge performance.

1.3 Why OFDMA for mobile broadband?

Orthogonal Frequency-Division Multiple Access (OFDMA) was chosen as the spectrum access technology of the 4G cellular systems. What are the fundamental advantages of OFDMA over CDMA, the dominant technology of 3G?

1.3 Why OFDMA for mobile broadband?

A simple answer is that OFDMA offers higher capacity because its *orthogonality* eliminates intra-cell interference. Recall from [168] that the capacity of a CDMA system, C_{CDMA}, is limited by the total interference, which consists of intra-cell interference and inter-cell interference. Denote by f the relative inter-cell interference factor equal to the ratio of mean inter-cell and intra-cell interference powers. Since the intra-cell interference is eliminated in OFDMA, as a first-order approximation,[2] the capacity is increased to

$$C_{\text{OFDMA}} = \frac{1+f}{f} C_{\text{CDMA}}. \tag{1.1}$$

The value of f depends on the wireless channel and how a user is connected to the base stations. In a typical scenario, $f \approx 0.5$ to 1, thereby translating into a factor of 2 to 3 capacity increase.

It should be pointed out that the capacity analysis in [168] is for voice traffic. OFDMA furthermore possesses salient features for mobile broadband. We highlight a few major ones in the following and leave details to Chapter 3.

The first salient feature is that the orthogonality enables OFDMA to multiplex users with *large signal dynamic range*. Because of the nature of the wireless channel, a user nearby a base station has a much higher channel gain than a faraway user does. In OFDMA, the received uplink signal from the nearby user can be much stronger than that from the faraway user, thereby achieving a high data rate without costing much system resource. However, without the orthogonality, the received signals are power controlled to arrive at the same level, because otherwise the strong signal from the nearby user would overwhelm the weak one from the faraway user. As a result, the performance is limited by the worst-case channel. A similar observation can be made in the downlink where the base station can allocate only a small transmit power to the nearby user without the concern that it would be overwhelmed by a large transmit power allocated to the faraway user.

Second, the *fine granularity of bandwidth resource* enables OFDMA to allocate and utilize bandwidth resource flexibly. The bandwidth resource unit is a tone in an OFDMA symbol. The flexibility allows packet-switched scheduling to quickly respond to time-varying traffic needs and channel fluctuation. The resultant benefits are low latency for delay sensitive applications and multiuser diversity that harnesses, rather than averaging out, channel fading. Fast ARQ is also facilitated to improve link reliability. This is particularly important to ensure upper layer protocols, such as TCP originally designed for the wired world, to work transparently over the error-prone wireless channel.

Third, OFDMA decomposes the wireless channel into a number of *parallel flat subchannels in frequency*, each of which is characterized by a single-tap channel response. This simplifies the implementations of advanced signal processing techniques, such as MIMO, and also helps exploit frequency selectivity of both channel and interference. In addition, a single tone signal can be easily detected without fine-grained time

---

[2] The analysis in [168] is for a 2G (IS-95) CDMA system. More advanced CDMA systems attempt to reduce or eliminate intra-cell interference by employing sophisticated receivers such as multiuser detectors in the uplink and equalizers in the downlink, or using CDMA/TDMA in the downlink. A comprehensive capacity comparison is outside of the scope of this book.

and frequency synchronization, and thereby facilitate handoff initiation and system determination.

1.4 Systems approach and outline of the book

This book focuses on the airlink system design of OFDMA mobile broadband communications with an emphasis on a *systems approach*. By systems approach, we mean the following:

- Different design objectives and choices often greatly affect each other. This book highlights those interactions so as to appreciate the system as a whole. Cross layer design is one such example.
- This book starts with first principles to understand the thought process and rationale that led to the evolution from 3G to 4G, rather than accepting the existing LTE design as gospel.
- Point-to-point communications are relatively well understood in the literature. This book emphasizes design choices in a multiuser multicell environment.
- A performance gain usually comes with a cost, for example, signaling overhead or processing complexity. An important aspect of system design is to understand the cost–benefit tradeoff.

This book attempts to answer the following questions:

- What are strategies to deal with channel fading and interference in mobile broadband?
- Why does OFDMA fit well with mobile broadband and how are OFDMA advantages best utilized to maximize system performance?
- Why is IP-based network architecture a desired choice for mobile broadband and how does the network architecture choice affect the airlink system design?
- How can communications theory be applied to practice by taking into account real world impairments?

To this end, the remaining chapters of this book are organized as follows.

Chapter 2 introduces the basic concepts of OFDM and OFDMA, and presents a special variation called SC-FDMA, which is designed to address the peak-to-average power ratio problem, a well known drawback of OFDM. The key feature of OFDMA is its orthogonality. To study the robustness of that feature, we quantify the level of orthogonality under a variety of real world impairments. This study leads to the cross interference and self-noise models, which explain the reason that in the real world the SINR is saturated even when C/I is sufficiently high.

With the basic OFDMA concepts in place, Chapter 3 presents qualitatively the system design principles of OFDMA mobile broadband communications. We first outline the system benefits of OFDMA, and then elaborate on how to apply them to a number of system operations. Chapter 3 serves as a high level preview of the insights to be developed in Chapters 4 to 9, which quantitatively study those applications.

Chapters 4 to 7 aim to create a big data pipe. Chapter 4 studies the wireless channel and characterizes multipath fading. The specific communications strategies presented are to mitigate fading via feedback or diversity to improve reliability in the single-user case, and to opportunistically exploit fading via multiuser diversity in the multiuser case. Finally, we combine those two ideas and discuss the notion of reliable opportunistic communications.

Chapter 5 studies power and bandwidth allocation of user multiplexing within a cell. We start with an intra-sector problem; in OFDMA, orthogonal multiplexing is an obvious choice, and we present the benefit of multiplexing users with large dynamic range of channel gains. Next, we study capacity-achieving non-orthogonal multiplexing with superposition coding; we find that the performance drops quickly in the presence of self-noise. To achieve robust non-orthogonal multiplexing, we then present superposition-by-position coding. Finally, we extend our study to a sectorized cell and investigate inter-sector interference management. The real world challenge is sector antenna leakage, which would significantly reduce the user multiplexing gain if not dealt with appropriately.

Chapter 6 studies interference between cells. We present stochastic geometry as a tool to compare the SINR distributions of different spectrum access schemes (FDMA, CDMA, and OFDMA). Focusing on OFDMA, we first assume inter-cell interference to be averaged over the entire bandwidth. Inter-cell interference management is relatively simple in the downlink, but more complex in the uplink as it is tied closely with power control and SINR target setting for which neighboring cells are coupled. Next, we present a scheme, called fractional frequency reuse, to coordinate, rather than to simply average, inter-cell interference so as to improve cell edge SINR as well as overall spectral efficiency.

Chapter 7 studies the use of multiple antennas in the OFDMA mobile broadband system, in particular the benefit of multiplexing gain with MIMO, which depends on the rank of the wireless channel matrix. We model both linear antenna arrays and polarized antennas and find that the channel environment in which they achieve the maximum multiplexing gain is complementary. We study the MIMO gain in the single user, multiuser, and multicell cases.

Chapter 8 studies scheduling techniques to share the data pipe among users in a cell. Scheduling takes place in two distinct time scales. In the fast time scale (of the order of milliseconds), the base station schedules users' traffic in a packet-switched manner. We study scheduling algorithms designed for different traffic types including infinitely backlogged traffic, elastic traffic, inelastic traffic, and flow level traffic. In the slow time scale (of the order of seconds), the base station manages the users' MAC states so as to allow power saving and low-latency contention-free access to a large user population. The key idea of MAC states lies in the very different time scale requirements of power control and timing control in OFDMA. Semi-persistent scheduling is another slow time scale scheduling mechanism that handles periodic traffic such as Voice over IP.

Chapter 9 argues that the network architecture of mobile broadband should be IP-based, and shows how such an architecture is supported with the OFDMA airlink. We focus on handoff. First, we present make-before-break handoff as an alternative way of replacing soft handoff to obtain macrodiversity. Make-before-break handoff fits well with IP-based networks. We present (expedited) break-before-make handoff as a backup when make-before-break handoff is not feasible. Second, we propose a single-tone signal, called beacon, to significantly reduce the complexity for users to search handoff candidate base stations.

Finally, Chapter 10 explores ideas beyond the conventional cellular framework. There are three main ideas. The first idea is to change communications topology and get a much denser network. We focus on relays, small cells such as femtocells, and as an extreme, direct device-to-device path without going through a base station. The second is to treat concurrent transmissions as part of a distributed network MIMO system rather than interference. We investigate user cooperation assisted with relays and network cooperation with multicell processing. The third is to access spectrum cognitively and opportunistically rather than following a static allocation rule. We focus on two communications aspects of spectrum sensing and sharing.

2 Elements of OFDMA

2.1 OFDM

2.1.1 Tone signals

OFDM (Orthogonal Frequency Division Multiplexing) is a multicarrier communication scheme where the entire system bandwidth is divided into N_c equally spaced tones. Denote by Δ_f the frequency spacing between two adjacent tones and $f[k]$ the baseband tone frequency given by $f[k] = k\Delta_f$, for $k = 0, \ldots, N_c - 1$. In one OFDM symbol duration, a block of complex symbols $X[0], \ldots, X[N_c - 1]$ are modulated on tones $f[0], \ldots, f[N_c - 1]$, respectively. The complex symbols can be QPSK or QAM symbols obtained from coding and modulation. In this case, the complex symbols are also referred to as modulation symbols. In the time domain, the baseband OFDM signal $x(t)$ in one OFDM symbol duration can be described by

$$x(t) = \sum_{k=0}^{N_c-1} X[k]\exp(j2\pi f[k]t), \, t \in [0, T_s), \quad (2.1)$$

where $j = \sqrt{-1}$, and T_s is the OFDM symbol duration. Each complex sinusoid $X[k]\exp(j2\pi f[k]t)$ in (2.1) is called a tone signal at frequency $f[k]$. The OFDM signal $x(t)$ is the sum of those tone signals.

Let

$$T_s = \frac{1}{\Delta_f}. \quad (2.2)$$

(2.1) becomes

$$x(t) = \sum_{k=0}^{N_c-1} X[k]\exp\left(j2\pi k \frac{t}{T_s}\right). \quad (2.3)$$

It is easy to see that the time interval of $t \in [0, T_s)$ contains an integer number of periods for each tone signal.

Note that two tone signals with distinct $f[k_1]$ and $f[k_2]$ are orthogonal because

$$\frac{1}{T_s}\int_0^{T_s} \left[\exp\left(j2\pi k_1 \frac{t}{T_s}\right)\right]\left[\exp\left(j2\pi k_2 \frac{t}{T_s}\right)\right]^* dt = \delta(k_1 - k_2)$$

$$= \begin{cases} 1, & \text{if } k_1 = k_2; \\ 0, & \text{otherwise.} \end{cases}$$

Sinusoids are special because they are the only eigenfunctions of linear time-invariant systems. If the OFDM symbol duration T_s is sufficiently shorter, that is, much shorter than the coherence time in time-selective fading, for practical purposes the wireless channel can be treated as a linear time-invariant system. Denote by $h(\tau)$ the impulse response of the wireless channel, which characterizes one or multiple propagation paths from the transmitter to the receiver. In a multipath channel, the delays of those paths may not be all identical and the difference results in *delay spread*. Assume $h(\tau) = 0$ when $\tau > T_{\max}$, where $T_{\max}$ is the maximum delay spread. For a tone signal $\exp(j2\pi f[k]t)$ transmitted over the wireless channel, the received signal is given by

$$y(t) = \int_{-\infty}^{t} \exp(j2\pi f[k]\tau) h(t-\tau) d\tau$$

$$= \int_{0}^{T_{\max}} \exp(j2\pi f[k](t-\tau)) h(\tau) d\tau$$

$$= \exp(j2\pi f[k]t) H[k], \qquad (2.4)$$

where complex number

$$H[k] = \int_{0}^{T_{\max}} \exp(-j2\pi f[k]\tau) h(\tau) d\tau \qquad (2.5)$$

is the channel response coefficient at tone frequency $f[k]$.

In Equation (2.4), the received signal $y(t)$ reaches the steady state after the tone signal has been transmitted for at least as long as $T_{\max}$. We make the following two observations about the steady state $y(t)$.

- After two orthogonal tone signals travel over the wireless channel, the received signals are also sinusoids at the exactly same tone frequencies and therefore still orthogonal. This property is called no ICI (inter-carrier interference). While many orthogonal function families can be possibly used as the basis for signal multiplexing, only sinusoids are the eigenfunctions, thereby making OFDM a unique multiplexing scheme that preserves signal orthogonality in the wireless channel.
- The channel response at any tone frequency is just a complex coefficient, which may vary with tone in a frequency selective channel. In OFDM the wideband channel is in effect converted into a number of parallel flat channels. This feature can be used to exploit frequency selectivity and to simplify signal processing in OFDM.

2.1.2 Cyclic prefix

Now let us consider the transient response of the received signal. Since the maximum channel response is $T_{\max}$, the received signal $y(t)$ takes up to $T_{\max}$ before it reaches the steady state $\exp(j2\pi f[k]t) H[k]$. In order for the received signal to further stay at the steady state for a time interval of T_s, the transmitted tone signal needs to be at least as long as $T_s + T_{\max}$. This motivates the use of a *cyclic prefix*: the transmitted OFDM signal $x(t)$ should be extended to time interval $t \in [-T_{cp}, T_s)$, a little longer than $t \in [0, T_s)$

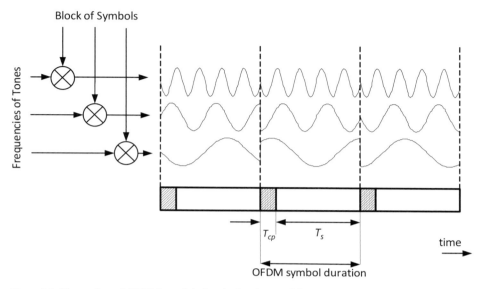

Figure 2.1 Illustration of OFDM modulation in the time and frequency domains.

originally defined in Equation (2.1), where T_{cp} represents the length of the cyclic prefix:

$$x(t) = x(t + T_s), t \in [-T_{cp}, 0). \tag{2.6}$$

Figure 2.1 illustrates the OFDM modulation scheme where the total OFDM symbol duration is now $T_s + T_{cp}$. Note that the extended portion, obtained by cyclically extending $x(t)$ into $t \in [-T_{cp}, 0)$, is the exact copy of the portion in $t \in [T_s - T_{cp}, T_s)$. We will discuss the choice of parameter T_{cp} later in this chapter. For now we assume $T_{cp} = T_{\max}$.

Instead of just one OFDM symbol shown in Figure 2.1, now suppose that a sequence of OFDM symbols are transmitted back-to-back. In the wireless channel delay spread causes interference between successive symbols, known as inter-symbol interference (ISI). As illustrated in Figure 2.2, the cyclic prefix acts as a guard interval and in effect eliminates the ISI between two successive OFDM symbols.

We make the following comments on the use of a cyclic prefix:

- A cyclic prefix allows an OFDM symbol to reach its steady state when it passes through the wireless channel such that the receiver only needs to deal with the steady state response.
- A cyclic prefix allows the channel response of the previous OFDM symbol to die down and thus not to interfere with the steady state period of the present symbol. This property is called no ISI.
- A cyclic prefix is an overhead to the system, and is meant to be discarded at the receiver before further processing. The system cost is a factor of T_{cp}/T_s reduction in the bandwidth efficiency. Because a block of N_c data symbols are transmitted simultaneously in OFDM, the symbol duration T_s is N_c times longer than what would

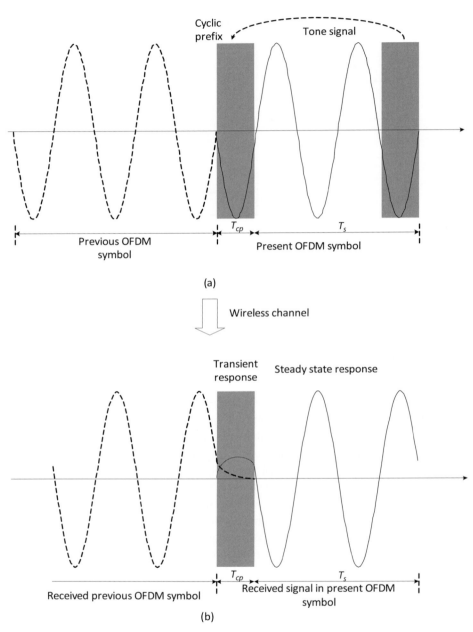

Figure 2.2 Illustration of a cyclic prefix. A transmitted tone signal of two successive OFDM symbols is shown in (a) and the channel responses of the two OFDM symbols are shown in (b) as dotted and solid curves, respectively. The actual received signal is the sum of the two curves. The received signal in the present OFDM symbol duration consists of two portions: the transient portion in which the previous OFDM symbol dies down and the present OFDM reaches its steady state, and the steady state portion that contains only the steady state channel response of the present OFDM symbol.

2.1 OFDM

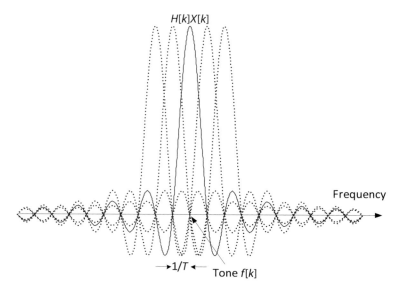

Figure 2.3 Frequency domain illustration of OFDM demodulation. Ignoring the phase factor, the figure shows $H[k]X[k]\text{sinc}(fT_s - k)$ of various k.

be in single-carrier communications with the same total bandwidth. A longer T_s reduces overhead T_{cp}/T_s for a given T_{cp}.

The receiver can discard the transient response and use only the steady state portion to recover the original data symbols $X[0], \ldots, X[N_c - 1]$. In this case, the receiver in effect uses a rectangular window of size T_s in the time domain to take the appropriate portion out of the total received signal. The frequency domain representation is

$$H[k]X[k]\text{sinc}((f - f[k])T_s)\exp(-j\pi(f - f[k])T_s). \tag{2.7}$$

The resultant signal is thus the sum of N_c sinc functions in the frequency domain, as shown in Figure 2.3. The sinc function associated with tone $f[k]$ reaches its peak at tone $f[k]$ and crosses zero at all other tones. This is another way of viewing the OFDM orthogonality.

2.1.3 Time-frequency resource

The bandwidth resource in OFDM can be viewed in a time-frequency grid as shown in Figure 2.4. The y-axis represents frequency tones and the x-axis OFDM symbols. Each small square represents the basic unit of bandwidth resource, referred to as *tone-symbol*, which is a tone in an OFDM symbol.

In summary, the property of no ISI and ICI means that the tone-symbols are orthogonal with each other. At any tone-symbol the wireless channel is constant, and the channel response is represented by a multiplicative complex coefficient. The coefficient varies

14 Elements of OFDMA

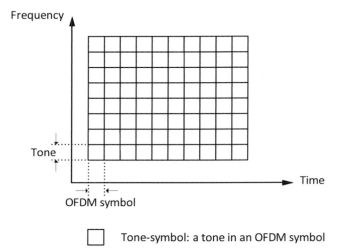

Figure 2.4 Illustration of bandwidth resources in OFDM.

over time and frequency characterizing the time and frequency selectivity of the wireless channel.

2.1.4 Block signal processing

In a practical communications system, the continuous OFDM signal $x(t)$ is generated with a chain of digital and analog signal processing modules. The specific design of those modules depends on implementation choices. We next provide a general description.

At the 1× Nyquist sampling rate with the sampling frequency equal to $\left(\frac{T_s}{N_c}\right)^{-1} = N_c \Delta_f$, $x(t)$ of Equation (2.3) at discrete time instant $t = \frac{l}{N_c} T_s$ is given by

$$x[l] = x\left(\frac{l}{N_c} T_s\right) = \sum_{k=0}^{N_c-1} X[k] \exp\left(j 2\pi \frac{kl}{N_c}\right), l = 0, \ldots, N_c - 1. \quad (2.8)$$

Equation (2.8) says that $x[0], \ldots, x[N_c - 1]$ is the inverse discrete Fourier transform (IDFT) of $X[0], \ldots, X[N_c - 1]$. Thus the block processing shown in Figure 2.5 is a much more efficient implementation than calculating $x[l]$ sample by sample.

At the transmitter, in every OFDM symbol, coding and modulation generates a block of complex symbols $X[0], \ldots, X[N_c - 1]$ to be sent. The IDFT module transforms them to $x[0], \ldots, x[N_c - 1]$ as a block. $\frac{N_c T_{cp}}{T_s}$ samples of the cyclic prefix are then added. The total $\frac{N_c T_{cp}}{T_s} + N_c$ baseband complex samples are filtered to control the out-of-band spectrum emission. The filtered time domain samples are subsequently converted to the analog domain, mixed up to the carrier frequency, and transmitted through the transmit antenna. At the receiver, in every OFDM symbol the received signal from the receive antenna is down-converted to the baseband digital and filtered to remove the adjacent band interference. $\frac{N_c T_{cp}}{T_s}$ samples of the cyclic prefix are removed. The

2.1 OFDM

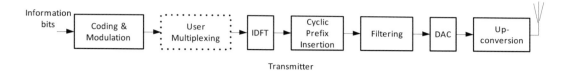

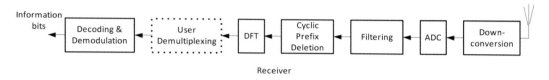

Figure 2.5 Simplified block diagram of signal processing modules in OFDM or OFDMA transmitter and receiver. The "user multiplexing" and "user demultiplexing" modules are only used in OFDMA.

DFT module transforms the remaining N_c time domain samples $y[0], \ldots, y[N_c - 1]$ to a block of complex symbols in the frequency domain. The output of the DFT module is

$$Y[k] = \frac{1}{N_c} \sum_{l=0}^{N_c-1} \exp\left(-j2\pi \frac{kl}{N_c}\right) y[l]$$
$$= H[k]X[k] + W[k], k = 0, \ldots, N_c - 1, \qquad (2.9)$$

where $W[k]$ represents an additive noise. Similar to (2.5), $H[k]$ is the frequency response coefficient of the wireless channel as well as the filters used in the transmitter and the receiver. The demodulated symbols $Y[0], \ldots, Y[N_c - 1]$ are then passed to demodulation and decoding for further processing.

We next highlight a few notes on digital signal processing of the diagram in Figure 2.5 that are important in practical implementation. First-time readers can skip these notes and return to them later if so desired.

Discussion notes 2.1 FFT/IFFT

If N_c is a power of 2, then the DFT and IDFT modules can be very efficiently implemented using Fast Fourier Transform (FFT) and inverse Fast Fourier Transform (IFFT). Otherwise, we can choose another integer N'_c that is a power of 2 and greater than N_c. Now the 1x sampling rate is increased to $\left(\frac{T_s}{N'_c}\right)^{-1} = N'_c \Delta_f$. The time domain samples to be generated are

$$x[l] = x\left(\frac{l}{N'_c}T_s\right) = \sum_{k=0}^{N_c-1} X[k] \exp\left(j2\pi \frac{kl}{N'_c}\right), l = 0, \ldots, N'_c - 1. \qquad (2.10)$$

One way of viewing the time domain sample calculation in Equation (2.10) is to pretend that there are N'_c tones except that the symbols carried on the top $N'_c - N_c$

tones are all equal to zero. To calculate $x[l], l = 0, \ldots, N'_c - 1$, the transmitter first pads $N'_c - N_c$ zeros to the end of the original block of complex data symbols to form a new symbol block of size N'_c: $X_0, \ldots, X_{N_c-1}, \underbrace{0, \ldots, 0}_{N'_c - N_c}$. An IFFT module of size N'_c transforms it to $x_0, \ldots, x_{N'_c-1}$. The number of the cyclic prefix samples now becomes $\frac{N'_c T_{cp}}{T_s}$.

The receiver uses a similar approach to calculate a block of N'_c complex symbols from the time domain samples $y[l] = y\left(\frac{l}{N'_c} T_s\right)$ with an FFT module:

$$Y[k] = \frac{1}{N'_c} \sum_{l=0}^{N'_c-1} y[l] \exp\left(-j2\pi \frac{kl}{N'_c}\right), k = 0, \ldots, N'_c - 1. \quad (2.11)$$

The receiver simply discards the symbols in the frequency domain corresponding to the top $N'_c - N_c$ tones $Y_{N_c}, \ldots, Y_{N'_c-1}$.

This technique makes use of efficient FFT/IFFT and yet allows the system parameter N_c to be flexible.

Discussion notes 2.2 Filtering

While the FFT/IFFT modules run at the $1\times$ sampling rate, the filtering modules in both the transmitter and the receiver often operate at a higher rate, for example, $4\times$ or $8\times$ sampling rate, to be more effective in rejecting aliasing. This book does not address multirate signal processing and filter design. However, a close look at the frequency and time responses of a typical filter, as shown in Figure 2.6, helps us to understand the system impact of the filtering modules.

First, note that the time response of the transmitter and the receiver filters are cascaded with the wireless channel response. The total effective delay spread to be covered by a cyclic prefix includes the delay spread due to the wireless channel and that introduced with the filters. Thus, even if the wireless channel itself is flat, the filters still cause delay spread.

Second, the sharper the filter transition from the passband to the stopband, the longer the time response. In the frequency domain, a sharper transition roll-off means more usable spectrum to pack tones. However, in the time domain, a longer filter response requires a larger cyclic prefix to cover. Therefore, we have to trade off the efficiency between the frequency domain and the time domain. It is noted that for a given transition roll-off, as the total bandwidth increases, the length of the time response decreases proportionally. Therefore the efficiency tradeoff is more favorable with larger bandwidth.

Third, note that the in-band ripple in the passband adds fluctuation to the frequency selectivity of the steady state wireless response $H[k]$. The receiver has to track the fluctuation as part of channel estimation. On the other hand, since the filter response is known, the transmitter can easily compensate for it (or at least the phase): the block of complex symbols to be sent is $X[0]/B[0], \ldots, X[N_c - 1]/B[N_c - 1]$ instead of

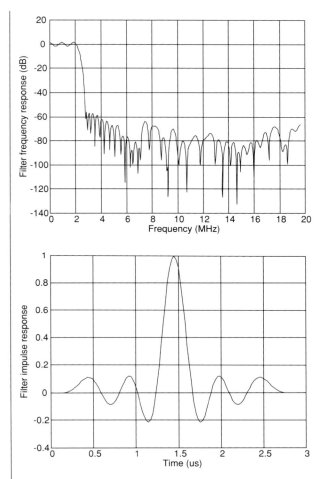

Figure 2.6 Frequency and time responses of a baseband linear phase FIR filter. The system bandwidth is 5 MHz (from -2.5 to 2.5 MHz). The filter operates at $8\times$ sampling rate. The passband of the filter is up to 2.15 MHz with ± 1.5 dB in-band ripple. Therefore, the usable spectrum $N_c \Delta_f = 2.15 \times 2$ MHz. The filter rejection is -44 dB at about 2.7 MHz. The time response spans about 3 µs.

$X[0], \ldots, X[N_c - 1]$, where $B[k]$ is the filter frequency response at tone $f[k]$. From the water-filling principle, such a pre-emphasis scheme is suboptimal. However, as long as the in-band ripple is small, the loss of optimality is minor.

Discussion notes 2.3 Equalization

Suppose that $X[0], \ldots, X[N_c - 1]$ are the modulation symbols of a code block. How do we recover the original code block from the output of the DFT module $Y[k]$? From Equation (2.9), $Y[k]$ differs from $X[k]$ by channel coefficient $H[k]$. The receiver first estimates $H[k]$ (see Section B.3 for channel estimation). Once the channel estimate $\hat{H}[k]$ is known, the receiver can equalize the channel, for example,

$$\hat{X}[k] = \frac{1}{\hat{H}[k]} Y[k] \qquad (2.12)$$

with zero-force equalization, or

$$\hat{X}[k] = \frac{\hat{H}[k]^*}{|\hat{H}[k]|^2 + \sigma^2/P} Y[k] \qquad (2.13)$$

with MMSE equalization where σ^2 is the estimated power of noise $W[k]$ and P the signal transmit power, and then feed $\hat{X}[k]$ to a decoder to recover the original code block. Sometimes this linear equalization may enhance noise because $Y[k]$ with smaller $\hat{H}[k]$ causes larger boost in the decoding metric. In effect, this rewards less reliable tones and penalizes more reliable tones.

An alternative strategy is to add weights to the received symbols in the decoding metric according to their reliability. An often-used decoding metric is the so-called log-likelihood ratio [127], which measures the ratio of likelihoods of an information bit being 1 versus 0. In Equation (2.9), suppose that $W[k]$ is white Gaussian and $X[k]$ is a BPSK symbol with $X[k] = +1$ or -1 of equal probability. Then the log-likelihood ratio is

$$L(X[k]|Y[k]) = \log\left(\frac{\mathbb{P}(Y[k]|X[k]=+1)}{\mathbb{P}(Y[k]|X[k]=-1)}\right)$$

$$= \frac{2}{\sigma^2} \Re(\hat{H}^*[k]Y[k]), \qquad (2.14)$$

where $\Re(\hat{H}^*[k]Y[k])$ represents the real part of $\hat{H}^*[k]Y[k]$. $\hat{H}^*[k]$ in effect acts as a reliability weight such that $L(X[k]|Y[k])$ associated with smaller $\hat{H}^*[k]$ contributes less to the decoding metric based on which the entire code block is to be decoded. A benefit of this scheme is that when the wireless channel fades at one tone, the loss to decoding is localized to that tone. In the extreme case where $\hat{H}^*[k] = 0$, the corresponding symbol $Y[k]$ is in effect erased but the damage does not spread to the entire code block.

2.2 From OFDM to OFDMA

2.2.1 Basic principles

OFDM is a point-to-point modulation scheme between a transmitter and a receiver. Applying the concepts of tone signals and cyclic prefixes to the multiuser scenario results in a new multiple access scheme, namely OFDMA (Orthogonal Frequency Division Multiple Access). The basic idea is to share the time-frequency resource, namely the tone-symbols shown in Figure 2.4, among multiple users. Let us focus on OFDMA in the cellular system, although the OFDMA scheme can be used in other types of systems, such as ad hoc ones.

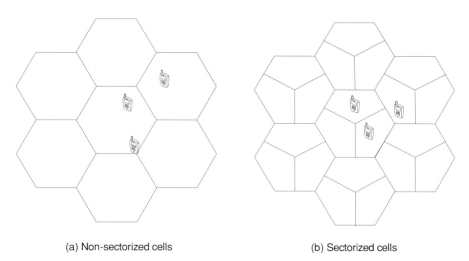

(a) Non-sectorized cells (b) Sectorized cells

Figure 2.7 Illustration of typical cellular deployments. (a) and (b) show a hexagonal deployment without and with sectorization, respectively.

Before we get into the details of OFDMA, we first review a few basic concepts in the cellular system. A typical cellular deployment is shown in Figure 2.7 where a base station is placed in the center of each hexagon. A hexagon represents the coverage area of a base station and is usually referred as a *cell*. An economic way to boost the network capacity without incurring too much additional deployment cost is to introduce *sectorization* by using multiple directional antennas in a cell. As shown in Figure 2.7(b), one hexagonal cell can be divided into three non-overlapping *sectors* which are covered by different directional antennas.

A *downlink* transmission is a transmission from a base station to a user, and an *uplink* transmission is from a user to a base station. In the multiuser scenario, a base station transmits to multiple users in the downlink, and multiple users transmit to a base station in the uplink. A signal transmission may interfere with another signal transmission within the same sector, in different sectors of the same cell, or in neighboring cells, when the two transmissions use the same time-frequency bandwidth resource. We refer to those types of interference as intra-sector, inter-sector, and inter-cell interference, respectively. Combined intra-sector and inter-sector interference is called *intra-cell interference*.

We next study how the OFDMA bandwidth resources are shared in the cellular system. We start with two simple principles in this section and will refine them as we address various interference issues in Chapters 5 and 6.

- *Zero intra-cell interference*. Within a sector, the tone-symbols allocated to different users are different to eliminate intra-sector interference. In a sectorized cell, all tone-symbols are reused in every sector of the cell. If sectorization is perfect, inter-sector interference is zero.

20   Elements of OFDMA

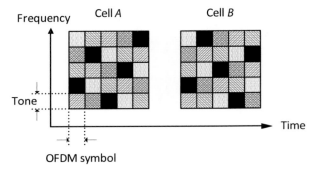

Figure 2.8 OFDMA tone hopping in cells A and B. In the time-frequency grid, the x-axis represents OFDM symbols and the y-axis represents frequency tones. A small square represents a tone-symbol. A sequence of small squares with the same filling pattern represents a subset of tone-symbols allocated to a user in either cell A or B. There are five users in cell A and another five in cell B.

- *Averaged inter-cell interference*. With universal frequency reuse, the same set of tone-symbols are reused among all the cells. This maximizes the usable spectrum available in every cell, but meanwhile results in inter-cell interference. As described later in this section, randomized resource allocation such as tone hopping ensures inter-cell interference is averaged.[1]

The allocation of tone-symbols in OFDMA is flexible. For example, all the tones of an OFDM symbol or one tone of a sequence of OFDM symbols can be allocated to a given user. In general, any subset of tone-symbols can be allocated to a user. Tone hopping means that the tone-symbols allocated to a given user hop across different frequency tones over time, thereby providing *frequency diversity*. In the example shown in Figure 2.8, a user in either cell A or B is allocated one tone-symbol in every OFDM symbol, and the frequency tone of the allocated tone-symbol varies from one OFDM symbol to another. In a given cell, the tone-symbols allocated to two different users never collide. Moreover, cells A and B uses distinct tone hopping patterns such that a user in cell A interferes with different users in cell B in different OFDM symbols. This is an example of *interference diversity* or *interference averaging*. As a result, the inter-cell interference seen by the user is not limited by the worst case co-channel interferer but depends on the average.

Tone hopping over time is one way of achieving frequency diversity and interference averaging. When a number of tones are allocated to a user simultaneously, those tones can spread over a large frequency bandwidth in a pseudo-random manner, thereby achieving the same goal. In the following discussion, we use the term "tone hopping" to mean "randomized resource allocation" in general, over time and/or frequency, for frequency diversity and interference averaging.

---

[1] Here we rely on resource allocation for inter-cell interference averaging. Chapter 6 will expand the idea to active interference control and coordination in the context of fractional frequency reuse.

Table 2.1. A brief comparison of FDMA, CDMA and OFDMA.

	FDMA	CDMA	OFDMA
Intra-cell interference	Zero	Non-zero	Zero
Inter-cell interference	Not averaged	Averaged	Averaged
Frequency reuse factor	> 1	1	1

2.2.2 Comparison: OFDMA, CDMA, and FDMA

A brief comparison of OFDMA with CDMA and narrowband FDMA provides interesting insights.

In FDMA, the entire system bandwidth is divided into a number of narrowband carriers. A carrier is allocated to one user in a cell and reused in another cell that is a minimum reuse distance apart. The minimum reuse distance determines that a carrier is reused in one out of every K cells, and $K > 1$ is called *frequency reuse factor*. Within a given cell, different narrowband carriers do not interfere with each other. Co-channel interference occurs between two users in the cells that reuse the same carrier, and is not averaged among many users in those cells. The value of frequency reuse factor is usually determined based on the worst case co-channel scenario where the interfering users are located most disadvantageously with each other (i.e., near cell edges).

In CDMA, a transmitter uses a unique spreading code to spread its signal over the entire system bandwidth. Different users use different spreading codes and their signals all occupy the same total bandwidth to achieve *universal frequency reuse*. Knowing the spreading code used by the desired transmitter, the intended receiver despreads the received signal to recover the original signal. If the cross correlations between the spreading code of the desired user and those of others are small, the interference from other transmitters is significantly rejected after despreading, by a factor equal to a spreading gain that depends on the length of the spreading code. Since all the signals share the same total bandwidth, a receiver sees averaged interference from all other transmitters. Universal frequency reuse is an important advantage of CDMA over FDMA because the available spectrum per cell is increased by a factor of K. In CDMA, interference averaging is crucial to make universal frequency reuse possible, because now the inter-cell interference is not limited by the worst-case co-channel interference scenarios. On the other hand, because spreading codes are not completely orthogonal in the multipath wireless channel, users within the same cell interfere with each other.

We summarize the above comparison in Table 2.1 and show that OFDMA combines the salient features of both FDMA and CDMA.

2.2.3 Inter-cell interference averaging: OFDMA versus CDMA

As we have just presented, both CDMA and OFDMA are designed to average out the inter-cell interference. However, there is a subtle yet important difference. CDMA uses code spreading while OFDMA uses tone hopping, thereby resulting in different statistical characteristics of inter-cell interference. To elaborate, consider a basic communication scenario where a transmitter sends a block of modulation symbols. For the sake of comparison, we ignore intra-cell interference in the CDMA case.

In OFDMA, a modulation symbol is sent in one tone-symbol. The inter-cell interference is caused by transmissions in other cells that share the same tone-symbol. With tone hopping, the interference is averaged over the entire block. However, for a given modulation symbol, the interference is caused by a small number of users in adjacent cells that share the same tone-symbol, and therefore the power may vary drastically from one modulation symbol to another. For a given modulation symbol, we cannot apply the Central Limit Theorem here and the interference may be non-Gaussian and heavy tailed.

In CDMA, a modulation symbol is spread to the entire bandwidth according to a spreading code, and the inter-cell interference from other transmissions in an adjacent cell depends on the cross correlations between that spreading code and the ones used by the interfering transmissions. For a given modulation symbol, every user in other cells contributes a fraction to the total interference, and therefore the power fluctuation between modulation symbols tends to be less drastic. When the number of users in other cells is large, the Central Limit Theorem applies and the interference can be reasonably approximated by a complex Gaussian.

To get a quantitative sense, consider an uplink example in the sectorized cellular system shown in Figure 2.7. Suppose N_c users are distributed in every sector. In OFDMA, each user is allocated one tone-symbol in every OFDM symbol. In CDMA, each user is allocated a spreading code of length N_c in every CDMA symbol. An OFDM symbol and a CDMA symbol have the same duration. Suppose that in CDMA or OFDMA, the transmit power of any user is set so that the received power is constant at its serving base station for all users.

Let us model the inter-cell interference for an intended modulation symbol received at a reference base station 0 as follows:

$$W = \sum_{m>0} \sum_{i=1}^{N_c} \rho_{m,i} \frac{|H_{m,i,0}|}{|H_{m,i}|} \exp(j\theta_{m,i}), \qquad (2.15)$$

where m is the index of a neighboring cell, and i is the index of a user. $|H_{m,i}|$ and $|H_{m,i,0}|$ represent the channel amplitude gain from user i in neighboring cell m to its own base station m and to base station 0, respectively, and $\theta_{m,i}$ is a phase parameter assumed to be random uniformly in $[0, 2\pi)$. $\rho_{m,i}$ is the cross correlation coefficient between the intended modulation symbol and an interfering modulation symbol transmitted by user i in cell m. In OFDMA,

$$\rho_{m,i} = \begin{cases} 1, & \text{if user } i \text{ occupies the same tone-symbol}; \\ 0, & \text{otherwise}. \end{cases} \qquad (2.16)$$

Because only one user in cell m occupies a given tone-symbol, $\rho_{m,i} = 0$ for all i except for the user sharing the same tone-symbol as the intended modulation symbol. In CDMA,

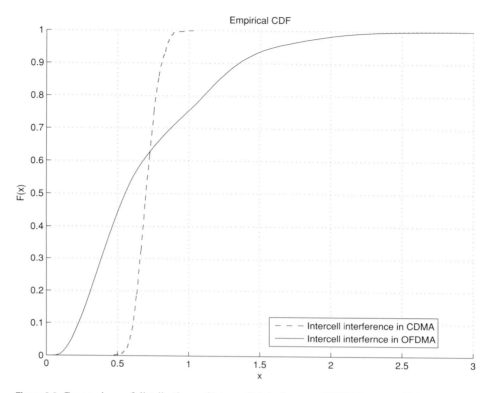

Figure 2.9 Comparison of distributions of inter-cell interference: OFDMA versus CDMA. Both systems have 128 users in each sector in a wrap-around, two-ring, 57-sector deployment. Users are uniformly dropped over the network. The interference strength is normalized to the received power of the signal from the intended user.

$\rho_{m,i}$ is modeled as

$$\rho_{m,i} = \frac{1}{N_c} \sum_{j=1}^{N_c} c_j \qquad (2.17)$$

with $c_j = \pm 1$ with equal probability. Thus, $\rho_{m,i}$ is a random variable distributed in $[-1, 1]$.

Figure 2.9 shows the distribution of the inter-cell interference power. We observe that while the inter-cell interference in CDMA is fairly accurately modeled as Gaussian, the inter-cell interference in OFDMA is clearly non-Gaussian and indeed heavy tailed.

It should be pointed out that if time and frequency are not completely aligned between the users in cell m and the base station receiver at cell 0, then more than one modulation symbol from the users in cell m may leak into the tone-symbol of the intended modulation symbol. In that case, the distribution is less heavy tailed than what is shown in Figure 2.9.

A practical implication of the above observation is that we should be very careful when we make a Gaussian assumption about the inter-cell interference in OFDMA. Take decoding for example. Consider a very simple model $Y = X + W$ where Y is the

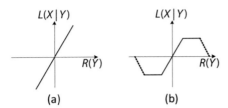

Figure 2.10 Illustration of decoding metrics when the interference is Gaussian (a) or heavy-tailed non-Gaussian (b). In (a), $L(X|Y)$ increases linearly with $\Re(Y)$. In (b), $L(X|Y)$ initially increases, but then saturates and even drops for large $\Re(Y)$ in order to discount the impulsive impact of interference outliers.

received modulation symbol, X is the transmitted BPSK symbol of a code block with $X = \pm 1$ with equal probability, and W the interference of unit power. If we assume W to be Gaussian, then the log-likelihood ratio is

$$L(X|Y) = \log\left(\frac{p(Y|X=+1)}{p(Y|X=-1)}\right)$$
$$= 2\Re(Y), \qquad (2.18)$$

where $\Re(Y)$ represents the real part of Y. In other words, when $\Re(Y) \gg 1$, the probability of $X = +1$ is much greater than that of $X = -1$. However, if W is heavy-tailed, then a large $\Re(Y)$ may simply indicate that Y has been hit by an interference outlier, in which case there is not much information from Y about the probability of $X = +1$ versus $X = -1$. One simple algorithm for handling heavy-tailed interference is to saturate $L(X|Y)$ as $\Re(Y)$ grows too large or even discount it entirely, for example, treat Y as an erasure. Figure 2.10 contrasts the decoding metrics used when interference is Gaussian versus heavy-tail non-Gaussian distributed.

Moreover, now that we anticipate that the inter-cell interference may be non-Gaussian, the coding and modulation strategy may also be different from the Gaussian interference scenario. For example, rate 1/4 coding with 16-QAM modulation can tolerate more erasure on modulation symbols and thus work better than rate 1/2 coding with QPSK in such an interference environment, although the latter works better in the AWGN channel.

2.2.4 Tone hopping: averaging versus peaking

As we have seen earlier in this chapter, OFDMA can be viewed as another spread spectrum technique similar to CDMA. Therefore the purpose of tone hopping is to achieve interference averaging and frequency diversity, meaning that a user experiences averaged inter-cell interference and averaged wireless channel condition. Averaging improves link reliability and thus is desirable for circuit-switched voice.

On the other hand, because of averaging, the best opportunities for communications may also be lost when the interference is low and/or the channel is good. Without averaging, those opportunities would have occurred because of the statistical nature of channel fluctuation or careful interference coordination of base stations. We will study opportunistic or coordinated communications in Chapters 4 and 6 with the idea

2.2 From OFDM to OFDMA

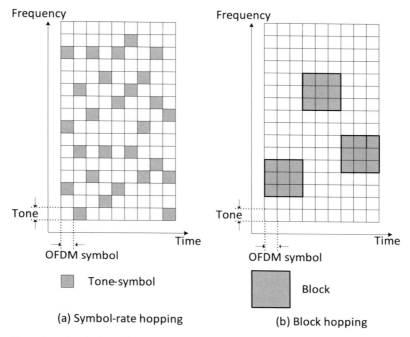

(a) Symbol-rate hopping (b) Block hopping

Figure 2.11 Symbol-rate hopping and block hopping. Shaded tone-symbols are the pilot.

of exploiting that selectivity of channel or interference and achieving a "peaking" effect as opposed to "averaging." Peaking improves instantaneous SINR and thus data rate. An important design goal of mobile broadband communications is to strike a good balance between averaging and peaking. Tone hopping makes OFDMA more flexible than CDMA to achieve such a goal.

Tone hopping design is quite different if the goal is to achieve averaging versus peaking. Roughly speaking, for averaging, tone hopping covers the time-frequency grid as widely as possible in a pseudo random manner. One such an example is shown in Figure 2.8. In this case, tones hop every OFDM symbol (symbol-rate hopping). In contrast, for peaking, tones concentrate in a much smaller footprint in the grid and usually hop as a block (block hopping) or do not hop at all. Figure 2.11 compares symbol-rate hopping and block hopping.

As a side note, from the viewpoint of point-to-point communications, concentrating tones in a block (as opposed to spreading everywhere) makes channel estimation more efficient, because the receiver only needs to estimate the channel in a smaller time-frequency region. Channel estimation is usually based on pilots, which are known signals sent by the transmitter. From Section B.3, in OFDMA enough pilots should be sent to cover the time-frequency region. Everything else being equal, block hopping requires a smaller number of pilots, thereby being more attractive to be used in MIMO (where the pilot overhead is already high in order to estimate the MIMO channel matrix) and uplink (where unlike the downlink there are no common pilots to be shared among users).

Elements of OFDMA

Practical example 2.1 Physical resource block allocation and hopping in LTE data channels

We use the LTE system as real world system examples to illustrate various design principles studied in this book. In those examples, we will use the LTE terminology. In particular, "eNB" is a base station, "UE" a user, and "RE" a tone-symbol. The 3GPP LTE standards specify a variety of deployment scenarios such as different bandwidth (from 1.4 MHz to 20 MHz), FDD or TDD operation, and different cyclic prefix length (normal versus extended). To keep the description simple and not get lost in the details, we focus on a common scenario of FDD with the normal cyclic prefix. All the LTE examples, except for Example 10.1, are based on LTE Release 8.

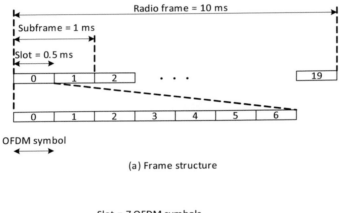

(a) Frame structure

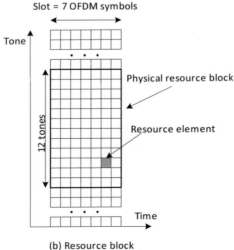

(b) Resource block

Figure 2.12 LTE frame structure and physical resource block. Frame type 1 (FDD) and normal cyclic prefix.

2.2 From OFDM to OFDMA

In LTE, most of the time-frequency resources (PRBs) are dynamically allocated to different UEs using the downlink and uplink data channels, called Physical Downlink Shared Channel (PDSCH) and Physical Uplink Shared Channel (PUSCH). In this example, we examine how LTE achieves frequency diversity and interference averaging in those two channels with resource block allocation and hopping.

Figure 2.12(a) depicts the frame structure (type 1) in the time domain, which is the same for both the downlink and uplink. A block of 12 contiguous tones over a slot is called a physical resource block (PRB) as shown in Figure 2.12(b). The basic tone spacing is 15 kHz. Thus a PRB occupies 180 kHz. The total number of PRBs available in a LTE system depends on the system bandwidth (from 1.4 MHz to 20 MHz) and ranges from 6 to 110. A PRB is the basic unit of resource allocation. A subframe consists of two slots and is the minimum time unit for scheduling. A UE is always allocated one or more pairs of PRBs over two slots of a subframe. Therefore, the LTE time structure allows block hopping in PDSCH and PUSCH. Symbol-rate hopping is also employed in LTE for reference signals (see Figure 7.20).

In PDSCH, there are three types of resource allocation. In type 0, P PRBs consecutive in frequency are grouped to be a resource block group (RBG), as shown in Figure 2.13(a), where P is the RBG size and depends on and increases with the system bandwidth. The basic unit of resource allocation becomes an RBG instead of a PRB. A resource allocation is expressed with a bitmap where each bit indicates whether the corresponding RBG is allocated to a given UE or not. Clearly, the resource allocation is less granular in frequency; however, the benefit is the reduction in the overhead of signaling a resource allocation. A drawback of type 0 allocation is reduced interference averaging (assuming neighboring cells use type 0 as well). To solve this problem, as shown in Figure 2.13(b), type 1 allocation divides all the RBGs into a number of dispersed subsets with the number of subsets equal to P. Within a subset, a bitmap indicates whether a PRB is allocated.

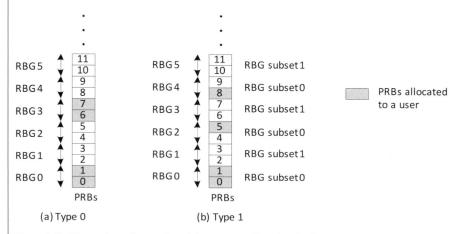

Figure 2.13 Illustration of types 0 and 1 resource allocation in the downlink PDSCH channel. $P = 2$.

Unlike type 0 or type 1, type 2 does not use a bitmap but employs the notion of virtual resource blocks (VRBs). A VRB is of the same size as a PRB. The difference is that the frequency position of a VRB is not fixed beforehand as in the case of a PRB, but can vary according to a mapping function. The idea is to allocate resources in terms of VRBs and then map VRBs to PRBs for actual transmissions. Specifically, a resource allocation indicates the start position and the length of a set of consecutive VRBs. Two types of VRBs are specified in LTE, namely localized VRBs and distributed VRBs. Localized VRBs are identical to PRBs and the mapping is direct. Distributed VRBs are mapped to PRBs in two steps, as shown in Figure 2.14: *interleaving* such that consecutive VRBs are mapped to non-consecutive PRBs that spread over a large frequency bandwidth, and *splitting* by keeping the PRBs unchanged in the first slot and modularly shifting the PRBs in the second slot to different frequencies by an offset. Splitting can be seen as block hopping at the slot boundary. The goal of distributed VRBs is to achieve frequency diversity and interference averaging, while local VRBs are suitable for frequency selective scheduling.

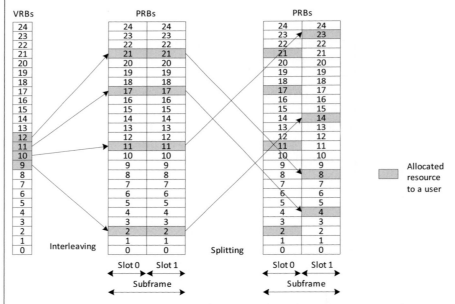

Figure 2.14 Illustration of type 2 resource allocation in the downlink PDSCH channel with distributed VRBs. The total number of PRBs is 25 and the offset is equal to 12.

The idea of VRB-to-PRB mapping can be applied to the uplink. Similar to type 2 allocation in PDSCH, resources of PUSCH are allocated in terms of VRBs, which are then mapped to PRBs. Note from Section 2.3 that the LTE uplink uses localized SC-FDMA and the PUSCH sent by a UE occupies a set of PRBs contiguous in frequency. Thus the step of interleaving in Figure 2.14 is skipped and only slot-based block frequency hopping takes place.

There are two types of hopping, as shown in Figure 2.15. The first type is *cell-specific hopping*. The bandwidth of PUSCH is divided into a few subbands. The VRBs are shifted by a number of subbands to a set of PRBs according to a cell-specific hopping pattern. Different shifts can be used in the two slots of a subframe. In addition, a cell-specific mirroring pattern can be applied to the second slot such that the PRBs in each subband are numbered in a reverse order. The second type is *explicit offsetting*. The PRBs in the first slot are identical to the allocated VRBs and move an offset of $1/2$, $1/4$, or $-1/4$ of the PUSCH bandwidth in the second slot.

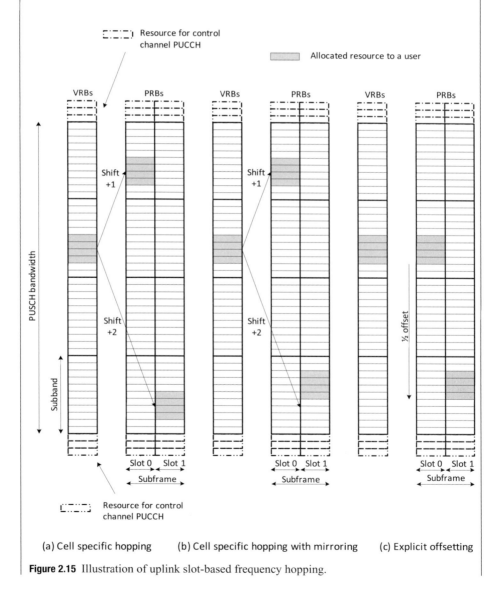

(a) Cell specific hopping (b) Cell specific hopping with mirroring (c) Explicit offsetting

Figure 2.15 Illustration of uplink slot-based frequency hopping.

> In addition to slot-based (intra-subframe) hopping, LTE also supports inter-subframe hopping where the PRBs allocated to a UE change from one subframe to another. Inter-subframe hopping is applicable to multiple H-ARQ transmissions to achieve frequency diversity.

2.2.5 Time-frequency synchronization and control

OFDMA involves multiple transmitters in the uplink and multiple receivers in the downlink. This leads to an important issue of time and frequency synchronization.

In a typical communications system, it is the receiver's responsibility to adjust its time and frequency to be synchronized with the received signal (see Appendix B). The same idea is applicable in the OFDMA downlink. In particular, the base station transmitter uses fixed symbol timing and carrier frequency to generate orthogonal tone signals to all the users in the sector, and each user receiver carries out time and frequency synchronization to the received downlink signal.

However, the scenario is very different in the uplink where multiple user transmitters are distributed in the sector. It would be impossible for the base station receiver to be *simultaneously* time synchronized to the received signals from *all* the user transmitters, if the user transmissions are not carefully coordinated. In other words, synchronization is no longer solely a receiver issue; a control mechanism is required at the transmitter side. Specifically, the base station receiver uses fixed symbol timing and carrier frequency to receive tone signals from all the users in the sector. Each user transmitter adjusts its time and frequency such that all the tone signals from different users arrive at the base station receiver synchronously in time and in frequency.

Let us consider time synchronization first. Figure 2.16 illustrates the downlink and uplink time synchronization within a sector where base station A communicates with users A_1 and A_2. In the downlink, the base station sends two tone signals to A_1 and A_2, respectively, with the same symbol time. After propagation delay, the signals arrive synchronously at the nearby user A_1 and the faraway user A_2, which adjust their respective receiver symbol times to be synchronized with the received signals. In the uplink, since A_1 and A_2 have different propagation delays to the base station, the faraway user A_2 needs to transmit earlier than the nearby user A_1 such that their tone signals arrive synchronously at the base station receiver.

We next examine time synchronization in the uplink. A user first synchronizes its receiver symbol time to the downlink signal. With *open-loop timing control*, the user sets its initial transmitter symbol time slaved to the receiver symbol time. The initial symbol time setting is not very accurate as the arrival time of the uplink signal at the base station varies with the round-trip propagation delay between the user and the base station. The base station further corrects the user's transmit symbol time using *closed-loop timing control* in which the base station measures the arrival time of the signal from the user and sends timing control instructing the user to advance its transmitter symbol time. Figure 2.17 illustrates the concepts of open-loop and closed-loop timing

2.2 From OFDM to OFDMA

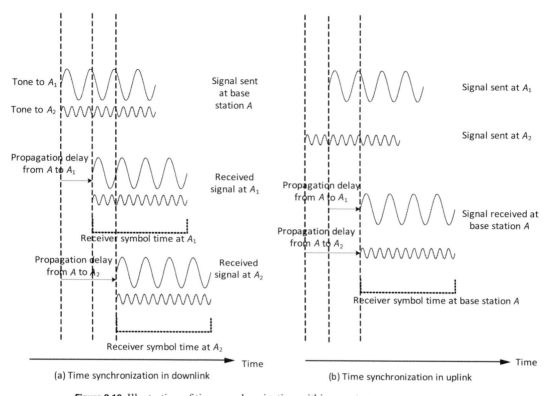

Figure 2.16 Illustration of time synchronization within a sector.

control in the uplink. When the user moves, closed-loop timing control (advance or delay transmitter symbol time) is needed periodically to compensate for changing propagation delay.

Note that because the cyclic prefix absorbs any residual timing misalignment, the accuracy requirement of closed-loop timing control is somewhat relaxed. The arrival time of an uplink signal does not have to be aligned with the base station receiver symbol time within a single sample. For example, in a 5 MHz system, a sample duration is only 0.2 μs. Suppose that the cyclic prefix is of 10 μs. Then the cyclic prefix can accommodate a few μs residual timing misalignment, assuming the delay spread of the wireless channel itself does not take the whole 10 μs. It should be pointed out that in this case, while the use of the cyclic prefix helps preserve orthogonality among the uplink signals, the residual timing misalignment causes phase rotation across tones of a given uplink signal; a residual timing misalignment of 1 μs causes a 2π phase rotation between two tones 1 MHz apart, which is in addition to the channel response of the wireless channel. The base station receiver needs to estimate and compensate for such a phase rotation in frequency prior to decoding and demodulation.

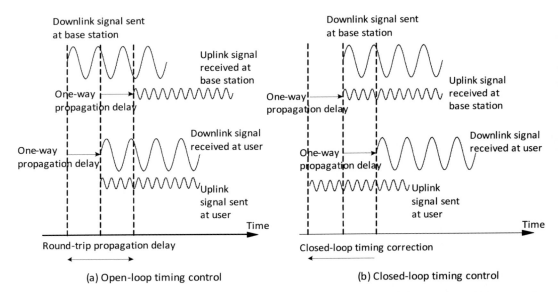

Figure 2.17 Illustration of open-loop and closed-loop timing control. In open-loop timing control (a), the uplink transmitter symbol time is slaved to the downlink receiver symbol time at the user. This results in misalignment of arrival time of the uplink signal equal to the round-trip propagation delay. In (b), the base station uses closed-loop timing control to instruct the user to advance or delay its transmitter symbol time such that the arrival time of the uplink signal is aligned with the base station receiver symbol time. The total amount of the timing correction is in effect equal to the round-trip propagation delay.

Now let us consider frequency synchronization. Similar to time synchronization, in the downlink, the user receiver synchronizes to the frequency of the dominant path of the received signal. In the uplink, the user may slave its transmit carrier frequency to the synchronized receive frequency, similar to open-loop timing control. In addition, the base station may further instruct the user to adjust its transmit frequency in a closed-loop manner. However, the key difference is that there are no equivalent "cyclic prefixes" in frequency to absorb frequency misalignment. When the frequency at the receiver is not completely synchronized with that of a received tone signal, the tone signal power leaks to other tones. Closed-loop frequency control may correct carrier frequency offset and Doppler shift, but cannot overcome multipath Doppler spread. One way to reduce the ICI impact of multipath Doppler spread is to make the tone spacing Δ_f much greater than the maximum Doppler spread. Moreover, similar to the residual timing misalignment issue, any residual frequency misalignment causes phase rotation across OFDM symbols of a given uplink signal. A residual frequency misalignment of 100 Hz causes the phase of channel coefficient $H[k]$ to rotate 2π between two OFDM symbols 10 ms apart, which is in addition to the channel response of the wireless channel. The base station receiver needs to estimate and correct such a phase rotation in time prior to decoding and demodulation.

2.2 From OFDM to OFDMA

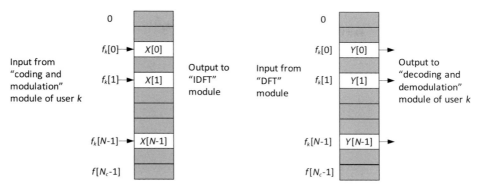

(a) User multiplexing at transmitter (b) User demultiplexing at receiver

Figure 2.18 Block diagrams of user multiplexing module at the transmitter (a) and user demultiplexing module at the receiver (b). A user is allocated N tone-symbols, indexed as $f_k[0], \ldots, f_k[N-1]$, at an OFDM symbol. The user multiplexing module takes N modulation symbols $X[0], \ldots, X[N-1]$ from the coding and modulation module of the user and inserts them in the corresponding positions in the N_c-element vector to the IDFT module. The gray elements represent the modulation symbols from other users at the base station transmitter (downlink) or zeros at the user transmitter (uplink). The user demultiplexing module takes N modulation symbols $Y[0], \ldots, Y[N-1]$, corresponding to $f_k[0], \ldots, f_k[N-1]$, of the N_c-element vector from the DFT module and provides them to the decoding and demodulation module of the user. The gray elements represent the modulation symbols to be provided to the decoding and demodulation module of other users at the base station receiver (uplink) or be discarded at the user receiver (downlink).

2.2.6 Block signal processing

The block diagram of signal processing of OFDMA is similar to that of OFDM shown in Figure 2.5, except for an additional "user multiplexing" module at the transmitter and an "user demultiplexing" module at the receiver. Those two modules are shown in Figure 2.18.

In the downlink, when the base station transmitter is to send the signals to multiple users, the coding and modulation module generates complex modulation symbols separately for each of the users. It then multiplexes the modulation symbols of those users to the corresponding tone-symbols according to the OFDMA tone-symbol allocation. Consider the tone hopping example shown in Figure 2.8 where the base station sends to five users in any OFDM symbol. The base station transmitter multiplexes one modulation symbol of each user to the tone-symbol allocated to that user as the input to the IDFT module. At the receiver side, from the output of the DFT module each user demultiplexes one symbol in its allocated tone-symbol, discards all other symbols, which are intended for other users, and further processes the demultiplexed symbol in the decoding and demodulation module.

Similarly, in the uplink, a user transmitter multiplexes the complex modulation symbols generated from the coding and modulation module to the tone-symbols allocated to it, and pads zeros to the remaining tone-symbols. At the receiver side, from the output of the DFT module the base station demultiplexes symbols that belong to different users according to their corresponding tone-symbols, and further processes them separately in the decoding and demodulation module.

We next add a few short notes on the block signal processing in Figure 2.18.

Discussion notes 2.4 Block front-end processing at the base station

While the coding and modulation module and the decoding and demodulation module are processed separately for individual users, the DFT/IDFT or filtering module is implemented as one single block. Those front-end sample rate processing modules are not done separately for different users. This helps simplify the base station implementation. In contrast, the CDMA base station needs to carry out sample rate processing, such as RAKE finger tracking, individually for different users. As a result, the complexity of the base station sample rate signal processing linearly increases with the number of users connected with the base station.

Discussion notes 2.5 Wideband processing at the user

Even if a user is allocated a subset of tone-symbols, the front-end processing at the user transceiver (DFT/IDFT and filtering) covers the entire bandwidth. This allows fast tone hopping with which the allocated subset of tone-symbols can change from OFDM symbol to another for frequency and interference diversity as shown in Figure 2.11(a). In addition, the size of the allocated subset may change rapidly with the scheduling needs. This wideband hopping is very different from narrowband FDMA hopping where a user only processes a narrowband carrier by employing a narrowband analog filter at the front-end. In FDMA, the carrier can hop from one frequency to another at a much slower rate than in wideband OFDMA hopping.

2.3 Peak-to-average power ratio and SC-FDMA

2.3.1 PAPR problem

High peak-to-average power ratio (PAPR) is a well known drawback for OFDM. What is PAPR and why is it important? Simply put, PAPR measures the power fluctuation of a signal $x(t)$ over time. For any OFDM symbol[2] $x(t), t \in [0, T_s)$, we define

$$\text{PAPR} = \frac{\max_{t \in [0, T_s)} |x(t)|^2}{\text{avg}_{t \in [0, T_s)} |x(t)|^2}. \tag{2.19}$$

High PAPR means large power fluctuation.

---

[2] Here we focus on one OFDM symbol. The cyclic prefix is ignored as it does not affect PAPR. The transition between successive OFDM symbols due to filtering is not taken into account in our PAPR analysis.

A signal with high PAPR requires large linear dynamic range of the power amplifier. If the power of the signal exceeds the linear region, the signal suffers from nonlinear distortion (e.g., clipping) and results in out-of-band spurious emissions and in-band corruption of the signal. Therefore, the operating point of the amplifier has to be set conservatively in the sense that the average transmit power is much lower than the peak power allowed by the maximum linearity. This large power backoff leads to two undesired results: low power efficiency defined as the ratio of the transmit power to the total power consumed, and low average transmit power given the linearity dynamic range of a power amplifier.

High PAPR is mostly a concern in the uplink as it affects the cost and efficiency of the power amplifier at the user.

2.3.2 PAPR of OFDMA

Suppose that in an OFDM symbol a user is allocated N tones $f[0], \ldots, f[N-1]$ ($N \leq N_c$) and transmits N i.i.d. QAM modulation symbols. Equation (2.1) is revised to be

$$x(t) = \sum_{k=0}^{N-1} X[k] \exp(j2\pi f[k]t), \, t \in [0, T_s), \quad (2.20)$$

The OFDMA signal at any time instant t is the sum of i.i.d. complex symbols. When N is large, the Central Limit Theorem says that $x(t)$ is approximately a complex Gaussian random variable. Then for a given constant PAPR_0,

$$\mathbb{P}\left(\frac{|x(t)|^2}{\text{avg}_{t \in [0,T_s)}|x(t)|^2} < \text{PAPR}_0 \right) = 1 - e^{-\text{PAPR}_0}. \quad (2.21)$$

When $N = N_c$, at the $1\times$ Nyquist sampling rate, N_c samples $x\left(\frac{l}{N_c}T_s\right), l = 0, \ldots, N_c - 1$ are i.i.d. Gaussian. Thus, $\mathbb{P}(\text{PAPR} > \text{PAPR}_0)$ is lower bounded by

$$\mathbb{P}(\text{PAPR} > \text{PAPR}_0) \geq 1 - (1 - e^{-\text{PAPR}_0})^{N_c}, \quad (2.22)$$

because the peak may occur in between two $1\times$ rate samples. The difference between $\mathbb{P}(\text{PAPR} > \text{PAPR}_0)$ and the right-hand side of this inequality is about 0.5 dB.

In the OFDMA uplink, it is more likely that $N < N_c$. In this case, the N_c samples are correlated. Figure 2.19 shows the complementary cumulative distribution curve of the PAPR. We consider two tone allocation schemes, with the N allocated tones contiguous and with them randomly selected. We observe that the PAPR in random tone allocation is strictly greater than that in contiguous tone allocation. Even with $N = 16$, the PAPR is about 10 dB with probability 10^{-4} to 10^{-3}.

2.3.3 SC-FDMA and PAPR reduction

The reason that OFDM has a large PAPR is because *independent* modulation symbols $X[0], \ldots, X[N-1]$ are modulated in the frequency domain to N tone signals,

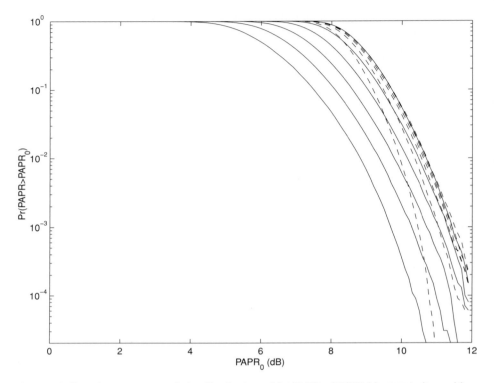

Figure 2.19 Complementary cumulative distribution of the PAPR of OFDMA: comparison with different tone allocation. $N_c = 512$. The group of solid curves is for contiguous tone allocation, and the group of dashed curves is for random tone allocation. In each group, from left to right, $N = 16, 32, 64, 128, 256, 512$.

respectively. When the N independent tone signals are added together, the signal at time t, $x(t)$, becomes a complex Gaussian random variable, which is known to have large amplitude fluctuation. The idea of single-carrier FDMA (SC-FDMA) is to modulate the symbols in the time domain instead, such that the resultant signal is no longer Gaussian and therefore has much a smaller PAPR.

To understand how SC-FDMA reduces the PAPR, first consider contiguous tone allocation where

$$f[k] = f[0] + \frac{k}{T_s}, k = 0, \ldots, N - 1, \tag{2.23}$$

where $f[0]$ is the index of the starting tone. This scheme is referred to as *localized* SC-FDMA.

Define a DFT-precoded vector

$$\tilde{X}[k] = \frac{1}{\sqrt{N}} \sum_{m=0}^{N-1} X[m] \exp\left(-j2\pi \frac{mk}{N}\right), \tag{2.24}$$

where constant $\frac{1}{\sqrt{N}}$ is added to ensure that when $X[m]$ is zero mean i.i.d., $\mathbb{E}(|\tilde{X}[k]|^2) = \mathbb{E}(|X[m]|^2)$. The transmitted signal in localized SC-FDMA, denoted by $x_1(t)$, is similar to Equation (2.20) except that $X[k]$ is replaced by $\tilde{X}[k]$:

$$x_1(t) = \sum_{k=0}^{N-1} \tilde{X}[k] \exp(j2\pi f[k]t), \, t \in [0, T_s), \quad (2.25)$$

The cyclic prefix in SC-FDMA is constructed in the same way as defined in Equation (2.6). To simplify the terminology, we treat SC-FDMA as a special form of OFDMA and still refer to $t \in [-T_{cp}, T_s)$ as an OFDM symbol.

It follows:

$$x_1(t) = \exp(j2\pi f[0]t) \sum_{k=0}^{N-1} \left(\frac{1}{\sqrt{N}} \sum_{m=0}^{N-1} X[m] \exp\left(-j2\pi \frac{mk}{N}\right) \right) \exp\left(j2\pi \frac{kt}{T_s}\right)$$

$$= \exp(j2\pi f[0]t) \sum_{m=0}^{N-1} z_m(t) X[m], \quad (2.26)$$

where

$$z_m(t) = \frac{1}{\sqrt{N}} \sum_{k=0}^{N-1} \exp\left(j2\pi k \left(\frac{t - \frac{m}{N}T_s}{T_s}\right)\right)$$

$$= \frac{1}{\sqrt{N}} \frac{\sin\left(\pi N \frac{t - \frac{m}{N}T_s}{T_s}\right)}{\sin\left(\pi \frac{t - \frac{m}{N}T_s}{T_s}\right)} \exp\left(j\pi(N-1) \frac{t - \frac{m}{N}T_s}{T_s}\right). \quad (2.27)$$

Figure 2.20 plots $|z_m(t)|$ for $m = 0, 1$. $z_0(t)$ peaks at $t = 0$ and is equal to zero at $t = \frac{1}{N}T_s, \ldots, \frac{N-1}{N}T_s$. For $m > 0$, $z_m(t)$ is obtained by cyclically shifting $z_0(t)$ in time by $\frac{m}{N}T_s$. $z_m(t)$ are orthogonal with each other at time instants $t = 0, \frac{1}{N}T_s, \ldots, \frac{N-1}{N}T_s$ and thus can be viewed as the *basis pulse functions* in SC-FDMA.

Equation (2.26) shows that $X[m]$ are modulated at time instants $t = \frac{m}{N}T_s$, because

$$x_1\left(\frac{m}{N}T_s\right) = \exp\left(j2\pi f[0]\frac{m}{N}T_s\right) \sqrt{N} X[m], \, m = 0, \ldots, N-1. \quad (2.28)$$

In between those time instants, the basis pulse functions smoothly connect between the symbols, and the amplitude of the transmitted signal $x_1(t)$ fluctuates a little (but not too much) for the following reason. Note that $|z_m(t)|$ decays quickly as $|t - \frac{m}{N}T_s|$ increases. For any $m = 0, \ldots, N-1$, in interval $t \in (\frac{m}{N}T_s, \frac{m+1}{N}T_s)$, $x_1(t)$ is dominated by $z_m(t)X[m] + z_{(m+1)\bmod N}(t)X[(m+1)\bmod N]$. This is very different from the OFDM case where N independent tones are added to form $x(t)$. As a result, $x_1(t)$ is not Gaussian and the PAPR is significantly smaller. Figure 2.21 compares the complementary cumulative distribution of the PAPR of OFDMA and SC-FDMA. At probability 10^{-3}, SC-FDMA reduces the PAPR by 2.5 to 4 dB.

Figure 2.22 compares the PAPR of different modulation schemes. We observe that the difference in the modulation schemes makes virtually no change in the PAPR in OFDMA. In SC-FDMA, the PAPR increases by about 1 dB when the modulation scheme

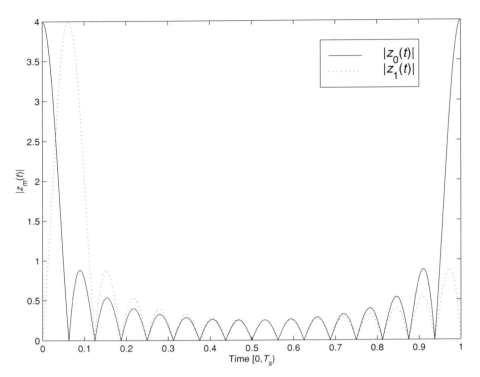

Figure 2.20 Amplitude plot of basis pulse functions $z_m(t)$ of SC-FDMA. $N = 16$.

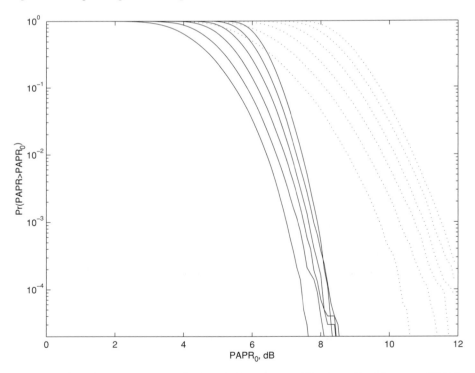

Figure 2.21 Complementary cumulative distribution of the PAPR: comparison between OFDMA and SC-FDMA. $N_c = 512$. QPSK symbols are modulated. The group of solid curves is for SC-FDMA, and the group of dotted curves is for OFDMA. In each group, from left to right, $N = 16, 32, 64, 128, 256, 512$.

2.3 Peak-to-average power ratio and SC-FDMA

Table 2.2. Duality of OFDMA and SC-FDMA.

	Frequency domain	Time domain
OFDMA	$\{X[k]\}$	$N \cdot \text{IDFT}\{X[k]\}$
SC-FDMA	$\frac{1}{\sqrt{N}} \cdot \text{DFT}\{X[k]\}$	$\sqrt{N} \cdot \{X[k]\}$

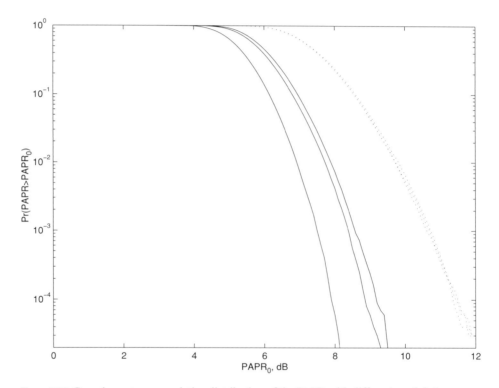

Figure 2.22 Complementary cumulative distribution of the PAPR with different modulation schemes. $N_c = 512$, $N = 64$. The group of solid curves is for SC-FDMA, and the group of dotted curves is for OFDMA. In each group, the modulation schemes from left to right are QPSK, 16-QAM, and 64-QAM.

changes from QPSK to 16-QAM or 64-QAM. Although the PAPR of 64-QAM symbols is about 1 dB higher than that of 16-QAM symbols, the difference of the corresponding SC-FDMA signals is much smaller.

It is interesting to note the time-frequency duality of OFDMA and SC-FDMA. Consider N contiguous allocated tones $f[k] = \frac{k}{T_s}$, $k = 0, \ldots, N - 1$. Table 2.2 summarizes the complex symbols modulated on frequency tones $f[k]$ and the signal values at time instants $\frac{n}{N}T_s$, $n = 0, \ldots, N - 1$ in OFDMA and in SC-FDMA.

So far we have considered the localized SC-FDMA scheme where the allocated tones are contiguous. In another scheme referred to as *distributed* SC-FDMA, the user is allocated equally spaced tones

$$f[k] = f[0] + \frac{kQ}{T_s}, k = 0, \ldots, N-1, \tag{2.29}$$

where $Q > 1$ is a positive integer. Using the same $\tilde{X}[k]$ as in (2.24), the distributed SC-FDMA transmitted signal, denoted by $x_Q(t)$, is given by

$$x_Q(t) = \sum_{k=0}^{N-1} \tilde{X}[k] \exp(j 2\pi f[k] t), \, t \in [0, T_s)$$

$$= \exp(j 2\pi f[0] t) \sum_{k=0}^{N-1} \tilde{X}[k] \exp\left(j 2\pi \frac{kQt}{T_s}\right)$$

$$= \exp(j 2\pi f[0] t) \sum_{m=0}^{N-1} z_m(Qt) X[m]. \tag{2.30}$$

Comparison of Equations (2.26) and (2.30) shows that other than the phase rotation $\exp(j 2\pi f[0] t)$, $x_Q(t)$ is obtained by compressing $x_1(t)$ from $t \in [0, T_s)$ to $t \in [0, T_s/Q)$ and then duplicating the resultant waveform $Q - 1$ times to fill up the remaining interval $t \in [T_s/Q, T_s)$. Figure 2.23 shows an example of the localized and distributed SC-FDMA waveforms. Therefore, from the perspective of the PAPR, there is no difference between the localized and distributed SC-FDMA schemes.

Is the SC-FDMA idea only limited to the above two tone allocation scenarios (contiguous and equally spaced)? In principle, for any set of allocated tones $f[k], k = 0, \ldots, N-1$, we select time instants $t[m] \in [0, T_s), m = 0, \ldots, N-1$, and $\tilde{X}[k]$ are solved by making sure that the resultant signal defined in Equation (2.25) is equal to modulation symbol $X[m]$ at $t[m]$:

$$X[m] = \frac{1}{\sqrt{N}} \sum_{k=0}^{N-1} \tilde{X}[k] \exp(j 2\pi f[k] t[m]). \tag{2.31}$$

The PAPR depends on how the continuous signal fluctuates between those time instants. One can show that if $t[m]$ is set to $t[m] = \frac{m}{N} T_s$ as before, the PAPR increases by about 2 dB with random tone allocation but by only 0.5 dB if the allocated tones consist of two equal-size contiguous tone blocks. In general, the following questions are still open:

- What is the optimal selection of $t[m]$ for any given tone allocation $f[k]$?
- What is the essential structure of tone allocation $f[k]$ to have a low PAPR?

In the following, we only focus on the localized and distributed SC-FDMA schemes.

2.3.4 Frequency domain equalization at the SC-FDMA receiver

Recall from Section 2.1 that in OFDM the receiver does not have to linearly equalize the channel to avoid noise enhancement. In SC-FDMA, when the channel is frequency

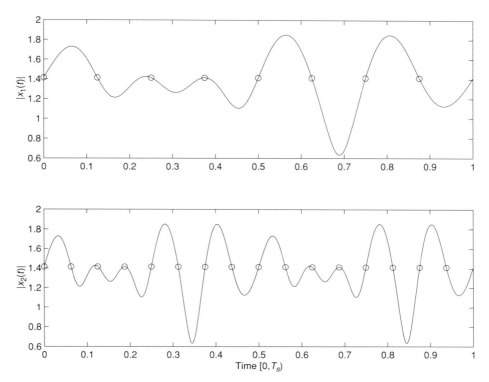

Figure 2.23 Comparison of localized SC-FDMA $x_1(t)$ and distributed SC-FDMA $x_Q(t)$. In the figure, $N = 8$, $Q = 2$, QPSK modulation. The upper portion shows localized SC-FDMA $|x_1(t)|$ where the circles represent the QPSK symbols modulated at time instants $t = 0, \frac{1}{N}T_s, \ldots, \frac{N-1}{N}T_s$. The lower portion shows distributed SC-FDMA $|x_2(t)|$ where the waveform in $t \in [0.5T_s, T_s)$ is identical to that in $t \in [0, 0.5T_s)$ and is the same as $|x_1(t)|$ in $t \in [0, T_s)$ compressed in time by a factor of 2.

selective, ISI arises among symbols modulated at time instants $t = \frac{1}{N}T_s, \ldots, \frac{N-1}{N}T_s$ within an OFDM symbol. Note that because of the cyclic prefix, there is no ISI between OFDM symbols. To deal with the ISI within an OFDM symbol, the SC-FDMA receiver has to equalize the channel. Figure 2.24 illustrates the block diagram of the SC-FDMA transmitter and receiver.

Comparison with Figure 2.5 shows that the SC-FDMA transmitter uses an additional module of size-N DFT precoding as defined in Equation (2.24). The SC-FDMA receiver first equalizes the channel in the frequency domain, and then applies size-N IDFT to recover $X[0], \ldots, X[N-1]$. A linear equalizer as in Equation (2.12) or (2.13) is often used for frequency domain equalization. As we have pointed out before, linear equalization suffers from enhanced noise and/or residual ISI. The following discussion notes and analysis illustrate SINR degradation in SC-FDMA when the channel is frequency selective.

Elements of OFDMA

Transmitter

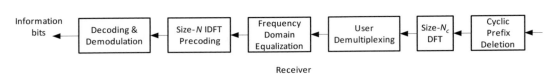

Receiver

Figure 2.24 Simplified block diagram of signal processing modules in SC-FDMA transmitter and receiver.

Discussion notes 2.6 SINR degradation in SC-FDMA

Similar to Equation (2.9) $Y[k]$ at the SC-FDMA receiver is given by

$$Y[k] = H[k]\tilde{X}[k] + W[k], k = 0, \ldots, N-1. \quad (2.32)$$

Denote by $G[k]$ the weight of the linear frequency domain equalizer. With frequency domain equalization and size-N IDFT, the estimate of $X[n]$ is calculated as follows:

$$\hat{X}_n = \sum_{k=0}^{N-1} G[k] Y[k] \exp\left(j2\pi \frac{nk}{N}\right) \quad (2.33)$$

$$= \sum_{k=0}^{N-1} G[k] \left(\frac{H[k]}{\sqrt{N}} \sum_{m=0}^{N-1} X[m] \exp\left(-j2\pi \frac{mk}{N}\right) + W[k]\right) \exp\left(j2\pi \frac{nk}{N}\right)$$

$$= \frac{1}{\sqrt{N}} \left(\sum_{k=0}^{N-1} G[k] H[k]\right) X[n]$$

$$+ \frac{1}{\sqrt{N}} \sum_{m=0, m \neq n}^{N-1} \left(\sum_{k=0}^{N-1} G[k] H[k] \exp\left(j2\pi \frac{(n-m)k}{N}\right)\right) X[m]$$

$$+ \sum_{k=0}^{N-1} G[k] \exp\left(j2\pi \frac{nk}{N}\right) W[k]. \quad (2.34)$$

In this equation, $\hat{X}[n]$ consists of the desired signal component from $X[n]$, the ISI from $X[m]$ ($m \neq n$), and the noise $W[k]$. To calculate the SINR, we assume the following:

- $X[m]$ are i.i.d.: $\mathbb{E}(X[m]X^*[n]) = P\delta(n-m)$, where P is the power of a modulation symbol.
- $W[k]$ are i.i.d.: $\mathbb{E}(W[k]W^*[l]) = \sigma^2 \delta(k-l)$, where σ^2 is the power of noise on each received tone.

- $X[m]$ and $W[k]$ are uncorrelated: $\mathbb{E}(X[m]W^*[k]) = 0$.

Hence,

$$\text{SINR} = \frac{\frac{P}{N}\left|\sum_{k=0}^{N-1} G[k]H[k]\right|^2}{\frac{P}{N}\sum_{m=0, m\neq n}^{N-1}\left|\sum_{k=0}^{N-1} G[k]H[k]e^{j2\pi\frac{(n-m)k}{N}}\right|^2 + \sigma^2 \sum_{k=0}^{N-1} |G[k]|^2}$$

$$= \frac{\left|\sum_{k=0}^{N-1} G[k]H[k]\right|^2}{N\sum_{k=0}^{N-1} |G[k]H[k]|^2 - \left|\sum_{k=0}^{N-1} G[k]H[k]\right|^2 + \frac{N}{\gamma}\sum_{k=0}^{N-1} |G[k]|^2},$$

(2.35)

where $\gamma = P/\sigma^2$. Note that in Equation (2.35), the SINR does not depend on n. This is because unlike OFDMA, SC-FDMA spreads $X[n]$ uniformly in all N tones.

To completely eliminate the ISI in Equation (2.34), $G[k]$ in zero-force equalization is set as in (2.12)

$$G[k] = \frac{1}{H[k]}. \tag{2.36}$$

Thus, Equation (2.35) reduces to

$$\text{SINR}_{\text{ZF}} = \left(\frac{1}{N}\sum_{k=0}^{N-1}\frac{1}{\gamma|H[k]|^2}\right)^{-1}, \tag{2.37}$$

where $\gamma|H[k]|^2$ represents the SINR of tone k. Equation (2.37) shows the effect of noise enhancement: if $|H[k]|^2$ on one of tones is much smaller than the rest, then SINR_{ZF} is dominated by the SINR on the worst tone.

To balance the ISI and the noise, we can set $G[k]$ according to the MMSE criterion $\min \mathbb{E}(|X[n] - \hat{X}[n]|^2)$. The optimal $G[k]$ is given by (2.13)

$$G[k] = \frac{H^*[k]}{|H[k]|^2 + \gamma^{-1}}. \tag{2.38}$$

Thus, Equation (2.35) reduces to

$\text{SINR}_{\text{MMSE}}$

$$= \frac{\left(\sum_{k=0}^{N-1}\frac{|H[k]|^2}{|H[k]|^2+\gamma^{-1}}\right)^2}{N\sum_{k=0}^{N-1}\left(\frac{|H[k]|^2}{|H[k]|^2+\gamma^{-1}}\right)^2 - \left(\sum_{k=0}^{N-1}\frac{|H[k]|^2}{|H[k]|^2+\gamma^{-1}}\right)^2 + \frac{N}{\gamma}\sum_{k=0}^{N-1}\frac{|H[k]|^2}{\left(|H[k]|^2+\gamma^{-1}\right)^2}}$$

$$= \frac{\sum_{k=0}^{N-1}\frac{\gamma|H[k]|^2}{\gamma|H[k]|^2+1}}{N - \sum_{k=0}^{N-1}\frac{\gamma|H[k]|^2}{\gamma|H[k]|^2+1}}$$

$$= \left(\frac{1}{N}\sum_{k=0}^{N-1}\frac{1}{\gamma|H[k]|^2+1}\right)^{-1} - 1. \tag{2.39}$$

From the convexity of function $\frac{1}{x+1}$, it can be shown that

$$\text{SINR}_{\text{MMSE}} \leq \gamma |\bar{H}|^2, \qquad (2.40)$$

where

$$|\bar{H}|^2 = \frac{1}{N} \sum_{k=0}^{N-1} |H[k]|^2. \qquad (2.41)$$

The above inequality (2.40) shows that linear equalization in the frequency selective channel causes loss in $\text{SINR}_{\text{MMSE}}$.

To quantify the loss, we consider a simple two-path channel model $h(\tau) = h_1 \delta(\tau) + h_2 \delta(\tau - \tau_0)$, where $|h_1| = |h_2|$ and the phase between h_1 and h_2 is uniformly distributed in $[0, 2\pi)$. For normalization, we adjust $|h_1|, |h_2|$ such that $|\bar{H}|^2 = 1$. In this case, γ would be the SINR if the channel were flat. Thus, $\gamma - \text{SINR}_{\text{MMSE}}$ measures the SINR loss in SC-FDMA. Figure 2.25 plots $\text{SINR}_{\text{MMSE}}$ versus γ for a variety of Q.

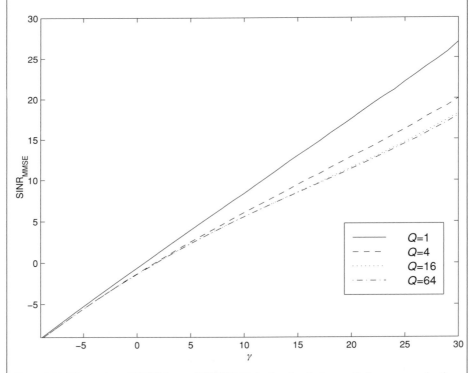

Figure 2.25 Illustration of SINR loss of SC-FDMA in the simple two-path frequency selective channel model. In this figure, $N = 16$, $\tau_0 = 1$ μs, $\Delta_f = 10$ kHz.

We observe that the SINR loss is negligible at low γ and increases with γ. At $\gamma = 30$ dB, the loss can be as large as 12 dB. The SINR loss increases with Q. For a given N, larger Q means that the allocated tones spread over larger bandwidth and may experience more severe frequency selectivity. The loss is most significant when Q increases from 1 to 4 and then saturates at $Q = 16$. From this perspective, localized SC-FDMA is superior to distributed SC-FDMA.

2.3.5 System aspects of SC-FDMA

SC-FDMA is similar to OFDMA in the sense that both are the sum of tone signals at the tones allocated to the user and employ the cyclic prefix. SC-FDMA has no ISI between two OFDM symbols. However, when the channel is frequency selective, ISI occurs among the symbols that are modulated in a given OFDM symbol. Furthermore, because of tone orthogonality, two SC-FDMA signals in an OFDM symbol do not interfere with each other as long as the allocated tones do not overlap. Indeed SC-FDMA signals and OFDMA signals can be orthogonally multiplexed using disjoint tone sets.

Constraint on resource allocation. The localized or distributed SC-FDMA scheme imposes a strict constraint on resource allocation: tones allocated to a user have to be either contiguous or equally spaced. So, what are the limitations?

First, the flexibility of scheduling and multiplexing users based on their frequency selective channel conditions is limited. For example, suppose the channel response of a user has a null in the middle of the bandwidth. Without the SC-FDMA constraint, the scheduler allocates to the user the tones at both bandwidth edges but not those in the middle. However, such tone allocation is not allowed in SC-FDMA.

Second, multiplexing control signaling and data traffic of a given user becomes complicated. Often a user needs to transmit control signaling such as channel quality indicator (CQI) reports to support downlink traffic scheduling, in addition to traffic. Without the SC-FDMA constraint, resource allocation for control signaling and traffic could be decoupled. For example, CQI reports are sent periodically and therefore a set of tones periodic in time is allocated for CQI, while traffic is generally much more bursty and thus scheduled dynamically every slot. Together, however, the tones allocated in traffic scheduling and those allocated for CQI may not satisfy the SC-FDMA resource allocation constraint.

Impact on inter-cell interference. The use of SC-FDMA affects the characteristics of inter-cell interference. Consider a simple scenario where every user in all cells is allocated N tones. In localized SC-FDMA, N tones of any user are contiguous. Thus, inter-cell interference from an adjacent cell seen by a given user mostly comes from one or two users in that cell. As a result, interference is not well averaged as compared with the OFDMA case where the N tones of the user may see interference from different users in the adjacent cell.

Moreover, in a given SC-FDMA signal, the transmit power tends to vary substantially from one tone to another. Equations (2.1) and (2.25) show that the complex symbol modulated on tone $f[k]$ is $X[k]$ in OFDMA or $\tilde{X}[k]$ in SC-FDMA. From Equation (2.24), if $X[0], \ldots, X[N-1]$ are i.i.d., then $\tilde{X}[k]$ is Gaussian when N is large. Suppose modulation symbols $X[k]$ are QPSK. Then the transmit power is the same on every tone in OFDMA, but varies according to the Gaussian distribution in SC-FDMA. This is evident from the time-frequency duality of OFDMA and SC-FDMA described in Table 2.2.

Localized versus distributed schemes. Finally we briefly compare the localized and distributed SC-FDMA schemes. From the perspective of SINR loss due to equalization in the frequency selective channel, the localized scheme is superior. In addition, since the distributed scheme covers wider bandwidth, it requires more pilot overhead for the receiver to estimate the channel.

By spreading the signal into a wider bandwidth, the distributed scheme has better frequency diversity. On the other hand, the localized scheme concentrates the signal on a narrower bandwidth and makes it easier to exploit frequency selectivity. Specifically, the base station scheduler tracks frequency selective channel quality and schedules the user in the most favorable portion of the bandwidth. We will further study frequency diversity and selective scheduling in Chapter 4.

Practical example 2.2 Uplink data and control channels in LTE

The LTE uplink supports the localized SC-FDMA scheme. An important constraint is that a UE only transmits in a set of contiguous tones. We briefly describe three major channels in the LTE uplink, namely PUSCH (Physical Uplink Shared Channel), PUCCH (Physical Uplink Control Channel), and SRS (Sounding Reference Signal), with the emphasis on how that constraint is met. We will not get into the details of physical channel processing of these channels.

PUSCH is used to carry traffic data and control signals for higher layers. PUCCH is used to transfer physical layer control information including:

- Downlink channel quality indicator (CQI) to assist adaptive modulation and coding and channel-aware scheduling in PDSCH.
- Downlink precoding matrix indicator (PMI) and Rank Information (RI) for downlink MIMO transmissions. RI indicates the maximum number of layers for spatial multiplexing and PMI the preferred precoding matrix.
- H-ARQ acknowledgments (H-ARQ-ACK) to indicate whether a corresponding PDSCH has been received successfully.
- Scheduling requests (SR) to request PUSCH channel resource.

As shown in Figure 2.28 later, PUSCH occupies the PRBs in the middle of the total bandwidth. PUCCH takes both edges so as to maximize frequency diversity (when combined with frequency hopping; see below) and not to fragment the bandwidth (thereby making it easy to allocate contiguous PRBs to PUSCH).

Let us first consider PUSCH. As described in Figure 2.24, information bits of a transport block are channel coded, mapped to modulation symbols, and processed with DFT precoding (ignoring other steps such as CRC insertion, segmentation, and scrambling). The output of DFT precoding is a block of complex symbols to be mapped to the PUSCH physical resource allocated to the UE. Example 2.1 describes the allocation and hopping of the PUSCH physical resource. In a slot, the middle

OFDM symbol is used to send DM-RS, which is for coherent demodulation at the eNB receiver. DM-RS is defined by a cyclic shift of some base sequence based on a Zadoff-Chu sequence (when the number of PRBs is at least 3) or a special QPSK sequence (when the number is 1 or 2). The complex symbols from DFT precoding are mapped to the remaining six OFDM symbols, as shown in Figure 2.26.

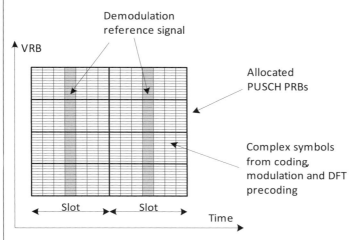

Figure 2.26 Illustration of PUSCH physical resource mapping in a subframe. This figure shows VRBs and does not include slot-based hopping shown in Figure 2.15 that maps VRBs to PRBs for frequency diversity.

Next we consider PUCCH. Roughly speaking, H-ARQ-ACK (1 or 2 bits, equal to one BPSK or QPSK symbol) and SR (ON-OFF keying) are sent with PUCCH format 1 and CQI/PMI/RI (20 coded bits, equal to 10 QPSK symbols) are sent with format 2. Figure 2.27 shows the physical resource mapping of PUCCH formats 1 and 2.

In format 1, the middle three OFDM symbols in each slot are for DM-RS. The BPSK/QPSK symbol of a H-ARQ-ACK or SR report is modulated in the remaining four OFDM symbols. Clearly sending only one BPSK/QPSK in a PRB is wasteful. To utilize the resource efficiently, multiple format 1 BPSK/QPSK symbols of different UEs orthogonally share the same PRB in a code division multiplexing (CDM) manner. Specifically, the BPSK/QPSK symbol of a UE is multiplied by a length-4 orthogonal cover sequence, and its reference signal is also multiplied by a length-3 orthogonal cover sequence. Each of the seven resultant symbols is then multiplied by a length-12 sequence to generate 12 symbols, which are sent in the 12 tones of the PRB in a corresponding OFDM symbol in the slot. The length-12 sequence is an orthogonal phase rotation of a CS-RS sequence. There are at most 12 orthogonal phase rotations given 12 tones in a PRB. As long as two UEs are assigned different orthogonal covers or phase rotations, their BPSK/QPSK PUCCH signals are orthogonal. Hence, up to $3 \cdot 12 = 36$ different UEs can ideally be multiplexed in the code space, although not all the 12 phase rotations are orthogonal in a frequency selective channel.

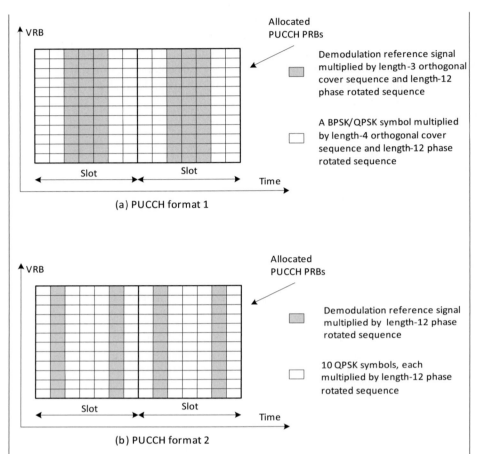

Figure 2.27 Illustration of PUCCH physical resource mapping in a subframe. This figure shows VRBs and does not include slot-based hopping shown in Figure 2.28 that maps VRBs to PRBs for frequency diversity.

In format 2, there is a total of four OFDM symbols in a subframe for DM-RS. Each of the 10 QPSK symbols of a CQI/PMI/RI report is sent in a remaining OFDM symbol. A QPSK symbol is multiplied by a phase rotated length-12 cell-specific sequence to generate 12 symbols, which are sent in the 12 tones of the PRB in a corresponding OFDM symbol.

Because both formats use phase rotated length-12 cell-specific sequences, format 1 and 2 PUCCH signals can be multiplexed in the same PRB using different orthogonal phase rotations. A phase rotation can be used either by three format 1 or one format 2 PUCCH signals since format 2 does not use orthogonal cover sequences to separate UEs of the same phase rotation. In addition to orthogonal multiplexing different UEs of the same cell in a PUCCH PRB, the phase rotations of the CS-RS sequence vary from one OFDM symbol to another according to a cell-specific hopping sequence so as to randomize inter-cell interference.

Furthermore, similar to Figure 2.15 shown for PUSCH, slot-based block hopping is used to maximize frequency diversity for PUCCH, as shown in Figure 2.28. Note that the number of PRBs needed for PUCCH, particularly the number of H-ARQ-ACK reports, may vary from one subframe to another. PRBs allocated for format 2 are at the bandwidth edge and for format 1 are next to PUSCH so that the PUSCH bandwidth can be adjusted dynamically to absorb such a variation.

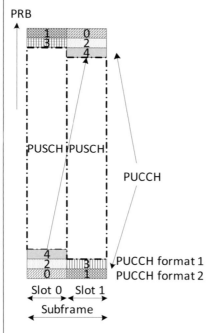

Figure 2.28 Illustration of PUCCH slot-based block hopping.

So far we have considered the basic case of a single control report (H-ARQ-ACK, SR, or CQI/PMI/RI). We next address more complex scenarios in practice. The first question is, how to send more than one report in a subframe? The PUCCH resource for CQI/PMI/RI and SR is determined beforehand and occurs in a periodic manner. In principle, the eNB can schedule the PUCCH resource such that a UE is never required to send CQI/PMI/RI and SR reports simultaneously. On the other hand, the PUCCH resource for H-ARQ-ACK is slaved to PDSCH in the sense that if a UE receives PDSCH, it should transmit H-ARQ-ACK in a corresponding subframe. Thus, H-ARQ-ACK may overlap in time with CQI/PMI/RI or SR for a given UE.

The general principle of handling this situation is to transmit in only one PUCCH PRB to keep the PAPR low. The specific processing scheme depends on the situation as exemplified below:

- H-ARQ-ACK and SR. Note that SR is signaled with ON-OFF keying. The UE does not transmit any signal if it has no SR to send. In this case, the UE transmits the H-ARQ-ACK report in the corresponding PUCCH resource of H-ARQ-ACK.

However, if the UE intends to send an SR report, it instead transmits the H-ARQ-ACK report in the PUCCH resource of SR without explicitly sending the SR report. The PUCCH resource of H-ARQ-ACK is left unused. By comparing the received energy in the SR and H-ARQ-ACK resource, the eNB receiver is able to determine which PUCCH resource actually carries a signal and therefore whether the UE transmits SR.
- H-ARQ-ACK and CQI. In this case, the PUCCH resource of H-ARQ-ACK is left unused and the PUCCH format 2 is modified slightly in the CQI resource. Specifically, in each slot the BPSK/QPSK symbol of a H-ARQ-ACK report is sent in the second OFDM symbol that is reserved for DM-RS in the original format 2 (see Figure 2.27b).

While the above processing schemes allow multiple reports to be sent simultaneously, they may degrade the decoding performance and lead to more subtle error events. For example, in the H-ARQ-ACK and CQI case, using only one OFDM symbol for DM-RS, as opposed to two, sacrifices the performance at high Doppler frequencies. Moreover, now it becomes hard to detect whether the UE actually sends H-ARQ-ACK. Note that there are indeed three states of a H-ARQ-ACK report: ACK, NAK, or no report. NAK indicates failed decoding of PDSCH while no report means the UE misses the scheduling grant. Clearly the error recovery mechanism is different for those two events. However, carrying the H-ARQ-ACK modulation symbol in the resource that would otherwise be used for DM-RS makes it hard for the eNB receiver to detect which error event actually takes place.

The next question is how to send both data and control in a subframe from a given UE? This can happen since data transmission in PUSCH is determined by the uplink scheduler, which may not always coordinate with the scheduling of H-ARQ-ACK and the PUCCH resource scheduling of CQI/PMI/RI and SR.

The principle of handling this situation is to leave the PUCCH resource unused and transmit control symbols in PUSCH. Together, the data and control symbols are time multiplexed using the SC-FDMA scheme.

In general, control symbols need to be better protected than data symbols because of the lack of H-ARQ and short codeword lengths. This can be done by boosting the transmit power of a control symbol as compared with a data symbol. However, such a scheme leads to power fluctuation and a high PAPR. Alternatively, one can keep the transmit power constant over the subframe and increase the number of REs of control symbols for better coding protection. In LTE, the number of REs for CQI/PMI, H-ARQ-ACK, or RI depends on the coding and modulation scheme of the PUSCH and an offset parameter, which is used to control the reliability of control symbols relative to data symbols. Figure 2.29 illustrates the time multiplexing of control and data symbols in PUSCH. Note that the H-ARQ-ACK symbols are mapped near the reference signal to benefit from better channel estimation at the eNB receiver.

Finally, SRS is used for the eNB to estimate the uplink channel quality. SRS is sent in the last OFDM symbol of a subframe. The subset of subframes in which SRS may be sent is specified in a downlink broadcast message. To avoid collision, PUSCH is restrained from transmitting in the last OFDM symbols in those subframes.

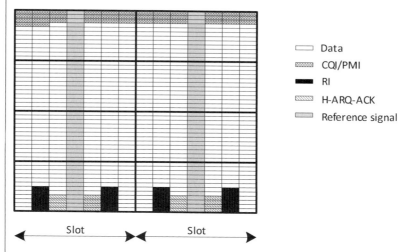

Figure 2.29 Illustration of multiplexing control and data symbols in PUSCH. Each column represents the block of symbols to be processed according to the SC-FDMA scheme, namely with DFT precoding and IFFT in an OFDM symbol. Although shown in a 2-D grid, the symbols in a column do not correspond to different tones but to time instants within an OFDM symbol.

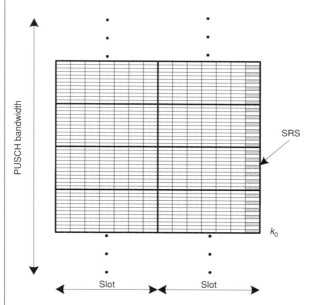

Figure 2.30 Illustration of an SRS signal.

Figure 2.30 shows an example of an SRS. In frequency, it starts from some tone k_0 and occupies every other tones for some bandwidth, thereby creating a comb-like spectrum. As far as modulation symbols are concerned, an SRS signal is similar to DM-RS described in Example 2.2. In particular, to generate orthogonal SRS signals, different cyclic shifts can be applied to the same base sequence derived as a cyclic extension of prime-length Zadoff-Chu sequences. Thus, multiple UEs can send their SRS signals in the physical resource shown in Figure 2.30 in a CDM manner. In addition, other UEs can share the same OFDM symbol in an FDM manner by utilizing other tones, particularly since an SRS occupies every other tones. A given UE is scheduled to transmit an SRS signal periodically, where the period ranges from 2 to 320 ms.

The SRS signal is intended to cover a wide bandwidth to facilitate a frequency selective scheduler. The eNB scheduler measures the channel quality of different portions of the PUSCH bandwidth and schedules a UE in favorable ones. To this end, the SRS signal can be wideband in one OFDM symbol or be narrowband (with the smallest bandwidth equal to 4 PRBs) but hop over frequency in a time duration of multiple scheduled subframes.

2.4 Real-world impairments

So far we have assumed that, excluding the cyclic prefix, the tone signals in OFDMA[3] are orthogonal. As will be elaborated in Chapter 3, orthogonality is a fundamental property of OFDMA, and a central theme of the OFDMA mobile broadband system design principles is to exploit orthogonality. However, in the real world no two tone signals are *perfectly* orthogonal because of imperfect wireless channel or implementation. The *level of orthogonality*, measured by the interference power between the tone signals, determines the extent to which the OFDMA system can exploit orthogonality in practice. In this section, we will model the ICI caused by a variety of impairments. To simplify the equations, we ignore the additive thermal noise.

2.4.1 Carrier frequency offset and Doppler effect

First, consider frequency offset $\delta_f \Delta_f$ between the transmitter and the receiver where $\delta_f \leq 0.5$. If $\delta_f > 0.5$, then most of the signal energy of tone k will show up in a different tone n, $n \neq k$. We ignore this integer tone shift scenario. The frequency offset can be caused by carrier frequency mismatch between the transmitter and receiver or by Doppler shift in the channel.

Consider a signal transmitted at tone k, $X[k] \exp(j2\pi f[k]t)$, where $X[k]$ is the complex symbol with normalized signal power $|X[k]|^2 = 1$. The output of the DFT module

---

[3] The discussion in this section is also applicable to SC-FDMA. We use "OFDMA" as a general term to include both OFDMA and SC-FDMA.

2.4 Real-world impairments

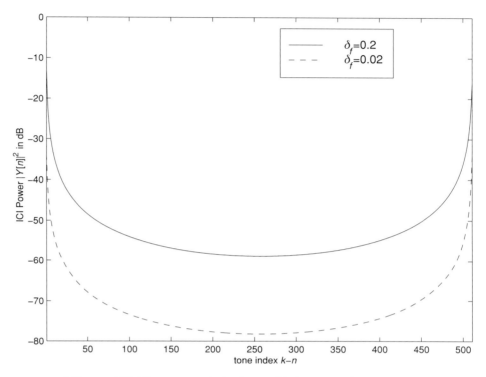

Figure 2.31 ICI power $|Y[n]|^2$ versus tone index $k - n$ with frequency offset δ_f.

at the receiver is

$$Y[n] = \frac{1}{N_c} \sum_{l=0}^{N_c-1} X[k] \exp\left(j2\pi \frac{(k-n+\delta_f)l}{N_c}\right)$$

$$= \frac{\sin(\pi(k-n+\delta_f))}{N_c \sin\left(\pi \frac{k-n+\delta_f}{N_c}\right)} \exp\left(j2\pi \frac{N_c-1}{N_c}(k-n+\delta_f)\right) X[k]. \quad (2.42)$$

Note that the desired signal component at tone k is distorted by factor $\frac{\sin(\pi \delta_f)}{N_c \sin\left(\pi \frac{\delta_f}{N_c}\right)} \exp\left(j2\pi \frac{N_c-1}{N_c} \delta_f\right)$, which is independent of k.

Figure 2.31 plots $|Y[n]|^2$, $n \neq k$ versus tone index $k - n$. We observe that the ICI power drops drastically as n moves away from k in a modular sense, or more precisely as $\left|\frac{N_c}{2} - ((k-n) \mod N_c)\right|$ decreases.

Now let us consider Doppler spread. For flat fading, the received signal is given as

$$y(t) = g(t)X[k]\exp(j2\pi f[k]t), \quad (2.43)$$

where $g(t)$ is a wide-sense stationary stochastic process with zero mean and unit variance. Define correlation function of $g(t)$ to be

$$r(\tau) = \mathbb{E}(g(t+\tau)g^*(t)). \quad (2.44)$$

Some commonly used models of $r(\tau)$ are listed below:

$$r(\tau) = J_0(2\pi f_d \tau), \text{ Jake's model} \quad (2.45)$$

$$r(\tau) = \text{sinc}(f_d \tau), \text{ uniform model} \quad (2.46)$$

$$r(\tau) = \cos(2\pi f_d \tau), \text{ two-path model}, \quad (2.47)$$

where f_d is the maximum Doppler shift, $J_0(\cdot)$ is the zeroth-order Bessel function of the first kind.

The output of the DFT module at the receiver is

$$Y[n] = \frac{1}{N_c} \sum_{l=0}^{N_c-1} g[l] \exp\left(j2\pi \frac{(k-n)l}{N_c}\right) X[k], \quad (2.48)$$

where $g[l] = g\left(\frac{l}{N_c} T_s\right)$. It follows:

$$\mathbb{E}(|Y[n]|^2) = \frac{1}{N_c^2} \sum_{l_1=0}^{N_c-1} \sum_{l_2=0}^{N_c-1} \mathbb{E}(g[l_1]g[l_2]^*) \exp\left(j2\pi \frac{(k-n)(l_1-l_2)}{N_c}\right)$$

$$= \frac{1}{N_c} \sum_{l=-N_c+1}^{N_c-1} r[l] \left(1 - \frac{|l|}{N_c}\right) \exp\left(j2\pi \frac{(k-n)l}{N_c}\right), \quad (2.49)$$

where $r[l] = r\left(\frac{l}{N_c} T_s\right)$. $\mathbb{E}(|Y[k-n]|^2)$ drops quickly with $k-n$, similar to what is shown in Figure 2.31. Furthermore,

$$\sum_{n \neq k} \mathbb{E}(|Y[n]|^2) = \frac{1}{N_c} \sum_{l=-N_c+1}^{N_c-1} r[l] \cdot \left(1 - \frac{|l|}{N_c}\right) \sum_{n \neq k} \exp\left(j2\pi \frac{(k-n)l}{N_c}\right)$$

$$= \frac{N_c - 1}{N_c} r[0] - \frac{1}{N_c} \sum_{l=-N_c+1, l\neq 0}^{N_c-1} r[l] \cdot \left(1 - \frac{|l|}{N_c}\right)$$

$$= \frac{N_c - 1}{N_c} r[0] - \frac{2}{N_c} \sum_{l=1}^{N_c-1} r[l] \cdot \left(1 - \frac{l}{N_c}\right). \quad (2.50)$$

The last step assumes that the correlation function $r(\cdot)$ is real and thus an even function.

Figure 2.32 plots the total expected ICI power $\sum_{n \neq k} \mathbb{E}(|Y[n]|^2)$ for a variety of $\delta_f = f_d T_s$ using the models in Equations (2.45) (2.46) and (2.47). We note that the ICI powers of the two-path model and the frequency offset model are the same and are more severe than that of the other two models.

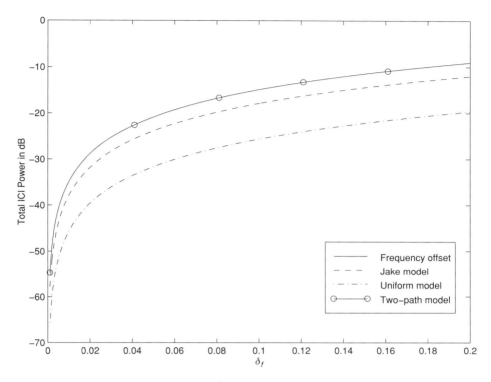

Figure 2.32 Total ICI power $\sum_{n \neq k} \mathbb{E}(|Y[n]|^2)$ versus δ_f using the frequency offset model and the Jake's, uniform, and two-path Doppler spread models.

2.4.2 Arrival time beyond the cyclic prefix

A signal that arrives at the receiver beyond the cyclic prefix interval will cause ISI and ICI. This may occur because of symbol synchronization mismatch between the transmitter and receiver or delay spread exceeding the cyclic prefix.

To study the effect of the time misalignment, suppose that the transmitter sends $X[-1, k]$ and $X[0, k]$ on tone k during two successive OFDM symbols, where $X[-1, k]$ and $X[0, k]$ are i.i.d. complex modulation symbols. The symbol 0 signal arrives beyond the cyclic prefix interval of the receiver symbol 0 time. As shown in Figure 2.33, the first N_G input samples of the DFT module come from the symbol -1 signal and the remaining $N_c - N_G$ samples from the symbol 0 signal.

The output of the DFT module at the receiver is

$$Y[n] = \frac{1}{N_c} \sum_{l=0}^{N_G-1} X[-1, k] \exp\left(j2\pi \frac{k(l + N_c - N_G) - nl}{N_c}\right)$$
$$+ \frac{1}{N_c} \sum_{l=0}^{N_c-N_G} X[0, k] \exp\left(j2\pi \frac{k(l + N_c - N_{cp}) - n(l + N_G)}{N_c}\right). \quad (2.51)$$

Elements of OFDMA

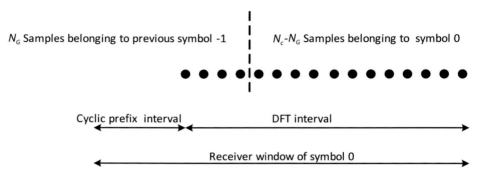

Figure 2.33 Illustration of arrival time beyond the cyclic prefix.

The first term of $Y[n]$ is the ISI from symbol -1. The second term of $Y[n]$, $n \neq k$, is the ICI from the current symbol 0. It follows

$$\mathbb{E}(|Y[n]|^2) = \frac{\sin^2\left(\pi \frac{(k-n)}{N_c}(N_c - N_G)\right) + \sin^2\left(\pi \frac{(k-n)}{N_c} N_G\right)}{\left(N_c \sin\left(\pi \frac{k-n}{N_c}\right)\right)^2}$$

$$= \frac{2\sin^2\left(\pi \frac{(k-n)}{N_c} N_G\right)}{\left(N_c \sin\left(\pi \frac{k-n}{N_c}\right)\right)^2}. \tag{2.52}$$

Figure 2.34 plots $|Y[n]|^2$, $n \neq k$ versus tone index $k - n$. We observe that similar to Figure 2.31 the ICI/ISI power drops drastically as n moves away from k.

The total ICI/ISI power is given by $\sum_{n \neq k} \mathbb{E}(|Y[n]|^2) + \left(\frac{N_G}{N_c}\right)^2$. The last term is the ISI on tone k from symbol -1. Figure 2.35 plots the total ICI/ISI power versus N_G/N_c. Note that the ICI/ISI is caused by the N_G samples of symbol -1 that enter the DFT module and the missing N_G samples of symbol 0. Therefore we approximate the total ICI/ISI power to be $\frac{2N_G}{N_c}$. Figure 2.35 shows that the approximation is quite close.

2.4.3 Sampling rate mismatch

In general, the sampling clock is usually different at the user and at the base station. To reduce cost, the crystal used at the user is usually less accurate than that used at the base station. When the user is communicating with a base station, it can detect the base station sampling rate from the downlink signal and use it to adjust its own sampling rate for both the downlink and the uplink. This way, the base station uses a fixed sampling rate and does not need to explicitly control the user. In the following, we study the interference from any residual sampling rate mismatch.

Suppose that the receiver sampling rate is N_c/T_s, as before, but the transmitter sampling rate is $N_c/(T_s(1 + \delta_s))$. Consider a signal transmitted at tone k, $X[k]\exp(j2\pi f[k]t(1 + \delta_s))$, where $X[k]$ is the complex symbol with normalized signal power $|X[k]|^2 = 1$. The received signal samples l are $X[k]\exp\left(j2\pi \frac{kl(1+\delta_s)}{N_c}\right)$. The output

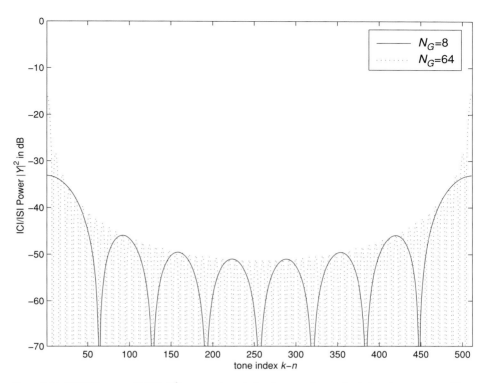

Figure 2.34 ICI/ISI power $\mathbb{E}(|Y[n]|^2)$ versus tone index $k - n$ when arrival time is not covered by the cyclic prefix. $N_c = 512$.

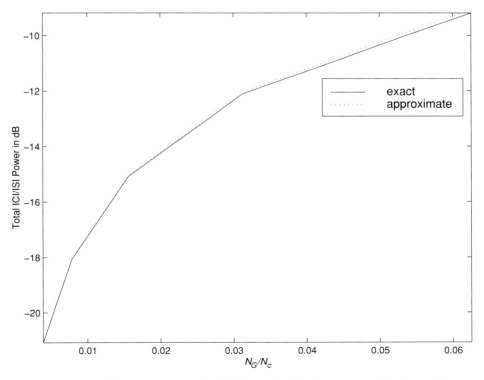

Figure 2.35 Total ICI/ISI power versus N_G/N_c when arrival time is not covered by the cyclic prefix.

of the DFT module at the receiver is

$$Y[n] = \frac{1}{N_c} \sum_{l=0}^{N_c} X[k] \exp\left(j2\pi \frac{kl(1+\delta_s)}{N_c}\right) \exp\left(-j2\pi \frac{nl}{N_c}\right)$$

$$= \frac{1}{N_c} \sum_{l=0}^{N_c} X[k] \exp\left(j2\pi l \frac{k-n+\delta_s k}{N_c}\right)$$

$$= \frac{\sin(\pi(k-n+\delta_s k))}{N \sin\left(\pi \frac{k-n+\delta_s k}{N_c}\right)} \exp\left(j\pi \frac{N_c - 1}{N_c}(k-n+\delta_s k)\right) X[k]. \quad (2.53)$$

It follows that

$$|Y[n]|^2 = \left(\frac{\sin[\pi(k-n+\delta_s k)]}{N_c \sin\left(\pi \frac{k-n+\delta_s k}{N_c}\right)}\right)^2 \quad (2.54)$$

Note that Equations (2.49) and (2.52) depend only on $k - n$ while Equation (2.54) depends on both $k - n$ and tone index k. In the previous two impairments, the total interference power does not depend on the tone index of which the signal is sent. However, this is not the case with the sampling rate mismatch impairment. Let the transmitted signal to be $X[k]$ on tone k for $k = 0, \ldots, N_c - 1$ and $X[k]$ are all i.i.d. Define $ICI_s(k)$ the total ICI power on all tone n's ($n \neq k$) where the ICI is caused by a signal on tone k, and $ICI_d(n)$ the total ICI power seen on the tone n where the ICI is caused by signals sent on all tone k's ($k \neq n$). Figure 2.36 plots $ICI_s(k)$ and $ICI_d(n)$. We observe that they are flat over a large range of n or k.

Phase noise

Phase noise is caused by the instability of phase in a local oscillator. Consider a baseband signal $x(t)$. The signal affected by phase noise can be modeled as

$$y(t) = x(t) \exp[j\phi(t)]$$
$$\approx x(t)[1 + j\phi(t)], \quad (2.55)$$

where $\phi(t)$ is assumed to be small. $\phi(t)$ is modeled as a stationary Gaussian process with zero mean, and the power spectrum density $L(f)$ is typically given as follows:

$$L(f) = 10^{-c} + \begin{cases} 10^{-a}, & |f| \leq f_1 \\ 10^{-(|f|-f_1)\frac{b}{f_2-f_1} - a}, & |f| > f_1. \end{cases} \quad (2.56)$$

Define autocorrelation function

$$R[l] = \int_{-\infty}^{\infty} L(f) \exp\left(j2\pi f T_s \frac{l}{N_c}\right) df. \quad (2.57)$$

2.4 Real-world impairments

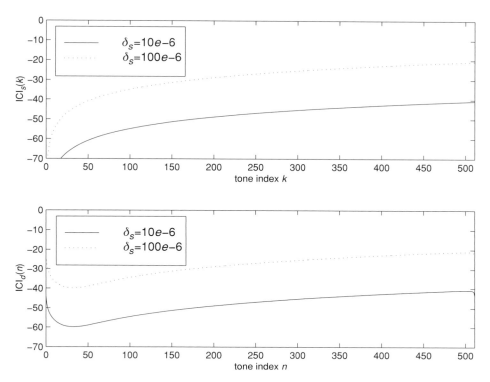

Figure 2.36 $\text{ICI}_s(k)$ versus k and $\text{ICI}_d(n)$ versus n in the presence of sampling rate mismatch. $N_c = 512$.

Consider a signal transmitted at tone k, $X[k]\exp(j2\pi f[k]t)$, where $X[k]$ is the complex symbol with normalized signal power $|X[k]|^2 = 1$. The output of the DFT module at the receiver is

$$Y[n] = \frac{1}{N_c} \sum_{l=0}^{N_c-1} X[k](1 + j\phi[l])\exp\left(j2\pi\frac{(k-n)l}{N_c}\right)$$
$$= X[k]\delta(k-n) + jX[k]\Phi(k-n), \quad (2.58)$$

where $\phi[l] = \phi\left(\frac{l}{N_c}T_s\right)$, and

$$\delta(k-n) = \begin{cases} 1, & \text{if } k = n \\ 0, & \text{otherwise} \end{cases} \quad (2.59)$$

$$\Phi(k-n) = \frac{1}{N_c} \sum_{l=0}^{N_c-1} \phi[l]\exp\left(j2\pi\frac{(k-n)l}{N_c}\right). \quad (2.60)$$

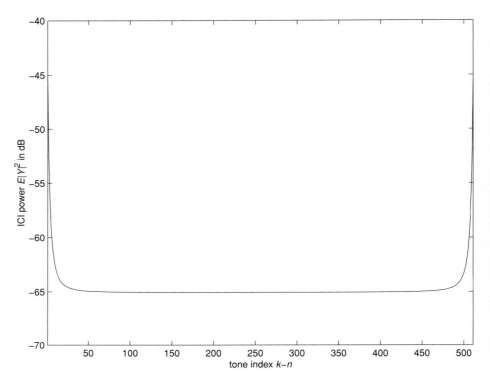

Figure 2.37 ICI power $\mathbb{E}(|Y[n]|^2)$ versus tone index $k - n$ in the presence of phase noise. $N_c = 512$, $T_s = 100$ µs. Phase noise parameters: $a = 6.5$, $b = 4$, $c = 10.5$, $f_1 = 1$ kHz, and $f_2 = 10$ kHz.

Note that $Y[k] = X[k][1 + j\Phi(0)]$, so the desired signal component at tone k is distorted by factor $1 + j\Phi(0)$. Moreover, when $n \neq k$,

$$\mathbb{E}(|Y[n]|^2) = \mathbb{E}(|\Phi(k-n)|^2)$$

$$= \frac{1}{N_c} R[0] + \frac{2}{N_c^2} \sum_{l=1}^{N_c-1} R[l](N_c - l) \cos\left(\frac{2\pi(k-n)l}{N_c}\right). \quad (2.61)$$

Figure 2.37 plots $\mathbb{E}(|Y[n]|^2)$, $n \neq k$ versus tone index $k - n$. Similar to Figure 2.31, the ICI power drops drastically as n moves away from k.

Table 2.3 shows the total ICI power as we vary the parameters of the phase noise model.

2.4.4 I/Q imbalance

I/Q imbalance occurs when the gain/phase is different along the in-phase and the quadrature signal processing path. Consider a signal transmitted at tone k, $A[k] \exp(j(2\pi f[k]t + \theta[k]))$, where $X[k] = A[k] \exp(j\theta[k])$ is the complex symbol with normalized signal power $|X[k]|^2 = 1$. The signal affected by I/Q imbalance can be

2.4 Real-world impairments

Table 2.3. Total ICI power in the phase noise model.

a	b	c	Total ICI power (dB)
5.5	4	10.5	−30
6.5	4	10.5	−36
7.5	4	10.5	−38
6.5	4	9.5	−28
6.5	4	11.5	−40

modeled as

$$\begin{aligned} y(t) &= A[k](1+\delta_g)\cos(2\pi f[k]t + \theta[k] + \delta_\theta) \\ &+ jA[k](1-\delta_g)\cos(2\pi f[k]t + \theta[k] - \delta_\theta) \\ &= X[k]\exp(j2\pi f[k]t)\left(\cos(\delta_\theta) + j\delta_g\sin(\delta_\theta)\right) \\ &+ X[k]^*\exp(-j2\pi f[k]t)\left(\delta_g\cos(\delta_\theta) - j\sin(\delta_\theta)\right), \end{aligned} \qquad (2.62)$$

where δ_g represents the imbalance in gain and δ_θ the imbalance in phase. Note that δ_g and δ_θ may vary with tone when I/Q imbalance is frequency selective.

The impact of I/Q imbalance is that a signal on tone k leaks energy in its image tone $-k$, that is, $N_c - k$, where the leakage power is given by

$$\mathbb{E}(|Y[N_c - k]|^2) = \delta_g^2\cos^2(\delta_\theta) + \sin^2(\delta_\theta). \qquad (2.63)$$

Figure 2.38 plots $\mathbb{E}(|Y[N_c - k]|^2)$ versus δ_θ for a few values of δ_g.

2.4.5 Power amplifier nonlinear distortion

When a transmitted signal exceeds the linear region of the power amplifier, the signal suffers nonlinear distortion. This can be modeled as an impulse noise added at the clipping time instant, which leads to a fairly flat power spectrum density function of in-band noise and interference to adjacent channels. While the adjacent channel interference can be mitigated with filtering, the in-band noise will uniformly affect all the tones.

Discussion notes 2.7 Determination of OFDMA parameters

Two key design parameters in OFDMA are tone spacing Δ_f and length of the cyclic prefix T_{cp}. Recall that the overhead due to the cyclic prefix is

$$\frac{T_{cp}}{T_s + T_{cp}} = \frac{1}{1 + \frac{1}{\Delta_f T_{cp}}}. \qquad (2.64)$$

Clearly the smaller $\Delta_f T_{cp}$, the smaller the overhead. On the other hand, larger Δ_f or T_{cp} in general helps mitigate the effect of the impairments. In particular, from Section 2.4, we learn that

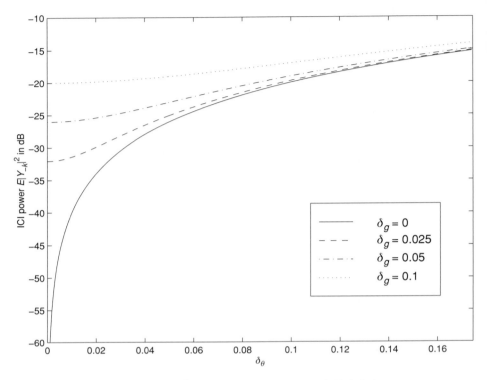

Figure 2.38 ICI power $\mathbb{E}(|Y[N_c - k]|^2)$ versus δ_θ in the presence of I/Q imbalance.

- Larger T_{cp} reduces the duration during which a signal may arrive beyond the receiver cyclic prefix interval.
- Larger Δ_f reduces the relative carrier frequency offset and Doppler effect as well as reducing the impact of phase noise.

In LTE, $\Delta_f = 15$ kHz, $T_{cp} \approx 5$ μs (normal mode) or 17 μs (extended mode).

Now consider the effect of delay spread. In many ITU channel models, the maximum delay spread is within 5 μs and can be completely covered by the cyclic prefix. If the maximum delay spread is 15 μs and the fraction of signal energy beyond the cyclic prefix interval is -25 dB, then the total ICI/ISI power due to delay spread is about -35 dB from Figure 2.35. However, it should be pointed out that the cyclic prefix needs to cover both delay spread and potential time misalignment between the transmitter and the receiver. As described in Section 2.2, in the uplink the base station controls the transmit symbol times of all the users within the sector in a closed-loop manner. Inaccuracy in timing control reduces the margin of the cyclic prefix to cover delay spread.

Next consider the effect of Doppler spread. Recall that at velocity v and carrier frequency f_c, the Doppler shift is given by $\frac{f_c v}{c}$ where $c = 3 \times 10^8$ meters per second is the speed of light. With a typical 900 MHz or 1.9 GHz carrier frequency for a cellular system, the Doppler shift is less than 200 Hz at $v = 70$ miles per hour. With the tone spacing of 15 kHz, $\delta_f < 0.02$. Therefore, the total ICI power due to Doppler spread is about -40 to -30 dB from Figure 2.32, and the total ICI power due to phase noise is about -36 dB from Table 2.3. However, we should point out that in open-loop frequency synchronization (Section 2.2), after the user synchronizes its receive carrier frequency with the downlink signal, the user applies the same frequency correction to its transmit carrier frequency. Then the worst-case Doppler shift in the uplink may be doubled and became $2\frac{f_c v}{c}$, twice as much as the previous estimate. The frequency discrepancy can be reduced if the user's transmit frequency is controlled in a closed-loop manner by the base station similar to closed-loop timing control.

2.5 Cross interference and self-noise models

2.5.1 Cross interference and self-noise due to ICI

So far we have studied a variety of real world impairments. When a tone signal is sent, the ICI it generates belongs to one of the following two types:

- non-uniform ICI: the ICI power mostly concentrates on nearby or image tones;
- uniform ICI: the ICI power uniformly distributes among all other tones.

The techniques of mitigating those impairments are out of the scope of this book. What we are interested in here is an *abstraction model* that captures the essence of ICI characteristics and yet is simple enough for system level design. To this end, we propose the following two ICI models. Consider a signal multiplexing scenario where two signals (X_i, $i = 1, 2$) are received in an OFDM symbol and occupy distinct tones. Denote by P the total received power. Suppose X_i occupies a fraction α_i of the total tones, with $\alpha_i \in [0, 1]$, $\alpha_1 + \alpha_2 = 1$, and is received at power $\beta_i P$, with $\beta_i \in [0, 1]$, $\beta_1 + \beta_2 = 1$.

The *cross interference model* assumes that the ICI power to X_i is

$$P_{ICI,i} \approx \epsilon \alpha_i P, \tag{2.65}$$

and the *self-noise model* assumes that the ICI power to X_i is

$$P_{ICI,i} \approx \eta \beta_i P, \tag{2.66}$$

where ϵ, η are small positive numbers. The cross interference model applies to uniform ICI or non-uniform ICI with the tones of X_1 and X_2 very much interleaved with each other. Thus, the ICI power seen by X_i is roughly proportional to the fraction of bandwidth it occupies. On the other hand, the self-noise model applies to non-uniform ICI with each of X_1 and X_2 occupying a contiguous tone block. In this case, since the ICI is dominated by the leakage from nearby tones, the ICI power between X_1 and X_2 tone blocks is much smaller than that within either tone block itself. Thus, the ICI power seen by X_i is roughly proportional to the fraction of power it receives.

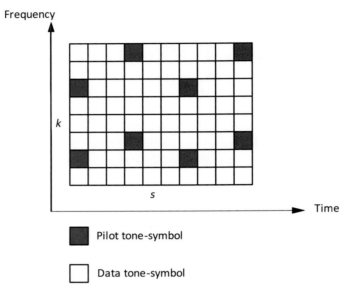

Figure 2.39 Illustration of pilot tone-symbols and data tone-symbols.

Qualitatively speaking, the leakage power of an OFDMA signal is spilled evenly to all the tones of the entire bandwidth in the cross interference model but contaminates only the fraction of bandwidth that the signal itself occupies in the self-noise model.

2.6 Self-noise due to imperfect channel estimation

A similar phenomenon of the self-noise model (2.66) can also be caused by imperfect channel estimation. In this case, unlike the ICI, the self-noise in a tone is caused by the signal power on that tone itself.

In Section 2.4, we see that under various real world impairments, the output of the DFT module at tone k of a transmitted tone signal $X[k]\exp(j2\pi f[k]t)$ is given by $Y[k] = H[k]X[k]$, where $H[k]$ is a complex number that depends on the models and parameters of the impairments. For example, in the phase noise model, Equation (2.58) says that $H[k] = (1 + j\frac{1}{N_c}\sum_{l=0}^{N_c-1}\phi[l])$. In general, $H[k]$ also includes the wireless channel and filter response from the transmitter to the receiver.

In coherent demodulation, the receiver first obtains the channel estimate of $H[k]$, referred to as $\hat{H}[k]$, and then demodulates $X[k]$ from $Y[k]$ and $\hat{H}[k]$. self-noise arises when the channel estimation is imperfect.

First let us see how channel estimation is typically done in an OFDMA system. Section B.3 reviews channel estimation techniques. As shown in Figure 2.39, a subset of tone-symbols are used to send known symbols, called pilots, and the remaining ones carry information-bearing data symbols. At a tone-symbol (s, k), where s and k are OFDM symbol index and tone index, respectively, the received symbol is given by

$$Y[s, k] = H[s, k]X[s, k] + W[s, k]. \qquad (2.67)$$

We drop index (s, k) to simplify notation in the following and use subscripts p, d to indicate pilot and data.

The receiver first estimates H_p at a pilot tone-symbol and then H_d at a data tone-symbol from $\hat{H}_p$. For simplicity, here we assume that H_d is estimated from one pilot. In reality, channel estimation uses several pilots, but the analysis following still holds in general.

Denote by $H^\delta = H - \hat{H}$ the channel estimation error. Suppose that the channel estimate at the pilot is

$$\hat{H}_p = \mathbb{E}(H_p | Y_p) = \frac{Y_p}{X_p} = H_p + \frac{W_p}{X_p}. \tag{2.68}$$

The channel estimation error is

$$\mathbb{E}(|H_p^\delta|^2) = \frac{\sigma^2}{\mathbb{E}(|X_p|^2)}, \tag{2.69}$$

where $\sigma^2 = \mathbb{E}(|W_p|^2)$ is the power of the additive noise.

The estimation of H_d from H_p assumes they are correlated in time and frequency. Next we assume the Gaussian Markov model (see Section 4.1.5 for details) for channel correlation. Specifically, suppose that the distance between the data and pilot tone-symbols are s_0 OFDM symbols and k_0 tones. Then

$$H_d = \sqrt{c} H_p + \sqrt{1-c} W_H, \tag{2.70}$$

where H_d, H_p, W_H are zero mean complex Gaussian random variables with the same variance equal to the expected channel power gain $\mathbb{E}(|H|^2)$, and

$$c = (1 - \eta_t)^{s_0} (1 - \eta_f)^{k_0}, \tag{2.71}$$

with parameters η_t, η_f defined in (4.39) and (4.40). The unbiased estimate is given by

$$\hat{H}_d = \mathbb{E}(H_d | \hat{H}_p) = \sqrt{c} \hat{H}_p. \tag{2.72}$$

It can be shown that assuming $\eta_t, \eta_f \ll 1$, the channel estimation error is

$$\mathbb{E}(|H_d^\delta|^2) = \left(s_0 \eta_t + k_0 \eta_f\right) \mathbb{E}(|H|^2) + \frac{\sigma^2}{\mathbb{E}(|X_p|^2)}. \tag{2.73}$$

When the receiver demodulates data symbol X_d from Y_d and $\hat{H}_d$, suppose we use zero-force equalization (2.12)

$$\hat{X}_d = \frac{1}{\hat{H}_d} Y_d = \frac{1}{\hat{H}_d} \left((\hat{H}_d + H_d^\delta) X_d + W_d\right)$$

$$= X_d + \frac{H_d^\delta}{\hat{H}_d} X_d + \frac{1}{\hat{H}_d} W_d. \tag{2.74}$$

This equation shows that the noise consists of two components. The power of the second one drops as $\mathbb{E}(|H|^2)$ increases, while that of the first one remains unchanged. The

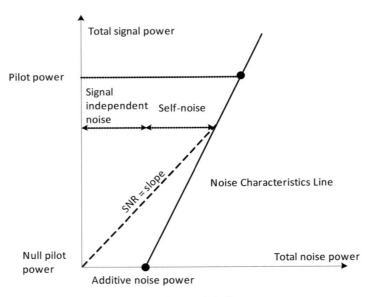

Figure 2.40 Illustration of noise characteristic line.

effective SNR is

$$\text{SNR} = \left(\left(s_0 \eta_t + k_0 \eta_f \right) + \frac{\sigma^2}{\mathbb{E}(|H|^2)} \left(\frac{1}{\mathbb{E}(|X_p|^2)} + \frac{1}{\mathbb{E}(|X_d|^2)} \right) \right)^{-1}, \quad (2.75)$$

which states that even if $\frac{\sigma^2}{\mathbb{E}(|H|^2)}$ goes to zero, SNR saturates at a value that depends on the inherent channel uncertainty characterized by η_t and η_f.

This residual channel estimation error leads to a self-noise term, which exhibits very different characteristics from the additive noise in the traditional channel model Equation (2.67), as its power is equal to $\left(s_0 \eta_t + k_0 \eta_f \right) \mathbb{E}(|HX|^2)$, which grows with the signal power. To emphasize the effect of the self-noise, we modify the channel model (2.67) to be

$$Y[s, k] = \sqrt{1 - \eta} H[s, k] X[s, k] + \sqrt{\eta} U[s, k] \mathbb{E}(|H[s, k] X[s, k]|) + W[s, k], \quad (2.76)$$

where $U[s, k]$ is zero mean unit variance complex Gaussian random variable. Hence, the total noise consists of signal independent noise and self-noise, which is signal dependent. From Equation (2.76), the power of the self-noise is equal to $\eta \mathbb{E}(|HX|^2)$ and grows linearly with the signal power. Figure 2.40 depicts the signal power versus the total noise power, which is modeled by a straight line, referred to as *noise characteristic line*, with slope equal to $1/\eta$ and x-axis intercept σ^2.

The effective SNR at tone-symbol (s, k) is thus

$$\gamma = \frac{(1 - \eta) \mathbb{E}(|HX|^2)}{\eta \mathbb{E}(|HX|^2) + \sigma^2}$$

$$= \frac{(1 - \eta) \gamma_0}{\eta \gamma_0 + 1}, \quad (2.77)$$

2.6 Self-noise due to imperfect channel estimation

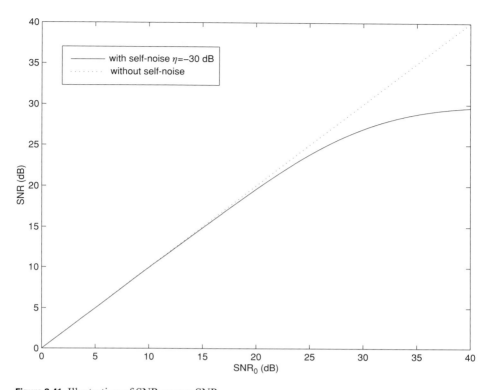

Figure 2.41 Illustration of SNR versus SNR_0.

where $\gamma_0 = \frac{\mathbb{E}(|HX|^2)}{\sigma^2}$ is the SNR in the absence of the self-noise. As shown in Figure 2.41, at high γ_0, SNR grows sub-linearly with γ_0 and eventually saturates at $\frac{1-\eta}{\eta}$.

Now consider the signal multiplexing scenario we discussed previously where two signals (X_i, $i = 1, 2$) are received in an OFDM symbol. Unlike ICI, the self-noise due to imperfect channel estimation does not spread into other tones, and therefore the noise power is represented by the self-noise model (2.66).

2.6.1 Self-noise measurement via null pilot

In scheduling and resource allocation, we often need to predict the SNR for a given signal power. The usual approach, without taking into account self-noise, is as follows. Suppose that the pilot is sent at power P per tone-symbol. The receiver measures the SNR at the pilot tone-symbols in Figure 2.39, denoted by $\gamma(P)$. Then, if the transmit power of the data signal is Q per tone-symbol, then the SNR is linearly predicted to be

$$\gamma(Q) = \frac{Q}{P}\gamma(P). \tag{2.78}$$

However, in the presence of self-noise, this linear prediction (2.78) may not be accurate. In particular, when $\gamma(P)$ is high and in the sublinear region shown in Figure 2.41, a reduction in transmit power from P to Q results in a smaller reduction in SNR than what is expected by the linear scaling (2.78). In other words, when a user is very close to the base station such that its pilot SNR in the downlink is saturated by self-noise, the base station can allocate a smaller transmit power to the user without hurting its SNR much.

We can use the noise characteristic line to more accurately predict the effective SNR. In Figure 2.40, $\gamma(Q)$ is equal to the slope of the dotted line that connects the origin and the point on the noise characteristic line corresponding to the signal power Q. Clearly, to fully define the noise characteristic line, we need two measurement points on the line. The two points should be made apart as much as possible to minimize the estimation error. The measurement at the pilot power is an obvious one to be used, as usually a pilot symbol is sent with higher power than a data symbol. The other measurement can come from a *null pilot*, which is similar to the usual pilot, except that zero signal power is transmitted on the null pilot tone-symbol. The two measurement points are illustrated in Figure 2.40.

The receiver measures σ^2, the power of signal independent noise W at the null pilot tone-symbol. At the pilot tone-symbol, the receiver measures $\gamma(P)$ and the total received power, denoted by $R(P)$. It follows that:

$$\frac{1}{\gamma(P)} = \frac{1}{1-\eta}\left(\eta + \frac{\sigma^2}{\mathbb{E}(|H|^2)P}\right)$$

$$R(P) = \mathbb{E}(|H|^2)P + \sigma^2.$$

After some algebra, we have

$$\gamma(Q) = \left(\frac{1}{\gamma(P)} + \left(1 + \frac{1}{\gamma(P)}\right)\frac{\sigma^2}{R(P)}\left(\frac{P}{Q}-1\right)\right)^{-1}. \tag{2.79}$$

2.7 Summary of key ideas

- Orthogonality, that is, no ISI and ICI, is the fundamental feature of OFDM. It is attributed to two enabling ideas, namely, sinusoid basis functions and cyclic prefixes. OFDM can be implemented efficiently with IFFT and FFT.
- The basic principles of OFDMA in a cellular system are to multiplex intra-cell users to distinct tones so as to make them orthogonal with each other, and to reuse the same tones in all cells to achieve universal frequency reuse. Inter-cell interference is averaged with tone hopping. In this sense, hopped OFDMA, similar to CDMA, belongs to the family of spread spectrum technologies.
- To maintain the orthogonality in OFDMA uplink, closed-loop time control is required to compensate for round-trip propagation delay between a user and a base station. Timing control is not required in CDMA. Closed-loop frequency control may not be required if the tone spacing is sufficiently large.
- SC-FDMA reduces the PAPR by about 2.5–4 dB at the cost of resource allocation not being flexible. Two resource allocation schemes are possible; the allocated tones have

2.7 Summary of key ideas

to be either contiguous or equally spaced. SC-FDMA also requires frequency domain equalization and incurs a loss in SNR in a frequency selective channel.

- The orthogonality in OFDMA is not perfect in practice. ICI arises in the presence of real world impairments. In a typical channel environment and implementation, ICI is about -35 to -25 dB. For user multiplexing, ICI leads to self-noise and cross interference that limit the SNR.
- Self-noise also arises from imperfect channel estimation with its power proportional to signal power. Self-noise is in addition to the additive noise in a conventional channel model, and causes the SNR to saturate in the high SNR regime. Self-noise does not leak to other tones.

3 System design principles

This chapter first summarizes the system level benefits of using OFDMA as the underlying multiple access technology, and then qualitatively presents the basic system design principles of OFDMA mobile broadband communications in a cellular network. This chapter serves as a preview of many topics to be covered in the remaining chapters of the book. We emphasize the concepts and insights here and leave the quantitative modeling and analysis to the subsequent chapters.

In order to better understand the system design principles, throughout this section we often contrast OFDMA with CDMA. CDMA has been widely-used in cellular wireless communications from the second generation circuit-switched voice system to the third-generation data system. A basic commonality between CDMA and OFDMA is that they are both wideband spread spectrum technologies. Therefore, comparisons between the two will help demonstrate the evolution of the system design ideas. CDMA is well studied in the literature. The readers can learn CDMA from textbooks such as [61, 159, 168].

3.1 System benefits of OFDMA

Recall that an OFDMA signal is the sum of tone signals, each being sinusoid at a given tone frequency. We next elaborate on the fundamental property of OFDMA, *orthogonality*, in three aspects.

When a receiver is synchronized to an OFDMA signal both in time and in frequency, the power of a tone signal is entirely contained on that tone and does not spill to others. Therefore two synchronized OFDMA signals on distinct sets of tones do not interfere with each other as long as the receiver is synchronized to them. This first aspect of orthogonality is also referred to as *no ICI or ISI* between tone signals. Orthogonality is unique to tone signals because sinusoids are the *only eigenfunction* of the linear time invariant system, a good approximation of the multipath fading channel within an OFDM symbol time. A sinusoid signal remains a sinusoid of the same frequency after passing through the wireless channel. As a consequence, the orthogonality between OFDMA tones remains intact at the receiver. This eigenfunction property only holds for sinusoid signals, and not for other orthogonal basis signals such as code sequences in CDMA. Figure 3.1 shows an illustration of this property.

Synchronization in a given cell is not difficult to achieve, because the base station is the only transmitter in the downlink and the only receiver in the uplink. In the downlink,

3.1 System benefits of OFDMA

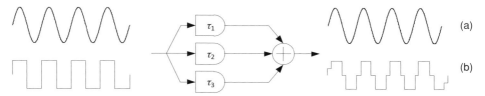

Figure 3.1 Illustration of the sinusoids' ability being the eigenfunction of a multipath fading channel. τ_is ($i = 1, 2, 3$) in the figure indicate the different delays associated with each path. (a) shows the received signal when a sinusoid is transmitted over the channel; (b) shows the received signal when a CDMA code sequence is transmitted. While (a) keeps the sinusoid waveform intact, (b) sees a distorted version of the code sequence due to the different delays at different paths.

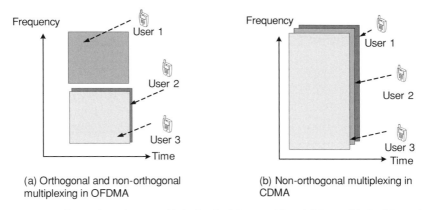

(a) Orthogonal and non-orthogonal multiplexing in OFDMA

(b) Non-orthogonal multiplexing in CDMA

Figure 3.2 Illustration of user multiplexing in OFDMA (a) and CDMA (b). In OFDMA, user 1 is allocated a different subset of tones from users 2 and 3, and therefore orthogonal in frequency with users 2 and 3. Users 2 and 3 are allocated the same subset of tones and are not orthogonal in frequency. Some other mechanism, such as spatial multiplexing or superposition coding, is used to further separate them. In CDMA, all the users are allocated different codes; they are separated not in frequency, but in code. As shown in Figure 3.1, different codes are not orthogonal in the presence of a multipath.

tone signals for different users arrive at any user receiver synchronously. In the uplink, every user transmitter adjusts its transmit symbol time such that all of their tone signals arrive at the base station receiver synchronously. Therefore different users within a cell can be made orthogonal.

Orthogonality is the key to the intra-sector system design. Take user multiplexing for example. A straightforward approach is orthogonal multiplexing where users are allocated to different tones such that there is no intra-sector interference. Perhaps a little more subtly, OFDMA can also take a non-orthogonal approach by multiplexing users in the same subset of tones, for example, using spatial multiplexing or superposition coding, with no interference to/from the remaining tones. Orthogonality makes user multiplexing more flexible and easier to control. Figure 3.2 shows that OFDMA uses a mixed orthogonal and non-orthogonal approach to multiplex users and CDMA is not as flexible.

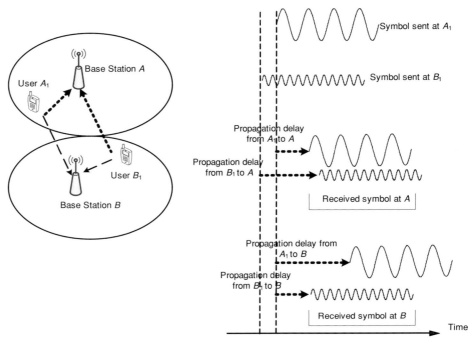

Figure 3.3 Illustration of uplink time synchronization issue in multiple cells. User A_1 is synchronized with base station A and user B_1 synchronized with base station B. However, A_1 is not synchronized with B and B_1 not synchronized with A.

However, synchronization in multiple cells may not be possible for all transmitter/receiver pairs, even if the base stations are all synchronized themselves. Figure 3.3 shows an uplink example of two adjacent cells A and B where no transit symbol times exist that allow users A_1 and B_1 to be synchronized to *both* base stations A and B. A similar observation can be made in the downlink. Note that, while it is possible to use a cyclic prefix to achieve synchronization in this case, the length of the cyclic prefix is usually chosen to cover the delay spread of the wireless channel, and therefore not long enough to cover the difference in propagation delay.

When the receiver is not synchronized to a tone signal, the signal power will spill to other tones. Section 2.4 shows that the spilled power does not distribute uniformly over all the tones, but instead mostly concentrates on neighboring tones. Figure 3.4 illustrates an example of how the signal power of a block of contiguous tones transmitted in a cell spills over tones at a receiver in an adjacent cell. Most of the signal power remains in the tone block with a small fraction of power leaking at the edges. Ignoring the small power leakage, the second aspect of orthogonality says that interference to the adjacent cell for the most part stays within the tone block. This greatly simplifies inter-cell interference coordination.

The third aspect of orthogonality is that in OFDMA the wideband wireless channel is converted into parallel narrowband channels such that the signals on different tones can

3.1 System benefits of OFDMA

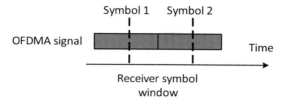

(a) Transmit and receive symbol time misalignment

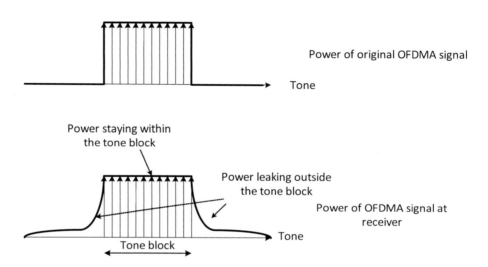

(b) Power spillage when receiver is not time synchronized with the OFDMA signal

Figure 3.4 Illustration of signal power spillage when the receiver is not time synchronized with an OFDMA signal. In a cellular network, the illustrated OFDMA signal may be transmitted by the base station in one cell and received by a user in an adjacent cell, or transmitted by a user in one cell and received by the base station in an adjacent cell. The OFDMA signal consists of a block of contiguous tones. (a) Shows two symbols of the incoming OFDMA signal when the receiver symbol window is not aligned with the signal. At the receiver, the signal power is spilled to other tones where the power spillage drops as tones becomes farther from the original ones. The power spillage decay can be approximated with a sine function. When the size of the tone block is large, the total power leaked outside of the block tends to be much smaller than the power that remains within, as shown in (b).

be separately processed. This simplifies implementation and facilitates exploitation of frequency selectivity. For example, if the transmitter knows the channel, it can allocate its transmit power budget to the tones according to the water-filling principle for the optimal performance. Frequency selectivity also adds a new dimension to multiuser diversity scheduling and spatial processing with multiple antennas.

The following system level benefits are attributable to the three fundamental property of OFDMA just discussed. We will explain these benefits qualitatively in more detail in the following sections.

- *No intra-sector interference.* When they are allocated distinct tones, users in the same sector do not interfere with each other. Eliminating intra-sector interference increases the sector capacity.
- *User multiplexing with large signal dynamic range.* The nature of the wireless channel creates a large dynamic range in path loss between different users. Orthogonality of OFDMA helps to maintain a large signal dynamic range, consistent with path loss dynamic range, when users are multiplexed. This means that in the downlink the base station can transmit to nearby and faraway users simultaneously with very different transmit powers, and in the uplink receive signals from them at very different powers. The sector capacity benefits from frequency multiplexing users with large path loss dynamic range.
- *Universal and fractional frequency reuse.* Similar to CDMA, OFDMA embraces universal frequency reuse via interference averaging, which can be achieved using distinct hopping patterns in adjacent cells. Thus spectrum reuse is not limited by the worst-case inter-cell interference. However, cell boundary users become the capacity bottleneck because they experience the most severe inter-cell interference. To improve spectral efficiency beyond universal reuse, OFDMA enables *fractional frequency reuse* (FFR), a flexible way of assigning different powers to different frequency subbands and coordinating power pattern in adjacent cells.
- *Flexible scheduling.* In OFDMA, the total degrees of freedom for communication can be seen as a time-frequency grid of tone-symbols. In principle, the scheduler can assign any set of tone-symbols to a user with any power, without much impact on other tone-symbols. This enables flexible scheduling in time, frequency, and power to jointly optimize spectral efficiency, QoS control, and fairness in response to rapidly varying traffic needs and time-frequency selective channel/interference conditions. Furthermore, mobile broadband is not just about creating high rate data links; it also requires efficient support of many low rate links for low latency control signaling. In OFDMA the minimum resource unit is a tone-symbol and fine resource granularity makes scheduling agile.
- *MIMO enhancements.* Some key MIMO channel characteristics, such as spatial correlation, are frequency selective in nature. The flexibility of scheduling in frequency helps OFDMA exploit frequency selectivity and enhances multiuser MIMO performance by scheduling users in the frequency bands with most favorable MIMO channel characteristics.

3.2 Fading channel mitigation and exploitation

In mobile communications, fading is a fundamental phenomenon where signals are attenuated over the wireless channel. Fading is a random process and introduces channel fluctuation over time, frequency, and space. Path loss, shadowing, mobility, and multipath propagation are the main causes of channel fading, large-scale or small-scale. When a signal fades, the received power may not be sufficient to recover the signal. A key question in a wireless system design is how to communicate *reliably* and *efficiently* in fading channels. Towards this end, we present the following two main ideas and elaborate on them in Chapter 4.

3.2.1 Fading mitigation

The first idea is *to make the signal more robust by mitigating channel fading*.

Diversity is an effective technique for improving the signal reliability. The basic idea is to send the same information through multiple signal paths, each of which preferably fades independently such that the transmission fails only if all of the signal paths fail. Common forms of diversity include time diversity, frequency diversity, and space diversity. For example, tone hopping and coding over multiple tones in OFDMA achieves frequency diversity, and multiple antennas can be used for space diversity. The diversity techniques here are mostly point-to-point between a transmitter and a receiver, and thus referred to as *single user diversity*.

The diversity techniques are often used in conjunction with forward error correction coding where the transmission of one coding block is designed to experience multiple independent fades. As long as a sufficient fraction is received with enough power, the coding block can be successfully decoded. ARQ is a feedback-based error control technique that can also exploit diversity. If the first transmission fails, the receiver requests the transmitter to retransmit. The retransmission may experience independent fading and therefore improve the chance of success. Hybrid ARQ combines the advantages of both channel coding and ARQ by utilizing the signals received in both transmissions in decoding.

Power control is another fading mitigation technique in which the transmitter adjusts the transmit power in response to the wireless channel fluctuation so as to maintain the required SINR target corresponding to a given data rate. If the data rate is flexible, then rate control (adaptive coding and modulation) is yet another mechanism to mitigate fading. The idea is to determine the coding and modulation scheme based on the wireless channel condition; increase the data rate when the channel condition is favorable and reduce it otherwise.

3.2.2 Fading exploitation

Rate control works in the packet data system because it in essence exploits the flexibility of scheduling. We can explore this idea further, and question why bother to schedule a user at all when it is in deep fading? This leads to the second idea: *to exploit channel fading with opportunistic scheduling to maximize system level spectral efficiency*. In other words, to exploit the time varying nature of the fading channel and schedule a user when its channel condition is favorable. However, if a user is not scheduled because of its bad channel condition, does the system resource go unused and wasted? Fortunately,

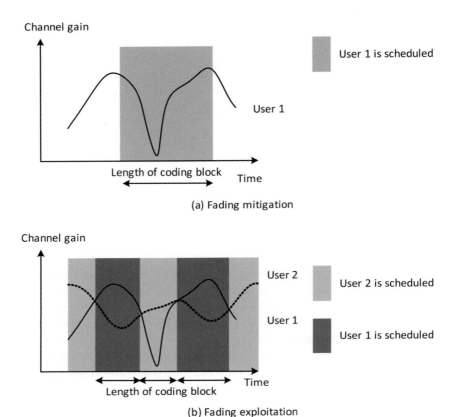

Figure 3.5 Illustration of channel fading mitigation (a) and fading exploitation (b). The channel gain varying over time of user 1 is shown in (a). The coding block is sufficiently longer than channel coherence time to obtain time diversity and combat channel fades. The channel gains varying over time of users 1 and 2 are shown in (b). User 1 or 2 is scheduled when its channel gain is relatively better. When a user is in deep fade, it is not scheduled. The coding block is usually shorter.

the answer is no, because other users can be scheduled at that time as long as their channel conditions are more favorable. Opportunistic scheduling [23, 90, 102] exploits multiuser diversity inherent in the multiuser system where the channel fluctuations of different users are assumed to be statistically independent and therefore it is very likely that *some* user experiences favorable channel condition. The coding block is sufficiently short to exploit multiuser diversity. Figure 3.5 contrasts the ideas of fading mitigation and exploitation.

Multiuser diversity differs from single user diversity that was mentioned earlier in that it is a system level technique and the benefit increases with the number of the users. It attempts to make the best use of, rather than mitigating, the effect of channel fading. For this reason, it would even be beneficial to artificially induce channel fading to amplify the multiuser diversity gain.

A potential cost of multiuser diversity is increased scheduling delays, as it may take a while for a user to see a good channel. OFDMA can exploit frequency selectivity to reduce scheduling delays. Specifically, in OFDMA, the wireless channel is divided into a set of parallel narrowband flat channels, which have different gains in a frequency selective channel. With opportunistic scheduling a user is assigned to narrowband channels of favorable channel gains, thereby achieving multiuser diversity gain. Taking advantage of both time and frequency selectivity, OFDMA allows more granular exploitation of multiuser diversity for higher spectral efficiency and smaller scheduling delays.

3.2.3 Mitigation or exploitation?

These ideas of fading mitigation and exploitation may appear contradictory to each other. How does one apply them in practice? Clearly, because of limited scheduling capability, the circuit-switched voice system has to rely on fading mitigation. However, the answer is not straightforward for the packet data system. One should take into account two aspects: first, the tradeoff of multiuser diversity gain versus control overhead and other performance metrics such as scheduling latency; and second, the possibility of combining the two ideas for reliable opportunistic communications.

3.3 Intra-cell user multiplexing

In OFDMA the entire bandwidth is divided into a number of tones. The users in the same sector use the tone-symbols in an orthogonal manner such that no two users collide on the same tone-symbols. In a sectorized cell, all the sectors of a given cell reuse the same set of tones. The orthogonality of OFDMA leads to the following two system benefits: no intra-sector interference, and user multiplexing with large signal dynamic range.

The first benefit is straightforward to appreciate. In contrast, CDMA is non-orthogonal: in the uplink, users are assigned non-orthogonal pseudo-random code sequences; in the downlink, the base station assigns orthogonal Walsh sequences to different users. However, the orthogonality of Walsh sequences is not preserved at a user receiver after the signal travels through the multipath fading channel. Even if the wireless channel itself is flat, the transmit/receive filters introduce delay spread.

The second benefit is worth some elaboration. Assuming ideal antenna isolation, the (inter-sector) interference between the sectors of the same cell is zero. In this case, intra-cell user multiplexing is mostly an intra-sector issue and each sector can independently multiplex users without inter-sector coordination. As will be elaborated in Chapter 5, this view is too simplistic as sectorization is never perfect in reality. However, for the purpose of this section, we focus on intra-sector user multiplexing.

In the wireless channel, nature often creates large path loss dynamic range because a signal decays rapidly as the distance increases. For example, in a macrocellular system, the path loss exponent of the wireless channel propagation is about 3.5. Consider a nearby user, for example, 50 m from the base station, and another faraway user, for example, 5000 m away; the path loss dynamic range, that is, the ratio of the path losses of the two users, is as high as $3.5 \times 10 \log_{10}(5000/50) = 70$ dB. Large path loss dynamic range leads to large signal power dynamic range. For example, in the uplink if the nearby user and the faraway user both transmit at the same power, the received signal strengths at the base station will be 70 dB apart; in the downlink, in order to achieve the same received power, the required transmit power to the faraway user has to be 70 dB higher than that to the nearby user.

We next show qualitatively how OFDMA broadband communications benefit from large signal dynamic range caused by nature. Quantitative analysis can be found in Chapter 5.

In practice the faraway user is usually power limited, meaning that the most effective way of increasing its capacity is to allocate more power because it operates in a regime where SINR affects capacity linearly and spending the same amount of power in a smaller bandwidth does not reduce the capacity much (see Appendix C). Meanwhile, the nearby user is bandwidth limited, meaning that the most effective way of increasing its capacity is to allocate more bandwidth because it operates in a regime where capacity increases linearly in bandwidth but only logarithmically in power (see Appendix C). This observation suggests the following power bandwidth tradeoff that benefits both the faraway and the nearby users when the base station serves them simultaneously:

- In the downlink, the base station allocates a large fraction of the transmit power but a small fraction of bandwidth to the faraway user, and a small fraction of the transmit power but a large fraction of bandwidth to the nearby user. The faraway user benefits more from high power allocation, and it is therefore beneficial to trade bandwidth for power. The nearby user benefits more from large bandwidth allocation, and it is therefore beneficial to trade power for bandwidth.
- In the uplink, the base station allocates a small fraction of bandwidth to the faraway user and a large fraction of bandwidth to the nearby user. Unlike the downlink, the uplink does not have a fixed total transmit power budget to be shared between the two users. Indeed, the received power target of the nearby user can be made higher than that of the faraway user. Given large path loss difference, the nearby user likely still transmits at much lower power, thereby not significantly causing interference to adjacent cells. The nearby user tends to be far from the base stations in adjacent cells. The inter-cell interference in the uplink is usually dominated by the faraway user.

Figure 3.6 illustrates the above power and bandwidth allocation ideas in the downlink and the uplink in a sector. In general, a larger fraction of bandwidth should be allocated to the nearby users, in either the downlink or the uplink. Figure 3.6(a) shows that in the downlink the per unit bandwidth transmit power is much lower for the nearby than for faraway users. Figure 3.6(b) shows that in the uplink the per unit bandwidth receive power is the same for the nearby and faraway users, which can be shown to maximize the sum rate in the power limited regime.

One consequence of the above power and bandwidth allocation scheme is large signal dynamic range of the signals being multiplexed. In the downlink, at the nearby user receiver, the desired signal is much weaker than the signal sent to the faraway user. In

3.3 Intra-cell user multiplexing

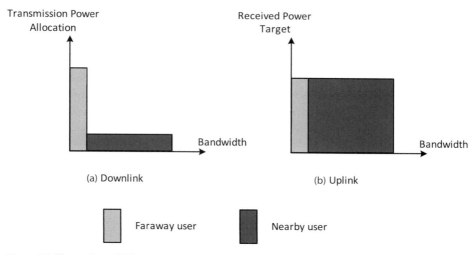

Figure 3.6 Illustration of joint power and bandwidth allocation in the (a) downlink and (b) uplink cases. The x-axis represents the fraction of bandwidth allocated to the faraway and nearby users. The y-axis represents per unit bandwidth transmit power in (a) or receive power in (b).

the uplink, at the base station receiver, the signal from the faraway user is possibly much weaker than that from the nearby user. OFDMA orthogonality makes it feasible to keep a large signal dynamic range.

As a comparison, consider the CDMA downlink, which is not orthogonal in the multipath channel. In the absence of the faraway user, the base station can serve the nearby user with low transmit power. When the base station transmits to both the faraway and nearby users simultaneously, cross interference occurs since the signals to the faraway and nearby users are not orthogonal. If the base station allocates most of the total transmit power to the faraway user and only a small fraction to the nearby user, then the nearby user will be overwhelmed by the cross interference from the strong signal power that the base station intends to send to the faraway user. As a result, the base station cannot effectively allocate very different powers to the two users.

Similarly, the CDMA uplink uses power control to mitigate the so-called near-far effect, in which a weak signal from a faraway user could be overwhelmed by the strong one from a nearby user because the CDMA uplink is not orthogonal. To address the near-far effect, open-loop and closed-loop power control is implemented to ensure that the received power from different users is roughly equal. Thus, rather than exploiting the signal dynamic range, CDMA power control works to *minimize* it. In circuit-switched voice systems, because the traffic rate of each user is about the same, there is not much system utility to have large signal dynamic range in the uplink. However, in broadband data systems, limiting the dynamic range by power control can significantly reduce the flexibility in user multiplexing and thus cause a loss in overall system spectrum efficiency.

In summary, the OFDMA orthogonality is the key to maintaining large signal dynamic range within a sector. However, orthogonality is never perfect in reality. The level of orthogonality is limited by the real world impairments as studied in Section 2.4, and in turn limits the signal dynamic range that can be realistically supported in OFDMA. The basic premise is that with proper choice of parameters (tone spacing and cyclic prefix), intra-sector interference is much smaller in OFDMA than in CDMA.

Moreover, because intra-sector interference dominates in CDMA, self-noise and inter-sector interference are of secondary importance in system design. In OFDMA, with intra-sector interference mostly eliminated, they become important. Chapter 5 will study how they would limit signal dynamic range in user multiplexing.

3.4 Inter-cell interference management

In universal frequency reuse, the same set of tones are reused in all the cells, and therefore the users experience inter-cell interference. In the absence of intra-cell interference, inter-cell interference is a key factor that determines the sector capacity.

First, we briefly review how the inter-cell interference issue is addressed in the FDMA and CDMA systems. The narrowband FDMA system is based on frequency reuse. A narrowband frequency channel is reused among cells that are at least a minimum distance apart. Co-channel interference occurs between the users using the same narrowband channel in two cells. The minimum reuse distance is such that the SINR is sufficient for successful communications at the *worst-case* co-channel interference. A narrowband frequency channel is reused in one of every K cells and K is called frequency reuse factor. For example, $K = 7$ in AMPS and $K = 4$ in GSM.

Frequency reuse reduces the amount of bandwidth available in a cell. To overcome this problem, CDMA uses the direct sequence spread spectrum technology such that a user sees interference not from just one user but from all the users in an adjacent cell. This way, the frequency reuse is limited not by the worst-case but by the *average* interference.

In general, there is a tradeoff between SINR and bandwidth; the smaller the reuse factor, the more bandwidth available in a cell but the lower the SINR is. In CDMA, the intra-sector interference is a significant part of the total interference. Since frequency reuse only reduces inter-cell interference but not intra-sector interference, universal frequency reuse where $K = 1$ is the preferred choice.[1]

Secondly, we consider the inter-cell interference issue in OFDMA. Similar to CDMA, OFDMA employs the tone hopping spread spectrum technology to achieve interference averaging. Specifically, distinct tone hopping patterns are used in adjacent cells such that over several OFDM symbols, one tone-hopping sequence in a cell sees interference from different physical tones in an adjacent cell. Provided that the information is coded across many tone-symbols, the decoding performance is determined by the average interference rather than the worst-case interference. Therefore, OFDMA can also use universal frequency reuse.

---

[1] Universal frequency reuse has other benefits such as soft handoff, which are not discussed in this section.

3.4 Inter-cell interference management

The following two points potentially make mobile broadband different from circuit-switched voice:

- To achieve low latency and high data rate in mobile broadband, scheduling needs to respond more rapidly (on the order of milliseconds) than what is typically required in circuit-switched voice (on the order of seconds). The number of scheduled users at a given time may be much smaller, and the set of scheduled users may be completely different from one slot to another. In that case, interference averaging tends to be less effective. Inter-cell interference needs to be more actively controlled by the scheduler.
- As mentioned before, inter-cell interference in OFDMA exhibits the orthogonality property where a tone in one cell mostly interferes with tones of nearby frequencies in another cell. This makes it possible to devise something better than universal frequency reuse by retaining the bandwidth benefit of universal frequency reuse and yet improving the SINR by coordinating power allocation across tones in adjacent cells.

We next elaborate on those two points and leave the detailed discussion to Chapter 6.

3.4.1 Interference averaging and active control

Interference averaging is mostly an uplink issue. This is because in the downlink, the locations of the interferers (the base stations in neighboring cells) are fixed. With cell-specific tone hopping, the effective inter-cell interference seen at a user receiver depends on the total transmit power of the interfering base stations, and hence can be easily managed by controlling the total transmit power.[2]

However, the situation is very different in the uplink, because the locations of the interferers vary depending on which interfering users in neighboring cells are transmitting at a given time. In a typical circuit-switched voice CDMA system, many users are transmitting simultaneously, each at low data rate and thus contributing a small fraction to the total interference. In that case, Gaussian approximation works well and the total interference does not fluctuate significantly. To a great extent, the system can rely on interference averaging without the need for a scheduler to actively control inter-cell interference.

In the broadband data system, to accommodate bursty traffic and to reduce latency, quite often only a small number of users are scheduled at a given time and some of them transmit at high data rate and account for a large fraction of the total interference. This potentially causes large fluctuation in inter-cell interference. Figure 3.7 shows an example where base station A measures the interference power at time t_1 and subsequently uses the measurement to determine the uplink data rate of its own users at time t_2. At time t_1, because the interfering users (B_1, B_2) in cell B and (C_1, C_2) in cell C are all far from cell A, the measured interference power is low. Based on that measurement, the base station predicts that it can receive high data rate from its users. However, at time t_2, a different set of interfering users (B_3, B_4, C_3, C_4) are scheduled in cells B and C,

---

[2] Strictly speaking, given the total transmit power, the power allocation across tones also affects the statistical characteristics of the inter-cell interference and thus the decoding performance. We ignore that second-order effect in this section.

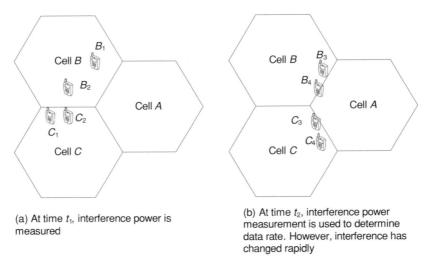

(a) At time t_1, interference power is measured

(b) At time t_2, interference power measurement is used to determine data rate. However, interference has changed rapidly

Figure 3.7 Illustration of large fluctuation of inter-cell interference in the uplink.

and cause rapid increase in interference. The uplink transmission in cell A will fail unless a large SINR margin has been added. In either case, large interference fluctuation causes capacity penalty. Alternatively, one can rely on (hybrid) ARQ or link adaptation to reduce capacity penalty at the cost of increased latency.

To reduce the fluctuation of inter-cell interference, it is desirable for the scheduler to actively control the inter-cell interference rather than passively relying on interference averaging. This means that the scheduler needs to allocate the power budget to the scheduled users in addition to the time-frequency bandwidth resource. We note that because of orthogonality, OFDMA does not suffer from the near-far effect as much as CDMA, and therefore does not require tight power control in the uplink. Rather, it uses power allocation to control inter-cell interference.

3.4.2 Universal versus fractional frequency reuse

OFDMA eliminates intra-sector interference. Any reduction in inter-cell interference by using a frequency reuse factor $K > 1$ directly translates to an increase in SINR. The SINR benefit would be smaller in CDMA because there intra-sector interference dominates and is not affected by the choice of frequency reuse K. Therefore, it is worthwhile to revisit the tradeoff of universal frequency reuse in the OFDMA context.

The benefit of universal frequency reuse is that more bandwidth is available per sector. For the nearby users, which are limited by the available bandwidth, universal reuse makes sense because increase in bandwidth directly translates into increase in capacity. However, for the faraway users, which are limited by the inter-cell interference, the increase in bandwidth does not noticeably help. Instead, they would rather prefer frequency reuse $K > 1$ so as to reduce the inter-cell interference. It would be desirable to devise a hybrid reuse scheme to satisfy the seemingly conflicting needs.

3.4 Inter-cell interference management

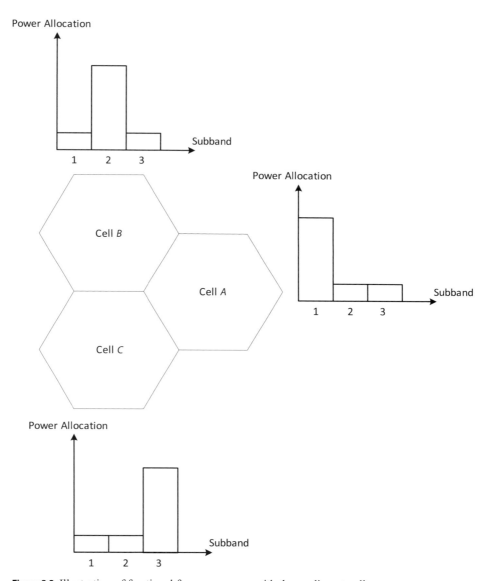

Figure 3.8 Illustration of fractional frequency reuse with three adjacent cells.

Recall from Section 3.3 that in the joint power and bandwidth allocation scheme, the faraway users are allocated high power over small bandwidth and the nearby users with low power over large bandwidth. Therefore, one idea of the hybrid reuse scheme is that the bandwidth over which high power is allocated to the faraway users does not overlap between adjacent cells so as to prevent high power signals from interfering with each other. Meanwhile, the bandwidth over which low power is allocated to the nearby users is free to be reused in every cell. This motivates so-called fractional frequency reuse.

Figure 3.8 illustrates the power and bandwidth allocation scheme of fractional frequency reuse in three adjacent cells. For the sake of simplicity, the example shows omnicells. The entire bandwidth is divided into three subbands. In cell A, the faraway users are allocated to subband 1 with high power. The nearby users are allocated to the remaining two subbands with low power. In cell B, the faraway users are allocated to subband 2, and subband 3 in cell C. The pattern can be extended to the entire cellular network.

We have introduced fractional frequency reuse from the perspective of the power and bandwidth tradeoff between faraway and nearby users. We can also motivate the idea from the viewpoint of inter-cell interference coordination. With universal frequency reuse, a user sees averaged inter-cell interference. By introducing power coordination, fractional frequency reuse aims at improving SINR while retaining the benefit of reusing the whole bandwidth in each cell. Continue the example of Figure 3.8, the faraway users in cell A now see reduced interference from cells B and C, and the SINR is close to what would be achieved with frequency reuse $K = 3$. The reduction in interference directly translates into capacity gain as the faraway users operate in low SINR regime. Meanwhile, the nearby users in cell A are allocated large bandwidth, close to what would be achieved with universal reuse. Although the nearby users now see an increase in the inter-cell interference as compared with the universal reuse scenario, the impact on the capacity would be insignificant as they are already in the high SINR regime.

Fractional frequency reuse is in essence static interference coordination agreed among adjacent cells beforehand. The idea can be extended to a more dynamic setting where the base stations exchange real-time control information of interference and traffic loading and coordinate the power and bandwidth allocation.

3.5 Multiple antenna techniques

The advanced use of multiple antennas has been a crucial enabling element for mobile broadband. In this section, we summarize the main ideas and the system benefits.

3.5.1 System benefits

Antenna diversity. Since the probability that all antennas deeply fade decreases as the number of antennas increases, antenna diversity improves link reliability. There are both receive diversity and transmit diversity. Space-time coding [4, 155] can be used for transmit diversity when the channel is unknown to the transmitter. Among other realizations, antenna diversity includes spatial diversity (multiple antennas of the same characteristics physically separated from one another), pattern diversity (co-located antennas with different radiation patterns), and polarization diversity (antennas with orthogonal polarizations).

Beamforming. When the channel is known to the transmitter, transmit beamforming achieves power gain by transmitting in the spatial direction of the intended receiver. However, the receiver has to feed back the channel information to the transmitter in a frequency-division duplexing (FDD) system because the uplink and downlink channels are not symmetric. In the absence of channel information at the transmitter, effective transmit beamforming is not possible from a point-to-point perspective. However, in a system with multiple users separated in different locations, multiple transmit antennas

3.5 Multiple antenna techniques

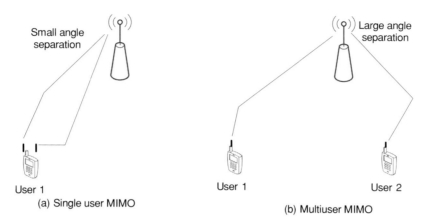

Figure 3.9 Comparison of single user MIMO (a) and multiuser MIMO (b). Single user MIMO requires the user to have multiple antennas. As the base station antennas are high, the angle separation of the multiple signal paths from a given user tends to be small.

can be used to induce channel variation and thus increase multiuser diversity [169]. On the other hand, receive beamforming is relatively easy because the receiver can estimate the channel and coherently combine signals to boost signal power, null out interference, or strike an optimal balance in the MMSE sense.

Spatial multiplexing. While beamforming achieves power gain, spatial multiplexing brings degree-of-freedom gain, and is more attractive in the high SINR regime. Spatial multiplexing can be between the base station and a user (single user MIMO). The spatial multiplexing gain depends on the rank (or more precisely the condition number) of the MIMO channel matrix. For example, in a line-of-sight environment, the channel matrix has rank 1 and the system has no spatial multiplexing gain to exploit. In general the achievable gain depends on the richness of scattering as multiple global spatially separable paths are needed at *both* the base station and the user. Since the base station is usually placed in a tower higher than the surroundings (≈ 30 m), it may not have multiple spatially separable paths because the angle separation for a given user tends to be small (see Figure 3.9a). In this case, reflection and scattering in the local environment of the user do not improve the rank. The situation becomes very different if we take a system, instead of point-to-point, view. The idea is not to be limited to single user MIMO, but to exploit the geographic separation between multiple users and achieve multiuser spatial multiplexing gain (multiuser MIMO). A key difference between single user MIMO and multiuser MIMO is that multiuser MIMO does not require a rich scattering environment, since the geographic separation between users is sufficient for full multiplexing gain. Another advantage is that each user is not required to have multiple antennas. The requirement of multiple antennas in single user MIMO is a big burden on the user because of cost, power consumption, and other reasons. The benefit of multiuser MIMO is on the system capacity rather than the peak data rate of a given user. Figure 3.9 contrasts single user MIMO and multiuser MIMO.

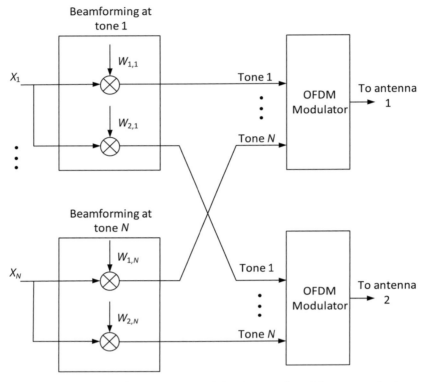

Figure 3.10 Illustration of transmitter spatial signal processing of frequency selective beamforming. There are two TX antennas. X_i is a scalar number transmitted on tone k. $[W_{1,k}, W_{2,k}], k = 1, \ldots, N$ is the beamforming weight vector.

3.5.2 OFDMA advantages

OFDMA has two main advantages. First, it simplifies implementation of spatial signal processing in a wideband channel. The wideband channel is converted into a number of parallel narrowband subchannels, each of a flat channel response characterized with a single-tap channel response. In the MIMO case, the wideband MIMO channel can be expressed by a number of narrowband MIMO channel matrices. To implement MIMO in the OFDMA system, the receiver/transmitter can apply single-tap channel-based spatial signal processing techniques individually in each subchannel. Therefore extending the narrowband MIMO techniques to the wideband channel is conceptually straightforward. As an example, consider a MISO downlink where the base station transmitter is equipped with multiple TX antennas and each user has only one RX antenna. Figure 3.10 illustrates a spatial signal processing diagram of the base station transmitter.

Second, as will be shown in Chapters 4 and 7, OFDMA adds the frequency dimension to spatial signal processing so as to achieve frequency selective diversity, beamforming, and spatial multiplexing. In a frequency selective wideband channel, if the base station has the full information of the channel of a user, it can use frequency selective beamforming in which the beamforming weight varies with tone. Frequency selective beamforming achieves the power gain benefit in every tone.

Without the channel information, the base station cannot effectively form beams targeting any given receiver. Continuing the MISO example in Figure 3.10, the base station can employ a delay diversity scheme where the weight of transmit antenna i at tone k is given by $W_{i,k} = e^{j\phi_i k}$ for some delay constant ϕ_i. By introducing distinct delays among the transmit antennas, the delay diversity scheme artificially creates/amplifies channel frequency selectivity. For a given user, the channel gain is enhanced at some tones while degraded at others. The tones where the channel gain is enhanced are likely different for different users. What makes the delay diversity scheme interesting is that on the system level the base station can exploit multiuser diversity over frequency. It can schedule users in the tones where the channel gains are enhanced. In essence, this is opportunistic frequency selective beamforming, because although the base station cannot target any specific user, a user is scheduled when its signal paths from different antennas are constructively added at the scheduled tones. The idea is similar to the dumb antenna scheme in [169] where multiple TX antennas are used to artificially create/amplify channel time selectivity.

3.6 Scheduling

The sector capacity is to be shared among many users. In mobile broadband, a key idea is to apply *statistical multiplexing* principles to share system resources (time, bandwidth, and power) instead of partitioning them in a fixed manner among users. The reason is that the resource demand of a specific user varies rapidly due to the following reasons:

- *Traffic variation.* Traffic arrivals are bursty for most data applications. A user may be silent most of the time and yet require a significant amount of resources over a short period of time when it becomes active.
- *Channel variation.* Channel fading leads to significant channel variation. Communications would consume much more system resources when a user is in deep fade than when in a good channel condition.
- *Interference variation.* Dependent on the activity of other users in the system, a user may see significant interference variation.

Thus, it is highly inefficient to pre-allocate a fixed portion of system resources to a user.

OFDMA is flexible in assigning different users to different symbol-tones with different power, due to orthogonality of system resources within the same sector. We have also seen that opportunistic scheduling can bring multiuser diversity and capacity gain. Thus, to fully exploit the advantage of OFDMA, a natural way to share system resources in OFDMA mobile broadband is to use *scheduled statistical multiplexing* by explicitly *scheduling* users to system resources based on their current traffic demands and channel/interference conditions.

OFDMA can be viewed as a special case of multichannel systems where each channel exhibits different time-frequency channel and interference characteristics. By exploiting the multichannel nature, an OFDMA scheduler can not only increase the capacity with opportunistic scheduling but also reduce delay as studied in [17, 18, 84, 143].

It should be pointed out that scheduling is not necessary for the sake of achieving statistical multiplexing per se. For example, in the circuit-switched voice CDMA system, statistical multiplexing is achieved via voice activity without explicit base station scheduling. Specifically, when a user is admitted, it is assigned a code sequence that is sufficient to support the voice traffic rate when the user is active. In the uplink, if the user is inactive, it simply suspends its transmission, thereby not causing any interference; in the downlink, if the voice circuit of the user is temporarily inactive, the base station simply stops transmitting to the user and saves the unused power for other active users.[3]

Unscheduled statistical multiplexing is relatively simple because it does not involve scheduling and needs no additional signaling and control mechanisms. It works well for circuit-switched voice or low rate data, but becomes less appealing for broadband data where traffic becomes much more bursty and resource allocation needs to quickly respond to rapidly varying traffic needs. With unscheduled statistical multiplexing, the base station relies on admission control to allocate the system resource and subsequently adjusts resource allocation by exchanging control messages between the base station and the users. The timescale in which the base station exerts its control of system resources is on the order of seconds, much longer than what is required in mobile broadband (on the order of milliseconds).

In contrast, scheduled statistical multiplexing tightly controls the system resources (power and bandwidth) and enables rapid re-allocation of the system resources among different users. Flexibility and granularity are the key advantages of using OFDMA to harness statistical multiplexing gains via scheduling. The bandwidth resource in OFDMA is divided in tone-symbols. In principle, the scheduler can assign any set of tone-symbols to a user with any power. Multiplexing may occur in time; for example, the scheduler devotes all the sector capacity to best effort traffic most of the time but quickly switches to service users with delay sensitive traffic when it arrives. Multiplexing may occur in frequency; for example, the scheduler assigns a small fraction of tones to faraway power limited users and yet gives a large fraction to nearby bandwidth limited users. Associated with each assigned set of tone-symbols, the scheduler determines appropriate transmit power in the downlink and receive power target in the uplink.

Several system level design issues are important to develop an efficient scheduling system:

- *IP aware base stations*. Maximizing the physical layer sector capacity is not the only objective; other performance objectives such as latency and delay jitter are also important. Architecturally, it makes sense that the base station processes IP packets *natively* such that the scheduler is aware of both IP QoS and physical layer channel and interference information. IP QoS-aware scheduling makes it possible to treat QoS consistently end-to-end.
- *Low system overhead and agile operation*: Scheduling is not just about an optimization algorithm. Scheduling comes at a signaling cost of collecting information from

---

[3] The actual mechanism of taking advantage of voice activity is more sophisticated. But for the purpose of our discussion, the simple description given here suffices.

users and announcing scheduling decision. Scheduling optimization is as effective as signaling is agile. This requires low latency control signaling. Having high rate *data* pipes is not sufficient; it is also important to efficiently support many low rate *control* pipes. The fine granularity of MAC resource in OFDMA is a key to designing such thin control pipes.

- *MAC state scheduling*: The duty cycle of data traffic is often low. Usually, the users are placed into different MAC states, depending on the traffic activities, to conserve battery power and to manage control overhead. QoS should not be provided only to a small set of users that are already in an active state for traffic scheduling. Rather, the idea is to treat traffic scheduling and MAC state scheduling in a unified QoS framework so that QoS is provided to the entire user population and cover the entire session when a user is within the system. Again, the OFDMA MAC resource granularity plays an important role in achieving this goal.

3.7 Network architecture and airlink support

The preceding sections of this chapter discuss the design principles for achieving large data capacity and sharing it efficiently among users. The airlink that connects the base station and users needs to be integrated into an end-to-end network architecture to form the mobile Internet.

Given the dominance of IP-based wired networks, the mobile broadband system should also be IP-based because such a network architecture greatly simplifies network installation, operation and maintenance. Figure 3.11 illustrates an IP-based network architecture. An IP base station is in effect a combination of an IP access router and a wireless radio base station. IP traffic goes directly to an IP base station and then to a user, which is an IP device, with no need for wireless-specific proprietary subnetwork protocols that may conflict with mechanisms already at work in the IP layers. This architecture permits a mobile device to automatically use all existing IP features and capabilities, while keeping additional link layer development to a minimum.

The emphasis in this book is not on the IP-based network architecture itself. Rather we focus on how the OFDMA airlink supports the IP-based network architecture. Airlink and network architecture should not be treated as two separate subjects in system design. While OFDMA mobile broadband mostly focuses on the physical and MAC/link layers, it should work well with the IP network layer and the IP-based network architecture. IP protocols work on the network layer and deliver packets from the source to the destination host through intermediate network nodes that may use a variety of link layer protocols. In principle, the airlink is just another link layer protocol, like any other form of broadband Internet access, for example, DSL, cable, leased lines, or Ethernet. However, the wireless link is inherently different from the wired link due to the channel and interference characteristics. To make existing IP applications work transparently over wireless, the airlink link layer has to combat the challenges in wireless and deliver performance similar to the wired link in terms of throughput, latency, and reliability.

Section 3.6 already argues that base stations should be IP aware. We next elaborate on two additional system ideas.

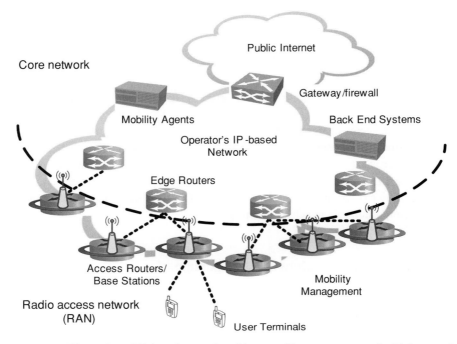

Figure 3.11 Illustration of IP-based network architecture. Users are connected with base stations via the OFDMA mobile broadband airlink. The base stations are also IP routers and connected to edge routers of the operator's IP-based network, which also includes mobility agents and back end systems for the purposes of mobility management, authentication, authorization, accounting. The operator's network is connected with the public Internet via gateway servers.

3.7.1 Unplanned deployment of base stations

Base stations are deployed in a plug-and-play manner similar to wireless local area networks, resulting in a flexible and scalable architecture and therefore lower system planning, deployment, and maintenance costs. On the other hand, unplanned deployment imposes new challenges on the OFDMA airlink regarding cell search, handoff, and inter-cell interference management.

As an example, consider the inter-cell interference problem in a deployment shown in Figure 3.12, where a high power base station A provides large cell coverage and within the coverage a low power base station B is installed to provide high capacity for users B_1, B_2, B_3 in the hot spot area. User A_1 is a little outside of the hot spot and is therefore connected to base station A because the downlink from A is stronger. User A_1 is far from base station A and has to transmit at a high power to reach A. Users B_1, B_2, B_3 are close to base station B and therefore can transmit at a low power. As a result, A_1 causes excessive interference to B.

This and other interference questions will be addressed in Section 10.1.2.

3.7 Network architecture and airlink support

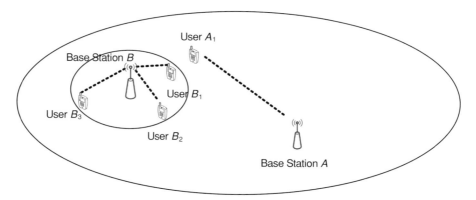

Figure 3.12 Illustration of potential inter-cell interference issues when a low power base station is installed within the coverage of a high power base station.

3.7.2 Mobile IP-based handoff

Handoff is a unique phenomenon in the mobile cellular system. As the user moves away from its serving base station, the link quality degrades while the signal from an adjacent base station grows stronger, thereby triggering handoff from the old base station to the new one. In CDMA soft handoff, multiple base stations receive the same uplink signal from the user. A base station controller is responsible for selecting the best decoding result from the base stations. In the downlink, the base station controller forwards the data to the base stations, which then transmit them to the user such that the user combines the signals from all the base stations with the RAKE receiver, similar to multipath reception from one base station. The base station controller is a key element in the CDMA-specific access networks. However, it is not IP-based, making it difficult for the base station to fully control the processing of native IP packets. In addition, the hierarchical structure of base station controller/base station is hard to implement in an unplanned deployment.

Mobile IP is an Internet Engineering Task Force (IETF) standard communications protocol (particularly IETF RFC 4721, 5944, and 6275) that builds on the IP protocols to address the issue of mobility management. The protocol allows the user to change its topological point of attachment to an IP internetwork without changing its IP address, thus making mobility transparent to applications and higher level protocols like TCP. As opposed to soft handoff in CDMA, Mobile IP provides a common denominator for a variety of IP-based handoff solutions that can be used in OFDMA mobile broadband as well as across diverse technologies.

Two issues need to be addressed. The first issue is airlink support for Mobile IP-based handoff protocols. One such protocol is make-before-break (MBB) handoff in which a user maintains two parallel links with two base stations at the same time to make handoff more seamless (minimizing packet loss and service disruption). MBB handoff can also be used for macrodiversity. MBB handoff in OFDMA eliminates the need of the non IP-based base station controller. We need to study how to design the airlink to achieve MBB handoff. The second issue is handoff initiation. A user keeps on searching for new candidate base stations frequently as a background process. The search complexity is especially high in an unplanned deployment and directly relates to battery power consumption. We need to study how to design the signaling and protocol of handoff initiation to minimize the search complexity.

More handoff related topics are discussed in Chapter 9.

3.8 Summary of key ideas: evolution of system design principles

To conclude this section, we contrast the new design principles of OFDMA mobile broadband with conventional wisdom and practice, labeled under "conventional view" in the table on the next page. The conventional view is best represented by 2G circuit-switched voice CDMA systems. It should be pointed out that some of the ideas labeled under "OFDMA mobile broadband" have already been adopted in 3G high speed data CDMA systems.

The evolution of the design principles is attributable not only to different characteristics of OFDMA versus CDMA but also to different requirements of broadband data versus circuit-switched voice. Certain salient features in CDMA, such as unscheduled statistical multiplexing and soft handoff, work very well for voice or relatively modest rate data but become less effective for mobile broadband.

Summary of new design principles

- Create a big data pipe (sector capacity) for transporting data traffic

Design topics	Conventional view	OFDMA mobile broadband
Channel fading and interference fluctuation	Mitigation of channel fades via single user diversity	Opportunistic exploitation of channel peaks via multiuser diversity
Intra-cell user multiplexing	Non-orthogonal multiplexing as signal power spreads into entire bandwidth	Flexible mix of orthogonal and non-orthogonal multiplexing
	Processing gain or multiuser detection to overcome intra-sector interference	Intra-sector interference eliminated due to signal orthogonality
	Power control to minimize signal dynamic range	Joint power and bandwidth allocation to exploit large signal dynamic range
	Self-noise and inter-sector interference considered secondary to intra-sector interference	Self-noise and inter-sector interference to be carefully managed to preserve large signal dynamic range
Inter-cell interference	Universal frequency reuse based on inter-cell interference averaging	Fractional frequency reuse based on inter-cell interference coordination
	Interference averaging	Active inter-cell interference control
Use of multiple antennas	Sectorization to reduce average interference	Frequency selective beamforming and multiuser spatial multiplexing

- Share the data pipe among users

Design topics	Conventional view	OFDMA mobile broadband
Packet scheduling	Relatively slowly-varying scheduling especially in uplink	Packet switched scheduling in both downlink and uplink in response to rapidly changing traffic needs
	Unscheduled statistical multiplexing through voice activity	Scheduled statistical multiplexing based on channel, interference and traffic conditions; more control overhead required to support sophisticated scheduling
MAC state scheduling	ACTIVE or SLEEP state	In addition, HOLD state to exploit data traffic duty cycle for power saving and QoS control of a large user population

- Support an IP-based end-to-end network architecture and handle mobility

Design topics	Conventional view	OFDMA mobile broadband
Network architecture	Adaptation of access network architecture to wireless	Adaptation of airlink so as to have consistent IP-based end-to-end network
	Hierarchical network architecture	Flat network architecture
	Networking planning required	Unplanned network deployment
Handoff and macro-diversity	Soft handoff	Mobile IP-based make-before-break and (expedited) break-before-make handoff

4 Mitigation and exploitation of multipath fading

The medium of wireless communications is the wireless radio frequency channel. We are interested in the characteristics of the wireless channel, in particular, how the channel response varies over time and frequency, as well as over the distance between a transmitter and a receiver. The variation in the channel response is usually referred to as *channel fading*. For a given signal, channel fading depends on the particular propagation environment, such as buildings, walls, ground, vehicles, between the transmitter and the receiver, as well as the carrier frequency of the signal. To characterize channel fading, we often use a statistical approach based on measurements made in a large variety of environments. Statistically, channel fading can be characterized by the following two different types of behaviors:

- *Large-scale fading*, which varies in a slow time scale (on the order of seconds) or in a large distance of many wavelengths. Large-scale fading is mainly caused by *path loss* and *shadowing*. Path loss is caused by signal strength degradation as the electromagnetic (EM) wave of the signal propagates through space. Shadowing results from penetration or reflection of objects much larger than the wavelength of the EM wave.
- *Small-scale fading*, which varies in a fast time scale (on the order of tens of milliseconds depending on mobility) or in a distance on the same order of the wavelength. Small-scale fading is mainly caused by multipath, as multiple copies of the transmitted signal add constructively or destructively at the receiver. Thus, small-scale fading is also referred to as *multipath fading*.

Note the wavelength is determined by $\lambda = c/f_c$, with $c = 3 \times 10^8$ m/s (the speed of light) and f_c, the carrier frequency of the signal. With a typical carrier frequency of cellular systems on the order of 1 GHz, λ is about 0.3 m.

Define *channel (power) gain* G the power ratio of the received and transmitted signals, and denote by G_{dB} the channel gain in the dB scale with $G_{dB} = 10 \log_{10}(G)$. Combining both large-scale and small-scale fading, we model G as follows:

$$G(\mathbf{r}_t, \mathbf{r}_r) = \bar{G}(d) S F, \qquad (4.1)$$

or in the dB scale,

$$G_{dB}(\mathbf{r}_t, \mathbf{r}_r) = \bar{G}_{dB}(d) + S_{dB} + F_{dB}, \qquad (4.2)$$

where $\bar{G}, S, F$ indicate the path loss, shadowing, and multipath fading components, respectively, of the overall channel gain, $\bar{G}_{dB}, S_{dB}, F_{dB}$ their counterparts in the dB scale, $\mathbf{r}_t$ and $\mathbf{r}_r$ the locations of the transmitter and receiver, respectively, and $d = |\mathbf{r}_t - \mathbf{r}_r|$ the distance between the two.

Small-scale fading F and F_{dB} can be removed by averaging the received signal strength over a few hundred wavelengths. What remains is the large-scale fading components of path loss $\bar{G}$ and shadowing S.

The shadowing component S can be modeled as a *log-normal* random variable, or equivalently, S_{dB} as a normal random variable. This is because as the EM wave propagates from the transmitter to the receiver, S results from penetration or reflection of main different objects, each of which causes a multiplicative loss to the strength of the signal. It follows that

$$S = \prod_i s_i, \tag{4.3}$$

where each s_i is assumed to be an i.i.d. random variable. Thus,

$$S_{dB} = \sum_i 10 \log_{10} s_i. \tag{4.4}$$

With a large number of the objects along the propagation path, we can apply the Central Limit Theorem here and approximate S_{dB} as a zero mean Gaussian random variable.

The shadowing effect occurs in a distance on the order of 10–100 m in an outdoor and in a shorter distance in an indoor environment. Once the shadowing effect S_{dB} is averaged out, we are left with the path loss component $\bar{G}$, which defines the variation of long-term median value as a function of the distance and can be considered the area average channel gain. The path loss effect occurs in a large distance on the order of the radius of a cell, for example, a few hundred or thousand meters. Many factors affect the path loss model, such as indoor or outdoor; rural, suburb, or urban; line-of-sight or non–line-of-sight; antenna heights; and carrier frequency. In general, G decreases logarithmically with the distance d from the transmitter,

$$\bar{G}_{dB}(d) = \bar{G}_{dB}(d_0) - 10r \log_{10}\left(\frac{d}{d_0}\right), \tag{4.5}$$

where d_0 is a reference distance close to the transmitter determined from measurements, and r is called the path loss exponent. In free space, $r = 2$. More commonly, r is between 3 and 5.

Figure 4.1 shows an example of the channel gain as a function of distance, with multipath fading, shadowing, and path loss taking effect at different distance scales.

Large-scale fading occurs over a time interval of several seconds or more. In such a slow time scale, the transmitter and/or receiver can easily track the channel variation, and compensate for it, for example, by increasing transmit power when the channel gain deteriorates. *Outage* occurs if the channel capacity is below the required data rate even when the transmitter operates at the maximum transmit power. The maximum transmit power is constrained by device capability (the peak output of a power amplifier without distorting the signal) and/or interference budget (the maximum amount of

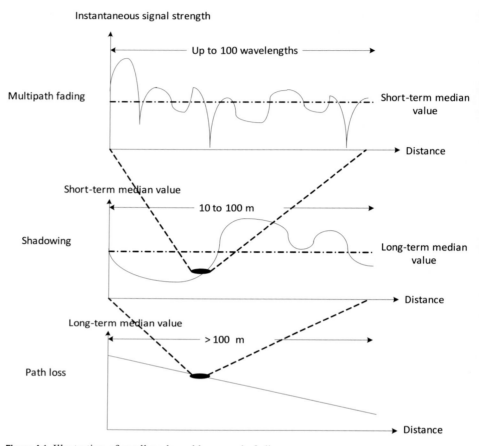

Figure 4.1 Illustration of small-scale and large-scale fading.

inter-cell interference allowed to be generated without damaging system stability). In circuit-switched voice, the required data rate is determined by the voice codec. The consequence of outage is call drops and is thus a crucial performance measure. In packet switched data, the data rate can be adapted to channel variation, and the data transmission can even be suspended temporarily if necessary. In general, outage is less critical unless data traffic imposes stringent QoS requirements on data rate and latency.

Small-scale fading corresponds to a time interval of a few milliseconds. As the time scale is on the order of a typical duration of physical layer communications, for example, for the transmission of a code block, multipath fading is more interesting and challenging, because of its following characteristics:

- *Channel attenuation*: with a certain probability, the channel gain drops so low that the instantaneous channel capacity is close to zero.
- *Channel amplification*: fading does not mean the channel gain always deteriorates; sometimes the channel gain actually improves. The capacity is increased if one can capture the opportunity when the channel gain is favorable.

- *Channel uncertainty*: it becomes harder to closely track the channel gain and requires more signaling overhead.

The focus of this chapter is to address the challenges and opportunities in multipath fading.

4.1 Multipath fading channel

4.1.1 Impulse response model

Suppose that the transmitter sends a baseband signal $u(t)$ modulated at carrier frequency f_c. The real signal transmitted over the wireless channel is

$$s(t) = \Re \left\{ u(t) e^{j2\pi f_c t} \right\}. \qquad (4.6)$$

The transmitted signal $s(t)$ arrives at the receiver via multiple propagation paths. A single path can be characterized with three time-varying parameters: amplitude gain $\alpha(t)$, delay $\tau(t)$, and Doppler frequency shift $f_D(t)$, where $f_D(t) = v\cos(\theta)/\lambda$ with v the transmitter/device speed and θ the angle of signal wave arrival relative to the direction of motion. Denote by $f_{Dm} = v/\lambda$ the maximum Doppler shift.

The received real signal is

$$r(t) = \Re \left\{ \sum_n \alpha_n(t) u(t - \tau_n(t)) e^{j2\pi(f_c + f_{D,n}(t))(t - \tau_n(t))} \right\}$$

$$= \Re \left\{ \sum_n \alpha_n(t) e^{-j\phi_n(t)} u(t - \tau_n(t)) e^{j2\pi f_c t} \right\}, \qquad (4.7)$$

where

$$\phi_n(t) = 2\pi \left((f_c + f_{D,n}(t)) \tau_n(t) - f_{D,n}(t) t \right) \qquad (4.8)$$

represents the phase shift on the baseband signal and n is path index. In some propagation environments, the above sum includes a special line-of-sight (LOS) path. The LOS path has the smallest delay τ, and often, but not always, the largest amplitude gain α.

Comparison of the baseband transmitted and received signals in (4.6) and (4.7) shows that the wireless channel can be modeled by a time-varying linear system with impulse response

$$h(t, \tau) = \sum_n \alpha_n(t) e^{-j\phi_n(t)} \delta(\tau - \tau_n(t)), \qquad (4.9)$$

where $h(t, \tau)$ is the baseband channel response at time t to an impulse input at time $t - \tau$.

In reality, the number of propagation paths is potentially very large. However, for a given bandwidth W which the signal occupies, not all the paths are *resolvable*. In particular, two paths are resolvable if their delay difference $|\tau_1 - \tau_2|$ is at least W^{-1}.

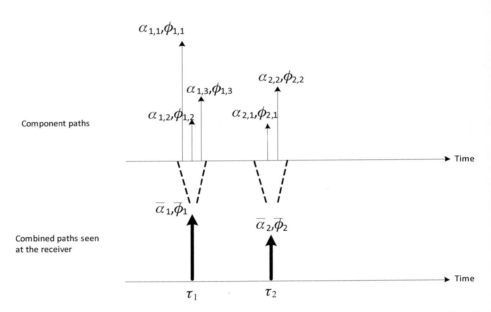

Figure 4.2 Illustration of multipath channel. Given bandwidth W, the receiver sees two combined paths. The first combined path is actually the sum of three component paths, which are non-resolvable because their delays are very close to each other. The second combined path comprises two component paths.

The receiver is unable to separate out the two paths if the delays are too close to each other, and thus in effect treats the non-resolvable paths as a single combined path, as illustrated in Figure 4.2.

Hence, the channel seen at the receiver is a *discrete path model*, an approximation of (4.9):

$$h(t, \tau) = \sum_{l=1}^{L(t)} \bar{\alpha}_l(t) e^{-j\bar{\phi}_l(t)} \delta(\tau - \tau_l(t)), \tag{4.10}$$

where $L(t)$ is the total number of resolvable paths and all $\tau_l(t), l = 1, \ldots, L(t)$, satisfy the multipath resolvability criteria. A combined path is referred to as a tap. The amplitude and phase of combined path l, $\bar{\alpha}_l(t)$, $\bar{\phi}_l(t)$, depend on $\alpha_{l,k}(t)$, $\phi_{l,k}(t)$ of the non-resolvable path components that have roughly the same delay $\tau_l(t)$:

$$\bar{\alpha}_l(t) e^{-j\bar{\phi}_l(t)} = \sum_k \alpha_{l,k}(t) e^{-j\phi_{l,k}(t)}. \tag{4.11}$$

The set of $\{\alpha_{l,k}(t), \phi_{l,k}(t)\}$ is the same as that of $\{\alpha_n(t), \phi_n(t)\}$ in (4.9) with different indexing. Next we briefly comment on the parameters in the discrete path channel model (4.10).

First, $L(t)$ and the parameters of each combined path depend on not only the underlying wireless channel itself but also the signal bandwidth W. As W increases, $L(t)$

increases while each combined path comprises fewer non-resolvable path components. The channel is called *narrowband fading* if $L(t) = 1$ and *wideband fading* if $L(t) > 1$. $L(t)$ changes slowly. A combined path l appears or disappears when $\sum_k \alpha_{l,k}^2(t)$ rises above or drops below the noise floor. This is caused by a significant change in the propagation environment along the component paths. Usually, the propagation environment consists of isolated objects of reflection, diffraction, or scattering, and the component paths of a given combined path correspond to a cluster of objects. Therefore, the rate of change in $L(t)$ is on the order of large-scale fading (shadowing and path loss) and can thus be ignored in the study of multipath fading.

Second, $\tau_{l,k}(t)$ also changes slowly. Because τ is determined by the propagation distance, the rate of its change is equal to $v/c \cos(\theta)$, with θ the angle of signal wave arrival relative to the direction of motion. At a vehicle speed of $v = 60$ mph, for example, τ changes about 0.1 µs every second. Because f_c is very large, even a nanosecond change in τ makes phase $2\pi f_c \tau$ rotate 2π. However, this effect is already modeled as Doppler shift f_D in (4.8). Hence, in the following study of the channel response autocorrelation function caused by phase evolution with time, we rely on $f_{D,l,k}(t)$ and treat $\tau_{l,k}(t)$ as fixed.

Third, although the the delay difference of any two component paths of a given combined path l is all within $\mathcal{W}^{-1}$, because f_c is very large, a very small delay difference causes large difference in phase $2\pi f_c \tau_{l,k}(t)$. Note that $\mathcal{W}$ is on the order of 1 MHz and f_c 1 GHz. Phase $\phi_{l,k}(t)$ of any component path k can be assumed to be independent with each other and uniformly distributed over $[-\pi, \pi]$. Therefore, $\bar{\alpha}_l(t), \bar{\phi}_l(t)$ may vary rapidly, because those component paths constructively or destructively combine in (4.11). The distribution of $\bar{\alpha}_l(t)$ is given in Section 4.1.2.

When the number of propagation paths is large, from the Central Limit Theorem, the channel impulse response in (4.10) is a Gaussian random process and therefore can be characterized with the mean and autocorrelation.

In the following sections, we study the combined paths in (4.11) more closely. We assume that the channel is *wide-sense stationary* such that the time variable t can be omitted in parameters α, τ, f_D when taking expectations.

4.1.2 Amplitude statistics

For a given l, in the absence of a dominant component and assuming a large number of components in the sum, the complex channel response of the combined path l, $\bar{\alpha}_l e^{-j\bar{\phi}_l}$, is zero-mean circular symmetric complex Gaussian, and the amplitude $\bar{\alpha}_l$ is a Rayleigh random variable with probability density function

$$p(x) = \frac{2x}{\Omega_l} e^{-\frac{x^2}{\Omega_l}}, \qquad (4.12)$$

where Ω_l represents the average power of path l:

$$\Omega_l = \sum_k \mathbb{E}(\alpha_{l,k}^2). \qquad (4.13)$$

Sometimes, a LOS path exists and dominates other non-LOS paths. In this case, (4.11) can be modeled as

$$\bar{\alpha}_l e^{-j\bar{\phi}_l} = \sqrt{\frac{\kappa}{\kappa+1}} \sqrt{\Omega_l} e^{-j\phi} + \sqrt{\frac{1}{\kappa+1}} \mathcal{CN}(0, \Omega_l), \qquad (4.14)$$

where ϕ is a uniform phase. The parameter κ, called K-factor, measures the power ratio of the LOS path and all the remaining non-LOS paths and is given by

$$\kappa = \frac{\mathbb{E}(\alpha_{l,0}^2)}{\sum_{k \neq 0} \mathbb{E}(\alpha_{l,k}^2)}, \qquad (4.15)$$

with the LOS component indexed as 0. The channel response of path l has non-zero mean, and the amplitude follows a Ricean distribution:

$$p(x) = \frac{2x(\kappa+1)}{\Omega_l} e^{-\kappa - \frac{(\kappa+1)x^2}{\Omega_l}} I_0 \left(2x \sqrt{\frac{\kappa(\kappa+1)}{\Omega_l}} \right), \qquad (4.16)$$

where $I_0(\cdot)$ is the zeroth order modified Bessel function of the first kind. When $\kappa = 0$, the Ricean distribution is reduced to the Rayleigh distribution. When $\kappa = \infty$, the channel does not exhibit any fading.

4.1.3 Channel variation in time

Rayleigh or Ricean fading characterizes the channel response at any time t. We next study channel variation in time by examining the autocorrelation of time domain channel response. The important concepts are *Doppler spread* and *coherence time*.

The autocorrelation function of the time domain channel response:

$$A_h(t, t + \Delta t; \tau, \tau + \Delta \tau) = \mathbb{E}(h^*(t, \tau) h(t + \Delta t, \tau + \Delta \tau)). \qquad (4.17)$$

We assume that the channel has *uncorrelated scattering*. Specifically, the channel response is uncorrelated if the path delay is different and the autocorrelation function is equal to zero for any $\Delta \tau \neq 0$. The rationale of this assumption is that different path delays correspond to different scatterers. Hence, (4.17) can be simplified to

$$A_h(t, t + \Delta t; \tau, \tau + \Delta \tau) = A_h(\Delta t; \tau) \delta(\Delta \tau), \qquad (4.18)$$

and

$$A_h(\Delta t; \tau) = \sum_{l=1}^{L} \left(\sum_k \mathbb{E} \left(\alpha_{l,k}^2 e^{j 2\pi f_{D,l,k} \Delta t} \right) \right) \delta(\tau - \tau_l)$$

$$= \sum_{l=1}^{L} \left(\sum_k \mathbb{E} \left(\alpha_{l,k}^2 e^{j 2\pi f_{Dm} \Delta t \cos \theta_{D,l,k}} \right) \right) \delta(\tau - \tau_l). \qquad (4.19)$$

In a dense scattering environment, Clarke's model assumes that the direction angle $\theta_{D,l,k}$ is independent and uniformly distributed over $[-\pi, \pi]$. To calculate that expectation over $\theta_{D,l,k}$, note that

$$\frac{1}{2\pi}\int_{-\pi}^{\pi}\cos(2\pi f_{Dm}\Delta t\cos\theta)\,d\theta = J_0(2\pi f_{Dm}\Delta t), \quad (4.20)$$

$$\frac{1}{2\pi}\int_{-\pi}^{\pi}\sin(2\pi f_{Dm}\Delta t\cos\theta)\,d\theta = 0, \quad (4.21)$$

where $J_0(\cdot)$ is the zeroth order Bessel function of the first kind. Hence, under the uniform θ assumption, it follows that

$$A_h(\Delta t;\tau) = \left(\sum_{l=1}^{L}\Omega_l\delta(\tau-\tau_l)\right)J_0(2\pi f_{Dm}\Delta t) \quad (4.22)$$

where Ω_l is given in (4.13).

Each term in the above sum represents the autocorrelation of the time domain channel responses, observed at t and $t+\Delta t$, of a combined path corresponding to delay τ_l. Figure 4.3(a) plots $J_0(2\pi f_{Dm}\Delta t)$. We observe that in Clarke's model, the autocorrelation function J_0 drops below 0.5 when $f_{Dm}\Delta t$ exceeds $1/4$, but then as Δt increases further, oscillates between positive and negative and does not monotonically decrease.

Taking the Fourier transform of $A_h(\Delta t;\tau)$ with respect to Δt, we get the power spectral density (psd):

$$S_h(f_D;\tau) = \int_{-\infty}^{\infty}A_h(\Delta t;\tau)e^{-j2\pi f_D\Delta t}\,d\Delta t. \quad (4.23)$$

The psd represents the power density as a function of Doppler shift. In Clarke's model, $S_h(f_D;\tau)$ is given by

$$S_h(f_D;\tau) = \begin{cases} \sum_{l=1}^{L}\frac{\Omega_l}{\pi f_{Dm}}\frac{1}{\sqrt{1-(f_D/f_{Dm})^2}}\delta(\tau-\tau_l), & \text{if } |f_D|\leq f_{Dm}; \\ 0, & \text{otherwise.} \end{cases} \quad (4.24)$$

Figure 4.3(b) plots $\frac{1}{\pi}\frac{1}{\sqrt{1-(f_D/f_{Dm})^2}}$ and shows that the psd concentrates heavily around the maximum Doppler shift $f_D = \pm f_{Dm}$ in Clarke's model.

In reality, the autocorrelation function and the psd depend on the propagation environment and may not strictly follow the patterns shown in Figure 4.3. Rather than relying on specific models, we can use two parameters, *coherence time* and *Doppler spread*, to characterize the autocorrelation function and the psd, respectively.

Coherence time T_c is defined as the length of a time interval beyond which the channel response undergoes a significant change. For example, suppose that the "significant change" means an autocorrelation of 0.5 or below. Then, in Clarke's model,

$$T_c = \frac{1}{4f_{Dm}}. \quad (4.25)$$

Doppler spread B_D is defined as the range of a Doppler frequency interval over which the psd is significant. In Clarke's model,

$$B_D = 2f_{Dm}. \quad (4.26)$$

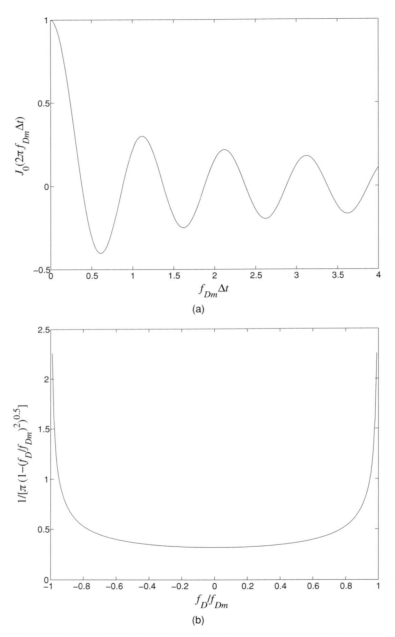

Figure 4.3 Autocorrelation function and Doppler power spectrum in Clarke's model.

While the coefficients in (4.25) and (4.26) are still based on the specific model, the following relationship between T_c and B_D holds in general:

$$T_c \propto \frac{1}{B_D}. \tag{4.27}$$

4.1.4 Channel variation in frequency

So far we have studied how the channel response evolves over time. We next study the channel variation over frequency by examining the autocorrelation of frequency domain channel response. The important concepts are *delay spread* and *coherence bandwidth*.

The channel impulse response (4.10) is defined in the time domain. Taking the Fourier transform of $h(t, \tau)$ with respect to τ, we get the channel response in the frequency domain at time t:

$$H(t, f) = \int_{-\infty}^{\infty} h(t, \tau) e^{-j2\pi f \tau} d\tau$$

$$= \sum_{l=1}^{L(t)} \bar{\alpha}_l(t) e^{-j\bar{\phi}_l(t) - j2\pi f \tau_l(t)}. \quad (4.28)$$

In narrowband fading with $L(t) = 1$, the phase of $H(t, f)$ rotates with f but the amplitude remains the same. Thus, it is called *flat fading*. In wideband fading with $L(t) > 1$, the amplitude also varies and the fading is called *frequency selective*. Similar to (4.17), define the autocorrelation of the channel response in the frequency domain

$$A_H(t, t + \Delta t; f, f + \Delta f) = \mathbb{E}(H^*(t, f) H(t + \Delta t, f + \Delta f)). \quad (4.29)$$

It can easily be shown that under the wide-sense stationary and uncorrelated scattering assumptions,

$$A_H(t, t + \Delta t; f, f + \Delta f) = A_H(\Delta t; \Delta f)$$

$$= \int_{-\infty}^{\infty} A_h(\Delta t; \tau) e^{-j2\pi \Delta f \tau} d\tau. \quad (4.30)$$

At $\Delta t = 0$, $A_H(0; \Delta f)$ is the correlation of the channel responses at two frequencies Δf apart at a given time.

$A_h(0; \tau)$ specifies the power of the channel impulse response corresponding to delay τ, and is thus called *power delay profile*. Letting $\Delta t = 0$ in (4.19), it follows that

$$A_h(0; \tau) = \sum_{l=1}^{L} \Omega_l \delta(\tau - \tau_l). \quad (4.31)$$

The power delay profile consists of L taps, with power Ω_l at delay τ_l, $l = 1, \ldots, L$. Thus,

$$A_H(0; \Delta f) = \sum_{l=1}^{L} \Omega_l e^{-j2\pi \Delta f \tau_l}. \quad (4.32)$$

There are no dominantly popular models of the power delay profile. To consistently evaluate system performance, the standards organizations adopt certain channel models. For example, the so-called ITU discrete path channel models have been used in developing the third generation (3G) "IMT-2000" family of radio access standards. Figure 4.4 plots $A_h(0; \tau)$ and $A_H(0; \Delta f)$ of two ITU models [138]. Clearly the two ITU models exhibit very different $A_h(0; \tau)$ and $A_H(0; \Delta f)$.

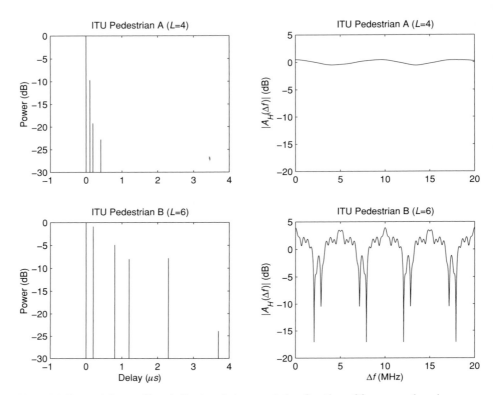

Figure 4.4 Power delay profiles $A_h(0; \tau)$ and autocorrelation function of frequency domain channel response $A_H(0; \Delta f)$ of ITU pedestrian A and B models.

To characterize the power delay profile and the autocorrelation function, we use two parameters, *coherence bandwidth* and *delay spread*, respectively.

Coherent bandwidth B_c is defined as the length of a frequency interval beyond which the frequency domain channel response undergoes a significant change, for example, an autocorrelation of 0.5 or below. Delay spread σ_τ is defined as the range of an interval of delay over which the power delay profile is significant. For example, we can define the mean delay

$$\mu_\tau = \frac{\int_0^\infty \tau A_h(0; \tau) \, d\tau}{\int_0^\infty A_h(0; \tau) \, d\tau}, \tag{4.33}$$

and the delay spread to be

$$\sigma_\tau = \sqrt{\frac{\int_0^\infty (\tau - \mu_\tau)^2 A_h(0; \tau) \, d\tau}{\int_0^\infty A_h(0; \tau) \, d\tau}}. \tag{4.34}$$

The mean delay μ_τ introduces a linear phase rotation across frequency to the frequency domain channel response. The delay spread σ_τ causes frequency selectivity in fading.

Similar to (4.27), the following general relationship holds:

$$B_c \propto \frac{1}{\sigma_\tau}. \qquad (4.35)$$

Finally, recall that in the above discussion, coherence time T_c is defined from $A_h(\Delta t; \tau)$, the autocorrelation of the time domain channel response. One can similarly define T_c from $A_H(\Delta t; \Delta f)$, the autocorrelation of the frequency domain channel response. In Clarke's model, for example, it follows from (4.22) that

$$A_H(\Delta t; \Delta f) = \left(\sum_{l=1}^{L} \Omega_l e^{-j2\pi \Delta f \tau_l} \right) J_0 \left(2\pi f_{Dm} \Delta t \right). \qquad (4.36)$$

We define T_c to measure how fast $A_H(\Delta t; \Delta f)$ evolves with Δt for any given Δf. Moreover, as in (4.23), we take the Fourier transform of $A_H(\Delta t; \Delta f)$ with respect to Δt and get the psd:

$$S_H(f_D; \Delta f) = \int_{-\infty}^{\infty} A_H(\Delta t; \Delta f) e^{-j2\pi f_D \Delta t} \, d\Delta t. \qquad (4.37)$$

We define Doppler spread using psd $S_H(f_D; \Delta f)$ instead of psd $S_h(f_D; \tau)$. The resultant T_c and B_D are the same in either definition.

4.1.5 Gaussian-Markov model

Recall from Section (2.1), the wireless channel in OFDM can be modeled with a one-tap frequency domain channel response (2.9), where the channel coefficient in tone-symbol (s, k) is denoted by $H[s, k]$. $H[s, k]$ is equal to $H(t, f)$ of (4.28) with t equal to OFDM symbol s and f equal to tone frequency f_k. $A_H(\Delta t; \Delta f)$ is the two-dimensional autocorrelation in time and in frequency. For example, in Clarke's model, (4.36) determines $A_H(\Delta t; \Delta f)$ for any power delay profile.

The *Gaussian-Markov model* is another simple model where the autocorrelation function is assumed to monotonically and exponentially decrease as Δf or Δt increases and the rates of the exponential decrease are given by parameters η_f and η_t, respectively. Specifically, the time-frequency variation of the channel coefficients is modeled by a Gaussian-Markov process [47]. Denote by $V_1[s, k]$ i.i.d. $\mathcal{CN}(0, \mathbb{E}(|H|^2))$ random variable for all s and k, where $\mathbb{E}(|H|^2)$ is the expected power of $H[s, k]$.

For any s, and $k = 1, \ldots, N_c - 1$, let

$$V_2[s, 0] = V_1[s, 0], \qquad (4.38)$$

$$V_2[s, k+1] = \sqrt{1 - \eta_f} V_2[s, k] + \sqrt{\eta_f} V_1[s, k+1]. \qquad (4.39)$$

Then, let $H[0, k] = V_2[0, k]$, and

$$H[s+1, k] = \sqrt{1 - \eta_t} H[s, k] + \sqrt{\eta_t} V_2[s+1, k]. \qquad (4.40)$$

Note that $\mathbb{E}(|H[s, k]|^2) = \mathbb{E}(|H|^2)$ for all s, k. $V_1[s, k]$ represents the random effect that cannot be predicted beforehand, and η_t and η_f reflect the amount of channel variation, over time and over frequency, respectively, caused by $V_1[s, k]$.

We can further determine η_t and η_f from the channel coherence time and coherence bandwidth and the OFDMA parameters. As before, channel coherence time T_c is defined as

$$\mathbb{E}\left(H\left[s + \frac{T_c}{T_s + T_{cp}}, k\right] H^*[s, k] \right) = 0.5, \quad (4.41)$$

and channel coherence bandwidth B_c such that

$$\mathbb{E}\left(H\left[s, k + \frac{B_c}{\Delta_f}\right] H^*[s, k] \right) = 0.5. \quad (4.42)$$

It follows that

$$\eta_t = 1 - 2^{\frac{-2}{T_s + T_{cp}}} \approx \frac{2\log(2)}{\frac{T_c}{T_s + T_{cp}}} \quad (4.43)$$

$$\eta_f = 1 - 2^{\frac{-2}{B_c}{\Delta_f}} \approx \frac{2\log(2)}{\frac{B_c}{\Delta_f}}, \quad (4.44)$$

where the above approximations assume $\frac{T_c}{T_s+T_{cp}}, \frac{B_c}{\Delta_f} \gg 1$. We note that

$$\eta_t \eta_f \approx \frac{(2\log(2))^2}{T_c B_c}, \quad (4.45)$$

where it is assumed that $T_s \gg T_{cp}$. It follows that

$$\max\{\eta_t, \eta_f\} \geq \frac{2\log(2)}{\sqrt{T_c B_c}}. \quad (4.46)$$

As an example, suppose $T_c = 10$ ms and $B_c = 500$ kHz. With the OFDMA parameters defined in Discussion notes 2.7, it follows $\eta_t = 0.0152$ and $\eta_f = 0.0277$.

4.2 Communications over a fading channel: the single-user case

So far we have presented the channel models of multipath fading. The fading channel poses unique challenges and opportunities. We will first study the single user case between a transmitter and a receiver, and then move on to the multiuser case where the wireless channel is shared among multiple links.

4.2.1 Performance penalty due to multipath fading

Consider a single user channel subject to multipath fading. The purpose of our study is not to provide a comprehensive performance analysis on the fading channel but to understand how the fading channel behaves differently from the AWGN channel from first principles. The insight will help us find ways of improving the performance.

We study the problem of sending a block of information bits over the channel. The information bits are encoded and modulated to form coded modulation symbols X_k, $k = 1, 2, \ldots$, which are transmitted in a block of tone-symbols.

We assume a block fading scenario where the channel response is the same for all the tone-symbols in the block. We furthermore assume that the coding and modulation is contained within the block. This assumption makes sense when the fading is flat and the coherence time is much longer than the time scale of typical physical layer communications (on the order of a few milliseconds). From (2.9), the received symbol Y_k in tone-symbol k is given by

$$Y_k = HX_k + W_k, \qquad (4.47)$$

where noise $W_k \sim \mathcal{CN}(0, \sigma^2)$. Denote by $P = \mathbb{E}(|X_k|^2)$ average transmit power. Define the average SNR $\bar{\gamma} = P\mathbb{E}(|H|^2)/\sigma^2$, and $z = |H|^2/\sqrt{\mathbb{E}(|H|^2)}$ relative channel gain variation due to fading experienced in a given code block. $z\bar{\gamma}$ is the instantaneous SNR. Denote by $p_z(x)$ the pdf of z.

Conditional on z, $P_{B|z}$ is the block error probability under the AWGN model. In Appendix Section B.4, Figure B.11 depicts the "waterfall" effect in $P_{B|z}$ versus instantaneous SNR of certain codes, where $P_{B|z}$ drops very rapidly once the instantaneous SNR exceeds a threshold γ_0. The behavior of error probability is very well characterized with a two-part approximation: error probability is very high (close to 1) if the instantaneous SNR is below a threshold $z\bar{\gamma} \leq \gamma_0$ or very low (around or below 10^{-4}) otherwise. The SNR threshold γ_0 depends on the coding and modulation scheme; it decreases when the coding rate or modulation order gets lower as depicted in Figure B.13.

The overall block error probability over the fading channel is thus given by

$$P_B = \int_0^\infty P_{B|z} p_z(x) \, dx. \qquad (4.48)$$

When $\bar{\gamma} \ll \gamma_0$, typically $z\bar{\gamma} \leq \gamma_0$, and the error probability is high. In practice, the coding and modulation scheme is usually chosen such that $\bar{\gamma} > \gamma_0$ to reduce the error probability. In this regime, typically $z\bar{\gamma} > \gamma_0$ and the error event caused by the additive noise is negligible thanks to the waterfall effect. However, *with a small probability* the channel is in a deep fade such that $z\bar{\gamma} \leq \gamma_0$, in which case the error event likely occurs. This event is called outage. Therefore,

$$P_B \approx \mathbb{P}\{z\bar{\gamma} \leq \gamma_0\} = \int_0^{\gamma_0/\bar{\gamma}} p_z(x) \, dx. \qquad (4.49)$$

In the fading channel, the deep fade event (as opposed to the additive noise) dominates the overall error probability. This behavior is very different from the AWGN channel.

$\bar{\gamma}/\gamma_0$ can be viewed as a margin required to cover the unknown channel fading z so as to avoid the outage event of $z\bar{\gamma} \leq \gamma_0$. For a given target of outage probability, (4.49) shows that the required margin depends on the tail distribution around $z \approx 0$. For example, with Rayleigh fading, z is an exponential random variable $p_z(x) = e^{-x}$. In this case, it follows that

$$\mathbb{P}\{z\bar{\gamma} \leq \gamma_0\} = 1 - e^{-\gamma_0/\bar{\gamma}}$$

$$\approx \gamma_0/\bar{\gamma}, \text{ when } \bar{\gamma} \gg \gamma_0. \qquad (4.50)$$

(4.50) shows that asymptotically, as $\bar{\gamma}$ increases, P_B drops only *inversely proportionally* to $\bar{\gamma}$, much slower than the waterfall effect observed in the AWGN channel where P_B drops *exponentially* fast as $\bar{\gamma}$ increases. To get the same P_B as in AWGN, a much higher $\bar{\gamma}$ would be required, simply because the probability of a deep fade does not drop with $\bar{\gamma}$ as rapidly.

So far we have studies the error probability of a given coding and modulation scheme as a function of average SNR $\bar{\gamma}$. How should the coding and modulation scheme be determined? We assume that the instantaneous channel response H is known to the receiver but not to the transmitter. The transmitter only knows the average SNR $\bar{\gamma}$ and the statistics of fading, but not the instantaneous z.

If a low coding rate or modulation order is selected, the corresponding γ_0 is small, and so is the error probability; however, the data rate is too conservative as the instantaneous SNR exceeds γ_0 most of the time. To quantify the tradeoff, suppose that the goal is to maximize the average data rate R_f. We assume that when the instantaneous SNR exceeds γ_0, the achieved data rate is equal to $\log(1+\gamma_0)$, the AWGN channel capacity at SNR γ_0. It follows that

$$\max_{\gamma_0} R_f = \log(1+\gamma_0)\mathbb{P}\{z\bar{\gamma} > \gamma_0\} \qquad (4.51)$$

$$= \log(1+\gamma_0)e^{-\frac{\gamma_0}{\bar{\gamma}}}, \text{ for Rayleigh fading} \qquad (4.52)$$

Setting the derivative of (4.52) with respect to γ_0 to 0, we get the following equation to determine the optimal γ_0:

$$(1+\gamma_0)\log(1+\gamma_0) = \bar{\gamma} \qquad (4.53)$$

As a comparison, the data rate with SNR $\bar{\gamma}$ in AWGN is

$$R_0 = \log(1+\bar{\gamma}) \qquad (4.54)$$

We next compare R_f and R_0 to see the performance penalty due to multipath fading. At low SNR, (4.53) leads to $\gamma_0 \approx \bar{\gamma}$, and

$$R_f \approx \gamma_0 e^{-1} \approx e^{-1} R_0. \qquad (4.55)$$

Apparently, the performance penalty is that the average data rate in fading is only a fraction of that in AWGN. At high SNR, $\gamma_0 \log(\gamma_0) \approx \bar{\gamma}$, and

$$R_f \approx \log(\gamma_0)e^{-1/\log(\gamma_0)} \approx e^{-\frac{1}{\log \gamma_0}}(R_0 - \log\log\gamma_0). \qquad (4.56)$$

We next show two approaches to combat channel fading.

4.2.2 Mitigation of fading via channel state feedback

The analysis in Section 4.2.1 assumes that only the receiver knows the channel. Suppose that from the feedback from the receiver, the transmitter also knows the instantaneous channel gain variation z. We assume that the knowledge at the transmitter is perfect: no error or latency in estimation or feedback. Now the transmitter can adjust transmission based on the feedback. The transmitter applies different approaches that depend on the performance and cost tradeoff: power control only, rate control only, and joint power and rate control.

Power control only

First, consider power control and fix the data rate to a constant target. The power control scheme is *channel inversion*; the transmit power P is set to be inversely proportional to the channel gain so as to make the instantaneous SNR equal to a constant target corresponding to the fixed data rate. With exact channel inversion, the fading channel is converted back to AWGN; however, the transmit power would be enormous when the channel gain is small. For Rayleigh fading, it is easy to show that $\mathbb{E}(P)$ would have to be infinite to achieve exact channel inversion. Clearly, such a scheme is not feasible. To limit the transmit power, we use *truncated channel inversion*; invert the channel only when $z\bar{\gamma}$ exceeds some threshold γ_0. Otherwise, declare outage and do not transmit:

$$P = \begin{cases} \frac{\beta}{z}, & \text{if } z \geq \frac{\gamma_0}{\bar{\gamma}}; \\ 0, & \text{otherwise.} \end{cases} \tag{4.57}$$

The outage probability is $\mathbb{P}\{z\bar{\gamma} \leq \gamma_0\}$. When the channel is not in outage, the instantaneous SNR is $\beta\bar{\gamma}$.

To keep the average transmit power unchanged, we set

$$\mathbb{E}(P(z)) = \beta \int_{\gamma_0/\bar{\gamma}}^{\infty} \frac{p_z(x)}{x} dx = 1. \tag{4.58}$$

In Rayleigh fading, it follows that

$$\beta = \left(\int_{\gamma_0/\bar{\gamma}}^{\infty} \frac{e^{-x}}{x} dx \right)^{-1}. \tag{4.59}$$

Equation (4.58) shows the tradeoff of design parameter γ_0. While the outage probability decreases with γ_0, the instantaneous SNR also drops when the channel is not in outage. Also note that we assume the constraint is the average transmit power. In practice, the transmitter is also constrained by the maximum instantaneous transmit power, which is not taken into account here.

Continuing the previous maximization of the average data rate, we need to optimize γ_0:

$$\max_{\gamma_0} R_f = \log(1 + \beta\bar{\gamma})\mathbb{P}(z\bar{\gamma} > \gamma_0) \tag{4.60}$$

$$= \log\left(1 + \frac{\bar{\gamma}}{\int_{\gamma_0/\bar{\gamma}}^{\infty} \frac{e^{-x}}{x} dx}\right) e^{-\gamma_0/\bar{\gamma}}, \text{ for Rayleigh fading.} \tag{4.61}$$

At low SNR $\bar{\gamma}$, (4.61) becomes

$$R_f \approx \frac{e^{-\gamma_0/\bar{\gamma}}}{\int_{\gamma_0/\bar{\gamma}}^{\infty} \frac{e^{-x}}{x} dx} R_0. \tag{4.62}$$

The multiplier in (4.62) is greater than 1 because letting $\gamma_0 > \bar{\gamma}$, we have

$$\int_{\gamma_0/\bar{\gamma}}^{\infty} \frac{e^{-x}}{x} dx < \int_{\gamma_0/\bar{\gamma}}^{\infty} e^{-x} dx = e^{-\gamma_0/\bar{\gamma}}. \tag{4.63}$$

The truncated channel inversion scheme achieves a higher average data rate R_f in the fading channel than R_0 in the AWGN channel. This result is somewhat surprising, as the fading channel not only suffers no loss but also outperforms the AWGN channel. What happened?

Given an average transmit power budget, the transmitter can spread the power over the entire bandwidth (time and frequency) or concentrate it to only a fraction (α) of the total bandwidth. In the AWGN channel, this choice does not make much difference because in the low SNR regime, $\alpha \log(1 + \bar{\gamma}/\alpha) \approx \log(1 + \bar{\gamma}) \approx \bar{\gamma}$. However, the situation is different in the fading channel. As the channel fluctuates, sometimes the channel gain is better than the average. This creates the possibility of *opportunistic truncation*: only transmit in those favorable instants and do not waste power when the channel is unfavorable. At low SNR, the bandwidth loss due to truncation has insignificant impact on the data rate. β represents the power boost when the transmission is on. Here, $\beta > 1$.

At high SNR $\bar{\gamma}$, we set $\gamma_0 \ll \bar{\gamma}$ to keep the outage probability and thus bandwidth loss low. (4.59) becomes

$$\beta \approx \frac{1}{\log(\bar{\gamma}/\gamma_0)} < 1. \tag{4.64}$$

It follows that

$$R_f \approx \log(\beta \bar{\gamma}) e^{-\gamma_0/\bar{\gamma}}$$
$$\approx e^{-\gamma_0/\bar{\gamma}} (R_0 - \log\log(\bar{\gamma}/\gamma_0)). \tag{4.65}$$

In this regime, $e^{-\gamma_0/\bar{\gamma}}$ is close to 1; however, the SNR is reduced to a fraction β of $\bar{\gamma}$. The capacity loss is thus $\log(\beta)$, equal to the offset term in (4.65). It is practically constant representing a fixed loss from the AWGN capacity R_0.

Rate control only

Next, consider rate control. In this scheme, the transmit power is fixed and the data rate of the transmission block is set equal to the AWGN channel capacity that corresponds to the instantaneous SNR. Clearly the data rate varies with the channel. If there is a minimum data rate requirement, then outage occurs if the instantaneous data rate is below the required minimum data rate. In the study below, however, we assume that no minimum data rate is required. We are interested in the expected data rate:

$$R_f = \mathbb{E}\left(\log\left(1 + z\bar{\gamma}\right)\right) \tag{4.66}$$

$$= \int_0^{\infty} \log(1 + x\bar{\gamma}) e^{-x} dx, \text{ for Rayleigh fading.} \tag{4.67}$$

Since log(·) is strictly concave, from Jensen's inequality,[1] it follows that R_f is always less than R_0. At low SNR,

$$R_f \approx \mathbb{E}(z)\bar{\gamma} = \bar{\gamma} \approx R_0. \tag{4.68}$$

At high SNR,

$$R_f \approx \mathbb{E}(\log(z\bar{\gamma})) = \log(\bar{\gamma}) + \mathbb{E}(\log(z))$$

$$\approx R_0 + \int_0^\infty \log(x) e^{-x}\, dx, \text{ for Rayleigh fading.} \tag{4.69}$$

The second term represents a constant capacity loss of the fading channel as compared with the AWGN channel.

Joint power and rate control

Finally, consider joint power and rate control. The transmitter jointly optimizes the instantaneous transmit power and data rate to maximize the expected data rate. The transmit power P depends on the instantaneous channel gain variation z, and as in the rate control only scheme, the data rate is set equal to the AWGN channel capacity corresponding to the instantaneous SNR. Hence, the optimization problem is stated as follows:

$$\max_P \ R_f = \mathbb{E}(\log(1 + z\bar{\gamma}P(z))) \tag{4.70}$$

$$\text{subject to } \mathbb{E}(P(z)) = 1. \tag{4.71}$$

The problem is solved by forming the Lagrangian

$$\mathcal{L}(P) = \mathbb{E}(\log(1 + z\bar{\gamma}P(z))) + \lambda\left(\mathbb{E}(P(z)) - 1\right) \tag{4.72}$$

Setting the derivative of $\mathcal{L}(P)$ to 0, it follows that for any fading realization z,

$$\frac{z\bar{\gamma}}{1 + z\bar{\gamma}P(z)} - \lambda = 0. \tag{4.73}$$

Hence,

$$P(z) = \begin{cases} \frac{1}{\lambda} - \frac{1}{z\bar{\gamma}}, & \text{if } z \geq \frac{\lambda}{\bar{\gamma}}; \\ 0, & \text{otherwise.} \end{cases} \tag{4.74}$$

Parameter λ is solved by plugging (4.74) into (4.71). For Rayleigh fading, λ solves the following equation:

$$\int_{\frac{\lambda}{\bar{\gamma}}}^\infty \left(\frac{1}{\lambda} - \frac{1}{x\bar{\gamma}}\right) e^{-x}\, dx = \frac{1}{\lambda} e^{-\frac{\lambda}{\bar{\gamma}}} - \frac{1}{\bar{\gamma}} \int_{\frac{\lambda}{\bar{\gamma}}}^\infty \frac{e^{-x}}{x}\, dx = 1. \tag{4.75}$$

The power allocation scheme (4.74) is referred to as *water-filling*. Note the similarity of (4.74) and (4.57) in that both expressions exhibit a form of opportunistic truncation. However, when truncation does not take place, the power allocation strategy is very

---

[1] Jensen's inequality: if $f(\cdot)$ is strictly concave, then $\mathbb{E}(f(x)) \leq f(\mathbb{E}(x))$ for any random variable x with equality if and only if x is deterministic.

different: the water-filling scheme allocates more power when the channel is good, while the channel inversion scheme does exactly the opposite. The rate control-only scheme can be viewed as operating in the middle where the power allocation is invariant of the channel.

Figure 4.5 compares the data rates of the Rayleigh fading channel in the four schemes studied so far and compares them with those in the AWGN channel. We observe that

- Power or rate control based on feedback significantly increases the data rate as compared with the scheme without feedback.
- The water-filling scheme is the optimal solution. At high SNR, the rate control only scheme performs almost the same as water-filling. At low SNR, the power control only scheme is close to water-filling. Hence, a practically simple alternative to water-filling is to use the two schemes at the low and high SNR regimes, respectively, with a switching point around 0 to 5 dB.
- At high SNR, the data rate in the fading channel is always lower than that in AWGN. However, the loss is insignificant. At low SNR, the fading channel with either truncated channel inversion or water-filling outperforms the AWGN channel. Indeed, R_f/R_0 grows significantly as the average SNR decreases.

Discussion notes 4.1 Practical consideration of feedback-based approaches

The main cost of the feedback-based schemes is the signaling overhead, which depends on the rate of feedback and the amount of information in every feedback.

We have assumed that the feedback provides a full description of the instantaneous channel response. However, such a fine granular feedback may not be needed. A feedback can be quantized to reduce the overhead. As an example, consider the following simple on-off power control scheme. For a given average SNR $\bar{\gamma}$ and the fading realization z, define a 1-bit feedback:

$$b = \begin{cases} 1, & \text{if } z\bar{\gamma} \geq \gamma_0; \\ 0, & \text{otherwise}. \end{cases} \quad (4.76)$$

The feedback indicates whether the channel is good or bad. If $b = 1$, the transmitter uses fixed power β and fixed data rate $\log(1 + \beta\gamma_0)$, where

$$\beta^{-1} = \mathbb{P}(z\bar{\gamma} \geq \gamma_0). \quad (4.77)$$

Parameter γ_0 is selected to maximize the expected data rate

$$\max_{\gamma_0} R_f = \log(1 + \beta\gamma_0)\mathbb{P}(z\bar{\gamma} \geq \gamma_0). \quad (4.78)$$

Note the difference between (4.60) and (4.78). With only 1-bit feedback, the data rate is set conservatively. It turns out that at low SNR, this simple on-off power control scheme achieves much of the gain of the truncated channel inversion or water-filling scheme. The latter requires much more detailed feedback.

4.2 Communications, fading channel: single-user case

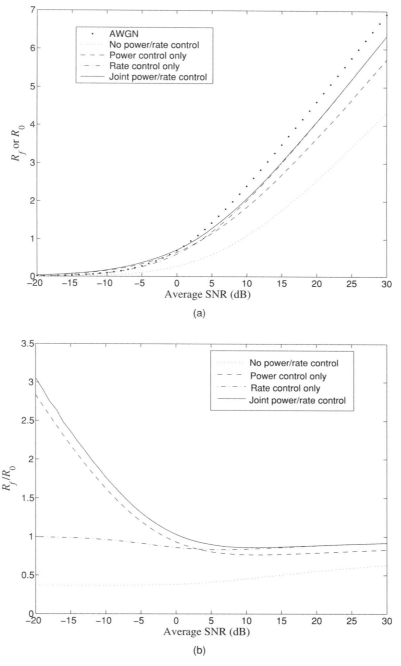

Figure 4.5 Comparison of data rate R_f of the Rayleigh fading channel in the schemes with no feedback (4.52), power control only (4.61), rate control only (4.67) and joint power/rate control (4.70) and R_0 of the AWGN channel. In (a), R_f and R_0 are plotted. In (b), R_f/R_0 is plotted.

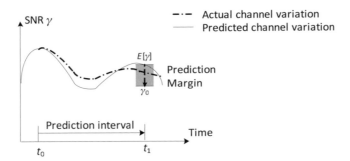

Figure 4.6 Illustration of channel state prediction and margin.

The rate of feedback depends on the coherence time and coherence bandwidth of the channel. To capture the channel variation in time and in frequency, at a minimum, the receiver sends one feedback for every block of coherence time and coherence bandwidth. More frequent feedback is desirable to keep good track of channel variation. In practice, the channel knowledge at the transmitter is fundamentally constrained by the delay required for channel measurement and feedback signaling. Figure 4.6 illustrates an example where the receiver sends a feedback at time t_0 and the transmitter predicts the channel state at time t_1. Even if the feedback at t_0 precisely reflects the channel state, the prediction is not perfect.

Uncertainty in channel state prediction has an adverse effect similar to channel fading. The instantaneous SNR γ at t_1 can be treated as a random variable conditional on the observation at t_0. The transmitter predicts SNR to be γ_0 for the power/rate control purpose. γ_0 is usually set conservatively, $\gamma_0 < \mathbb{E}(\gamma(t_1)|\gamma(t_0))$, to reduce the outage probability. Prediction margin is defined as

$$\mathrm{MG} = 1 - \frac{R(\gamma_0)}{R(\mathbb{E}(\gamma(t_1)|\gamma(t_0)))}, \tag{4.79}$$

where $R(x)$ represents the data rate corresponding to SNR x. The prediction margin MG increases with the conditional variance of $\gamma(t_1)|\gamma(t_0)$. Clearly, the conditional variance is smaller than the unconditional variance of $\gamma(t_1)$ itself. Therefore, feedback helps reduce the required prediction margin. MG > 0 is indicative of the loss in the data rate due to prediction uncertainty. In the rate control scheme, for example, the transmitter can determine the optimal MG that maximizes the expected data rate

$$\max_{\gamma_0} R_f = \log(1+\gamma_0)\mathbb{P}(\gamma(t_1) > \gamma_0|\gamma(t_0)). \tag{4.80}$$

Note the similarity between (4.51) and (4.80). Conditional on the observation at t_0, the uncertainty of $\gamma(t_1)$ should be reduced, and the achieved rate R_f is higher in (4.80) than that in (4.51). We will elaborate on the uncertainty of $\gamma(t_1)$ in Section 4.2.4.

4.2 Communications, fading channel: single-user case

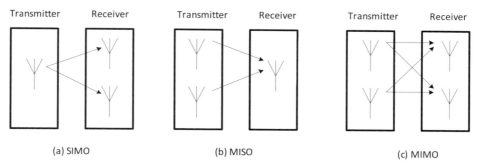

Figure 4.7 Illustration of scenarios of using multiple antennas: single input multiple output (SIMO), multiple input single output (MISO), and multiple input multiple output (MIMO).

4.2.3 Mitigation of fading via diversity

The analysis in Section 4.2.1 assumes that the receiver only sees one realization of the channel fading. As a result, the performance is limited by a single bad realization. The idea of diversity is to allow the receiver to see multiple, hopefully independent, realizations of the channel so as to improve the reliability: communication will probably succeed if the channel is good in *some* of the realizations. We next elaborate on different forms of diversity.

Antenna diversity

Antenna diversity is obtained by placing multiple antennas at the transmitter and/or receiver. One method is to separate them far apart such that the fading at each antenna is roughly uncorrelated. A second method is to use different (vertical or horizontal) polarization. In the following, we ignore the implementation details of obtaining antenna diversity and focus on the diversity gain in three different scenarios as shown in Figure 4.7, namely single input multiple output (SIMO), multiple input single output (MISO), and multiple input multiple output (MIMO).

First, consider receive diversity where a signal sent from one transmit antenna is received at K_r receive antennas as illustrated in Figure 4.7(a). Suppose that in a tone-symbol, the received symbol at every receive antenna is

$$Y_l = H_l X + W_l, l = 1, \ldots, K_r, \quad (4.81)$$

where l is the receive antenna index. Because W_l are i.i.d. $\mathcal{CN}(0, \sigma^2)$, the optimal receiver scheme is to coherently combine all Y_l:

$$Y = \sum_{l=1}^{K_r} \frac{H_l^*}{\sqrt{\sum_{l=1}^{K_r} |H_l|^2}} Y_l \quad (4.82)$$

$$= \left(\sqrt{\sum_{l=1}^{K_r} |H_l|^2} \right) X + \tilde{W}.$$

The noise term $\tilde{W} \sim \mathcal{CN}(0, \sigma^2)$. The linear processing of (4.82) is called *maximal ratio combining*.

Suppose that $\mathbb{E}(|H_l|^2) = \mathbb{E}(|H|^2)$ is the same for all l. As before, define $z_l = |H_l|^2/\sqrt{\mathbb{E}(|H|^2)}$. The instantaneous SNR is thus

$$(K_r \bar{\gamma}) \left(\frac{1}{K_r} \sum_{l=1}^{K_r} z_l \right).$$

The benefits of receive diversity are twofold:

- *Power gain*: coherently adding the signals from multiple receive antennas makes the first term (the mean of SNR) increase linearly with K_r.
- *Diversity gain*: because z_l are independent fading realizations, the variance of the second term decreases with K_r. However, the incremental benefit of diversity gain diminishes as K_r increases.

We next quantify the above benefits by calculating the probability of a deep fade in the case where each received signal is Rayleigh faded. The pdf of $\sum_{l=1}^{K_r} z_l$ is given by $\frac{1}{(K_r-1)!} x^{K_r-1} e^{-x}$. Similar to (4.49), the outage probability that the instantaneous SNR does not exceed threshold γ_0 is

$$\mathbb{P}\left(\bar{\gamma} \left(\sum_{l=1}^{K_r} z_l \right) \leq \gamma_0 \right) = \int_0^{\gamma_0/\bar{\gamma}} \frac{1}{(K_r - 1)!} x^{K_r-1} e^{-x} \, dx$$

$$= \sum_{l=K_r}^{\infty} \frac{(\gamma_0/\bar{\gamma})^l}{l!} e^{-\gamma_0/\bar{\gamma}}$$

$$\approx \frac{(\gamma_0/\bar{\gamma})^{K_r}}{K_r!}, \text{ when } \bar{\gamma} \gg \gamma_0. \quad (4.83)$$

Comparison of (4.83) with (4.50) shows that given $K_r > 1$, the outage probability decays with $\bar{\gamma}/\gamma_0$ at a faster rate than when there is no diversity ($K_r = 1$), since the exponent of $\gamma_0/\bar{\gamma}$ increases to K_r. K_r is thus called the *degree of diversity* and represents the diversity gain. Figure 4.8(a) plots the outage probability as a function of margin $\bar{\gamma}/\gamma_0$ for a variety of K_r. Figure 4.8(b) views the outage probability from a different angle to see how it drops with K_r given margin $\bar{\gamma}/\gamma_0$.

Recall that receive diversity benefits from both power gain and diversity gain. To demonstrate those two benefits separately, Figure 4.8(b) in addition plots the outage probability assuming only diversity gain, in which case the power of each independent signal is reduced by a factor of K_r to keep the total received power unchanged. Then, the instantaneous SNR becomes

$$\bar{\gamma} \left(\frac{1}{K_r} \sum_{l=1}^{K_r} z_l \right),$$

or equivalently, replace $\bar{\gamma}$ with $\bar{\gamma}/K_r$ in (4.83). We observe that with only diversity gain, the slope of the decay of outage probability becomes much less steep. When the margin is small (0 dB), the outage probability does not noticeably drop as K_r increases.

4.2 Communications, fading channel: single-user case

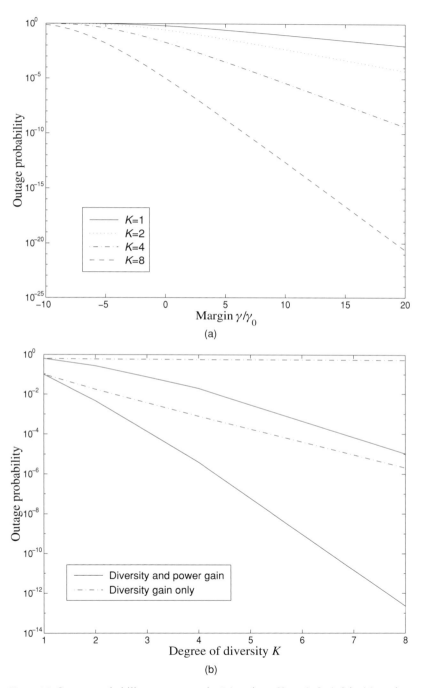

Figure 4.8 Outage probability versus margin $\bar{\gamma}/\gamma_0$ given $K_r = 1, 2, 4, 8$ in (a), and outage probability versus K_r given $\bar{\gamma}/\gamma_0$ in (b). In (b), two groups of curves are shown: $\bar{\gamma}/\gamma_0 = 0$ dB for the upper group and 10 dB for the lower group.

Similar to (4.51), we next determine the optimal choice of γ_0 that maximizes the expected data rate

$$\max_{\gamma_0} R_f = \log(1 + \gamma_0) \mathbb{P}\left(\sum_{l=1}^{K_r} z_l \bar{\gamma} > \gamma_0\right) \quad (4.84)$$

$$= \log(1 + \gamma_0) \sum_{l=0}^{K_r - 1} \frac{(\gamma_0/\bar{\gamma})^l}{l!} e^{-\gamma_0/\bar{\gamma}}, \text{ for Rayleigh fading.} \quad (4.85)$$

Figure 4.9(a) plots R_f of Rayleigh fading with a variety of diversity order K_r and compares with R_0 of the AWGN channel. With both diversity gain and power gain, R_f at $K_r = 4$ exceeds R_0. In contrast, Figure 4.9(b) shows the similar plots, but with diversity gain only, by replacing $\bar{\gamma}$ with $\bar{\gamma}/K_r$. In that case, R_f is always inferior to R_0. This is not surprising because R_f converges to R_0 as $K_r \to \infty$.

Next, consider transmit diversity where the signals sent from K_t transmit antennas are received at one single receive antenna as illustrated in Figure 4.7(b). Continue the coherent symbol detection example. The received symbol is

$$Y = \sum_{l=1}^{K_t} H_l X_l + W, l = 1, \ldots, K_t, \quad (4.86)$$

where l is the transmit antenna index.

To keep the total transmission energy unchanged, the average power at each transmit antenna is $\mathbb{E}(|X_l|^2) = 1/K_t$. We assume that the transmitter is unaware of the instantaneous channel H_l. Otherwise, the transmitter should set X_l such that they combine over the air in the maximal ratio combining manner:

$$X_l = \frac{H_l^*}{\sqrt{\sum_{l=1}^{K_t} |H_l|^2}} X. \quad (4.87)$$

This is also referred to as *beamforming*.

A naïve scheme may be to repeat the same symbol in all the antennas: $X_l = X$ for all l. The combined channel response would be equal to $\frac{1}{\sqrt{K_t}} \sum_{l=1}^{K_t} H_l$. If H_l is i.i.d. complex Gaussian, then the combined channel is still complex Gaussian with the same mean and variance. Using multiple transmit antennas in this way does not result in any diversity gain. The transmitter needs to be smart. *Space-time coding* is a family of such smart techniques ([122]). For the sake of illustration, we next present the so-called Alamouti scheme ([4]) designed to work in the $K_t = 2$ case. It is possible to generalize the idea to the $K_t > 2$ cases to some extent.

Suppose that the transmitter intends to send two symbols X_1 and X_2 using two tone-symbols. In the first tone-symbol, $X_1/\sqrt{2}$ is sent at antenna 1 and $X_2/\sqrt{2}$ at antenna 2. In the second tone-symbol, $-X_2^*/\sqrt{2}$ is at antenna 1 and $X_1^*/\sqrt{2}$ at antenna 2. Assume that the channel response from a given antenna remains unchanged in the two tone-symbols. Then from (4.86), the received symbol vector is given by

$$\begin{bmatrix} Y_1 \\ Y_2^* \end{bmatrix} = \frac{1}{\sqrt{2}} \begin{bmatrix} H_1 & H_2 \\ H_2^* & -H_1^* \end{bmatrix} \begin{bmatrix} X_1 \\ X_2 \end{bmatrix} + \begin{bmatrix} W_1 \\ W_2 \end{bmatrix}. \quad (4.88)$$

4.2 Communications, fading channel: single-user case

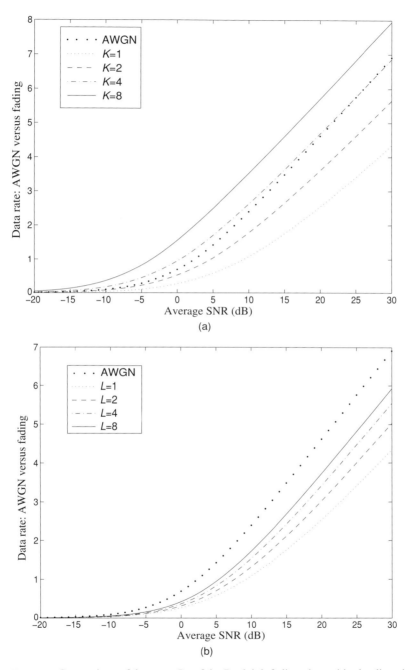

Figure 4.9 Comparison of data rate R_f of the Rayleigh fading channel in the diversity schemes with R_0 of the AWGN channel, with diversity gain and power gain in (a) and with diversity gain only in (b).

The receiver calculates a new vector from received Y_1, Y_2

$$\begin{bmatrix} Z_1 \\ Z_2 \end{bmatrix} = \frac{1}{\sqrt{|H_1|^2 + |H_2|^2}} \begin{bmatrix} H_1 & H_2 \\ H_2^* & -H_1^* \end{bmatrix}^H \begin{bmatrix} Y_1 \\ Y_2^* \end{bmatrix}$$

$$= \sqrt{\frac{|H_1|^2 + |H_2|^2}{2}} \begin{bmatrix} X_1 \\ X_2 \end{bmatrix} + \begin{bmatrix} \tilde{W}_1 \\ \tilde{W}_2 \end{bmatrix}, \qquad (4.89)$$

with noises $\tilde{W}_1$, $\tilde{W}_2$ being i.i.d. $\mathcal{CN}(0, \sigma^2)$. The instantaneous SNR for X_1 and X_2 is therefore $\bar{\gamma}(z_1 + z_2)/2$, which is 3 dB worse than that in the receive diversity case, meaning that the Alamouti scheme achieves $K_t = 2$ diversity gain but no power gain. This is true in general; without the knowledge of instantaneous channel, transmit diversity obtains only diversity gain. Therefore, the curves labeled as "diversity gain only" in Figures 4.8(b) and 4.9(b) are indicative of the performance of the transmit diversity schemes.

Finally, consider the MIMO scenario. For simplicity, assume two transmit antennas and two receive antennas as illustrated in Figure 4.7(c). To maximize the diversity gain, we combine the above transmit and receive diversity approaches. Specifically, the transmitter sends two symbols X_1 and X_2 in two tone-symbols as in the Alamouti scheme and the receiver performs maximal ratio combining. The effective channel becomes

$$\frac{1}{2} \left(\sum_{i=1}^{2} \sum_{j=1}^{2} |H_{i,j}|^2 \right), \qquad (4.90)$$

where i, j are transmit and receive antenna indices, respectively. Hence, in the 2×2 MIMO scenario, a four-fold diversity gain and 3 dB power gain are obtained. In addition to achieving diversity or power gain, MIMO can also be used to increase the degrees of freedom and therefore the data rate with spatial multiplexing, as will be studied in Chapter 7.

Time and frequency diversity

So far we have assumed a block fading scenario where all the tone-symbols of a code block are in one fading block such that the channel response coefficients H_ks are the same. From the channel modeling in Section 4.1, roughly speaking, a fading block comprises one coherence time and coherence bandwidth. Now consider a different scenario where H_ks are spread into multiple fading blocks and each block sees independent fading from each other, thereby achieving time and frequency diversity. OFDMA is flexible in mapping a code block to a subset of tone-symbols so as to achieve a desired tradeoff of time and frequency diversity. Figure 4.10 shows a few options for obtaining time and frequency diversity.

A simple scheme for achieving diversity gain is to use repetition coding: transmit the same symbol in multiple tone-symbols. The analysis of such a scheme is identical to that of receive antenna diversity, except that the former does not achieve power gain, since the transmitter does spend transmit power multiple times for a given symbol.

4.2 Communications, fading channel: single-user case

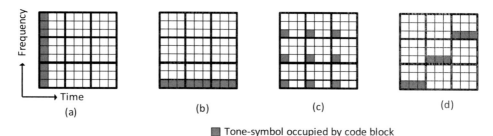

Figure 4.10 Comparison of choices of time and frequency diversity in an OFDMA system. A small block represents a tone-symbol. The channel responses of neighboring tone-symbols are correlated, as reflected in coherence time and coherence bandwidth. A big block with thick border line consists of tone-symbols within a fading blocking of a coherence bandwidth and a coherence time. Choice (a) obtains frequency diversity where the code block is on multiple frequency tones at an OFDM symbols. Choice (b) obtains time diversity where the code block is on the same tone over multiple OFDM symbols. Choices (c) and (d) both obtain time and frequency diversity over the same range of time and frequency.

The diversity gain in this case is due to diversity in time and frequency as opposed to diversity in space. Moreover, compared with receive antenna diversity, the simple scheme based on repetition coding consumes more time and frequency resource. Note that repetition coding does not utilize the time and frequency resource efficiently. For the same amount of increase in resource, a more sophisticated coding scheme achieves coding gain in addition to diversity gain, thereby resulting in a higher data rate. It has been shown in [159] and [61] that as the block size goes to infinity, the achievable data rate is given by

$$R_f = \mathbb{E}\left(\log(1 + z\bar{\gamma})\right). \tag{4.91}$$

In practice, (4.91) holds as long as the fading process is ergodic and the number of independent fading realizations seen by the code block, L_D, is sufficiently large such that

$$\frac{1}{L_D} \sum_{k=1}^{L_D} \log(1 + z_k \bar{\gamma}) \approx \mathbb{E}\left(\log(1 + z\bar{\gamma})\right). \tag{4.92}$$

Note that (4.91) is identical to the data rate with the rate control only scheme (4.67) in Section 4.2.2. However, the communication scheme to achieve the same R_f is very different, as explained below.

In the diversity scheme, the transmitter is only aware of the statistics of the channel, that is, the average SNR $\bar{\gamma}$ and distribution of z, but not the instantaneous channel realization. The transmitter sends a single code block with the fixed data rate R_f and the code block spans over multiple coherence time/bandwidth so as to experience a sufficient number of independent realizations of the fading channel. One drawback is that in an environment where the channel fading varies slowly, in order to pursue time diversity the code block has to be sent in a long time interval, thereby incurring large delay and making time diversity less attractive.

On the other hand, in the rate control scheme, the transmitter is aware of the instantaneous channel realization, and sends multiple code blocks at variable data rates with each code block staying within a coherence time/bandwidth. The drawback is the signaling overhead required for the transmitter to keep track of the channel variation. Clearly, when the channel fading varies rapidly, the overhead increases, making rate control less attractive.

Hence, the idea of diversity or rate control seems to work in a complementary fading environment: diversity for fast fading and rate control for slow fading. This observation naturally suggests making the choice between the diversity and rate control schemes based on the underlying channel fading characteristics and the traffic delay requirement. We will further elaborate on this idea in Section 4.2.4.

> **Discussion notes 4.2** Tradeoff considerations for achieving diversity
>
> For antenna diversity, there are the hardware cost of multiple antennas and the associated RF circuitry and the power cost of operating them. Ignoring those costs, receive diversity is *always* beneficial because of the power gain; using multiple receive antennas, the receiver in essence collects more energy from the environment. The story for transmit diversity or time and frequency diversity is, however, not that simple. The reason is that when a signal with a given total energy is spread over multiple blocks of coherence time/bandwidth or multiple transmit antennas, the number of independent channel parameters to be estimated grows linearly with the degrees of diversity. (Non-coherent communication does not require explicit channel estimation, but faces a similar cost in dealing with varying channel.) For example, in Figure 4.10, while choices (c) and (d) both obtain time and frequency diversity, the code block in (c) spreads over more blocks of coherence time and coherence bandwidth and therefore requires more power and bandwidth for channel estimation. Since channel estimation consumes power and bandwidth, increasing the degrees of diversity is not always beneficial, especially when the incremental benefit of diversity gain diminishes.
>
> In addition to the tradeoff between diversity gain and channel estimation cost, another tradeoff exists between diversity gain and transmission delay. In Figure 4.10, choice (a) is preferable from the delay perspective, because it has a smaller transmission delay than the other choices. However, other constraints may limit the use of choice (a). First, assuming the same total transmission energy for the code block in all the choices, choice (a) requires a higher transmit power because of the short transmission time, which becomes a problem if the transmitter is already constrained by the maximum transmit power. Second, choice (a) relies entirely on frequency diversity, which depends on the delay spread of the channel and the channel bandwidth as explained in Section 4.1.4. If the channel does not provide sufficient frequency diversity, we have to pursue time diversity and end up with choice (c) or (d).

4.2.4 Feedback or diversity

We have investigated the effect of fading on the performance of a single-user communication channel and the two basic approaches of communication over fading channels, namely diversity and power/rate control via channel state feedback. In particular, we have shown that joint power and rate control leads to an optimal water-filling solution as shown in (4.74), which takes advantage of the channel variation and assigns more power to the frequency band or time slot with better channels. Similar behavior is observed in the truncated power control algorithm. Both algorithms show significant gain over the AWGN channel when the SNR is low, as seen in Figure 4.5. On the other hand, the diversity scheme focuses on mitigating the channel variation by averaging or summing over multiple independently faded paths, in frequency, time or space.

In a sense the two approaches contradict each other and deal with fading in fundamentally different ways. In a practical system, one has to make a tradeoff between the two. In particular, how much diversity should we use in data transmission and how much can we take advantage of opportunistic transmission via feedback information?

We have shown that either diversity or power/rate control via channel state feedback increases the average data rate in the fading channel. But there are some subtle differences. In a diversity scheme, transmit power is spent irrespective of whether the channel is good or bad. At low SNR, by opportunistically transmitting only when the channel is good, the truncated power control scheme achieves a higher average data rate. The fundamental reason is that in this regime the power/bandwidth tradeoff is such that power has a more dominant impact on capacity than bandwidth. With opportunistic truncation, it is not desirable to have much diversity in a code block, because diversity averages good and bad channel conditions and thus makes it harder to harness good channel conditions opportunistically. At high SNR, rate control outperforms truncated power control, and achieves the same average data rate as the time and frequency diversity scheme does.

The above lessons indicate that the SNR regime is a key consideration in determining a diversity or feedback strategy. At low SNR, we should lean on using opportunistic communication via feedback. Moreover, combining opportunistic truncation and diversity, if possible, is desirable to balance the tradeoff between increasing the average data rate and reducing the scheduling delay (see below). At high SNR, we should maximize the degrees of diversity to the extent possible so as to mitigate the fading effect. Fully pursue frequency diversity and antenna diversity if available, as well as time diversity within the budget of the traffic delay requirement. If significant channel variation still remains, introduce rate control to further maximize the average data rate.

In addition to the average data rate studied so far, the following performance metrics are also important in practice:

- First, *delay*, which includes transmission delay and scheduling delay. Transmission delay is the time duration over which a code block is sent. In order to pursue time diversity, the code block is sent over multiple coherence time and the transmission delay depends on the length of channel coherence time. Scheduling delay measures time duration over which the transmitter has to wait for its transmission opportunities

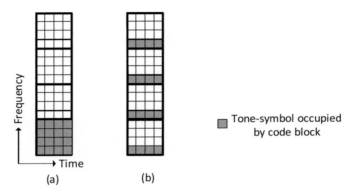

Figure 4.11 Two configurations of an OFDMA transmission block. A small block represents a tone-symbol. A big block with thick border line consists of tone-symbols within a coherence bandwidth and a coherence time. A transmission block is contained within one coherence bandwidth in (a) without diversity and spread into four coherence bandwidth in (b) with diversity.

to come, and is mostly relevant in the multiuser case to be addressed in Section 4.3. In the truncated power control scheme, the transmitter suspends the transmission when the channel is bad, thereby also causing scheduling delay.

- Second, *variation in data rate*. Scheduling delay measures the extreme case where the data rate drops to zero. In the fading channel, the data rate generally fluctuates from one code block to another. In addition to the average data rate, another important measure is the "minimum" rate, usually defined to be the bottom x percentile of its cumulative density function where x depends on the traffic requirement. In general, the smaller difference between the average and the minimum rates, the better for consistent and predictable data communication service. With diversity, the probability of a deep fade, and thus outage, decreases. In general, the minimum data rate is higher with a diversity scheme than with a feedback scheme. For example, although time and frequency diversity and rate control have the same average data rate, the rate variation is lower with the diversity scheme, and thus the minimum data rate is higher. Similarly, at low SNR, despite a higher average data rate, the truncated power control scheme results in bursty transmission and the issue of scheduling delay as noted before.
- Third, *performance sensitivity to channel prediction*. Power/rate control relies heavily on the accuracy of the channel feedback to make the right decision on the basis of channel prediction. On the contrary, the diversity schemes are not as much dependent on channel prediction, because diversity reduces the uncertainty in channel prediction thanks to the reduction in channel variation. To see this, continue the rate prediction example in Figure 4.6 where the channel SNR is measured at time t_0 and fed back to the transmitter, which is to select the rate of a transmission block at time t_1. Consider two configurations of an OFDMA transmission block: with or without diversity, shown in Figure 4.11. First, consider Figure 4.11(a) without frequency diversity. Denote the channel capacity at time t_i

$$R(t_i) = \log(1 + z(t_i)\bar{\gamma}), \qquad (4.93)$$

with $i = 0, 1$ and $z(t_i)$ the instantaneous channel gain variation of a coherence bandwidth at t_i. Assume that the channel at a given frequency is correlated between t_0 and t_1 and the correlation coefficient of $R(t_0)$ and $R(t_1)$ is given by

$$r = \frac{\mathbb{E}(R(t_1)R(t_0)) - \mathbb{E}(R(t_1))\mathbb{E}(R(t_0))}{\sigma_{R(t_1)}\sigma_{R(t_0)}}, \quad (4.94)$$

with $\sigma_{R(t_i)}$ the unconditional variance of $R(t_i)$. The transmitter knows the rate $R(t_0)$ from the feedback and uses a linear MMSE estimator to predict $R(t_1)$ from $R(t_0)$. It can be shown that the minimum squared prediction error is given by

$$e_1 = \sigma^2_{R(t_1)} \left(1 - r^2\right). \quad (4.95)$$

Now consider Figure 4.11(b) with frequency diversity. Assume frequency diversity of degree L_D. For coherence bandwidth k, $k = 1, \ldots, L_D$, denote by $z(t_i, k)$ the instantaneous channel gain variation, and the channel capacity at time t_i is

$$R(t_i, k) = \log(1 + z(t_i, k)\bar{\gamma}). \quad (4.96)$$

As before, assume that $R(t_0, k)$ and $R(t_1, k)$ are correlated with the same correlation coefficient r for all frequency k. For simplicity, we assume that there is no correlation between different coherence bandwidth pieces. From the feedback, the transmitter knows the rate at t_0,

$$\bar{R}(t_0) = \frac{1}{L_D} \sum_{k=1}^{L_D} R(t_0, k),$$

and uses a linear MMSE estimator to predict the rate at t_1,

$$\bar{R}(t_1) = \frac{1}{L_D} \sum_{k=1}^{L_D} R(t_1, k).$$

To compare with the above case with no diversity, assume that

$$\sigma_{R(t_i,k)} = \sigma_{R(t_i)},$$

$$\mathbb{E}(R(t_1, k)R(t_0, k)) - \mathbb{E}(R(t_1, k))\mathbb{E}(R(t_0, k)) = \mathbb{E}(R(t_1)R(t_0)) - \mathbb{E}(R(t_1))\mathbb{E}(R(t_0))$$

for all i, k. It can be shown that the correlation coefficient between $\bar{R}(t_i)$ is equal to r and the variance of $\bar{R}(t_1)$ is reduced by a factor of L_D. Therefore, it follows that the minimum squared prediction error is also reduced by a factor of L_D

$$e_{L_D} = \frac{e_1}{L_D}. \quad (4.97)$$

As a result, with diversity, the required prediction margin is significantly reduced in the rate control scheme. The reduction in the required prediction margin is a gain due to diversity.

4.3 Communications over a fading channel: the multiuser case

4.3.1 Fading channel and multiuser diversity

So far we have studied a variety of forms of diversity, which we collectively call *single user diversity*, as they focus on improving the reliability of a single user link. To this end, single user diversity pursues averaging effect. Single user diversity in general helps, with the exception that in the low SNR regime averaging is not always desirable if the goal is to maximize the average data rate. Is the same insight applicable to a multiuser case?

An important difference between the multiuser and single user cases lies in the notion of *scheduling* in mobile broadband communication, because a given link is not always allocated with a fixed amount of power/bandwidth resource. Rather, scheduling decisions are made dynamically by comparing the real time traffic needs and channel conditions of all the competing links. Scheduling in a multiuser scenario leads to a new form of diversity: *multiuser diversity*, in which the goal is to pursue not an *averaging* but a *peaking* effect, as explained next.

Downlink

First consider the downlink in a single cell where the base station sends data traffic to the users in the cell with a fixed transmit power P. Assume that the fading channel is flat and that the base station schedules at most only one user at any time. Denote by

$$\gamma_i = \frac{P|H_i|^2}{\sigma^2} = z_i \bar{\gamma}_i, i = 1, \ldots, M, \quad (4.98)$$

the instantaneous SNR of user i, with $|H_i|^2$ the channel gain and σ^2 noise power. As before, denote by $\bar{\gamma}_i = P\mathbb{E}(|H_i|^2)/\sigma^2$ the average SNR and $z_i = |H_i|^2/\sqrt{\mathbb{E}(|H_i|^2)}$ the relative channel gain variation. We consider flat fading in this section and will extend to frequency selective fading in the next section. The users feed back their SNRs to the base station so that it knows γ_i for all i.

Suppose that $\bar{\gamma}_i = \bar{\gamma}$ is the same for all i. To maximize the expected data rate in the downlink, the base station schedules the user i^* of the best SNR at any given time:

$$i^* = \arg\max_i \gamma_i. \quad (4.99)$$

The expected data rate is

$$R_f = \mathbb{E}\left(\log\left(1 + \bar{\gamma} \max_i z_i\right)\right). \quad (4.100)$$

Clearly, the scheduling scheme (4.99) achieves a higher total data rate than a round robin scheduling where the users are scheduled according to a fixed order irrespective of their channel conditions. If the channel fading is independent for different users, not all the users experience a bad channel at the same time. The base station can avoid scheduling a user if it is in a deep fade.

Not only (4.99) protects against deep fades, it also exploit the opportunity where sometimes the instantaneous γ_i exceeds its mean because of constructive addition in multipath fading. As the number of users M increases, the likelihood that at a given time, at least one of the users is in such a favorable channel condition becomes increasingly significant. This makes the total data rate R_f potentially higher than R_0, the rate in AWGN, a phenomenon we have never seen previously in the single user diversity scenario.

For single user diversity, the performance of interest is the reliability of a single link characterized by the probability of independent diversity branches all experiencing a deep fade which drops with the number of branches. The performance depends on the *low end* tail probability of z_k being small at any given branch k. Single user diversity pursues averaging effect to mitigate the damage of channel attenuation. On the other hand, for multiuser diversity, the performance of interest is the total data rate. The best SNR among independent users increases with the number of users. It depends on the *high end* tail probability of z_i being large of any given user i. Multiuser diversity pursues peaking effect to exploit the opportunity of channel amplification. Therefore, for the same average SNR, the reliability of a given link is higher in Ricean fading than in Rayleigh fading, but the total data rate of a multiuser system is higher in Rayleigh fading than in Ricean fading.

Figure 4.12(a) ($b = 0$, Rayleigh and Ricean) compares R_f in Rayleigh and Ricean fading as M increases. In either case, the total data rate increases with the number of users, an effect called multiuser diversity. We note that, even as small as $M = 2$, R_f in Rayleigh fading exceeds that in Ricean fading and the difference increases with M. The observation can be explained by studying the pdf of $\max_i z_i$, which is plotted in Figure 4.12(b) ($b = 0$, Rayleigh and Ricean) for a given $M = 8$. The high end tail probability is significantly heavier in Rayleigh fading than in Ricean fading.

The careful reader may recall that at low SNR the opportunistic truncation scheme exhibits a similar spirit of pursuing peaking, instead of averaging, effect: spend transmit energy only when the channel is good. When the transmission is truncated, the channel is left idle in the single user case but can be allocated to another link in the multiuser case so that no bandwidth is wasted.

So far we have assumed i.i.d. γ_i. In reality, when the users are distributed at different locations in a cell, while it is reasonable to assume γ_i to be independent, $\mathbb{E}(\gamma_i)$ is usually not all equal, as users close to the base station have greater average SNR than those at cell edges. To accommodate that asymmetric case, a fair scheduling algorithm normalizes γ_i in (4.99):

$$i^* = \arg\max_i \frac{\gamma_i}{\mathbb{E}(\gamma_i)}. \tag{4.101}$$

The idea is that a user is scheduled when the channel is near to its own peaks. One of the first algorithms that explores multiuser diversity, proportional fair scheduling uses relative data rate instead of relative SNR to pick a user:

$$i^* = \arg\max_i \frac{R_i}{T_i}, \tag{4.102}$$

where R_i is the data rate corresponding to γ_i, and T_i the average throughput user i has already obtained and averaging is done in an exponentially weighted window.

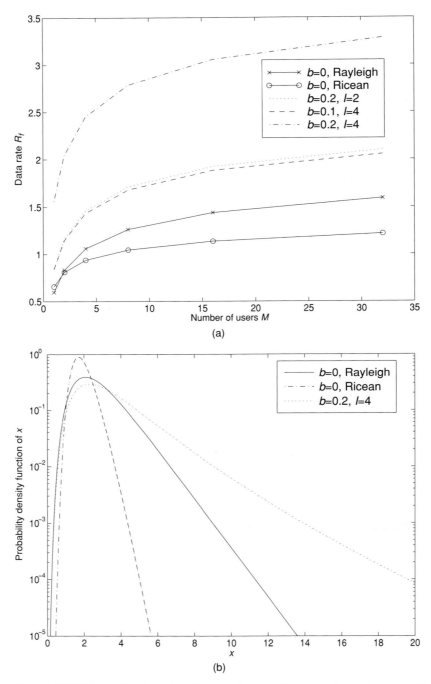

Figure 4.12 Multiuser diversity gain: (a) shows data rate R_f versus the number users M, and (b) shows the pdf of SNR or SINR γ for a given $M = 8$. Comparison is made among single cell ($b = 0$, Rayleigh and Ricean fading) and multi-cell scenarios ($b > 0$, Rayleigh fading only with different interference parameters). $\bar{\gamma}$ is determined such that $\frac{\bar{\gamma}}{1+bI\bar{\gamma}} = 0$ dB for all the choices of b and I.

The empirical throughput T_i is updated as follows:

$$T_i(t+1) = (1-\epsilon)T_i(t) + \epsilon R_i(t)\mathbf{1}_{\{i^*(t)=i\}}, \qquad (4.103)$$

Indicator function $\mathbf{1}_{\{i^*(t)=i\}}$ is equal to 1 if user i is scheduled and 0 otherwise. $\epsilon \in (0,1)$ is a filtering parameter. The proportional fairness metric $R_i(t)/T_i(t)$ attempts to favor users with either high instantaneous rate $R_i(t)$, or users who have not been recently scheduled and have low recent throughput $T_i(t)$.

The reason that a relative metric of (4.101) or (4.102) is preferred to an absolute metric of (4.99) is fairness. Always scheduling the user with the best absolute SNR will starve users at cell edges. Maximization of the total data rate is not the only performance goal. As noted in Section 4.2.4, other performance metrics such as delay and minimum rate are also important in reality.

The above study can be generalized to the downlink of a multi-cell system by replacing noise with interference plus noise. The SINR of user i of cell 0 is given by

$$\gamma_i = \frac{P_0|H_{0,i}|^2}{\sigma^2 + \sum_{n>0} P_n|H_{n,i}|^2}, \quad i=1,\ldots,M, \qquad (4.104)$$

where P_n is the transmit power of base station n. User i is connected with base station 0 with channel gain $|H_{0,i}|^2$ and all other base stations $n > 0$ are interferers with $|H_{n,i}|^2$. The SNR γ_i in the user selection algorithm (4.99) or (4.101) should be replaced with the SINR γ_i (4.104).

As noted above, the multiuser diversity benefit depends on the high end tail probability of γ_i. Compared with SNR, SINR tends to exhibit heavier high end tail. Specifically, when the noise power σ^2 dominates in the denominator in (4.104), it reduces to the single cell case. However, when the interference dominates, the instantaneous SINR can be boosted because of either channel boost in the desired signal $|H_{0,i}|^2 \gg \mathbb{E}(|H_{0,i}|^2)$, or channel fade in the interference $\sum_{n>0} P_n|H_{n,i}|^2 \ll \sum_{n>0} P_n\mathbb{E}(|H_{n,i}|^2)$, or both. In a typical fading channel, the probability of a channel gain being boosted by a large factor is smaller than the probability of it being faded by the same factor. Therefore, to get a large SINR, interference fade contributes more than signal boost, thereby making multiuser diversity more significant in the multi-cell than the single cell case. This is especially true when the number of dominant interfering base stations is small.

To provide a concrete example, assume that P_n is the same for all $n = 0, 1, \ldots$. As before, denote by $z_{n,i} = |H_{n,i}|^2/\mathbb{E}(|H_{n,i}|^2)$. Furthermore, assume $\mathbb{E}(|H_{n,i}|^2) = b\mathbb{E}(|H_{0,i}|^2)$ for all $n > 0$ with $b < 1$ indicating that the user is connected to the base station of the best average channel gain. (4.104) is rewritten as

$$\gamma_i = \frac{\bar{\gamma} z_{0,i}}{1 + b\bar{\gamma}\sum_{n=1}^{I} z_{n,i}}, \qquad (4.105)$$

where I is the number of dominant interfering base stations and average SNR $\bar{\gamma}$ defined to be $P\mathbb{E}(|H_{0,i}|^2)/\sigma^2$ as before.

Figure 4.12(a) plots the expected data rate

$$R_f = \mathbb{E}\left(\log\left(1 + \max_i \gamma_i\right)\right), \qquad (4.106)$$

for several possibilities of b and I. For the sake of comparison, we set $\bar{\gamma}$ to keep $\frac{\bar{\gamma}}{1+bI\bar{\gamma}}$ unchanged. $\bar{\gamma}$ increases with bI, making the interference more dominant than the noise. Figure 4.12(b) further plots the pdf of $\max_i \gamma_i$ for $M = 8$ and shows that the high end tail becomes significantly heavier in the multi-cell case.

The study so far addresses the problem of allocating all the power and bandwidth to only one user based on the SNR or SINR condition. The real-world problem is potentially more complex:

- Multiple users may share power and bandwidth. The problem of user multiplexing will be addressed in Chapter 5.
- SNR or SINR is not the only consideration. The base station needs to take into account other factors, such as traffic latency and queue length. The problem of scheduling will be addressed in Chapter 8.

Practical example 4.1 Multiuser diversity in the downlink: EV-DO

EV-DO ([12]) is arguably the first cellular system that exploits multiuser diversity in the downlink. It is an evolution from the IS-95 CDMA technology and is optimized for data traffic.

IS-95 is basically a CDMA voice system where an active user keeps a *continuous* circuit-switched connection with the base station. Simply put, a connection uses a spreading code, which is an orthogonal Walsh code in the downlink or a pseudo-random code in the uplink. The data rate of a connection is equal to full, 1/2, 1/4, or 1/8 voice codec rate, and varies only with the voice activity, but not with the wireless channel condition. Closed-loop power control is employed to combat against channel fluctuation in the uplink and power allocated is used in the downlink.

A significant evolution of EV-DO is that rate control replaces power control in the downlink. Specifically, the downlink becomes time multiplexed packet-switched as shown in Figure 4.13(b). At a given time slot, the base station transmits to at most one user at the fixed full power. Instead of power control, the data rate and the transmission length vary with the wireless channel condition of the scheduled user. Rate control is a form of adaptive modulation and coding (AMC), which adjusts the modulation and coding scheme in response to channel variations (see Appendix B.4). When the base station traffic queue is empty, the base station only transmits short pilots and control signals without scheduling any user so as to reduce interference to neighboring cells.

Every user estimates the wireless channel condition using the downlink pilots, and predicts the SINR of a future time slot. The value of the SINR is then mapped to a data rate that the SINR can support for a target error probability. The user sends that request data rate using a 4-bit value in an uplink control channel frequently at a rate of once per 1.67 ms. After receiving the request data rates from all the users, the base station schedules at most one user whose traffic is not empty and whose channel condition is the most favorable from the perspective of multiuser diversity.

A popular scheduling algorithm is proportional fair scheduling (4.102). The base station broadcasts the index of the scheduled user at the beginning of the time slot.

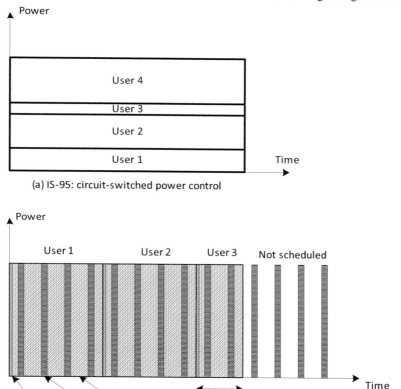

Figure 4.13 Comparison of IS-95 (a) and EV-DO (b) downlink. In (b), the index of the scheduled user is sent at the beginning of a slot, and the pilot and traffic are time-multiplexed. If no user is scheduled, then only the pilot is sent.

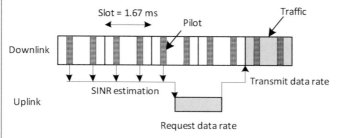

Figure 4.14 Time sequence of SINR measurement, request data rate, and transmit data rate.

Figure 4.14 depicts the time sequence of SINR measurement, request data rate, and transmit data rate. Apparently, a key to maximizing the multiuser diversity gain

> is to predict the SINR and thus data rate accurately and rapidly. Therefore, EV-DO minimizes the latency from the request data rate transmission in the uplink to the corresponding data traffic transmission in the downlink to only 0.5 slot.
>
> Adaptive modulation and coding, multiuser channel-aware scheduling, and fast hybrid-ARQ (to be discussed in Appendix Section B.4) make EV-DO perform very well in a data system. Those ideas help promote the evolution of UMTS WCDMA to HSDPA and then HSUPA, and are eventually adopted in various OFDMA mobile broadband systems.

Uplink

Now consider the uplink. To focus on multiuser diversity without complication of user multiplexing or scheduling, we continue the problem of scheduling only one user at any time based on the channel gain condition.

First, consider the power limited scenario where the constraint is that the maximum transmit power of a user not exceed a fixed value, P_m. In this case, to maximize the data rate, the transmit power of user i is set to $P_i = P_m$. Similar to the downlink case, denote by

$$\gamma_i = \frac{P_m |H_i|^2}{\sigma^2}, i = 1, \ldots, M, \quad (4.107)$$

the SNR of user i, with $|H_i|^2$ the channel gain and σ^2 the noise power at the base station receiver. Furthermore, assume that $\bar{\gamma} = \mathbb{E}(\gamma_i)$ is the same for all i. Then the problem is identical to the single cell downlink case. We know that the optimal scheme of maximizing the expected data rate is to schedule the user of the best instantaneous SNR (4.99) and multiuser diversity is the result of channel gain fluctuation to the desired base station.

Next, consider a multi-cell scenario where the SINR is given by

$$\gamma_i = \frac{P_i |H_{0,i}|^2}{\sigma^2 + Q}. \quad (4.108)$$

Q represents the total inter-cell interference from the users scheduled in all other cells. Meanwhile, the transmission of user i causes inter-cell interference to the base stations in those cells. Therefore, an additional constraint of analyzing P_i and γ_i comes from inter-cell interference.

To focus on multiuser diversity, in the following we simplify the scenario and assume that the system consists of only two cells, cells 0 and 1. Chapter 6 will address a more general problem of multi-cell inter-cell interference. The SINR is given by

$$\gamma_i = \frac{P_i |H_{0,i}|^2}{\sigma^2 + P_{i'} |H_{0,i'}|^2}, \quad (4.109)$$

where i' is the index of the user scheduled in cell 1. $P_{i'}$ is the transmit power of user i' and $|H_{0,i'}|^2$ is the channel gain from the user i' to the base station of cell 0.

How should the base station select the user? Intuitively, the scheduled user should have a large channel gain to its own base station and a small channel gain to the neighboring base station so as to reduce the inter-cell interference. We next derive the scheme more precisely.

Although it is possible to still let $P_i = P_{i'} = P_m$ for all users, the inter-cell interference seen at cell 0 would fluctuate drastically with $|H_{0,i'}|^2$ of the scheduled user in cell 1. Unless the base stations in the two cells *jointly* determine i and i', the random fluctuation in the inter-cell interference makes it hard for the base station to predict the SINR and therefore assign an appropriate coding and modulation scheme for user i. Alternatively, we can fix a target value, Q_m, of the inter-cell interference, called the interference budget

$$P_{i'}|H_{0,i'}|^2 = Q_m. \tag{4.110}$$

User i' in cell 1 sets

$$P_{i'} = \frac{Q_m}{|H_{0,i'}|^2}. \tag{4.111}$$

User i in cell 0 follows the same principle to set its transmit power. The SINR of user i is given by

$$\gamma_i = \frac{|H_{0,i}|^2}{|H_{1,i}|^2} \frac{Q_m}{\sigma^2 + Q_m}, \tag{4.112}$$

where $|H_{1,i}|^2$ is the channel gain from user i to the base station at cell 1. Hence, the base station schedules the user i^* such that

$$i^* = \arg\max_i \frac{|H_{0,i}|^2}{|H_{1,i}|^2}. \tag{4.113}$$

Note that the user selection scheme (4.113) does not depend on σ^2 or Q_m. Comparing (4.113) with (4.104) we note that multiuser diversity in this two-cell uplink case is the same as the multi-cell downlink case with $I = 1$ and $\sigma^2 = 0$; the gain comes from not only the channel fluctuation to the desired base station but also that to the interfering base station. The effect of the latter is more predominant and multiuser diversity is potentially much more significant.

Practical example 4.2 Multiuser diversity in the uplink: Flash-OFDM and LTE

Example 4.1 uses EV-DO as an example to study multiuser diversity in the downlink. The EV-DO uplink is, however, circuit-switched, similar to IS-95, and therefore does not support multiuser diversity scheduling. In the following we use Flash-OFDM and LTE, instead, to illustrate multiuser diversity in the uplink.

We note that despite the similarity between (4.113) and (4.101), implementing (4.113) to obtain multiuser diversity in the uplink is usually more difficult than in the downlink (4.101). The reason is that in the downlink, the user readily measures the SINR γ_i with the pilot signals from the desired and interfering base stations (4.104). In the uplink $|H_{0,i}|^2$ and $|H_{1,i}|^2$ need to be measured.

One possibility is that the user measures the downlink channel gain and uses it to approximate the uplink channel gain. This approach works well in a TDD system because of channel reciprocity. However, the downlink and uplink in FDD exhibit independent multipath fading because of large carrier frequency difference. Using the downlink measurement to approximate the uplink channel gain captures large-scale but not small-scale fading, and is therefore not particularly helpful here. Given a typical delay constraint, small-scale multipath fading is the key to harnessing multiuser diversity.

Ultimately, only the base station receiver can measure the receive power of the uplink signal from the user and derive the channel gain if the transmit power of the user is known. Meanwhile the uplink transmit power varies according to the power control commands from the base station when closed-loop power control is in place (see Example 6.1). The base station usually does not explicitly track the exact transmit power of the user, and is therefore unable to determine the uplink channel gain. Fortunately, knowing the variation in its transmit power under closed-loop power control, the user can tell whether its channel gain increases or decreases. Specifically, assuming closed-loop power control keeps a fixed received power target, the channel gain increases when the required transmit power drops, and vice versa.

In Flash-OFDM, when the user is close to the boundary between two cells, it is connected to two base stations in an MBB manner for macro-diversity and power controlled independently by the base stations (see Section 9.3). Assuming the same received power target at the base stations, the ratio of the required transmit powers with respect to the two base stations is equal to the ratio of the uplink channel gains

$$\frac{|H_{0,i}|^2}{|H_{1,i}|^2} = \frac{P_{1,i}}{P_{0,i}}, \quad (4.114)$$

where $P_{0,i}$, $P_{1,i}$ represent the required transmit powers to base stations 0 and 1, respectively, and $|H_{0,i}|^2$, $|H_{1,i}|^2$ the uplink channel power gains. The above equation can easily be modified if the received power targets are different. Hence, the user reports to the base station $|H_{0,i}|^2/|H_{1,i}|^2$ to support uplink multiuser diversity scheduling, the same way as reporting SINR γ_i for downlink multiuser diversity scheduling.

When the user is located in the interior area of a cell, it is connected only to one base station. Since the user is away from any neighboring base station, the uplink multiuser diversity scheduler ignores the effect of $|H_{1,i}|^2$ and uses only $|H_{0,i}|^2$ for user selection (4.113). In this case, except for the serving base station, no other base stations measure the signal from the user. As noted before, the user estimates

$$|H_{0,i}|^2 \propto \frac{1}{P_{0,i}}. \quad (4.115)$$

Therefore, it reports to the base station $1/P_{0,i}$. Equivalently, in Flash-OFDM, the user reports the power headroom $P_m/P_{0,i}$, where P_m is the maximum transmit power of the user device. In addition to supporting uplink multiuser diversity scheduling, the power headroom reports allow the base station to determine the maximum coding and modulation scheme that can be supported in the present channel condition of the user.

4.3 Communications, fading channel: multiuser case

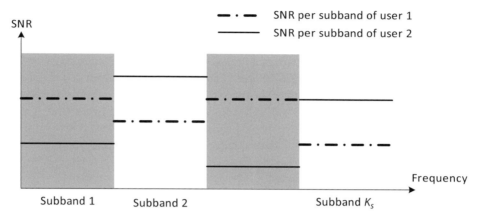

Figure 4.15 Illustration of frequency selective and frequency diversity scheduling in a wideband OFDMA channel where SNR is piecewise constant in each subband. In frequency diversity scheduling, all the subbands are allocated to user 1 because its average SNR is greater even though in subbands 2 and 4 it is inferior to user 2. In frequency selective scheduling, each subband is allocated depending on the SNR in that subband. As a result, user 1 is allocated with subbands 1 and 3 and user 2 with subbands 2 and 4.

> Now consider the LTE uplink. As studied in Example 9.3, in LTE a user is connected only to one base station (cell 0). No other base station tracks the channel gain of the user. As a result, user selection (4.113) has to be based on $|H_{0,i}|^2$ alone instead of the channel gain ratio $|H_{0,i}|^2/|H_{1,i}|^2$. Clearly, the effect of $|H_{1,i}|^2$ is missing as far as uplink multiuser diversity is concerned. As we noted before, in a fading channel, the effect of $|H_{1,i}|^2$ may be much more significant than the effect of $|H_{0,i}|^2$, and is especially important for the users at the cell boundary.

4.3.2 Exploring multiuser diversity in frequency and space

Unlike single user diversity, multiuser diversity facilitates opportunistic communication in which channel fading is exploited, rather than mitigated, via channel-aware scheduling with the goal of pursuing peaking, rather than averaging, effect. The amount of multiuser diversity depends on channel fluctuation. So far we have assumed flat fading and studied channel fluctuation in time. We next explore multiuser diversity in frequency and space (antennas). Our study will focus on the single cell downlink.

Frequency selective multiuser diversity

We continue to approximate the wideband channel with a block fading model. In OFDMA, the total bandwidth W is divided into $K_S = W/B_c$ non-overlapping subbands. Each subband is a coherence bandwidth B_c. The channel gain is constant within a subband and independently changes from one subband to another. From Section 4.2.3, when $K_S > 1$, the single user channel can obtain a frequency diversity of order K_S. How does frequency diversity interact with multiuser diversity? Figure 4.15 shows two basic ideas of allocation subbands to the users: frequency diversity scheduling and frequency selective scheduling.

In *frequency diversity scheduling*, all the K_S subbands are allocated to one user. Moreover, assume that the transmit power is uniformly distributed in the K_S subbands.[2] Denote by $\gamma_i(k)$ the SNR of user i in subband k. Assume that $\bar{\gamma} = \mathbb{E}(\gamma_i(k))$ for all i and k. Denote by $z_i(k) = \gamma_i(k)/\bar{\gamma}$ the i.i.d. fading realizations. The user selection algorithm (4.99) becomes

$$i^* = \arg\max_i \sum_{k=1}^{K_S} \frac{1}{K_S} \log(1 + \bar{\gamma} z_i(k)). \quad (4.116)$$

The expected data rate is

$$R_f = \mathbb{E}\left(\sum_{k=1}^{K_S} \frac{1}{K_S} \log(1 + \bar{\gamma} z_{i^*}(k)) \right). \quad (4.117)$$

Alternatively, in *frequency selective scheduling*, each of the K_S subbands can be allocated to a different user. Again, assume that the transmit power is uniformly distributed in the K_S subbands. Then, the user selection algorithm (4.99), with γ_i replaced by $\gamma_i(k)$, is run independently for each subband k. For i.i.d. fading, the expected data rate is the same as (4.100).

The averaging effect of frequency diversity indeed hurts the benefit of multiuser diversity. In frequency diversity scheduling, user selection is based on the average SNR in the entire bandwidth. A user may be selected even though its SNR in certain subbands is not the best. On the other hand, frequency selective scheduling is more flexible and always selects the user with the best SNR in each subband. Frequency selective scheduling achieves a higher expected data rate than frequency diversity scheduling does.

Figure 4.16 compares the data rate of the two scheduling schemes in Rayleigh fading with several possibilities of the number of users (M) and the number of subbands (K_S). We observe that in general as K_S increases, the data rate of frequency diversity scheduling drops indicating that the multiuser diversity gain diminishes. The reason is that the average SNR fluctuates much less significantly. As K_S goes to infinity, the average SNR is equal to $\bar{\gamma}$, same for all users, and the multiuser diversity gain completely disappears. However, frequency selective scheduling preserves multiuser diversity.

Finally, it should be pointed out that frequency selective scheduling is easy to implement in OFDMA, since a group of contiguous tones naturally comprise a subband. Frequency selectivity depends on the wireless channel of a given user. OFDMA allows flexible partition of the total bandwidth into subbands in that the base station can apply different subband partition patterns for different users according to their channel characteristics.

With frequency selective multiuser scheduling, a user is scheduled to a favorable subband. Within the subband, the instantaneous channel variation z of each tone tends

---

[2] One can use the optimal water-filling scheme to allocate power in the subbands. However, the conclusion about the interplay between frequency diversity and multiuser diversity remains the same.

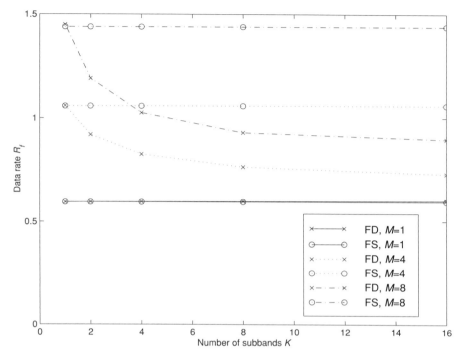

Figure 4.16 Data rate of frequency selective (FS) and frequency diversity (FD) scheduling as the number of independent subbands increases $\bar{\gamma} = 0$ dB.

to be strong and the channel variation across tones tends to be small. Therefore, there is not much gain of performing a sophisticated water-filling scheme within the subband. Allocating the transmit power uniformly across the subband is already close to the optimum. This is very different from the optimal water-filling solution in the single-user case (4.74).

Multiuser diversity with multiple transmit antennas

Now suppose that the base station is equipped with multiple transmit antennas. We know from Section 4.2.3 that space-time coding obtains transmit diversity where the goal is to improve the reliability of a single user link by averaging the channel from multiple antennas. Transmit diversity reduces the multiuser diversity gain, similar to what we have studied with frequency diversity in the previous section. Indeed, as far as multiuser diversity is concerned, the base station would be better off concentrating its transmit power on only one antenna rather than applying space-time coding. This does not seem to be an exciting way of employing multiple antennas. Naturally we wonder whether multiple transmit antennas can be used, in a constructive way, to enhance multiuser diversity. The following opportunistic beamforming idea proposed in [169] shows that multiple transmit antennas can indeed help.

Recall that multiuser diversity depends on channel fluctuation. What happens if the channel varies very slowly such that γ_i is essentially constant in the time scale of scheduling? Apparently, (4.99) is very unfair as it keeps on selecting the same user. The relative comparison of (4.101) of a fair scheduling is not useful, either. Multiuser diversity diminishes in the absence of channel fluctuation. The idea of opportunistic beamforming is to *artificially* induce channel fluctuation using transmit antennas.

Suppose that the base station is equipped with K_t transmit antennas. Suppose that the channel is flat in frequency and constant in time. Denote by $H_{i,k}$ the channel response from antenna k to user i. At symbol time t, the same signal symbol is multiplied by a complex number

$$g_k(t) = \sqrt{\alpha_k(t)} e^{j\theta_k(t)}, \tag{4.118}$$

and then sent from antenna k, with

$$\sum_{k=1}^{K_t} \alpha_k(t) = 1 \tag{4.119}$$

to keep the total transmit power unchanged. The SNR at user i becomes

$$\gamma_i = \frac{P \left| \sum_{k=1}^{K_t} g_k(t) H_{i,k} \right|^2}{\sigma^2}, i = 1, \ldots, M. \tag{4.120}$$

γ_i can be written as follows:

$$\gamma_i = \bar{\gamma}_i \Omega_i, \tag{4.121}$$

with

$$\bar{\gamma}_i = \frac{P}{\sigma^2} \frac{\sum_{k=1}^{K_t} |H_{i,k}|^2}{K_t}, \tag{4.122}$$

$$\Omega_i = \frac{\left| \sum_{k=1}^{K_t} g_k(t) H_{i,k} \right|^2}{1/K_t \sum_{k=1}^{K_t} |H_{i,k}|^2}. \tag{4.123}$$

As discussed before, because of averaging, the fluctuation of $\bar{\gamma}_i$ decreases as K_t increases. On the other hand, the combining power gain Ω_i ranges from 0 to K_t where $\Omega_i = K_t$ means that user i is in the coherent beamforming direction. To highlight the effect of opportunistic beamforming, we assume $\bar{\gamma}_i = \bar{\gamma}$ is the same for all i. Then, in the relative comparison (4.101), the user whose Ω_i is the greatest is the winner:

$$\Omega_{i^*} = \max_i \Omega_i. \tag{4.124}$$

The expected data rate is given by

$$R_f = \mathbb{E}\left(\log(1 + \gamma_{i^*})\right) = \mathbb{E}\left(\log(1 + \bar{\gamma} \Omega_{i^*})\right). \tag{4.125}$$

The beamforming direction corresponding to vector $[g_1(t), \ldots, g_{K_t}(t)]$ at time t is given by $[g_1^*(t), \ldots, g_{K_t}^*(t)]$. If the channel vector $[H_{i,1}, \ldots, H_{i,K_t}]$ of user i is along the beamforming direction, then the SNR reaches its own peak. As $g_k(t)$ varies dynamically, the beamforming direction to which the signal is sent is time-varying. At any time, the beamforming direction is not specifically targeted to a particular user. Whichever user happens to be near the beamforming direction stands out in the relative comparison (4.101) and gets scheduled. From that perspective, the beamforming scheme is opportunistic.

As a comparison, consider the single user beamforming scheme where the base station targets to a specific user i. The base station needs to know the channel $H_{i,k}$ of all the antennas. To this end, the base station sends separate pilot signals from individual antennas so as for the user to measure the individual channel $H_{i,k}$. The user feeds back all the channel responses, as opposed to a single overall SNR report γ_i in opportunistic beamforming. The beamforming parameters are then set according to (4.87)

$$g_k(t) = \frac{H_{i,k}^*}{\sqrt{\sum_{k=1}^{K_t} |H_{i,k}|^2}}. \qquad (4.126)$$

This direction is called the coherent beamforming direction to that user.

Opportunistic beamforming works in a multiuser system. If the number of users is sufficiently large, then at any given time, some user is near the beamforming direction. Thus, Ω_{i^*} increases with K_t.

Figure 4.17 plots R_f of opportunistic beamforming. We observed that as the number of users M increases, opportunistic beamforming approaches coherent beamforming, which obtains a factor of K_t power gain. Two groups of curves are plotted with $K_t = 2$ and 4, respectively. Two channel models are considered here. The first model assumes richly scattered Rayleigh fading environment such that $H_{i,k}$ is i.i.d. complex Gaussian random variable with zero mean and unit variance. $|H_{i,k}|$ is normalized to satisfy $\bar{\gamma}_i = \bar{\gamma}$. The second model assumes a single line-of-sight path from an antenna and user i and the angle between the signal path towards user i and the transmit antenna array is uniformly distributed in $[0, 2\pi]$. Figure 4.17 shows that opportunistic beamforming approaches coherent beamforming faster when K_t is smaller, and for a given K_t faster in the line-of-sight model than in the Rayleigh fading model.

To see the combined effect of both opportunistic beamforming and channel fluctuation denote by $|z_{i,k}|^2$ the instantaneous fading gain corresponding to $|H_{i,k}|^2$. Then the multiuser scheduling policy (4.99) becomes

$$i^* = \arg\max_i \Lambda_i, \quad \text{with } \Lambda_i = \left| \sum_{k=1}^{K_t} g_k(t) z_{i,k} \right|^2. \qquad (4.127)$$

The overall benefit of this opportunistic scheduling may not always come from opportunistic beamforming, since when user i gets scheduled, the fading direction $[z_{i,1}, \ldots, z_{i,K_t}]$ may not be along the present beamforming direction defined by $[g_1^*(t), \ldots, g_{K_t}^*(t)]$; it may simply be because the instantaneous fading gains $|z_{i,k}|^2$ are greater than those of other users. To see whether the artificially induced channel

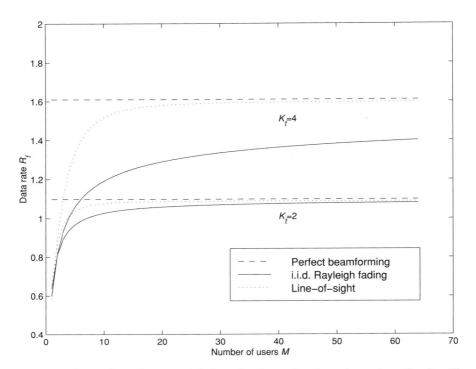

Figure 4.17 Comparison of opportunistic beamforming and perfect coherent beamforming. The data rate of opportunistic beamforming is given in (4.125). For coherent beamforming, set $R_f = \log\left(1 + \bar{\gamma} K_t\right)$ in (4.125) $\bar{\gamma} = 0$ dB.

fluctuation enhances multiuser diversity, we need to compare the high end tail probability of Λ_i in (4.127) with that of $|z_{i,k}|^2$.

Assume that $z_{i,k}$ is i.i.d. and circular symmetric (the phase is uniform over $[0, 2\pi]$). It follows that $\mathbb{E}(\Lambda_i) = \mathbb{E}\left(|z_{i,k}|^2\right)$. As an approximation, we use the variance to represent the high end tail probability. It can be shown that if $\mathbb{E}\left(|z_{i,k}|^4\right) > 2\left(\mathbb{E}\left(|z_{i,k}|^2\right)\right)^2$, then $\text{var}(\Lambda_i) > \text{var}\left(|z_{i,k}|^2\right)$ and likely multiuser diversity is enhanced. To be specific, consider the following two fading models as in [159]:

- Rayleigh fading. In this case, $z_{i,k}$ is complex Gaussian. $\mathbb{E}\left(|z_{i,k}|^4\right) = 2\left(\mathbb{E}\left(|z_{i,k}|^2\right)\right)^2$ and thus multiuser diversity is neither enhanced or degraded. The reason is that the term $\sum_{k=1}^{K_t} g_k(t) z_{i,k}$ in (4.127) is the sum of i.i.d. complex Gaussian and therefore has the same distribution as the original $z_{i,k}$ from any single antenna.
- Ricean fading. It follows that $\mathbb{E}\left(|z_{i,k}|^4\right) > 2\left(\mathbb{E}\left(|z_{i,k}|^2\right)\right)^2$, and the difference and thus multiuser diversity grow larger with parameter κ. The reason is that the fluctuation in the original channel only comes from the non-LOS paths. With time-varying g_k, the LOS paths from multiple transmit antennas add constructively or destructively, thereby increasing the overall channel fluctuation.

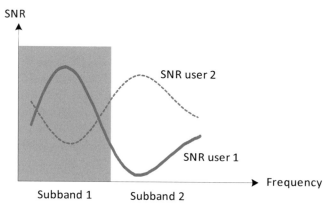

Figure 4.18 Illustration of frequency selective channel SNR induced by delay diversity and frequency selective multiuser scheduling where user 1 is scheduled in subband 1 and user 2 is scheduled in subband 2.

Let us get back to the case where the underlying wireless channel varies very slowly. How fast can the induced channel fluctuation be? To obtain multiuser diversity the base station needs to track the SNR of all the users, thereby imposing a practical limit on the rate of change. Specifically, the channel fluctuation should be slow enough for the user to reliably measure the SNR and feed it back to the base station and for the base station to schedule the users accordingly. A 2π phase change in about 20–50 ms is realistic currently [169]. Such a scheduling latency may be too long for certain delay sensitive traffic.

One idea for reducing the scheduling latency is to combine opportunistic beamforming in frequency and in space. In OFDMA, since the beamforming coefficients are applied to individual tones, $g_k(t)$ becomes $g_{l,k}(t)$ where l is tone index. As a simple example, consider delay diversity. Set

$$g_{l,k} = \sqrt{\frac{1}{K_t}} e^{j\tau_k l}, \qquad (4.128)$$

where τ_k is an artificial delay introduced at antenna k. Unlike (4.118) that leads to time-variation, (4.128) induces frequency selectivity. For a given user, the channel responses from different antennas are phase-rotated by $e^{j\tau_k l}$ at tone l. For example, assume channel response of user i at antenna k, $H_{i,k}$, itself to be constant in frequency. Then the combined channel response at tone l is $\sqrt{\frac{1}{K_t}} e^{j\tau_k l} H_{i,k}$. For a given user i, $H_{i,k}$s (for all k) add constructively or destructively depending on tone l, and the combining power gain is the same as in (4.123) where the channel responses and beamforming coefficients are now dependent on tone l, as shown in Figure 4.18.

To obtain multiuser diversity, the scheduler selects the best user according to (4.124) for every tone at the extreme. However, to reduce scheduling and signaling overhead, we can group tones into subbands, similar to Figure 4.15, and schedule users per subband as opposed to per tone. Moreover, we can combine (4.118) and (4.128) to achieve simultaneous time and frequency selective beamforming:

$$g_{l,k}(t) = \sqrt{\alpha_k(t)} e^{j\theta_k(t)} e^{j\tau_k l}. \tag{4.129}$$

In the above delay diversity scheme, delay τ_l's are usually small to make channel and SINR trackable. The combined channel response in (4.129) varies smoothly in the time-frequency grid. As a result, the scheduling latency may still be an issue to be resolved. We next present enhanced opportunistic beamforming ([89]) in which the idea is to induce *discontinuous* beamforming patterns from one subband to another.

Assume that the underlying wireless channel is flat in frequency. With multiple transmit antennas, we artificially induce different channel fluctuation patterns in frequency. Specifically, denote by K_S the total number of subbands and $\alpha_{l,k}(t)$, $\theta_{l,k}(t)$ the power allocation and phase direction in subband l at antenna k. All tones within a subband use the same beamforming coefficients so as to make the channel correlation high. Meanwhile, different beamforming coefficients are used in different subbands to induce rapid channel fluctuation across frequency. A user tracks the channel SNR of the subbands separately assuming no coherence between the subbands. The users report the SNRs in every subband and the base station schedules the one with the best SNR in that subband. To reduce the signaling overhead, a user may report only the subband in which its SNR is the highest among all the subbands, because the user is unlikely to be scheduled in other subbands when the number of users is large.

We next select $g_{l,k}(t) = \sqrt{\alpha_{l,k}(t)} e^{j2\pi \theta_{l,k}(t)}$ to reduce the scheduling latency. To provide an insight, consider $K_t = 2$. Let $g_{l,1}(t) = 1/\sqrt{2}$ and $g_{l,2}(t) = e^{j2\pi ft + v_l}/\sqrt{2}$, where the phase offset v_l is constant uniformly spaced in $[0, 2\pi]$. This particular choice results in multiple opportunistic beams, each rotating at frequency f and has a fixed phase offset from one another. For example, for $K_S = 2$, $v_1 = 0$, $v_2 = \pi/2$. The channel seen by user i is $H_{i,1}(t)/\sqrt{2} + e^{j2\pi ft} H_{i,2}(t)/\sqrt{2}$ and $H_{i,1}(t)/\sqrt{2} - e^{j2\pi ft} H_{i,2}(t)/\sqrt{2}$, shown in Figure 4.19. At any given time t, $H_{i,1}$ and $H_{i,2}$ add constructively in one of the two subbands thereby making it more likely to be scheduled in that subband.

Figure 4.20 illustrates the basic opportunistic beamforming idea and the enhancement with subband selection. In the basic scheme, there is only one beam rotating slowly over the space and a user has to wait until it is in the beamforming direction to get scheduled. In the enhanced scheme, two beams are simultaneously present, each in one subband. While the rate of rotation of either beam remains the same, a user is in the beamforming direction of one beam twice as often because of the phase offset between the two beams. Therefore the opportunistic scheduling latency is reduced by half. The benefit of subband selection increases with K_S. When $K_S = 4$, $v_1 = 0$, $v_2 = \pi/2$, $v_3 = \pi$, $v_4 = 3\pi/2$, and it is more likely than when $K_S = 2$ that one of the subbands is highly beamformed at any time.

To evaluate the benefit of enhanced opportunistic beamforming, we conduct a simple simulation with 16 users served by a base station equipped with two transmit antennas. Consider $K_S = 1$ and 2. In each subband, the time is divided into time slots of 1.4 ms, with one user scheduled in each time slot. The users' channel gains h_i are modeled as complex Gaussian variables, constant over time and frequency. All the users experience an average 10 dB SNR in each subband, and the instantaneous rate achievable is given

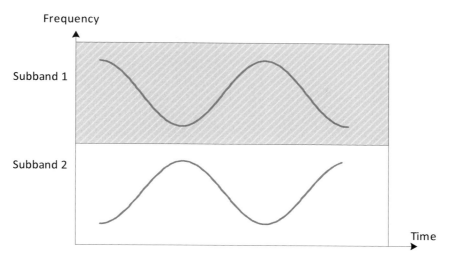

Figure 4.19 Illustration of channel response amplitude in enhanced opportunistic beamforming. There are two subbands. Channel fluctuation in a subband is induced using multiple transmit antennas, and follows a different (offset) pattern in a different subband.

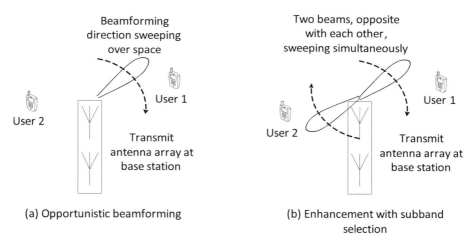

(a) Opportunistic beamforming (b) Enhancement with subband selection

Figure 4.20 Illustration of basic opportunistic beamforming and enhancement with subband selection. In basic opportunistic beamforming, one beam sweeps over the space and when a user happens to be in the beamforming direction gets scheduled. The scheduling latency depends on the sweeping rate. With $K_S = 2$ subbands, two beams with opposite beamforming directions simultaneously sweep and the scheduling latency is reduced by half.

by the Shannon capacity corresponding to the time-varying SNR. The base station uses a proportional fair scheduler (4.102).

The results of the simulation are shown in Figure 4.21. Here, throughput is the total throughput of the cell averaged over time and over 100 random user groups. The total throughput is normalized to the optimal throughput, labeled "optimal BF," given

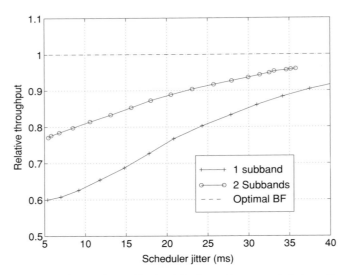

Figure 4.21 Comparison of basic and enhanced opportunistic beamforming. Rate versus jitter for 16 user, 2 antennas and $K_S = 1$ or 2 subbands.

by the rate that would be achieved if each of the 16 users were scheduled 1/16 of the time with their optimal beamforming configuration. Scheduling jitter refers to the standard deviation in inter-scheduling times for the users. For either one or two subbands, several different throughput versus jitter points are achievable by varying the scheduling parameter ϵ. As ϵ approaches zero, the rate achieved approaches a value close to the optimal rate, but at the price of increased scheduled jitter. The figure shows that the tradeoff of scheduling jitter is much better for two subbands in comparison to one. For the same total rate, the jitter is roughly cut by half from one subband to two subbands.

Multiuser diversity with SDMA

Using multiple antennas for beamforming achieves power gain. So far we have shown that multiuser diversity helps achieve opportunistic beamforming. Multiple antennas can also be used to obtain bandwidth gain with spatial multiplexing. One example is SDMA, an uplink multiuser MIMO scheme that allows multiple users to transmit to a base station with the same time-frequency resource. In Section 7.3.1 we demonstrate how SDMA benefits from multiuser diversity.

4.3.3 Multiuser or single-user diversity

Multiuser diversity is indeed very different from the single user diversity schemes as we discussed earlier, although they are all termed as types of *diversity*. Multiuser diversity exploits channel fluctuations and single-user diversity tends to mitigate channel fluctuation. From first principles, multiuser diversity can be viewed as a natural extension of the feedback schemes we discussed before, for example, truncated power control or water-filling schemes, since they all encourage *opportunistic* communications. In fact, multiuser diversity takes a step further. It not only takes advantage of channel fluctuation, but also utilizes the characteristic that different users experience channel fluctuation uncorrelated with each other. When a single user might refrain from transmitting at a given time slot due to undesired channel quality, there is a good chance that another user can utilize this slot with favorable channel quality. This simple idea can be further generalized to an opportunistic scheduling policy in which at each time slot the user with the most favorable SNR should be scheduled. This opportunistic scheduling policy has been widely adopted in mobile broadband cellular systems. We will discuss this and other scheduling policies in Chapter 8.

In Section 4.2.4, we discussed the tradeoff between single user diversity schemes and feedback-based power/rate control schemes. A similar tradeoff exists between multiuser diversity and single-user diversity. In particular, one has to decide whether to keep users on time-frequency selective channel resources so as to select the most favorable user on each channel resource or distribute users across all time-frequency resources for averaging. In the following, we continue this line of discussion by comparing the systems benefits and requirements of multiuser diversity and single-user diversity.

As far as the average rate is concerned, in principle multiuser diversity can always outperform single-user diversity since on all channel resources the user with the best channel is scheduled. In practice, multiuser diversity gain relies on the right rate of channel fluctuation. In Section 4.3.2, we observe that multiuser diversity diminishes in two extreme channel scenarios:

- Slow fading. The channel varies at a much slower time scale than what is acceptable to the latency requirement of the traffic. Within the latency tolerance, there is hardly any channel fluctuation to be exploited.
- Fast fading. When fading is so fast that the channel already experiences several independent fades even within a transmission block, there is not much channel fluctuation to exploit because of the averaging effect.

In addition, multiuser diversity gain depends on how accurate the scheduler can predict the SNR of all the users on all the scheduled transmission blocks. So far we have assumed that the multiuser scheduler is precisely aware of the channel state information. In practice, the scheduler uses a signaling protocol to track the SNR. In the downlink, for example, the users measure their SNRs and feed back to the base station scheduler. Such a protocol involves delay from measurement to reporting and to scheduling. In fast fading, the base station does not know the channel state precisely, because the SNR has changed significantly in the time interval of the protocol delay. In this case, multiuser diversity is hard to exploit and single-user diversity becomes attractive. In particular, it may be better off to code over multiple coherence times to achieve time diversity and for the base station scheduler to only track the average SNR across all the channel resources. Of course multiuser diversity gain diminishes in this case.

The above discussion suggests that multiuser diversity should be best harnessed in an intermediate channel fading scenario, "medium-speed fading" where the channel SNR remains unchanged within a transmission block, but fluctuates substantially in a time scale shorter than the latency tolerance; and yet the SNR fluctuation is still trackable at the multiuser scheduler.

The ability to accurately predict SNR is important to *reliably* exploit multiuser diversity. The scheduler should pursue multiuser diversity for peaking effect if it is able to accurately predict SNR; otherwise, the scheduler needs to be conservative by pursuing single-user diversity. The error in SNR prediction causes a performance loss in multiuser diversity; in that case, single-user diversity becomes more attractive, because single user diversity mitigates channel variations and is thus less sensitive to channel prediction errors. Here we note an inherent tension between pursuing averaging and peaking effects. From (4.97), averaging independent fading channel realizations reduces channel uncertainty and therefore the required prediction margin; however, multiuser diversity also diminishes.

To better demonstrate this point, we continue our example of the frequency selective and diversity schemes, and compare two OFDMA transmission blocks shown in Figure 4.11. Furthermore, we continue to use the notion of prediction margin defined in (4.79) to characterize the performance loss due to channel prediction uncertainty. In particular, we assume that in the frequency selective scheme, a transmission block is within one coherence bandwidth over one coherence time so as to pursue peaking effect. In the frequency diversity scheme, a transmission block is spread into a large frequency range of L_D coherence bandwidth over the same time duration and thus exhibits a diversity order of $L_D > 1$.

We make the same assumptions as in Section 4.3.2, except that user selection (4.116) and data rate (4.117) are revised to take into account the loss due to the prediction margin

$$i^* = \arg \max_{i=1,\ldots,M} \sum_{k=1}^{L} \frac{1}{L} \log(1 + \bar{\gamma} z_i(k)(1 - \mathrm{MG})), \qquad (4.130)$$

$$R_f = \mathbb{E} \left(\sum_{k=1}^{L} \frac{1}{L} \log(1 + \bar{\gamma} z_{i^*}(k)(1 - \mathrm{MG})) \right), \qquad (4.131)$$

where $L = L_D$ in the frequency diversity scheme and $L = 1$ in the frequency selective scheme. From (4.97) we assume that the required prediction margin MG in the frequency diversity scheme is only $1/\sqrt{L_D}$ of that in the frequency selective scheme.

Figure 4.22 plots the cdf of the scheduled data rates collected in simulation before taking the expectation in (4.131). We treat the single user ($M = 1$) case without prediction margin (MG $= 0$) as a baseline, and show the impact of first prediction margin (MG > 0) and then multiuser selection ($M > 1$). We observe that:

- In either the frequency selective or diversity scheme, after introducing the prediction margin, the cdf curve of the single user data rate shifts to the left, indicative of the loss due to prediction uncertainty; and the cdf curve of the multiuser data rate then shifts to the right representing multiuser diversity gain.

4.3 Communications, fading channel: multiuser case

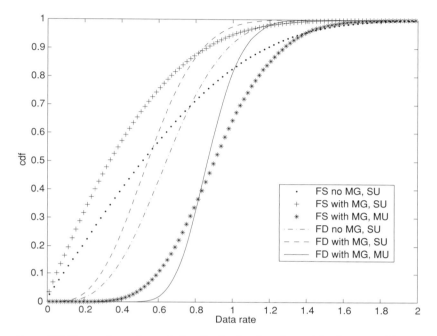

Figure 4.22 Comparison of cumulative distribution functions of the data rate of frequency diversity and selective transmission blocks. In either scheme, the figure plots the single user (SU) data rates with and without prediction margin (MG) and the multiuser (MU) data rate with prediction margin. $\bar{\gamma} = 0$ dB. MG $= 0.4$ in the frequency selective (FS) and MG $= 0.2$ in the frequency diversity (FD) cases. The order of frequency diversity is $L_D = 4$. $M = 8$ for multiuser and $M = 1$ for a single user.

- The leftward and rightward shifts are more significant in the frequency selective case than in the frequency diversity case, because in the latter case a smaller prediction margin is required but the multiuser diversity gain is also less.
- Comparison between frequency selective and diversity schemes shows that frequency selective scheme works better at the high end of the cdf curve (multiuser diversity benefit) but worse at the low end (prediction margin cost).

Hence, for reliable opportunistic communication, system design needs to balance the tension between pursuing averaging (frequency diversity) and peaking (frequency selective) effects by trading off prediction margin and multiuser diversity. As in Figure 4.22, we use the single-user case without prediction margin ($M = 1$, MG $= 0$) as the baseline and examine the impact of both multiuser selection ($M > 1$) and prediction margin (MG > 0). We divide the performance comparison into two steps:

$$\frac{\mathbb{E}\left(\sum_{k=1}^{L} \frac{1}{L} \log\left(1 + \bar{\gamma} z_{i^*}(k)(1 - \mathrm{MG})\right)\right)}{\mathbb{E}\left(\sum_{k=1}^{L} \frac{1}{L} \log\left(1 + \bar{\gamma} z_0(k)\right)\right)} = \Delta_1 \cdot \Delta_2, \quad (4.132)$$

where Δ_1, Δ_2 are defined next.

First, consider the single-user case and define the data rate loss due to the required prediction margin to be

$$\Delta_1 = \frac{\mathbb{E}\left(\sum_{k=1}^{L} \frac{1}{L} \log(1 + \bar{\gamma} z_0(k)(1 - \mathrm{MG}))\right)}{\mathbb{E}\left(\sum_{k=1}^{L} \frac{1}{L} \log(1 + \bar{\gamma} z_0(k))\right)} < 1. \quad (4.133)$$

The diversity benefit of the frequency diversity scheme is defined as the reduction in the rate loss as compared with the frequency selective scheme:

$$\Delta_{SU} = \frac{\Delta_{1,FD}}{\Delta_{1,FS}} > 1, \quad (4.134)$$

where the subscripts FD and FS represent the frequency diversity and selective schemes, respectively.

Next, consider the multiuser case and define the data rate increase due to multiuser diversity gain to be

$$\Delta_2 = \frac{\mathbb{E}\left(\sum_{k=1}^{L} \frac{1}{L} \log(1 + \bar{\gamma} z_{i^*}(k)(1 - \mathrm{MG}))\right)}{\mathbb{E}\left(\sum_{k=1}^{L} \frac{1}{L} \log[1 + \bar{\gamma} z_0(k)(1 - \mathrm{MG})]\right)} > 1. \quad (4.135)$$

Compared with the frequency selective scheme, the frequency diversity scheme has a smaller multiuser diversity gain. The difference between the two schemes is given by

$$\Delta_{MU} = \frac{\Delta_{2,FD}}{\Delta_{2,FS}} < 1. \quad (4.136)$$

The combined effect is therefore

$$\Delta_{total} = \Delta_{MU} \Delta_{SU}. \quad (4.137)$$

To illustrate the tradeoff, we next compare the frequency diversity and selective schemes over a range of required prediction margin MG.

Figure 4.23 plots Δ_{SU}, Δ_{MU}, and Δ_{total} with respect to MG and shows that the frequency selective scheme is favorable when MG is small, and as MG increases with the uncertainty in SNR prediction, the overall tradeoff becomes in favor towards the frequency diversity scheme. In OFDMA, it is flexible to adjust between the frequency selective and diversity schemes.

4.4 Summary of key ideas

- Multipath fading can cause drastic channel fluctuation in a short time interval. The channel is commonly described with a discrete path model consisting of a number of resolvable paths. Each path is associated with a complex channel gain and a delay. Path resolvability depends on the channel bandwidth.
- Channel variation in time depends on Doppler spread and is characterized with coherence time. Channel variation in frequency depends on delay spread and is characterized with coherence bandwidth.

4.4 Summary of key ideas

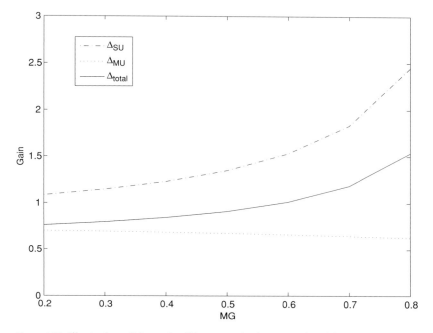

Figure 4.23 Illustration of the tradeoff between single user and multiuser diversity gains. The x-axis represents MG of the frequency selective scheme. The required margin is assumed to be reduced by a factor of $1/\sqrt{L_D}$ in the frequency diversity scheme. $\bar{\gamma} = 0$ dB. The order of frequency diversity is $L_D = 4$. $M = 8$ for the multiuser case. Gain equal to 0 means that the frequency selective and diversity schemes achieve the same rate.

- Multipath fading is a major challenge for reliable communications, and meanwhile a great potential for opportunistic communications.
- Reliable communications require mitigation of channel fading and therefore pursuit of averaging effect. Feedback-based rate/power control and time-frequency-space single-user diversity approaches are both effective and each possesses its own pros and cons. They can be combined to further enhance performance.
- Opportunistic communications exploit channel fading and therefore pursue peaking effect. The more drastically the channel fluctuates, the more significant the performance gain that can be achieved. Multiuser diversity is a system level diversity and the benefit increases with the number of users.
- Multiple antennas can be used to artificially induce channel fluctuation to amplify multiuser diversity and achieve the benefits of increased data rate and reduced scheduling delay. OFDMA adds the frequency dimension to allow flexible a design tradeoff between frequency diversity and frequency selectivity and to achieve reliable opportunistic communications.

5 Intra-cell user multiplexing

In a cellular system where a base station communicates with multiple users in every cell, a key question is how to share the system resources (power and bandwidth) among cells and users. We will study two aspects related to this question. The first aspect is about how to create a large data pipe between the base stations and users. We focus on the intra-cell issues in this chapter and leave the study of the inter-cell issues in Chapter 6. Specifically, in this chapter we will investigate how different user multiplexing schemes affect data rates and what the *Pareto-optimal user multiplexing schemes*, are. For simplicity, we decouple user multiplexing and channel fading by assuming that the wireless channel gain does not vary over time or frequency. Communications over the wireless fading channel has been studied in Chapter 4. The second aspect is about how to share the data pipe among users in a fair manner taking into upper layer considerations such as quality of service, and will be studied in Chapter 8.

We compare different multiplexing schemes, including *orthogonal* and *non-orthogonal* ones, in terms of *rate region*. In orthogonal multiplexing, different users are allocated system resources non-overlapping in time and frequency. Time division multiplexing (TDM) and frequency division multiplexing (FDM, such as OFDMA) are two examples. As noted in Chapter 3, an important advantage of OFDMA lies in its orthogonality and flexibility in user multiplexing. Specifically, with the granularity of bandwidth allocation as fine as a tone-symbol, users can be multiplexed in time and/or frequency. Different tone-symbols can be allocated different power. In non-orthogonal multiplexing, different users may share the same time and frequency resources with superposition coding. Advanced receiver algorithms, for example, successive interference cancellation (SIC), are required to decode the multiple layers of signals sharing the same bandwidth. In general non-orthogonal multiplexing requires more sophisticated coding and decoding techniques. Assuming perfect orthogonality, it has been well-known that superposition coding with SIC is capacity-achieving and thus superior to any orthogonal multiplexing schemes. Spatial multiplexing is another way of non-orthogonal multiplexing using multiple antennas and will be studied in Chapter 7. In this chapter, we only consider the single antenna case.

We would like to emphasize two points in this chapter. The first point is that the channel gains of different users exhibit large dynamic range (as large as 70 dB) because of small-scale multipath fading and large-scale cell geometry (path loss). The rate region increases when users with large dynamic range are multiplexed in frequency. The flexibility and orthogonality of OFDMA are key to achieving such a capacity improvement. The second point is the orthogonality in OFDMA degrades because of real-life impairments, which are characterized in Section 2.5 with the self-noise and cross interference models. We will show that self-noise and cross interference affect significantly the performance of user multiplexing schemes. In particular, superposition coding becomes even inferior to simple FDM in the self-noise model. We will then present a new coding scheme, called superposition-by-position, to combat the challenge of self-noise.

This chapter is organized as follows. We start with user multiplexing within a sector ignoring inter-sector interference, and study orthogonal multiplexing in Sections 5.1 and non-orthogonal multiplexing in 5.2. In both sections, we first assume perfect OFDMA orthogonality and then analyze the impact of loss of orthogonality using the self-noise and cross interference models. We then investigate user multiplexing between sectors of a cell in Section 5.3 to show how inter-sector interference due to imperfect sectorization can significantly reduce the advantages of OFDMA and present schemes to mitigate the adverse impact.

5.1 Orthogonal multiplexing

In this section, we first look at the performance of orthogonal multiplexing where no more than one user in a sector is allocated to the same tone-symbols.

5.1.1 Orthogonal multiplexing in the perfect model

Downlink multiplexing

Denote by P_m and $\mathcal{W}$ the total transmit power and bandwidth, respectively. Note that P_m here indicate the *maximum* power a base station can use at any time. There are two reasons why we assume the maximum, instead of average, total transmit power constraint. First, the power amplifier equipped at the base station has fixed maximum output power capability. In the coverage limited scenario, the base station is operated at the maximum output power to ensure the coverage. It is not feasible to "move" some power budget from one time instant to another. Second, fixing total transmit power at the base station allows the users in neighboring cells to easily track the inter-cell interference, thereby facilitating SINR measurement and downlink scheduling. However, in an interference limited deployment scenario, the base station does not have to always operate at a fixed maximum output power. As will be shown in Chapter 6, the base station can increase the capacity by varying the total transmit power in a coordinated manner among neighboring cells.

For simplicity, throughout this chapter, we consider a two-user case and study the power and bandwidth allocation between the two users served by the same base station. Let G_i be the downlink channel power gain, and σ_i^2 the total power of inter-cell interference plus noise of user i. Note that σ_i^2 is simply equal to $N_0\mathcal{W}$ if we ignore the inter-cell interference contribution, where N_0 the power density of the thermal noise.

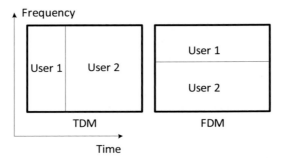

Figure 5.1 Time division multiplexing (TDM) and frequency division multiplexing (FDM) in OFDMA cellular systems.

Without loss of generality, we refer to user 1 as *inferior* user and user 2 as *superior* user, where $G_1/\sigma_1^2 \leq G_2/\sigma_2^2$. We now study the rate region achievable with different user multiplexing schemes.

Define the *nominal* SINR $\tilde{\gamma}_i$ of an inferior or superior user as the following:

$$\tilde{\gamma}_i = \frac{G_i P_m}{\sigma_i^2}, i = 1, 2. \tag{5.1}$$

The nominal SINR is the SINR when all the resources (power and bandwidth) are allocated to the user. In a cellular downlink, each user can estimate its nominal SINR based on the common pilots sent by the base station and report it back to the base station. In this section, we assume that the nominal SINR $\tilde{\gamma}_i$ is known *perfectly* to the base station and ignore the effect of SINR estimation errors and the delay caused by the SINR report loop.

Two basic schemes for orthogonal multiplexing are TDM and FDM shown in Figure 5.1. Since orthogonality among users is preserved in both TDM and FDM, it may appear that the two schemes would be equivalent. However, as shown next, FDM achieves a user multiplexing gain and outperforms TDM.

The data rates in TDM are given by

$$R_i = \psi_i \log(1 + \tilde{\gamma}_i), i = 1, 2, \tag{5.2}$$

where ψ_i is the fraction of time in which user i is scheduled with $\psi_1 + \psi_2 = 1$. The unit of R_i is nats per second per Hz. Denote by α_i and β_i the fractions of bandwidth and power assigned to user i in FDM, where $\alpha_i, \beta_i \in [0, 1]$, $\alpha_1 + \alpha_2 = 1$ and $\beta_1 + \beta_2 = 1$. The data rates in FDM are given by

$$R_i = \alpha_i \log\left(1 + \frac{\beta_i}{\alpha_i}\tilde{\gamma}_i\right). \tag{5.3}$$

Given $\tilde{\gamma}_1$ and $\tilde{\gamma}_2$, the rate region is obtained by varying ψ_i in (5.2) for TDM or α_i, β_i in (5.3) for FDM. Note that by setting $\alpha_i = \beta_i = \psi_i, i = 1, 2$, FDM can achieve the same rate as TDM. Thus, the achievable rate region using TDM is a *subset* of the rate region under FDM; FDM has additional flexibility in tuning α_i, β_i. Can this flexibility allow FDM to achieve higher rates for *both* users?

To compare the performance, we next numerically plot the data regions of the FDM and TDM schemes under various $\tilde{\gamma}_1$ and $\tilde{\gamma}_2$.

Figure 5.2(a) shows that when the nominal SINRs of the users are not equal, the FDM scheme outperforms the TDM scheme. The difference in the FDM and TDM rate regions represents FDM user multiplexing gain. The FDM user multiplexing gain increases with the ratio between $\tilde{\gamma}_1$ and $\tilde{\gamma}_2$. However, the benefit starts to diminish when the SINR ratio further increases; for example, the incremental improvement from increasing $\tilde{\gamma}_2/\tilde{\gamma}_1$ from 25 to 35 dB is less significant than that from an increase from 15 to 25 dB. In Figure 5.2(b), we vary $\tilde{\gamma}_2$ and $\tilde{\gamma}_1$, while keeping $\tilde{\gamma}_2/\tilde{\gamma}_1$ fixed to be 25 dB. We observe that the maximum user multiplexing benefit is achieved at $\tilde{\gamma}_1 = -9.5$ dB, that is, when $\tilde{\gamma}_1$ is not too low (-17 dB) or too high (5.5 dB).

The above observations indicate that the FDM multiplexing gain increases with the dynamic range of the nominal SINRs. What is the *practical* dynamic range of nominal SINR we are expecting in typical cellular deployment? As noted before, the dynamic range of channel gain can be as much as 70 dB in principle. A nearby user usually sees much less inter-cell interference than a faraway user does. Therefore, the dynamic range of nominal SINR is potentially substantial. At the low end, a cell edge user sees interference from multiple neighboring base stations. Nominal SINR is typically around -5 dB in an interference limited deployment, and even lower in a coverage limited deployment. At the high end, in theory, a user near the base station sees very high nominal SINR. However, real-world impairments, such as nonlinearity and phase noise, limit the maximum SINR in practice to about 35–40 dB. Further improvement of the maximum SINR, though feasible, significantly drives up implementation costs.

The reason that the FDM multiplexing gain depends on the SINR dynamic range is that FDM allows the tradeoff of power and bandwidth among the superior and inferior users operating in distinct SINR regimes. Recall that the Shannon capacity formula in an AWGN channel (C.2) is as follows:

$$C = W \log(1 + \tilde{\gamma}) = W \log\left(1 + \frac{GP}{\sigma^2}\right). \quad (5.4)$$

There are two SINR regimes, namely low SINR regime where $\tilde{\gamma} \ll 1$ such that $\log(1 + \tilde{\gamma}) \approx \tilde{\gamma}$ and high SINR regime where $\tilde{\gamma} \gg 1$ such that $\log(1 + \tilde{\gamma}) \approx \log(\tilde{\gamma})$. A user in the low SINR regime is power limited because capacity $C \approx \frac{GP}{N_0}$ changes linearly with power P. However, given power P, capacity C changes insignificantly with bandwidth W. Therefore, keeping the power unchanged, one can reduce the amount of utilized bandwidth with little impact on capacity. On the other hand, a user in the high SINR regime is bandwidth limited because capacity C changes linearly with bandwidth W. However, for a given W, C only changes logarithmically with P. Therefore, keeping the bandwidth unchanged, one can reduce the amount of utilized power without significantly reducing the capacity.

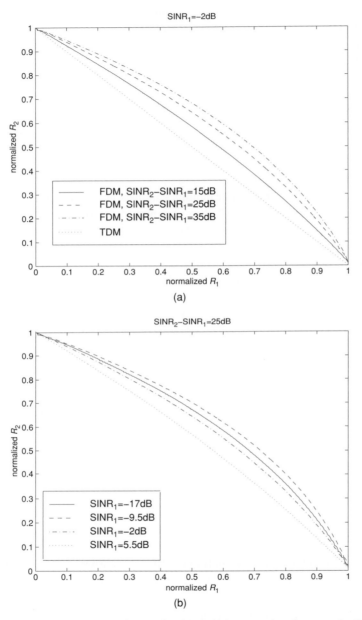

Figure 5.2 Comparison of normalized downlink rate regions between the TDM and FDM schemes. In (a), $\tilde{\gamma}_1$ is fixed to be -2 dB and $\tilde{\gamma}_2$ varies such that $\tilde{\gamma}_2/\tilde{\gamma}_1$ is equal to 15, 25 and 35 dB. In (b), $\tilde{\gamma}_2/\tilde{\gamma}_1$ is fixed to be 25 dB, and $\tilde{\gamma}_1$ varies from -17 dB to 5.5 dB. The rates are normalized with respect to the maximum R_1 and R_2 achieved given $\tilde{\gamma}_1$ and $\tilde{\gamma}_2$, when all bandwidth and power are assigned to either user 1 or 2.

5.1 Orthogonal multiplexing

In a typical cellular system, both power limited and bandwidth limited users are present. In FDM, it is desirable to multiplex power limited users and bandwidth limited users in frequency in the same time. Moreover, allocate most of the power to power limited users and most of the bandwidth to bandwidth limited users; this way, both types of users benefit from the improved utilization of power and bandwidth. In contrast, in TDM either bandwidth is under utilized when a power limited user is scheduled, or power is under utilized when a bandwidth limited user is scheduled. In general, the larger the dynamic range between the nominal SINRs, the greater the user multiplexing gain, as evident in Figure 5.2(a). However, a large dynamic range itself is not sufficient. If both users are either power or bandwidth limited, then the user multiplexing gain is insignificant, as can be inferred from Figure 5.2(b).

Next we take a step further to investigate the *Pareto-optimal* allocation of power and bandwidth (α_i and β_i) in FDM in the two-user example. From the above insight, it is desirable to allocate more power to the power limited (inferior) user and more bandwidth to the bandwidth limited (superior) user:

$$\frac{\beta_1}{\alpha_1} > 1, \frac{\beta_2}{\alpha_2} < 1. \tag{5.5}$$

The *operating* SINR γ_i, defined below, can be controlled or "shaped" with different choices α_i and β_i,

$$\gamma_i = \frac{\beta_i}{\alpha_i}\tilde{\gamma}_i, i = 1, 2. \tag{5.6}$$

The superior user has high nominal SINR and the inferior user low nominal SINR. The effect of (5.5) is that the operating SINR of the inferior user is shifted up to be greater than the nominal SINR such that it is less power limited; meanwhile, the operating SINR of the superior user is shifted down to be smaller than the nominal SINR such that it is less bandwidth limited. Hence, the utilization of power and bandwidth is improved for *both* the superior and inferior users. In contrast, in TDM the superior or inferior user uses the full bandwidth and power when scheduled. The downlink transmission alternates between $\tilde{\gamma}_1$ and $\tilde{\gamma}_2$ according to time sharing fraction ψ_i, but ψ_i does not shape the operating SINR from the nominal SINR.

Figure 5.3 shows the Pareto-optimal power and bandwidth allocation that achieves the points on one of the Pareto-boundary curves shown in Figure 5.2. We observe that except for the two end points where $\beta_1 = \alpha_1 = 1$ and $\beta_1 = \alpha_1 = 0$, (5.5) holds true in the entire Pareto-boundary. This confirms our insight.

One can make a tradeoff between R_1 and R_2 along the Pareto-boundary. Start from one end point of $R_1 = 0$ and R_2 at its maximum $\log(1 + \tilde{\gamma}_2)$. Increasing α_1 and β_1 moves the operating point towards the other end point of R_1 at its maximum $\log(1 + \tilde{\gamma}_1)$ and $R_2 = 0$. In the optimal power/bandwidth allocation, α_1, β_1 are not increased at the same rate. Specifically, when initially α_1, β_2 are small, $\frac{\beta_1}{\alpha_1}$ is large, indicating that to quickly increase R_1 the optimal power/bandwidth allocation gives the inferior user more power than bandwidth such that $\gamma_1 \gg \tilde{\gamma}_1$. As α_1, β_1 further increase towards 1, $\frac{\beta_1}{\alpha_1}$ gradually drops towards 1 and the power bandwidth allocation to the inferior user is very much

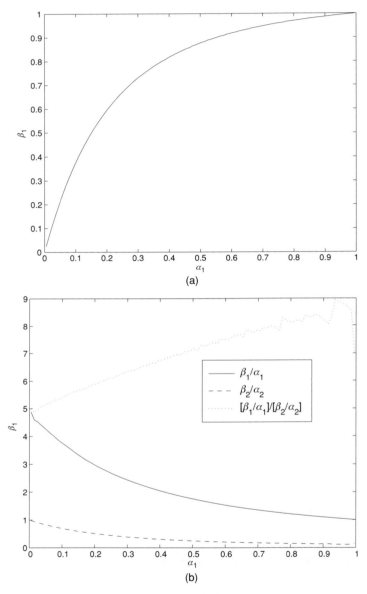

Figure 5.3 Pareto-optimal power and bandwidth allocation in the downlink: β_1 versus α_1 in (a) and $\frac{\beta_1}{\alpha_1}$, $\frac{\beta_2}{\alpha_2}$ and $\frac{\beta_1}{\alpha_1}/\frac{\beta_2}{\alpha_2}$ versus α_1 in (b). $\tilde{\gamma}_1 = -2$ dB and $\tilde{\gamma}_2 = 23$ dB.

balanced such that $\gamma_1 \approx \tilde{\gamma}_1$. Meanwhile, for the superior user, initially α_2, β_2 are close to 1, and thus $\gamma_2 \approx \tilde{\gamma}_2$. Then as α_2, β_2 decrease towards 0, $\frac{\beta_2}{\alpha_2}$ drops from 1, indicating that to avoid R_2 from dropping quickly, the optimal power/bandwidth allocation gives to the superior user more bandwidth than power such that $\gamma_2 \ll \tilde{\gamma}_2$. $\frac{\beta_1}{\alpha_1}$ and $\frac{\beta_2}{\alpha_2}$ represent the power per unit bandwidth allocation to the inferior and superior users, respectively. The ratio of $\frac{\beta_1}{\alpha_1}$ and $\frac{\beta_2}{\alpha_2}$ represent the dynamic range of power per unit bandwidth in the optimal power bandwidth allocation.

> **Discussion notes 5.1** An analysis of optimal power and bandwidth allocation in a cellular downlink
>
> So far, our main focus has been on the qualitative insight of optimal power and bandwidth allocation and we rely on simulation results to verify the insight. On the other hand, it is possible to characterize the optimal α_i and β_i in a more analytical way. In the following, we will summarize the main analytical results but leave out the detailed proof.
>
> First, while Figure 5.3 is based on a set of specific nominal SINRs, the observation holds in general. It can be proved that for any $\tilde{\gamma}_1$ and $\tilde{\gamma}_2$ with $\tilde{\gamma}_2 \geq \tilde{\gamma}_1$, a necessary condition to achieve any point on the Pareto-boundary of rate region (5.3) is that
>
> $$\frac{\beta_1}{\alpha_1} \geq 1 \geq \frac{\beta_2}{\alpha_2}. \quad (5.7)$$
>
> Second, we can calculate the Pareto-optimal power and bandwidth allocation α_i and β_i corresponding to each point on the Pareto-boundary. Define the following optimization problem: for a given parameter t
>
> $$\max_{\alpha_1, \beta_1} \mathcal{U} = t R_1 + R_2. \quad (5.8)$$
>
> Note that $-t$ is the slope of the tangent line at an optimal operating point on the Pareto-boundary. It follows that
>
> $$\frac{\partial \mathcal{U}}{\partial \beta_1} = t \frac{\tilde{\gamma}_1}{1 + \gamma_1} - \frac{\tilde{\gamma}_2}{1 + \gamma_2} = 0, \quad (5.9)$$
>
> $$\frac{\partial \mathcal{U}}{\partial \alpha_1} = t \left(\log(1 + \gamma_1) - \frac{\gamma_1}{1 + \gamma_1} \right) - \left(\log(1 + \gamma_2) - \frac{\gamma_2}{1 + \gamma_2} \right) = 0. \quad (5.10)$$
>
> Thus, a point on the Pareto-boundary of the rate region can be one-to-one mapped to a value of the slope $-t$. Given the value of $-t$ and the nominal SINRs $\tilde{\gamma}_1$ and $\tilde{\gamma}_2$, we can calculate the optimal operation SINRs, γ_1 and γ_2, which correspond to a specific choice of power and bandwidth allocation.
>
> Next we look at the range of the value of t. As shown in Figure 5.4, along the Pareto-boundary, t starts from some $t_{\min}$ at one end point ($R_1 = 0$, $R_2 = \log(1 + \tilde{\gamma}_2)$) and gradually increases to some $t_{\max}$ at the other end point ($R_1 = \log(1 + \tilde{\gamma}_1)$, $R_2 = 0$). At the former end point $\gamma_2 = \tilde{\gamma}_2$ and at the latter end point, $\gamma_1 = \tilde{\gamma}_1$. At those two end points, (5.9) and (5.10) are reduced to the following equations to solve $t_{\min}$ and $t_{\max}$:
>
> $$\left(1 - \log\left(\frac{(1 + \tilde{\gamma}_1)\tilde{\gamma}_2}{\tilde{\gamma}_1 t_{\max}} \right) \right) \frac{1}{t_{\max}} + \log(1 + \tilde{\gamma}_1) + \frac{\tilde{\gamma}_2 - \tilde{\gamma}_1}{(1 + \tilde{\gamma}_1)\tilde{\gamma}_2} = 1, \quad (5.11)$$
>
> $$\left(1 - \log\left(\frac{(1 + \tilde{\gamma}_2)\tilde{\gamma}_1 t_{\min}}{\tilde{\gamma}_2} \right) \right) t_{\min} + \log(1 + \tilde{\gamma}_2) + \frac{\tilde{\gamma}_1 - \tilde{\gamma}_2}{(1 + \tilde{\gamma}_2)\tilde{\gamma}_1} = 1. \quad (5.12)$$

Given any $t \in [t_{\min}, t_{\max}]$, one can calculate the operating SINR γ_1 and γ_2 from (5.9) and (5.10), and therefore the optimal power bandwidth allocation α_1, β_1.

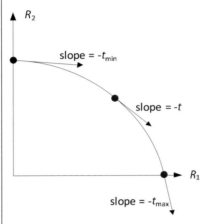

Figure 5.4 Tangent line slope $-t$ is bounded by $-t_{\min}$ and $-t_{\max}$, the tangent line slopes at the two end points of the Pareto-boundary.

From Figure 5.2, the more convex the Pareto-boundary, the greater the user multiplexing gain. The convexity can be studied using the slope of the tangent line at any point along the Pareto-boundary. Comparison of $t_{\min}$ and $t_{\max}$ tells how convex the FDM Pareto-boundary is. In one extreme case where $\tilde{\gamma}_1 = \tilde{\gamma}_2$, it follows that $t_{\max} = t_{\min} = 1$ and thus the Pareto-boundary is a straight line. The other extreme scenario is where $\tilde{\gamma}_2 \gg \tilde{\gamma}_1$. Fixing $\tilde{\gamma}_1$ and letting $\tilde{\gamma}_2$ increase to infinity, it follows that

$$t_{\max} \to \frac{\log(\tilde{\gamma}_2)}{\log(1+\tilde{\gamma}_1) + \frac{1}{1+\tilde{\gamma}_1} - 1} \qquad (5.13)$$

$$t_{\min} \to \xi(\tilde{\gamma}_2), \qquad (5.14)$$

where $\xi(\tilde{\gamma}_2)$ is the solution to $\xi^\xi = \tilde{\gamma}_2$. When $\tilde{\gamma}_2$ is large, $\xi(\tilde{\gamma}_2)$ is close to $\frac{\log(\tilde{\gamma}_2)}{\log\log(\tilde{\gamma}_2)}$. Thus,

$$\frac{t_{\max}}{t_{\min}} \approx \frac{\log\log(\tilde{\gamma}_2)}{\log(1+\tilde{\gamma}_1) + \frac{1}{1+\tilde{\gamma}_1} - 1}. \qquad (5.15)$$

When $\tilde{\gamma}_1$ is small, the ratio is large, implying significant convexity of the Pareto-boundary. In particular, when $\tilde{\gamma}_1$ gets close to zero, $t_{\max}$ becomes very large and the Pareto-boundary touches the R_1 axis almost vertically, meaning that when the inferior users are power limited, a significant rate R_2 can be achieved without sacrificing R_1 much. The above analysis is confirmed by Figure 5.5, which shows that as the disparity of nominal SINR (in dB) increases, both $t_{\max}$ and $t_{\min}$ increases roughly linearly with different slopes, and the gap between them grows wider.

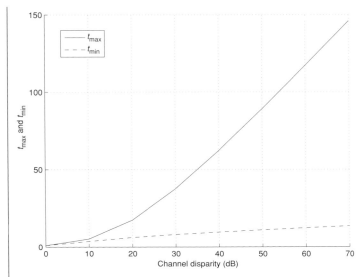

Figure 5.5 Plot of $t_{\max}$ and $t_{\min}$ versus disparity of nominal SINR $\tilde{\gamma}_2/\tilde{\gamma}_1$. $\tilde{\gamma}_1$ is fixed to be -3 dB.

Before we conclude the downlink user orthogonal multiplexing, we would like to make a few more notes:

- Although the results obtained in this section are based on a two-user model, the insight can be extended to a multiuser case.
- The advantage of FDM over TDM only exists when the constraint is on *maximum* power, rather than *average* power. Suppose that the total transmit power of the base station is allowed to vary over time and that P_m is the average total transmit power. Then, for any FDM scheme with α_i and β_i, in the TDM scheme we can set $\psi_i = \alpha_i$ and the total transmit power to be $\frac{\beta_i}{\alpha_i} P_m$ when user i is scheduled. It can be easily verified that the above TDM scheme achieves the same rates as FDM. Note that in the above TDM scheme, the total transmit power fluctuates around the average constraint P_m because one of $\frac{\beta_1}{\alpha_1}$ and $\frac{\beta_2}{\alpha_2}$ is greater than 1 and the other is less than 1, unless $\alpha_i = \beta_i$.
- For simplicity, we do not explicitly deal with channel fading in this chapter. It should be pointed out that channel fading does affect nominal SINR. In particular, when opportunistic scheduling is allowed, the nominal SINR of a scheduled user is higher that of an average user, thanks to the multiuser diversity effect described in Chapter 4. Thus the dynamic range of nominal SINRs would be smaller. However, if opportunistic scheduling is not possible, for example, because of delay constraint, then channel fading can significantly hurt nominal SINR, leading to a large dynamic range of nominal SINRs.

Practical example 5.1 Downlink user multiplexing: EV-DO, HSDPA, and LTE

Let us continue Example 4.1. In the EV-DO downlink, the base station transmits to at most one user. Thus, the users are multiplexed in time as depicted in Figure 4.13(b). From the above study, we know that such a TDM approach is inferior to FDM.

We can also reach the same conclusion by observing the data rate table and SINR CDF distribution provided in [12]. Specifically, in a 1.25 MHz downlink, the lowest data rate is 38.4 kbps operating at -12.5 dB SINR, in which case the channel coding rate is much lower than $1/2$. In such a low SINR regime, the downlink is very much power limited and could achieve the same data rate with a much reduced bandwidth. On the other hand, the highest data rate is 2457.6 kbps at 9.5 dB SINR. Indeed, the SINR CDF distribution reveals that the SINR saturates below 15 dB. However, we know that when a user is close to the base station, the C/I could be much higher than 15 dB. It seems that the observed SINR is saturated because of self-noise, an effect similar to what is demonstrated in Figure 2.41. Here, self-noise is partly attributed to the suboptimality of a RAKE receiver, which is the MMSE optimal in the low SINR regime. In the high SINR regime, one has to employ an equalizer to overcome multipath delay spread. When the SINR is already saturated, the base station could reduce the transmit power without noticeably hurting the SINR and data rate.

To avoid wasting power or bandwidth, the base station would have to transmit to nearby and faraway users simultaneously. However, in CDMA, simultaneous transmissions to multiple users cause intra-cell interference, precisely something EV-DO attempts to avoid. Hence, using a TDM transmission approach, EV-DO suffers from the loss of TDM as opposed to FDM in downlink user multiplexing.

HSDPA (for the downlink) is a major evolution of UMTS WCDMA to support high speed data traffic, followed by HSUPA (for the uplink). Similar to EV-DO, HSDPA adopts the ideas of adaptive modulation and coding, multiuser channel-aware scheduling, and fast hybrid-ARQ. An important difference is that the HSDPA downlink multiplexes users not only in time but also in code, thereby transmitting simultaneously to nearby and faraway users and allowing joint power bandwidth allocation.

Simultaneous transmissions in the downlink are separated by channelized codes, which are constructed with the Orthogonal Variable Spreading Factor technique ([1]). The use of channelized codes allows the spreading factor to be changed (bandwidth allocation) while maintaining orthogonality between spreading codes of different lengths. The codes are selected from a code tree illustrated in Figure 5.6. The number of spreading codes for HSDPA is configurable between 1 and 15 with spreading factor 16. HSDPA transmissions are code multiplexed with other channels such as for common control and circuit-switched traffic.

The channelized codes are no longer orthogonal in the presence of multipath delay spread. A conventional RAKE receiver yields suboptimal performance because of intra-cell multiple access interference. Channel equalization is needed at the receiver. Traditionally, equalization is done in the time domain. However, the complexity is high if the channel response is long. Frequency domain equalization can reduce complexity.

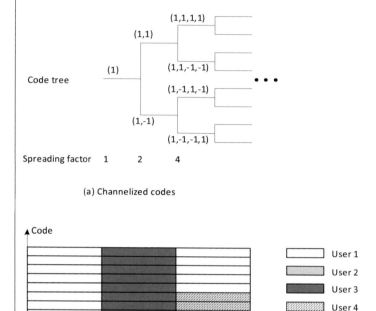

(a) Channelized codes

(b) User multiplexing in time and in code

Figure 5.6 Illustration of code tree and user multiplexing in HSDPA.

It should be pointed out that because users are multiplexed in code, an equalizer in HSDPA equalizes not only its own signal but also signals to other users, and therefore has to deal with the potentially large dynamic range of those signals. In practice, equalization is not perfect and the residual ISI post-equalization results in interference between users. As a comparison, an equalizer in EV-DO only deals with its own signal. Similarly, because of tone orthogonality, an LTE receiver first removes all other signals at the output of FFT before any equalization (see Figure 2.18b) and then only equalizes its own signal. Hence, without the concern of interference between users, an EV-DO or LTE equalizer can live with larger residual errors than the HSDPA equalizer can.

Now let us look at OFDMA-based LTE. In LTE, users are multiplexed in time and in frequency (tone), as shown in Figure 5.7. Specifically, the downlink includes the control and data channels. The control channels are for downlink/uplink scheduling grants, acknowledgment to uplink data channels, and uplink power control commands. The data channels are for downlink traffic. The control and data channels are TDM within a subframe where the control channels span the first few OFDM symbols and the data channels take the rest. The control channels always span over the entire bandwidth for frequency diversity. The data channels support localized transmissions for frequency selective scheduling and distributed transmissions for frequency diversity scheduling.

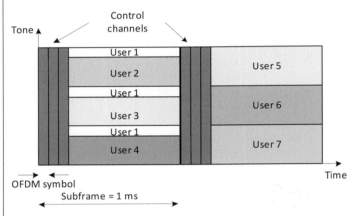

Figure 5.7 Illustration of user multiplexing in the LTE downlink. User 1 is for distributed transmissions and all the other users are for localized transmissions.

Finally, we briefly describe three control channels in the LTE downlink, namely PCFICH (Physical Control Format Indicator Channel), PHICH (Physical Hybrid-ARQ Indicator Channel) and PDCCH (Physical Downlink Control Channel). As shown in Figure 5.7, they all reside in the first few OFDM symbols of a subframe. The PCFICH is used to specify the number of OFDM symbols occupied by the control channels, which can be 1, 2, or 3. The PCFICH consists of 2 information bits, which are coded and scrambled to be sent in 16 REs in the first OFDM symbol. Those 16 REs spread in frequency for diversity and their frequency locations depend on the physical layer cell identity to avoid collision between neighboring cells. The PHICH is used to send acknowledgment to corresponding PUSCH in the uplink. The number of PHICH signals depends on the number of PUSCH transmissions to be acknowledged. The PHICH is typically transmitted in the first OFDM symbol. A PHICH consists of 1 information bit, which is coded and scrambled to be sent in 12 REs. Multiple PHICH signals can share the same set of 12 REs in a CDM manner with orthogonal codes. The PDCCH is used to send downlink scheduling assignments of corresponding PDSCH, uplink scheduling grants of corresponding PUSCH, and uplink power control commands. A PDCCH can use one of a variety of formats, and the number of information bits in the PDCCH depends on the format. A PDCCH is transmitted on a set of Control Channel Elements (CCEs), where each CCE consists of 36 REs. The number of CCEs of a given PDCCH can be 1, 2, 4, or 8 depending on the payload size and the required coding rate. For example, if a PDCCH is intended to send to a user with unfavorable channel conditions, then the required coding rate is low and more CCEs are used. Usually, whether the control channels need 1, 2, or 3 OFDM symbols depends on the size of the PDCCH, which can change dynamically from one subframe to another because the number of scheduling assignments or grants may be different. Thus, the use of PCFICH makes it flexible to partition the control and data channels. Note that in a given subframe, not all the REs in the first few OFDM symbols are occupied by the control channels. Unused REs help reduce inter-cell interference of the control channels.

Uplink multiplexing

Now consider user multiplexing in the uplink. Similar to the downlink, consider only two users in a sector, an inferior user and a superior user, with respective channel power gains G_1 and G_2. We use σ^2 to denote the total power of inter-cell interference plus noise. We are interested in the sum data rates R_1 and R_2 achieved for the inferior and superior users. In the uplink, the definition of the inferior and superior users is $G_1 \leq G_2$ since the interference plus noise power is common to both.

We assume that the constraints of user multiplexing in the uplink are the maximum transmit power of *each* user P_m and a maximum *total* inter-cell interference power Q_m caused by the users in the sector to all neighboring base stations. Denote by P_i the transmit power of an inferior or superior user. The constraints are given as

$$P_i \leq P_m, i = 1, 2 \tag{5.16}$$

$$V_1 P_1 + V_2 P_2 \leq Q_m. \tag{5.17}$$

The inter-cell interference from a user i to a neighboring base station n is proportional to its transmit power and the corresponding channel gain $G_{n,i}$. Thus in (5.17) the weight coefficient $V_i = \sum_{n>0} G_{n,i}$. As we will show later in Section 6.2.3, Q_m is closely related to the quantity *load* of a sector in the load-spillage mechanism, which is a distributed inter-cell interference management protocol and can be shown to be *Pareto-optimal*. The constraint (5.17) on Q_m follows directly from the load-spillage mechanism. Usually, an inferior user is close to cell edges and causes more inter-cell interference than a superior user does. Thus we assume $V_1 \geq V_2$ in the following.

Define $Q_i = V_i P_i$ and $Q_{m,i} = V_i P_m$. The above equations are rewritten as

$$Q_i \leq Q_{m,i}, i = 1, 2 \tag{5.18}$$

$$Q_1 + Q_2 \leq Q_m, \tag{5.19}$$

Denote by α_i the fraction of the bandwidth allocated to the inferior or superior users with $\alpha_1 + \alpha_2 = 1$, and $\beta_i = \frac{Q_i}{Q_m}$ the fraction of the *interference budget*, as opposed to the power budget in the downlink. The achieved rates are given as follows:

$$R_i = \alpha_i \log\left(1 + \frac{\beta_i}{\alpha_i}\tilde{\gamma}_i\right), \tag{5.20}$$

where the nominal SINR is defined to be

$$\tilde{\gamma}_i = \frac{G_i Q_m}{\sigma^2 V_i}, i = 1, 2. \qquad (5.21)$$

The nominal SINR in the uplink is the SINR when an inferior or superior user is allocated all the bandwidth and allowed to use up the total inter-cell interference budget Q_m *without* taking into account the maximum transmit power constraint P_m.

Without the individual constraint on Q_i (5.18), the uplink multiplexing problem would be almost identical to the downlink version by substituting P_i with Q_i. However, the individual constraint on Q_i can lead to different characteristics of user multiplexing, as we will show below. An obvious effect of having the additional constraint (5.18) is that a user may already be limited by its maximum transmit power before using up the total interference budget. As a result, it is possible that the power allocation parameter β_1 or β_2 cannot reach the entire range of [0, 1] and that $\beta_1 + \beta_2 < 1$.

Depending on values of Q_m, $Q_{m,1}$ and $Q_{m,2}$, there are four operating regimes as shown in Figure 5.8. Note that $Q_{m,1} \geq Q_{m,2}$, since $V_1 \geq V_2$. Regime I is the *interference limited* regime where the transmit power constraint has no impact. Regime IV is the *power limited* regime where the inter-cell interference constraint has no impact. Regime II and III are the transition regimes between the other two regimes where both constraints take effect. The range of α_1, α_2 is [0, 1] with $\alpha_1 + \alpha_2 = 1$ in all the four regimes. The range of β_1, β_2 depends on the operating regime. In regimes I, II, and III,

$$\beta_1 \in \left[\max\left(1 - \frac{Q_{m,2}}{Q_m}, 0\right), \min\left(\frac{Q_{m,1}}{Q_m}, 1\right) \right], \qquad (5.22)$$

$$\beta_2 \in \left[\max\left(1 - \frac{Q_{m,1}}{Q_m}, 0\right), \min\left(\frac{Q_{m,2}}{Q_m}, 1\right) \right], \qquad (5.23)$$

and $\beta_1 + \beta_2 = 1$. In regime IV, there is a single optimal operating point

$$\beta_1 = \frac{Q_{m,1}}{Q_m}, \qquad (5.24)$$

$$\beta_2 = \frac{Q_{m,2}}{Q_m}, \qquad (5.25)$$

and $\beta_1 + \beta_2 < 1$. In regime IV, since Q_m is no longer relevant, it is more convenient to define power limited nominal SINR

$$\tilde{\gamma}_i' = \frac{G_i P_m}{\sigma^2}, i = 1, 2 \qquad (5.26)$$

and refer to the one defined in (5.21) as interference limited nominal SINR.

When would user multiplexing gain be the most significant in the uplink? In the interference limited regime (regime I), the expression of the Pareto-boundary is the same as in the downlink. Everything we have learned in the downlink applies here. Specifically, multiplexing gain is high when $\tilde{\gamma}_1$ is low and $\tilde{\gamma}_2$ is high. The scenario

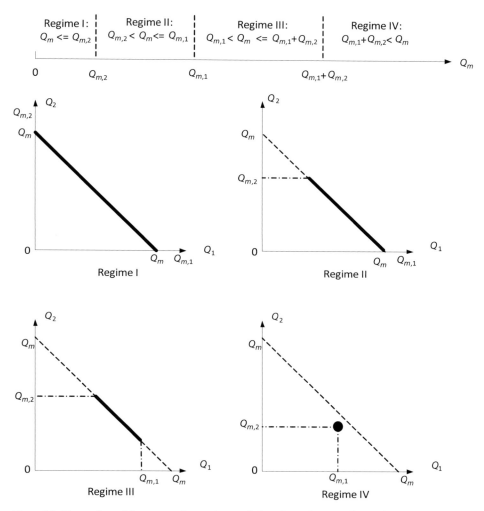

Figure 5.8 Illustration of four operating regimes of Q_m, $Q_{m,1}$, $Q_{m,2}$. In the perfect model, the thick solid line or single dot is the optimal choice of Q_1, Q_2 in the corresponding regime based on constraints (5.18) and (5.19) assuming $Q_{m,1} \geq Q_{m,2}$.

is, however, very different in the power limited regime (regime IV). In this regime, multiplexing users simply means combining their power budgets. Thus, multiplexing gain is high when $\tilde{\gamma}'_1$ and $\tilde{\gamma}'_2$ are both low.

Figure 5.9 compares the normalized Pareto-boundary of three different user multiplexing scenarios (two inferior users, two superior users, or one inferior user and one superior user) in the power limited regime (a) and in the interference limited regime (b). In (a), multiplexing gain is the most significant with "inferior-inferior," in which case the capacity increases almost linearly with the power budget increase. In (b), the Pareto-boundary of "inferior-superior" dominates as in the downlink.

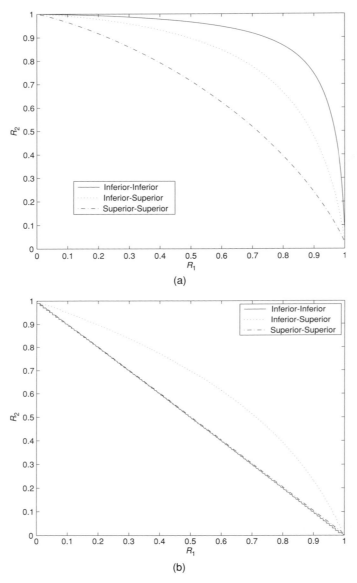

Figure 5.9 Comparison of normalized uplink rate regions in the power and interference limited regimes. "Inferior-Inferior" represents two inferior users that are multiplexed in frequency, "Superior-Superior" two superior users, and "Inferior-Superior" one inferior user and one superior user. For a superior user, power limited nominal SINR $\tilde{\gamma}' = -2$ dB, interference budget weight $V = 10$. For an inferior user $\tilde{\gamma}' = 13$ dB, $V = 1$. In the power limited regime (a), $Q_m = 20P_m$. In the interference limited regime (b), $Q_m = P_m$.

Table 5.1. Comparison of uplink and downlink orthogonal multiplexing.

		Uplink	
	Downlink	Interference limited	Power limited
General multiplexing strategy	Allocate greater transmit power per unit bandwidth to inferior users than to superior users	Same as in downlink with transmit power replaced by interference budget	Bandwidth tradeoff only; each user transmits at its maximum power
When is user multiplex gain most significant?	Inferior users in low SINR and superior users in high SINR regime; large SINR disparity	Same as in downlink	A large number of users in low SINR regime

Summary

We summarize the discussion on user multiplexing in the downlink and uplink in Table 5.1.

5.1.2 Orthogonal multiplexing in the cross interference model

So far we have assumed perfect OFDMA orthogonality to develop insight on user multiplexing. We next consider the impact of imperfect orthogonality. In Section 2.5, the loss of orthogonality in OFDMA is described with the cross interference model and self-noise model. We start with the cross interference model in this section and proceed to the self-noise model in the next section.

Recall that in the cross interference model, the desired power of a received signal is reduced by a factor of $1 - \epsilon$, and the remaining ϵ fraction leaks evenly to the entire bandwidth. Thus, the resultant intra-sector interference to a signal is proportional to the bandwidth the signal occupies.

Downlink multiplexing

The downlink achieved rates in FDM in the cross interference model are given by

$$R_i = \alpha_i \log\left(1 + \frac{\beta_i(1-\epsilon)G_i P_m}{\alpha_i \sigma_i^2 + \alpha_i \epsilon G_i P_m}\right), \quad (5.27)$$

where the cross interference is reflected in the additional noise term $\alpha_i \epsilon G_i P_m$. This term is proportional to the fraction of bandwidth assigned to user i, α_i, and is not dependent on β_i. We can rewrite (5.27) in the following form:

$$R_i = \alpha_i \log\left(1 + \frac{\beta_i}{\alpha_i}\tilde{\gamma}_{i,CI}\right), \quad i = 1, 2. \quad (5.28)$$

This expression is the same as in (5.3), except that the nominal SINR in the cross interference model is given by

$$\tilde{\gamma}_{i,CI} = \frac{(1-\epsilon)G_i P_m}{\sigma_i^2 + \epsilon G_i P_m}$$

$$= \frac{(1-\epsilon)\tilde{\gamma}_i}{1+\epsilon\tilde{\gamma}_i}, \tag{5.29}$$

where $\tilde{\gamma}_i$ is the nominal SINR in the absence of cross interference defined in (5.1).

When $\tilde{\gamma}_i$ is small, $\tilde{\gamma}_{i,CI}$ is close to $\tilde{\gamma}_i$. As $\tilde{\gamma}_i$ increases, $\tilde{\gamma}_{i,CI}$ is upper-bounded by $\frac{1-\epsilon}{\epsilon}$. Thus, it follows that

$$\tilde{\gamma}_{i,CI} \approx \min\left(\tilde{\gamma}_i, \frac{1-\epsilon}{\epsilon}\right). \tag{5.30}$$

From the study in the perfect model, user multiplexing gain is the most significant when the inferior users in the low SINR regime are multiplexed with the superior users in the high SINR regime. Cross interference factor ϵ does not affect the low SINR regime much but limits the maximum SINR achievable by the superior users and therefore the SINR dynamic range. For example, suppose that $\tilde{\gamma}_1 = -2$ dB. At $\epsilon = 10^{-3}$, $\tilde{\gamma}_2 < 30$ dB, and $\tilde{\gamma}_2/\tilde{\gamma}_1$ is at most 32 dB. At $\epsilon = 10^{-2}$, it drops to only 22 dB. Hence, we conclude that in the cross interference model, user multiplexing gain is less significant due to the limitation on the high nominal SINRs.

Uplink multiplexing

The uplink scenario is a little more complex than the downlink. Recall the four operating regimes in Figure 5.8 from the interference limited regime to the power limited regime. In the interference limited regime (regime I), with cross interference, the expression of the achieved rates in FDM is

$$R_i = \alpha_i \log\left(1 + \frac{(1-\epsilon)G_i P_i}{\alpha_i(\sigma^2 + \epsilon G_1 P_1 + \epsilon G_2 P_2)}\right)$$

$$= \alpha_i \log\left(1 + \frac{\beta_i(1-\epsilon)\tilde{\gamma}_i}{\alpha_i(1+\epsilon\beta_1\tilde{\gamma}_1 + \epsilon\beta_2\tilde{\gamma}_2)}\right), i = 1, 2, \tag{5.31}$$

where $\tilde{\gamma}_i$ is the nominal SINR in the absence of cross interference defined in (5.21). The above expression looks different from (5.28). However, if we let

$$\beta_i' = \frac{1+\epsilon\tilde{\gamma}_i}{1+\epsilon\beta_1\tilde{\gamma}_1 + \epsilon\beta_2\tilde{\gamma}_2}\beta_i, \tag{5.32}$$

it can be shown that in the interference limited regime where $\beta_i, \beta_i' \in [0, 1]$, (5.31) becomes

$$R_i = \alpha_i \log\left(1 + \frac{\beta_i'}{\alpha_i}\tilde{\gamma}_{i,CI}\right), \tag{5.33}$$

with

$$\tilde{\gamma}_{i,CI} = \frac{(1-\epsilon)G_i Q_m}{\sigma^2 V_i + \epsilon G_i Q_m}$$
$$= \frac{(1-\epsilon)\tilde{\gamma}_i}{1+\epsilon\tilde{\gamma}_i}, i = 1, 2. \quad (5.34)$$

Note the similarity between the above expression and (5.20). Therefore, as in the downlink case, the effect of cross interference is that the nominal SINR at the high end saturates due to ϵ, thereby reducing user multiplexing gain.

Now we consider the power limited regime (regime IV). In this regime,

$$R_i = \alpha_i \log\left(1 + \gamma_i(\alpha_1, \alpha_2, P_1, P_2)\right), i = 1, 2, \quad (5.35)$$

where $\alpha_1 + \alpha_2 = 1, \alpha_i \in [0, 1], P_i \in [0, P_m]$, and operating SINR

$$\gamma_i(\alpha_1, \alpha_2, P_1, P_2) = \frac{(1-\epsilon)G_i P_i}{\alpha_i(\sigma^2 + \epsilon(G_1 P_1 + G_2 P_2))}. \quad (5.36)$$

In the perfect model, the optimal choice for uplink multiplexing in the power limited regime is to let all users transmit at maximum power, that is, $P_1 = P_2 = P_m$, which is shown as the thick dot in Figure 5.8. However, this choice is not necessarily Pareto-optimal in the cross interference model. The reason is that when one user reduces its transmit power, the cross interference is also reduced, thereby benefiting the rate of the other user.[1] Obviously, both users need not back off from P_m simultaneously. Thus, the Pareto-boundary of the feasible rate region includes not only the rate pairs (R_1, R_2) achieved by $P_1 = P_2 = P_m$, but also those achieved with $P_1 = P_m, P_2 \in [0, P_m)$ (denoted by Γ_1) and $P_2 = P_m, P_1 \in [0, P_m)$ (denoted by Γ_2).

Figure 5.10(a) provides a conceptual characterization of the regions Γ_1 and Γ_2. Take Γ_1 for example. It can be shown that for any point on Γ_1, there exists one and only one allocation scheme $(\alpha_1, \alpha_2, P_1 = P_m, P_2)$ that achieves that point. Furthermore, for any two different points (R_1, R_2) and (R'_1, R'_2) on Γ_1, denote by $(\alpha_1, \alpha_2, P_1 = P_m, P_2)$ and $(\alpha'_1, \alpha'_2, P_1 = P_m, P'_2)$ the corresponding Pareto-optimal resource allocation choices. If $R'_2 > R_2$, we have $\alpha'_1 < \alpha_1, \alpha'_2 > \alpha_2$ and $P'_2 \geq P_2$. Figure 5.10(b) shows a numerical example of the uplink Pareto-optimal rate region in the presence of cross interference.

Discussion notes 5.2 An analysis of optimal power and bandwidth allocation for orthogonal uplink multiplexing with cross interference in the power limited regime

Next we provide a detailed characterization of the rate region of Figure 5.10.

It can be shown that Γ_1 is convex and that a necessary condition for allocation scheme $(\alpha_1, \alpha_2, P_1 = P_m, P_2)$ to be Pareto-optimal is

$$\frac{(1+\gamma_2)\log(1+\gamma_2) - \gamma_2}{(1+\gamma_1)\log(1+\gamma_1) - \gamma_1} = \frac{1-\epsilon}{\epsilon\gamma_1(1, 0, P_m, 0)}, \quad (5.37)$$

---

[1] This kind of behavior is much more dominant in the inter-cell interference management scenario, which we will discuss extensively as part of the topic of *uplink power control* in Chapter 6.

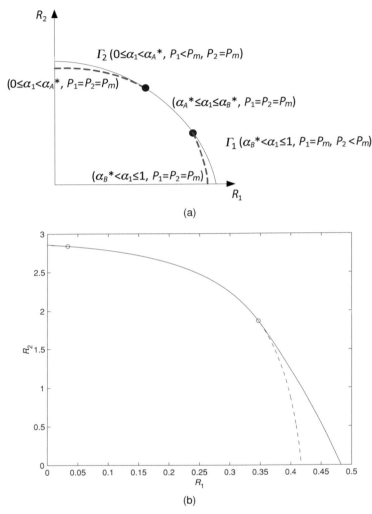

Figure 5.10 Pareto-boundary of the uplink rate region in the power limited regime. Part (a) is an illustration where the solid curve consists of three portions of the Pareto-boundary and the dotted curves represents the suboptimal rate pairs achieved by allocation schemes $\alpha_1 \in [0, \alpha_A^*)$, $P_1 = P_m$, $P_2 = P_m$ in the first portion and $\alpha_1 \in (\alpha_B^*, 1]$, $P_1 = P_m$, $P_2 = P_m$ in the third portion. Allowing $P_1 < P_m$ when $\alpha_1 \in [0, \alpha_A^*)$ and $P_2 < P_m$ when $\alpha_1 \in (\alpha_B^*, 1]$ improves the Pareto-boundary. Part (b) numerically plots a similar picture with the following parameters: $\epsilon = 10^{-2}$, $\tilde{\gamma}_1' = -2$ dB, $\tilde{\gamma}_2' = 13$ dB. Because of the large disparity in $\tilde{\gamma}_i'$, the first portion is very short and whether or not to allow $P_1 < P_m$ makes not much difference. However, as shown, varying P_2 between 0 and P_m in the third portion increases the Pareto-boundary remarkably.

where γ_1, γ_2 are defined in (5.36) with input variables $\alpha_1, \alpha_2, P_1 = P_m, P_2$, and $\gamma_1(1, 0, P_m, 0)$ is equal to γ_1 in (5.36) with $\alpha_1 = 1, \alpha_2 = 0, P_1 = P_m, P_2 = 0$. Hence, each point of the Pareto-boundary Γ_1 is obtained by fixing a P_2 with $P_2 < P_m$ and choosing α_i that satisfies (5.37).

Similar results can be obtained for rate region Γ_2. Specifically, a necessary condition for allocation scheme $(\alpha_1, \alpha_2, P_1, P_2 = P_m)$ to be Pareto-optimal is

$$\frac{(1+\gamma_2)\log(1+\gamma_2) - \gamma_2}{(1+\gamma_1)\log(1+\gamma_1) - \gamma_1} = \frac{\epsilon \gamma_2(0, 1, 0, P_m)}{1-\epsilon}, \quad (5.38)$$

where γ_1, γ_2 are defined in (5.36) with input variables $\alpha_1, \alpha_2, P_1, P_2 = P_m$, and $\gamma_2(0, 1, 0, P_m)$ is equal to γ_2 in (5.36) with $\alpha_1 = 0, \alpha_2 = 1, P_1 = 0, P_2 = P_m$.

Hence we reach the following full characterization of the Pareto-boundary of (5.35) under maximum transmit power constraint $P_i \leq P_m$. As R_1 increases and R_2 decreases, the Pareto-boundary has the following three portions:

1. Γ_2: $\alpha_1 \in [0, \alpha_A^*)$, $P_1 < P_m$, $P_2 = P_m$. Any Pareto-optimal allocation scheme satisfies (5.38) for a given P_1. α_A^* is the solution to (5.38) when $P_1 = P_2 = P_m$.
2. Middle portion: $\alpha_1 \in [\alpha_A^*, \alpha_B^*]$, $P_1 = P_m$, $P_2 = P_m$, where α_B^* is defined below.
3. Γ_1: $\alpha_1 \in (\alpha_B^*, 1]$, $P_1 = P_m$, $P_2 < P_m$. Any Pareto-optimal allocation scheme satisfies (5.37) for a given P_2. α_B^* is the solution to (5.37) when $P_1 = P_2 = P_m$. Note that $\alpha_B^* > \alpha_A^*$.

Figure 5.10(a) illustrates the three portions of the Pareto-boundary. The two circles represent where the Pareto-boundary is partitioned in to three portions.

Let us now consider two extreme cases. First, when ϵ is small, $\alpha_A^* \approx 0$ and $\alpha_B^* \approx 1$. The middle portion shown in Figure 5.10 dominates the Pareto-boundary and neither user should back off from P_m.

Second, when ϵ is large such that cross interference dominates noise $\epsilon \tilde{\gamma}_i' \gg 1$, both $\gamma_1(1, 0, P_m, 0)$ and $\gamma_2(0, 1, 0, P_m)$ are approximately equal to $\frac{1-\epsilon}{\epsilon}$. Combining (5.37) and (5.38) we have

$$\frac{(1+\gamma_2)\log(1+\gamma_2) - \gamma_2}{(1+\gamma_1)\log(1+\gamma_1) - \gamma_1} = 1. \quad (5.39)$$

It follows that $\gamma_2 = \gamma_1$ and therefore

$$\frac{G_1 P_1}{\alpha} = \frac{G_2 P_2}{1-\alpha}. \quad (5.40)$$

Furthermore, $\alpha_A^* = \alpha_B^*$, which can be obtained from (5.40) with $P_1 = P_2 = P_m$. Thus, the middle portion of the Pareto-boundary in Figure 5.10 becomes a single point. In Γ_2, $P_2 = P_m$ and $P_1 = \frac{\alpha G_2 P_m}{(1-\alpha)G_1}$ with $\alpha \in [0, \alpha_A^*]$. In Γ_1, $P_1 = P_m$ and $P_2 = \frac{(1-\alpha)G_1 P_m}{\alpha G_2}$ with $\alpha \in [\alpha_B^*, 1]$. From (5.40), the Pareto-optimal strategy is to have the same per unit bandwidth receive power for all users, which is illustrated in Figure 3.6(b) and similar to CDMA uplink power control.

In summary, in the interference limited regime, the optimal choice of Q_1, Q_2 in the cross interference model resides on the same solid line of the perfect model shown in Figure 5.8. In the power limited regime, in the perfect model the optimal choice is only one point $Q_1 = Q_{m,1}, Q_2 = Q_{m,2}$; in the cross interference model, however, the optimal choice in addition resides on lines $0 \leq Q_1 \leq Q_{m,1}, Q_2 = Q_{m,2}$ and $Q_1 = Q_{m,1}, 0 \leq Q_2 \leq Q_{m,2}$, which are shown as the two dash-dot lines in Figure 5.8.

The same idea is applicable to characterize the ICI impact in the transition regime (regime II or III). Specifically, recall that in the perfect model, the optimal choice of Q_1, Q_2 resides on the solid line in Figure 5.8. In the cross interference model, the optimal choice in addition resides on the dash-dot lines. The Pareto-boundary can be obtained by varying β_i along the above choice of Q_1, Q_2 and meanwhile α_1 from 0 to 1 and letting $\alpha_2 = 1 - \alpha_1$.

5.1.3 Orthogonal multiplexing in the self-noise model

Recall that in the self-noise model (see Section 2.5), the desired power of a received signal is reduced by a factor of $1 - \eta$, and the remaining η fraction becomes intra-sector interference to the signal itself. When multiple signals are multiplexed, self-noise does not spill over from one signal to another. In this section, we study the impact of self-noise on optimal user multiplexing schemes.

Downlink multiplexing

The downlink achieved rates in FDM in the self-noise model are given by

$$R_i = \alpha_i \log\left(1 + \frac{\beta_i(1-\eta)G_i P_m}{\alpha_i \sigma_i^2 + \beta_i \eta G_i P_m}\right), i = 1, 2. \quad (5.41)$$

Comparison of (5.41) and (5.28) shows that the only difference in the rate expression is that the interference component is proportional to α_i in the cross interference model and proportional to β_i in the self-noise model. From cross interference to self-noise, the interference component of the superior users reduces by a factor of $\frac{\beta_2}{\alpha_2}$, a number much smaller than 1 according to Figure 5.3. Meanwhile, the interference component for the inferior users increases by a factor of $\frac{\beta_1}{\alpha_1}$; however, its contribution to the overall SINR does not increase much as long as $\eta \tilde{\gamma}_1 \ll 1$. Recall from (5.30) that cross interference has negligible impact on the inferior users but saturates the SINR of the superior users. Thus we conjecture that compared with cross interference, self-noise would have a much smaller adverse impact on user multiplexing gain.

The above conjecture is confirmed in Figure 5.11(a) with $\eta = \epsilon$, which shows that when $\tilde{\gamma}_2$ is 13 dB, not as high as $\frac{1-\epsilon}{\epsilon} = 20$ dB, the performance difference between the cross interference and self-noise models is small. As $\tilde{\gamma}_2$ increases to beyond $\frac{1-\epsilon}{\epsilon}$, the Pareto-boundary in the cross interference model saturates; note that there is not much difference between $\tilde{\gamma}_2 = 23$ and 33 dB. However, the saturation effect is not noticeable in the self-noise model, where the Pareto-boundary exhibits remarkable improvement from $\tilde{\gamma}_2 = 13, 23$ to 33 dB.

Uplink multiplexing

In the interference limited regime (regime I in Figure 5.8), the expression of the achieved rates in FDM in the self-noise model is the same as in (5.41), except that P_m is replaced with $\frac{Q_m}{V_i}$ and σ_i^2 with σ^2. The range of α_i, β_i is [0, 1]. Therefore, the effect of self-noise is similar to the downlink case.

5.1 Orthogonal multiplexing

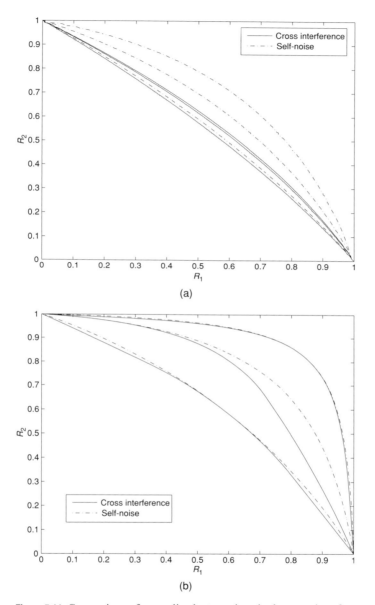

Figure 5.11 Comparison of normalized rate regions in the cross interference and self-noise models, downlink in (a) and uplink in (b). $\eta = \epsilon = 10^{-2}$. In (a), $\tilde{\gamma}_1$ is fixed to be -2 dB and $\tilde{\gamma}_2$ varies such that $\tilde{\gamma}_2/\tilde{\gamma}_1$ is equal to 15, 25, and 35 dB and the corresponding curves are from the left to the right in the figure. In (b), three groups of curves, from the left to the right, are $(\tilde{\gamma}_1', \tilde{\gamma}_2') = (13, 13)$, $(-2, 13)$, and $(-2, -2)$ dB, respectively.

Now consider the power limited regime (regime IV). The self-noise generated by one user does not leak to the received signal from the other user. Unlike in the case in the cross interference model, neither user needs to back off from the maximum transmit power to benefit the rate of the other user. The optimal scheme is therefore the same as in the perfect model where the superior and inferior users both transmit at the maximum power:

$$R_i = \alpha_i \log\left(1 + \frac{(1-\eta)G_i P_m}{\alpha_i \sigma^2 + \eta G_i P_m}\right), i = 1, 2. \quad (5.42)$$

Similar to the downlink, we are interested in how user multiplexing gain would be different in the cross interference or self-noise model. Figure 5.11(b) compares the normalized Pareto-boundary of three representative scenarios, all in the power limited regime, and finds that when both users are in the high (13 dB, 13 dB) or low SINR (-2 dB, -2 dB) regime, the performance difference between the two models is negligible. However, when the superior user is in the high SINR regime (13 dB) and the inferior user in the low SINR regime (-2 dB), the Pareto-boundary in the self-noise model significantly dominates, especially in the regime where the R_1 versus R_2 tradeoff favors R_1. In that regime, $P_1 = P_m$, and α_1 is large. In the cross interference model, the inferior user sees a large fraction of the total ICI power. To protect R_1, P_2 has to back off from P_m to reduce cross interference; R_2 decreases as a result. On the other hand, in the self-noise model, P_2 needs not back off from P_m and the intra-sector interference seen by the inferior user is smaller. Therefore, R_1 and R_2 are both better off in the self-noise model.

The self-noise impact in the transition regime (regime II or III) can be characterized similar to regime I, except that the range of β_i is limited as defined in the perfect model in Section 5.1.1.

5.2 Non-orthogonal multiplexing

In the preceding section, we have studied bandwidth and power allocation in the case of orthogonal multiplexing. The term "orthogonal" in OFDMA does not mean that orthogonal multiplexing is the only available choice. Indeed, the signals of different users can share the same tone-symbols, in which case the signals are no longer orthogonal with each other. This is referred to as non-orthogonal multiplexing and is the topic of this section.

Consider two users to be multiplexed. Let i be the user index, $i = 1, 2$. Superposition coding is an example of non-orthogonal multiplexing in the downlink. In classic superposition coding, in every tone-symbol, the transmitted complex symbol is given by

$$X = X_1 + X_2, \quad (5.43)$$

where X_1 and X_2 are the complex modulation symbols that the base station intends to send to users 1 and 2, respectively, in that tone-symbol, as illustration in Figure 5.12.

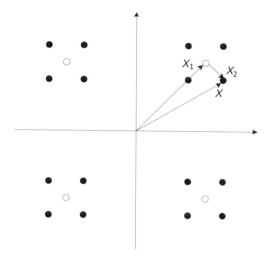

Figure 5.12 Illustration of encoding in the superposition coding scheme. X_1 is obtained from a large QPSK constellation and X_2 is from a small QPSK constellation and superposed on X_1 to generate the transmitted symbol X.

The symbol received at user i is thus

$$Y_i = H_i(X_1 + X_2) + W_i, \qquad (5.44)$$

where W_i is the additive noise and H_i the channel response coefficients of the wireless channel from user i.

The above superposition coding idea can be similarly applied in the uplink where both users transmit complex symbols in the same tone-symbol. The two signals X_1, X_2 are "superposed" in the air and the symbol received at the base station is given by

$$Y = H_1 X_1 + H_2 X_2 + W. \qquad (5.45)$$

From this reason, we loosely call the uplink scheme "superposition coding" as well, although neither transmitter does superposition coding.

By allowing both users to access the entire bandwidth, non-orthogonal multiplexing has bandwidth advantage over orthogonal multiplexing, but at the cost of additional interference between two users. Which scheme would perform better? In theory, assuming the perfect model, non-orthogonal multiplexing using classic superposition coding, together with successive interference cancellation (SIC) at the receiver as described below, is generally superior and indeed optimal in the sense that it achieves the capacity [159]. However, how would classic superposition coding perform in the presence of real-world impairments? We will start with the perfect model to see what is the key to the optimality of classic superposition coding, then compare superposition coding with orthogonal multiplexing in the cross interference and self-noise models, and finally present a new superposition coding scheme that is robust in the presence of self-noise.

5.2.1 Non-orthogonal multiplexing in the perfect model

As in Section 5.1, we consider the two-user case with an inferior user and a superior user, where $G_1/\sigma_1^2 \leq G_2/\sigma_2^2$ in the downlink or $G_1 \leq G_2$ in the uplink. To understand the bandwidth versus interference tradeoff in non-orthogonal multiplexing, we next examine the rate region.

Downlink multiplexing

Consider the downlink first. Similar to Section 5.1.1, suppose that the total transmit power P_m is shared between the inferior and superior users: $\beta_i P_m$ for user i, where $\beta_i \in [0, 1]$, $\beta_1 + \beta_2 = 1$. Unlike orthogonal multiplexing, X_1 and X_2 both occupy the entire bandwidth. How should user i recover X_i from Y_i in (5.44)?

The inferior user treats signal X_2 as noise, and the achievable rate is given by

$$R_1 = \log\left(1 + \frac{\beta_1 P_m G_1}{\sigma_1^2 + \beta_2 P_m G_1}\right)$$

$$= \log\left(1 + \frac{\beta_1 \tilde{\gamma}_1}{1 + \beta_2 \tilde{\gamma}_1}\right), \tag{5.46}$$

where $\tilde{\gamma}_i$ is defined in (5.1).

Similarly the superior user could treat signal X_1 as noise as well, in which case the rate would be given by

$$R_2 = \log\left(1 + \frac{\beta_2 \tilde{\gamma}_2}{1 + \beta_1 \tilde{\gamma}_2}\right). \tag{5.47}$$

However, the superior user has a better decoding strategy, called successive interference cancellation (SIC). It first decodes X_1 just like user 1. Since $\tilde{\gamma}_1 \leq \tilde{\gamma}_2$, the superior user can reliably decode X_1. It then reconstructs the received signal $H_2 \hat{X}_1$ from decoded $\hat{X}_1$, subtracts it from Y_2, and finally decodes X_2. This way, the interference from X_1 is eliminated, and the achievable rate now increases to

$$R_2 = \log(1 + \beta_2 \tilde{\gamma}_2). \tag{5.48}$$

Figure 5.13 plots the Pareto-boundary of the rate region of the superposition coding scheme with or without SIC, and compares it with that of orthogonal multiplexing for different nominal SINR conditions, with two superior users in (a), two inferior users in (b), and mixed superior and inferior users in (c). From (a) and (b), when the nominal SINRs are the same, the superposition coding scheme with SIC has the same performance as orthogonal multiplexing. However, when the nominal SINRs are different, as in (c), the superposition coding scheme is strictly superior, except for the two end points. The performance gap grows drastically as the disparity in nominal SINR increases. Without SIC, however, superposition coding can be much inferior to orthogonal multiplexing.

Uplink multiplexing

Next, consider the uplink case. The constraints are the same as in (5.18) and (5.19) in Section 5.1.1. As before, user 1 is inferior and user 2 superior. Define $\beta_i = \frac{Q_i}{Q_m}$. Unlike

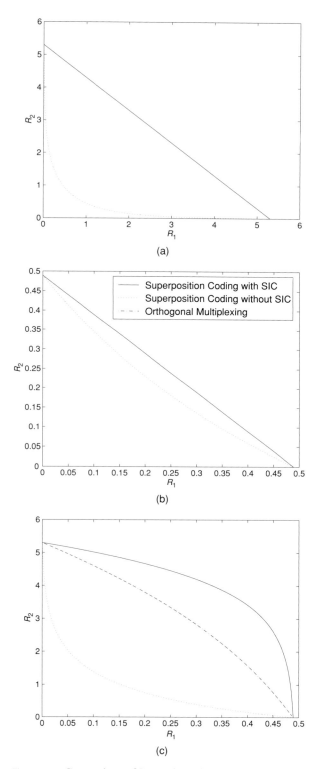

Figure 5.13 Comparison of Pareto-boundary of rate region: superposition coding with or without SIC and orthogonal multiplexing in the downlink. Three multiplexing scenarios are plotted: two superior users in (a), two inferior users in (b), and mixed superior and inferior users in (c). $\tilde{\gamma}_1 = -2$ dB and $\tilde{\gamma}_1 = 23$ dB.

orthogonal multiplexing, X_1 and X_2 both occupy the entire bandwidth. We next recover X_i from Y in (5.45).

The base station receiver can first decode X_1 treating X_2 as noise, and then decode X_2 after the decoded X_1 has been canceled out. The achieved rates are given as follows:

$$R_1 = \log\left(1 + \frac{G_1 P_1}{\sigma^2 + G_2 P_2}\right)$$

$$= \log\left(1 + \frac{\beta_1 \tilde{\gamma}_1}{1 + \beta_2 \tilde{\gamma}_2}\right) \tag{5.49}$$

$$R_2 = \log(1 + \beta_2 \tilde{\gamma}_2), \tag{5.50}$$

where $\tilde{\gamma}_i$ is defined in (5.21). Alternatively, the base station receiver can change the order of decoding – first X_2 and then X_1 – and the achieved rates are

$$R_1 = \log(1 + \beta_1 \tilde{\gamma}_1) \tag{5.51}$$

$$R_2 = \log\left(1 + \frac{\beta_2 \tilde{\gamma}_2}{1 + \beta_1 \tilde{\gamma}_1}\right). \tag{5.52}$$

Similar to orthogonal multiplexing, there are four operating regimes of Q_m, $Q_{m,1}$, $Q_{m,2}$ shown in Figure 5.8.

In the interference limited regime (regime I), $\beta_i \in [0, 1]$ and $\beta_1 + \beta_2 = 1$. It can be shown that the Pareto-boundary is the rate pair curve defined by (5.51) and (5.52), because it strictly dominates that of (5.49) and (5.50) except for the two end points with $\beta_i = 0, 1$. The optimal decoding order at base station receiver is therefore first X_2 and then X_1. Note the similarity and difference between (5.51) (5.52) in the uplink and between (5.46) (5.48) in the downlink. Figure 5.14(a) plots the Pareto-boundaries of the superposition coding scheme and orthogonal multiplexing in the interference limited regime. As a comparison, the rate pair curve of (5.49) and (5.50) is also plotted. Note that the rate pair curve of (5.51) and (5.52) is convex and dominates that of (5.49) and (5.50), which is concave.

In the transition regime (regime II or III) where both the interference and transmit power constraints take effect. Although $\beta_1 + \beta_2 = 1$ still holds, the range of β_i is only a strict subset of $[0, 1]$. As a result, (5.51) and (5.52) do not completely dominate (5.49) and (5.50). Indeed, it can be shown that the Pareto-boundary is the convex hull of the rate pair curve defined by (5.51) and (5.52), shown as the solid curve in Figure 5.14(b), and the end point defined by (5.49) and (5.50) at $\beta_1 = 1 - \frac{Q_{m,2}}{Q_m}$, shown as a square. The rate pair curve of (5.49) and (5.50), shown as the dotted line, does not contribute to the Pareto-boundary except for that square end point. The optimal decoding order depends on the tradeoff between R_1 and R_2. Specifically, on the solid curve, first X_2 and then X_1; on the dash-dot line, time-share between the two alternative decoding orders.

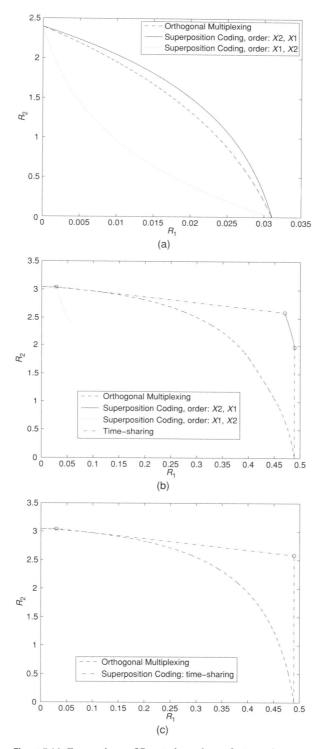

Figure 5.14 Comparison of Pareto-boundary of rate region: superposition coding and orthogonal multiplexing in the uplink. A superior user is multiplexed with an inferior user. $\tilde{\gamma}'_1 = -2$ dB, $\tilde{\gamma}'_2 = 13$ dB, $V_1 = 10$, $V_2 = 1$. $Q_m = 0.5P_m$, $10.5P_m$, $15P_m$ in (a), (b), (c), respectively, for the interference limited, transition, and power limited regimes, respectively.

Finally, in the power limited regime (regime IV), $\beta_1 + \beta_2 < 1$. Both users should transmit at the maximum power P_m. Equations (5.49) and (5.50) reduce to one end point

$$R_1 = \log\left(1 + \frac{G_1 P_m}{\sigma^2 + G_2 P_m}\right) \tag{5.53}$$

$$R_2 = \log\left(1 + \frac{G_2 P_m}{\sigma^2}\right), \tag{5.54}$$

and (5.51) and (5.52) reduce to another end point

$$R_1 = \log\left(1 + \frac{G_1 P_m}{\sigma^2}\right) \tag{5.55}$$

$$R_2 = \log\left(1 + \frac{G_2 P_m}{\sigma^2 + G_1 P_m}\right). \tag{5.56}$$

The two end points are marked as a circle and a square in Figure 5.14(c). Neither point dominates the other. The Pareto-boundary is the convex hull of those two end points, shown as the dash-dot line. A point on the line that connects the two end points is obtained by time sharing between the two alternative decoding orders. From (b) to (c), the solid curve is reduced to the single circle end point.

5.2.2 Non-orthogonal multiplexing in the cross interference and self-noise models

The above study shows that interference cancellation is the key to the optimality of classic superposition coding. Assuming perfect SIC, superposition coding is indeed capacity-achieving. However, without SIC superposition coding suffers from excessive interference between the two users and may even be inferior to orthogonal multiplexing. In the real world, interference cancellation is never perfect. We need to assess the robustness of the optimality of the superposition coding scheme in the presence of residual errors due to imperfect SIC. Similar to Section 5.1, we next consider the cross interference and self-noise model.

Downlink multiplexing
Since the signals in superposition coding occupy the entire bandwidth, the cross interference or self-noise model makes no difference. The rate expressions are given by

$$R_1 = \log\left(1 + \frac{(1-\epsilon)\beta_1 G_1 P_m}{\sigma_1^2 + \beta_1 \epsilon G_1 P_m + \beta_2 G_1 P_m}\right)$$

$$= \log\left(1 + \frac{\beta_1 \tilde{\gamma}_{1,CI}}{1 + \beta_2 \tilde{\gamma}_{1,CI}}\right) \tag{5.57}$$

$$R_2 = \log\left(1 + \frac{(1-\epsilon)\beta_1 G_2 P_m}{\sigma_2^2 + \epsilon G_2 P_m}\right)$$

$$= \log(1 + \beta_2 \tilde{\gamma}_{2,CI}), \tag{5.58}$$

where $\tilde{\gamma}_{i,CI}$ is defined in (5.29). In the self-noise model, ϵ is replaced by η.

Comparison of the above rate expressions with (5.46) and (5.48) in the perfect model shows that the only difference is $\tilde{\gamma}_{i,CI}$ versus $\tilde{\gamma}_i$. We have seen a similar comparison in the orthogonal multiplexing case between the perfect and cross interference model in Section 5.1.2, and can draw a similar conclusion. In the cross interference model, the superposition coding scheme outperforms orthogonal multiplexing. However, the performance gap between the two may become narrower as cross interference ϵ reduces SINR disparity.

The comparison is quite different in the self-noise model. To elaborate, compare the post-SIC residual interference caused by self-noise at the superior user in (5.58) and (5.41). In superposition coding, the residual interference $\eta P_m G_2$ at the superior user is proportional to the total transmit power P_m. In orthogonal multiplexing, the residual interference $\eta \beta_2 G_2 P_m$ only depends on the power allocated to the superior user. When self-noise dominates thermal noise ($\eta \tilde{\gamma}_2 \gg 1$), the ratio of the interference power in orthogonal multiplexing versus in superposition coding is given by

$$\frac{\alpha_2 + \eta \beta_2 \tilde{\gamma}_2}{1 + \eta \tilde{\gamma}_2} \ll 1. \tag{5.59}$$

The above inequality holds because when the inferior user is power limited and the superior user is bandwidth limited, the desired operating point is $\alpha_2 \approx 1$ and $\beta_2 \ll 1$ (most of the power is allocated to the inferior user and most of the bandwidth is allocated to the superior user). Thus, the residual interference is much smaller in orthogonal multiplexing. The difference can be large enough to compensate for the bandwidth loss in orthogonal multiplexing so that orthogonal multiplexing may even perform better than superposition coding.

This conjecture is confirmed by Figure 5.15(a) where we compare the Pareto-boundary of superposition coding and orthogonal multiplexing and show that in the self-noise model, superposition coding no longer always dominates orthogonal multiplexing. With $\eta = 0.01$, superposition coding is superior to orthogonal multiplexing. As η increases to 0.05, the two curves cross over.

Uplink multiplexing

Next extend the uplink study in the perfect model to include cross interference and self-noise. Consider the cross interference model first. Suppose that the decoding order is first X_2 and then X_1. The rate expressions are given by

$$R_1 = \log\left(1 + \frac{(1-\epsilon)G_1 P_1}{\sigma^2 + \epsilon G_1 P_1 + \epsilon G_2 P_2}\right)$$

$$= \log\left(1 + \frac{(1-\epsilon)\beta_1 \tilde{\gamma}_1}{1 + \epsilon \beta_1 \tilde{\gamma}_1 + \epsilon \beta_2 \tilde{\gamma}_2}\right) \tag{5.60}$$

$$R_2 = \log\left(1 + \frac{(1-\epsilon)G_2 P_2}{\sigma^2 + G_1 P_1 + \epsilon G_2 P_2}\right)$$

$$= \log\left(1 + \frac{(1-\epsilon)\beta_2 \tilde{\gamma}_2}{1 + \beta_1 \tilde{\gamma}_1 + \epsilon \beta_2 \tilde{\gamma}_2}\right). \tag{5.61}$$

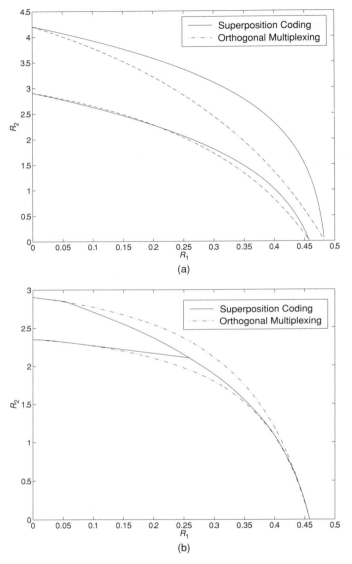

Figure 5.15 Comparison of Pareto-boundary of superposition coding and orthogonal multiplexing in the self-noise model. The downlink case is shown in (a) with $\tilde{\gamma}_1 = -2$ dB and $\tilde{\gamma}_2 = 23$ dB. The lower curves are with $\eta = 0.05$ and the upper ones are with $\eta = 0.01$. The uplink case of the power limited regime is shown in (b) with $\eta = 0.05$ and $\tilde{\gamma}'_1 = -2$ dB. The lower curves are $\tilde{\gamma}'_2 = 13$ dB and the upper ones are with $\tilde{\gamma}'_2 = 23$ dB.

Define β'_i as in (5.32). It can be shown that the above rate expressions (5.60) and (5.61) are equivalent to (5.51) and (5.52), respectively, with β_i replaced by β'_i and $\tilde{\gamma}_{i,CI}$ defined in (5.34). Hence, in the interference limited regime, the study in the perfect model is applicable here, and the effect of cross interference ϵ on the nominal SINR and therefore user multiplexing gain is similar to the downlink case studied earlier in this section.

The power limited regime can be investigated similarly to the study of orthogonal multiplexing from the perfect to the cross interference model. Specifically, recall that in the perfect model the Pareto-boundary is the convex hull of two end points given in (5.53) to (5.56). At the two end points, both users transmit at the maximum power P_m. The only difference between them is the decoding order. In the cross interference model, the Pareto-boundary still includes a line segment, which is obtained by time sharing two end points at which both users still transmit at P_m. Due to cross interference, the rates achieved at the two end points are lower than those in the perfect model. That line segment corresponds to the middle portion shown in Figure 5.10(a). From the study of orthogonal multiplexing in the cross interference model, the Pareto-boundary also includes two additional portions corresponding to $P_1 \in [0, P_m]$, $P_2 = P_m$ and $P_1 = P_m$, $P_2 \in [0, P_m]$, respectively. The former portion is connected with the line segment at one of the two end points whose decoding order is first X_1 and then X_2. The optimal decoding order in that portion is the same as the corresponding end point. The latter portion is connected with the segment at the other end point and the optimal decoding order reverses.

The transition regime can be studied similarly. Finally, we note that as in the perfect model, the superposition coding outperforms orthogonal multiplexing the cross interference model in all the operating regimes.

Next consider the self-noise model. The rate expressions are the same as (5.60) and (5.61) in the self-noise model with ϵ replaced by η. Hence, the Pareto-boundary of superposition coding is the same in either model.

In the interference limited regime, the impact of self-noise η is the same as in the downlink self-noise case studied earlier in this section. In particular, unlike in the cross interference model where superposition coding always outperforms orthogonal multiplexing, in the self-noise model orthogonal multiplexing may be superior in certain operating regimes.

The same can be said in the power limited regime as well. Figure 5.15(b) compares the Pareto-boundary of superposition coding and orthogonal multiplexing. Given $\eta = 0.05$, superposition coding is superior when the SINR disparity is 15 dB. As the disparity increases to 25 dB, the two curves cross over and orthogonal multiplexing outperforms in the right portion of the curves. The reason is that in superposition coding as P_2 increases from 0 to P_m, the residual interference from subtracting the superior user's signal grows significantly because $G_2 \gg G_1$, thereby limiting R_1 similar to the downlink case. On the other hand, in orthogonal multiplexing, because self-noise does not leak between the inferior and superior users, R_1 does not suffer from the self-noise caused by the superior user.

5.2.3 Superposition-by-position coding

So far we have learned that classic superposition coding, combined with successive interference cancellation (SIC) at the receiver, is *capacity-achieving* in the perfect model, which can significantly outperform orthogonal multiplexing. Furthermore, in the presence of cross interference, although the performance degrades, superposition coding still

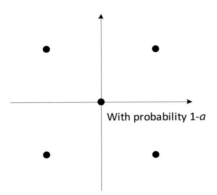

Figure 5.16 Illustration of QPSK$^+$ signaling scheme.

outperforms orthogonal multiplexing. However, this is no longer true when self-noise dominates, because in superposition coding post-SIC residual interference spreads over the entire bandwidth. In orthogonal multiplexing, users occupy disjoint subsets of tones and interference cancellation is thus not required. As a result, self-noise does not lead to interference leakage between users in orthogonal multiplexing. We have demonstrated that superposition coding and SIC can actually perform worse than simple orthogonal multiplexing when self-noise is present, despite the extra complexity required in coding and decoding.

Can we design a coding scheme that retains the benefit of superposition coding and yet is immune to self-noise? The superposition-by-position coding scheme proposed in [62] and introduced below is such a scheme that works in the downlink. The idea is that rather than spreading the power allocated to the inferior user to all the tones, it is concentrated on a small subset of tones. To the superior user, the residual interference caused by self-noise is limited to those tones only. To the inferior user, concentrating the power will not noticeably degrade the capacity when it operates in the low SNR regime.

To this end, superposition-by-position coding employs a signaling scheme where a fraction of symbols are zero:

$$X_1 = \begin{cases} 0, & \text{with probability } 1 - \alpha \\ \text{non-zero complex symbol}, & \text{otherwise.} \end{cases} \quad (5.62)$$

When $X_1 = 0$, it is called a zero symbol. The signal X_2 to the superior user is sent in the subset of tones where zero symbols are used in X_1. The non-zero complex symbols can be discrete points in a constellation such as QPSK. The signaling scheme is referred to as QPSK$^+$ scheme. Figure 5.16 illustrates the QPSK$^+$ signaling scheme, where a symbol is either 0 or one of the four QPSK constellation points.

Note that in the AWGN channel the *information-theoretic* optimal input signal follows a Gaussian distribution. QPSK constellation is a quite rough approximation of a Gaussian distribution. As an alternative to QPSK$^+$, the non-zero complex symbols can follow a Gaussian distribution and the signaling scheme is referred to as Gaussian$^+$. However, it

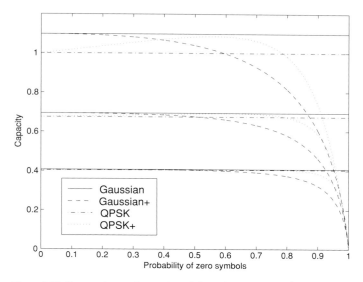

Figure 5.17 Comparison of capacity of Gaussian, Gaussian$^+$, QPSK, and QPSK$^+$ in the AWGN channel. The upper group of plots is for SNR $=3$ dB, the middle group is for SNR $=0$ dB, and the lower group is for SNR $=-3$ dB.

should be noted that both Gaussian and Gaussian$^+$ input signaling are not practical in real systems.

Dedicating a fraction of symbols to be zero in (5.62) makes the input signal, either QPSK$^+$ or Gaussian$^+$, deviate from the optimal Gaussian distribution. However, when the SNR is not high, the degradation is insignificant. Indeed, it is shown in [166] that all these signaling schemes are capacity-achieving in the low SNR limit, that is, $\gamma \to 0$. Figure 5.17 compares the capacity of Gaussian$^+$ or QPSK$^+$ versus Gaussian and QPSK as a function of the probability of zero symbols under different SNR condition. We observe that

- As the probability of zero symbols $1 - \alpha$ increases to 1, the capacity of Gaussian$^+$ or QPSK$^+$ drops to 0. However, it is remarkable that the capacity of Gaussian$^+$ or QPSK$^+$ does not degrade much from the Gaussian case over a wide range of $1 - \alpha$. This is particularly true at low SNR (-3 dB).
- It is interesting to note that the capacity of QPSK$^+$ increases slightly and exceeds that of QPSK as the probability of zero symbols increases over a quite wide range. This is because with a proper fraction of symbols being zero, QPSK$^+$ is a closer approximation to Gaussian than QPSK is. For the same reason, QPSK$^+$ outperforms Gaussian$^+$ where the probability of zero symbols is large.

Superposition-by-position coding bears a resemblance to orthogonal multiplexing in the sense that the energy of X_1 and X_2 concentrates on two disjoint subsets of tones. The probability of zero symbols $1 - \alpha$ controls the bandwidth partition of the two subsets. Indeed, from the perspective of the superior user, it is allocated $\alpha_2 = 1 - \alpha$ bandwidth,

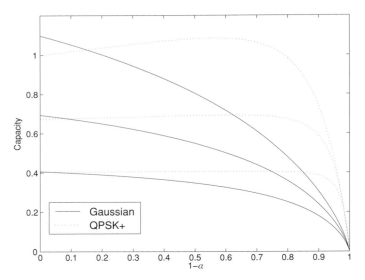

Figure 5.18 Comparison of R_1: Gaussian orthogonal multiplexing versus QPSK$^+$. The upper group of plots is for SNR $=3$ dB, the middle group is for SNR $=0$ dB, and the lower group is for SNR $=-3$ dB.

not different from orthogonal multiplexing. When the superior user is bandwidth limited and the inferior user is power limited, it is desirable that $1-\alpha$ be close to 1, in which case QPSK$^+$ outperforms Gaussian$^+$ according to Figure 5.17. In the following, we will use QPSK$^+$ for X_1 instead of Gaussian$^+$.

The superior user receiver first decodes X_1, and then decodes X_2 from the tones in which the decoded symbols of X_1 are zero symbols. As long as the receiver can correctly identify the tones of zero symbols in X_1, there is no residual interference leaking from X_1 to X_2. Hence, the rate R_2 is same as that in the orthogonal multiplexing case (5.41).

The key difference between superposition-by-position coding and orthogonal multiplexing lies in the signaling scheme of X_1. The rate R_1 is greater in superposition-by-position, because the information to the inferior user is conveyed not only in the non-zero symbols using phases/amplitudes (same as in orthogonal multiplexing), but also in the *positions* of the zero symbols. Using the zero symbols means that information is conveyed in the *positions* of the zero symbols. In other words, which subset of symbols are used as zero symbols carries information. The scheme of (5.62) can be viewed as a position-based signaling scheme.

To be more precise, we compare the following two schemes of constructing X_1 with the same total signal power budget P_m. The first scheme is Gaussian scheme in orthogonal multiplexing with bandwidth allocation fraction equal to α. The second scheme is QPSK$^+$ over the entire bandwidth and the fraction of non-zero symbols is equal to α. Note that the available bandwidth left for X_2 remains the same in either scheme.

Figure 5.18 compares the capacity of the two schemes as a function of $1-\alpha$ under different SNR condition. The Gaussian orthogonal multiplexing scheme outperforms

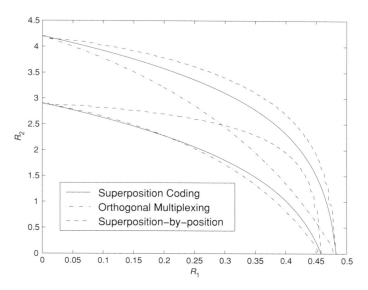

Figure 5.19 Comparison of downlink Pareto-boundary of superposition coding, superposition-by-position coding, and orthogonal multiplexing in the self-noise model. The upper set of three curves is with $\eta = 0.01$ and the lower set is with $\eta = 0.05$. $\tilde{\gamma}_1 = -2$ dB and $\tilde{\gamma}_2 = 23$ dB. The curves of superposition coding and orthogonal multiplexing are the same as in Figure 5.15(a).

only when $1 - \alpha$ is very small. As $1 - \alpha$ increases, the QPSK$^+$ scheme becomes much superior. The reason is that although the energy of X_1 only occupies a fraction of bandwidth α, the coding space of X_1 covers the entire bandwidth. Unlike in orthogonal multiplexing, QPSK$^+$ does not suffer from the penalty of bandwidth reduction α for a large range of $1 - \alpha$.

We now calculate the rate R_1. The received signal at the inferior user is given by

$$Y_1 = \begin{cases} \sqrt{1-\eta}H_1 X_1' + \sqrt{\eta}I_{1,1} + W_1, & \text{with probability } \alpha \\ \sqrt{1-\eta}H_1 X_2 + \sqrt{\eta}I_{1,2} + W_1, & \text{otherwise.} \end{cases} \quad (5.63)$$

Here, X_1' represents the non-zero symbols of X_1 and W_1 zero-mean complex Gaussian noise. We assume that the self-noise caused by X_1' is $\sqrt{\eta}I_{1,1} \sim \mathcal{CN}(0, \eta\beta_1 G_1 P_m)$, and the self-noise caused by X_2 is $\sqrt{\eta}I_{1,2} \sim \mathcal{CN}(0, \eta\beta_2 G_1 P_m)$, where β_1, β_2 are the fractions of power budget P_m allocated to X_1', X_2, respectively. The rate R_1 can be determined via the mutual information between X_1 and Y_1:

$$R_1 = I(X_1, Y_1) = h(Y_1) - h(Y_1|X_1), \quad (5.64)$$

where $h(Y_1)$ and $h(Y_1|X_1)$ are differential entropy of Y_1 and conditional differential entropy of Y_1 given X_1.

Figure 5.19 plots the Pareto-boundary of classic superposition coding, superposition-by-position coding, and orthogonal multiplexing. When the self-noise is modest ($\eta = 0.01$), superposition coding is superior to orthogonal multiplexing but slightly

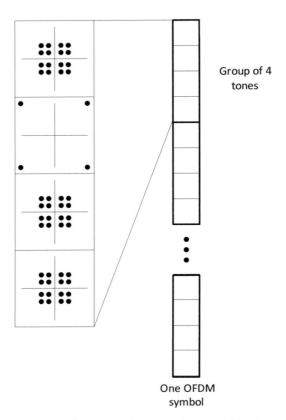

Figure 5.20 Illustration of a practical superposition-by-position coding scheme.

inferior to superposition-by-position coding. When the self-noise is severe ($\eta = 0.05$), the superposition-by-position coding scheme remarkably outperforms the other two schemes.

Strict use of the position-based signaling scheme (5.62) has certain drawbacks in practice. For example, if the codeword is spread over several OFDM symbols, then the distribution of the non-zero symbols may not be even over all the OFDM symbols, thereby causing fluctuation in the transmit power and causing certain power allocation β_i not feasible given the constraint of peak transmit power in every OFDM symbol. In addition, the decoding algorithm of the position information may be complex. A practical solution is as follows.

Assume that $\alpha = 1/4$ and that the number of tones in an OFDM symbol N_c is a multiple of 4. We partition the N_c tones in groups of size 4. In each group, one tone carries a non-zero symbol for X_1 and the other three tones are zero symbols. The non-zero symbol is a QPSK symbol. Hence, the position of the non-zero symbol carries 2 bits and the QPSK phase carries another 2 bits. The symbols of X_2 are transmitted in the remaining three tones. The scheme is illustrated in Figure 5.20, where X_2 is shown to use the 16-QAM constellation. The relative amplitudes of QPSK and 16-QAM constellations depend on the power allocation β_i.

Another benefit of this superposition-by-position coding scheme is that once a superior user decodes the signal intended for the inferior users, those QPSK symbols can be used as additional pilots to decode the remaining 16-QAM symbols. The QPSK symbols are usually denser than a regular pilot (the density equal to 25 percent in Figure 5.20 versus around 10 percent for a typical pilot), thereby helping enhance channel estimation.

So far we have studied the use of superposition-by-position coding in the downlink. The scenario is quite different in the uplink, because while one user can use the position-based signaling scheme and leave a fraction of "clean" tones to be used by the other user, the other user does not know where the clean tones are since their positions are not determined beforehand. One possible scheme is for the other user to spread its signal energy uniformly across all the tones and expect that the information is delivered only via the clean tones. However, the rate is probably lower than that with simple orthogonal multiplexing, since the signal energy transmitted in the other tones is not only wasted but also interferes with the position-based signal. Overall, such a scheme is likely to perform worse than orthogonal multiplexing in the uplink.

5.3 Inter-sector interference management

So far we have considered user multiplexing in a single sector. We next expand the study to multiple sectors of a cell.

5.3.1 Sectorization

Sectorization is widely used in cellular systems to reduce interference and improve cell capacity. The idea is to divide a cell into multiplex sectors, each of which is covered with a directional antenna, called a sector antenna. Sector antennas of a cell are co-located but have different orientations, concentrating signal transmission and reception towards the sectors of interest.

Figure 5.21 illustrates the pattern of a sector antenna showing the antenna gain in any direction. The beamwidth of the sector antenna is just wide enough to cover a 120° sector. The main lobe is relatively flat within the sector and decays quickly near the sector edges. The side and back lobes are attenuated. Three sector antennas deployed 120° offset with each other divide the cell into three sectors where the sector edges are marked as thick lines in the figure.

If sectorization is perfect, then the antenna gain is constant in any direction within the sector and drops immediately to zero outside of the sector. The sector antenna in Figure 5.21 does not achieve perfect sectorization. Similar to the cross interference and self-noise models of intra-sector interference, we assume the following simple *inter-sector interference model* to characterize imperfect sectorization for the purpose of system level design. Inter-sector interference is the interference between two sectors of a given cell. In this section, we assume that the two sectors reuse the same bandwidth.

Intra-cell user multiplexing

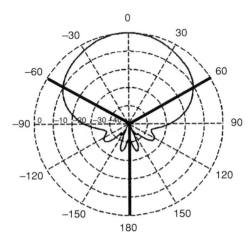

Figure 5.21 Illustration of the coverage of a sector antenna and sector antenna pattern of a three-sector cell.

Depending on its location relative to the sector antennas, a user can be either in the interior of a sector or at the edge between the two sectors. For either sector edge or interior user, the antenna gain from/to its own sector is constant G, the same as in the perfect model.

- For a sector interior user, the antenna gain from/to the interfering sector antenna is equal to $\zeta_0 G$. $\zeta_0 \in (0, 1)$ is called *front-to-back ratio* and characterizes the effect of back and side lobes. This is very different from inter-cell interference in that it affects all sector interior users uniformly irrespective of their locations within the sector.
- For a sector edge user, the antenna gain from/to the interfering sector antenna is location specific and can be as high as G. In this regard, inter-sector interference behaves similarly to inter-cell interference. The inter-sector interference is greater at sector edge than in sector interior.

We first consider the sector interior users in Section 5.3.2 and then include the sector edge users in Section 5.3.3.

5.3.2 Synchronized sectors

If the transmission in the two sectors is not synchronized, then the inter-sector interference is assumed to spread randomly and uniformly across the entire bandwidth. For a given user, the received power of the inter-sector interference is proportional to the bandwidth it occupies. For example, with orthogonal multiplexing in the downlink, the achieved rates are

$$R_i = \alpha_i \log\left(1 + \frac{\beta_i G_i P_m}{\alpha_i \sigma_i^2 + \alpha_i \zeta_0 G_i P_m}\right), \quad (5.65)$$

5.3 Inter-sector interference management

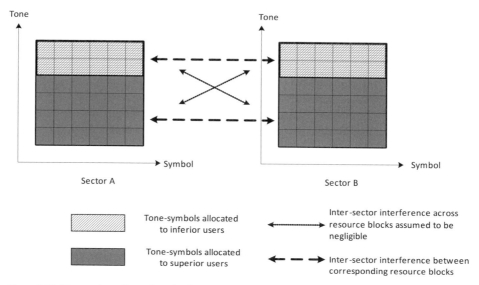

Figure 5.22 Illustration of synchronized sectors and the allocation of tone-symbols to the inferior and superior users. Tone hopping is not shown for simplicity.

where the same total transmit power P_m is assumed in both sectors. Typically, ζ_0 is about -20 dB depending on the local scattering and reflection around the sector antennas of the base station. Comparison with (5.28) shows that the inter-sector interference has the effect similar to the cross interference. The same observation can be made with orthogonal multiplexing in the uplink. From Section 5.1.2, it follows that the inter-sector interference reduces the SINR dynamic range and thus user multiplexing gain.

Fortunately, we can preserve the SINR dynamic range and user multiplexing gain by synchronizing the two sectors. Specifically, in the two sectors, the base station transmitters synchronize their symbol time and tone frequency for the downlink signal, and so do the base station receivers for the uplink signal. Because of the OFDMA orthogonality, a signal sent on any tone-symbol in one sector only causes inter-sector interference to the same tone-symbol in the other sector. Here, the inter-sector interference contributed by the ICI of the signal is considered negligible, since the power drops by a combined factor of $\zeta_0 \epsilon$. In orthogonal multiplexing of the inferior and superior users, the two sectors not only use the same bandwidth allocation α_i, but also allocate the same set of tone-symbols to the inferior (or superior) users. As a result, the inter-sector interference only occurs between the superior users and between the inferior users in the two sectors, and the inferior users in one sector do not interfere with the superior users in the other sector. Moreover, power allocation β_i is also the same or similar in both sectors. Figure 5.22 illustrates the synchronized sectors using orthogonal multiplexing.

For a given user, the received power of the inter-sector interference is proportional to the allocated power. For example, the downlink achieved rates are

$$R_i = \alpha_i \log\left(1 + \frac{\beta_i G_i P_m}{\alpha_i \sigma_i^2 + \beta_i \zeta_0 G_i P_m}\right), \qquad (5.66)$$

which is very similar to (5.41). Hence, in the synchronized sectors, the inter-sector interference now behaves like self-noise rather than cross interference. From Section 5.1.3, user multiplexing gain is greater than with self-noise (5.66) than with cross interference (5.65).

Why does sector synchronization make a difference here? From Section 5.1.1, the key to achieving user multiplexing gain is to allocate greater power per unit bandwidth to the inferior users than to the superior users. In the unsynchronized sectors, the superior users see randomized inter-sector interference from both the inferior and superior users in the other sector, which significantly reduces the dynamic range of the power per unit bandwidth allocation. On the other hand, in the synchronized sectors, the superior users only see inter-sector interference from the power allocated to the superior users in the other sector, which have the same or similar power profile. As a result, the inter-sector interference is not a significant limiting factor to the dynamic range of power allocation.

Superposition coding, including superposition-by-position coding, suffers from the same problem as the unsynchronized sectors. Unlike in the case of orthogonal multiplexing, the inter-sector interference is not confined only between the superior users even if the symbol time and tone frequency are synchronized in the two sectors.

5.3.3 Users at sector edge

Now consider multiplexing both sector interior and edge users. We study only the downlink in this section; the uplink case is similar. User index $i = 1$ or 2 represents the group of edge or interior users, respectively. Assume that among all the edge (or interior) users, the nominal SINR without inter-sector interference $\tilde{\gamma}_i$ is the same and defined in (5.1), and the inter-sector interference leakage is also the same. Denote by ζ_i the factor of inter-sector interference, where $\zeta_2 = \zeta_0$ the front-to-back ratio and ζ_1 can be as high as 0 dB. We are interested in the aggregate data rates R_1, R_2 in both sectors.

To focus on inter-sector user multiplexing, we assume that in either sector the sector interior and edge users are multiplexed in the TDM manner such that we do not have to address the power bandwidth allocation problem of intra-sector multiplexing. In practice, one has to jointly consider both intra-sector and inter-sector multiplexing.

Suppose that the two sectors transmit at the same power P_m. Then the total achieved rates are obtained by time sharing the following two end points of transmitting either to edge or to interior users in both sectors:

$$R_1 = 2\log\left(1 + \frac{\tilde{\gamma}_1}{1 + \zeta_1\tilde{\gamma}_1}\right), \ R_2 = 0 \tag{5.67}$$

$$R_1 = 0, \ R_2 = 2\log\left(1 + \frac{\tilde{\gamma}_2}{1 + \zeta_2\tilde{\gamma}_2}\right). \tag{5.68}$$

The factor is 2 because there are two sectors.

When $\tilde{\gamma}_1$ is large, ζ_1 close to 0 dB severely reduces the effective SINR of the sector edge users. One scheme of boosting R_1 is to coordinate the scheduling between the two sectors such that when one sector transmits to the sector edge users, the other sector transmits to the sector interior users at a reduced power βP_m, with $\beta \in [0, 1]$. The achieved rates in this reduced-power coordinated scheduling are

$$R_1 = \log\left(1 + \frac{\tilde{\gamma}_1}{1 + \zeta_1 \beta \tilde{\gamma}_1}\right), \; R_2 = \log\left(1 + \frac{\beta \tilde{\gamma}_2}{1 + \zeta_2 \tilde{\gamma}_2}\right). \quad (5.69)$$

The Pareto-boundary of the achieved rate pairs is the convex hull of the two end points of (5.67), (5.68) and the curve obtained by varying β in (5.69).

Note that at $\beta = 1$, (5.69) corresponds to the time sharing middle point between the two end points (5.67), (5.68). $\beta = 0$ represents the case of reuse factor 2. Specifically, when one sector is transmitting to the edge user, the other sector leaves the bandwidth unused. The tradeoff is a factor of 2 reduction in bandwidth in exchange of an increase in SINR from $\frac{\tilde{\gamma}_1}{1+\zeta_1\tilde{\gamma}_1}$ to $\tilde{\gamma}_1$. When $\tilde{\gamma}_1$ is large, or more precisely,

$$\log(1 + \tilde{\gamma}_1) > 2\log\left(1 + \frac{\tilde{\gamma}_1}{1 + \zeta_1 \tilde{\gamma}_1}\right), \; \text{or} \quad (5.70)$$

$$\zeta_1 > \frac{1}{\sqrt{1 + \tilde{\gamma}_1} - 1} - \frac{1}{\tilde{\gamma}_1}, \quad (5.71)$$

the tradeoff is beneficial. This is shown in Figure 5.23(a), where $\tilde{\gamma}_1 = \tilde{\gamma}_2 = 10$ dB. The Pareto-optimal scheme is to time share between the end point of (5.68) (transmitting to the interior users in both sectors) and the end point of (5.69) at $\beta = 0$ (transmitting to an edge user in only one of the two sectors). It is easy to check that inequality (5.71) is satisfied in Figure 5.23(a). However, reuse factor equal to 2 ($\beta = 0$) is not always included in the Pareto-optimal scheme. Figure 5.23(b) shows a different example where the Pareto-optimal scheme is to time share between transmitting to either the edge or interior users in both sectors and transmitting to the edge users at the full power in one sector and to the interior users at a reduced power ($\beta > 0$) in the other sector. When the nominal SINR $\tilde{\gamma}_1$ further drops, increasing the SINR of the edge users by using $\beta < 1$ becomes less effective and both sectors should use the full power P_m.

Finally, the following points are worth mentioning.

- In the case of reuse factor equal to 2 ($\beta = 0$), instead of letting one sector to transmit and the other one to be silent, both sectors can transmit the same signal to the edge users. The two signals are mixed over the air and treated as multipath copies at the user receiver. This is similar to softer handoff in CDMA from the perspective of combining signal energy. The difference is that because of sector synchronization and OFDMA, the sector edge user does not see inter-sector interference from other FDM signals.
- Taking a step further, assume that an edge user feeds back its channel state information to the base station. Then the two sectors can act as two transmit antennas of a MIMO channel provided that the user has multiple receive antennas. Comparing with the use of transmit antennas of just one sector to form a MIMO channel, using antennas of two sectors has the advantage of larger angular spread at the transmitter, thereby helping make the channel matrix full rank and well conditioned.
- With synchronized sectors, both sectors use the same way of partitioning tone-symbols into a number of subsets such that the inter-sector interference between any different

Intra-cell user multiplexing

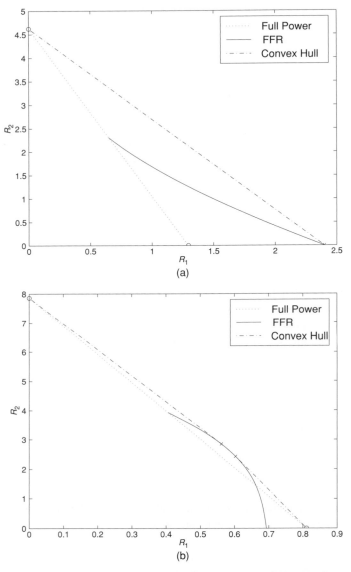

Figure 5.23 Pareto-boundary of inter-sector user multiplexing. The dotted curves labeled as "full power" are obtained by time sharing between two end points of (5.67), (5.68). The solid curves labeled "FFR" are from (5.69) by varying β from 0 to 1. The dash-dot curves are the convex hull. $\zeta_1 = 0$ dB, $\zeta_2 = -20$ dB. In (a), $\tilde{\gamma}_1 = \tilde{\gamma}_2 = 10$ dB. In (b), $\tilde{\gamma}_1 = 0$ dB, $\tilde{\gamma}_2 = 20$ dB.

two subsets is negligible (see Figure 5.22). Hence, different inter-sector user multiplexing schemes can be applied in different subsets. For example, one subset is used to multiplex the interior users with the full transmit power P_m in both sectors, while another subset is used to multiplex the edge users with the full power in one sector and the interior users with a reduced power βP_m in the other sector.

- The idea of coordinating the scheduling between sectors and reducing the transmit power for the interior users can be extended to manage inter-cell interference. In that case, the scheme is called "fractional frequency reuse," which is discussed in Chapter 6.

5.4 Summary of key ideas

- User multiplexing is about sharing power and bandwidth within a cell. In OFDMA, orthogonal and non-orthogonal multiplexing are the two basic forms of user multiplexing.
- In orthogonal multiplexing, FDM is superior to TDM. In the downlink or the interference limited regime of the uplink, the gain is achieved by multiplexing power limited inferior users and bandwidth limited superior users. The greater the SINR disparity, the greater the multiplexing gain. OFDMA orthogonality allows large signal dynamic range, the key to realizing the gain. In the power limited regime of the uplink, the gain is the most significant when inferior users are multiplexed, as FDM pools the power budget together.
- User multiplexing gain degrades in the presence of intra-sector interference because it reduces SINR disparity. Relatively speaking, self-noise is less detrimental than cross interference as intra-sector interference does not spread across different users.
- Without intra-sector interference, classic superposition coding (non-orthogonal multiplexing) is optimal and capacity-achieving. Interference cancellation is the key for classic superposition coding to outperform FDM orthogonal multiplexing.
- When cancellation is imperfect, residual self-noise degrades the data rate such that superposition coding may perform worse than orthogonal multiplexing. Superposition-by-position coding retains the benefit of superposition coding, and yet is robust against self-noise.
- Inter-sector interference due to imperfect antenna sectorization reduces OFDMA signal dynamic range. This issue can be resolved by synchronizing sectors such that inter-sector interference behaves similarly to self-noise rather than cross interference. Coordinated power and bandwidth allocation between sectors improves the rate of sector edge users.

6 Inter-cell interference management

In this chapter, we will study power and bandwidth allocation in a multi-cell scenario where inter-cell interference dominates. As in Chapter 5, we assume that the channel gain does not vary over time or frequency.

Recall from Section 1.1 that spectrum reuse among cells is the key to increasing overall spectrum utilization and that spectrum reuse leads inter-cell interference to be managed. In a conventional cellular deployment, there are two basic tools to manage inter-cell interference.[1] One is *cell planning*, including carefully choosing base station locations and fine tuning antenna patterns to maximize service quality. In an ideal world with homogeneous wireless channel propagation, base stations are placed in the hexagonal grids as shown in Figure 6.1. Practical considerations such as local terrain characteristics affect cell planning choices. The second tool is *handoff*. A user switches to a new base station as it moves across the boundary between two cells. Under so-called unrestricted association where the user is allowed to connect to any base station, handoff ensures that the user is always connected to the "best" base station, which is usually the closest one. As a result, the interference from an adjacent base station does not exceed the desired signal from a serving base station.

In the early generations of the cellular systems, those two tools were deemed insufficient, and a third tool, *frequency reuse*, was introduced to control inter-cell interference [159]. In such a system, the available bandwidth is divided into a number of non-overlapping narrowband frequency channels, and each base station is allocated $1/K$ of the channels such that a channel is reused in every K cells that are sufficiently apart from each other. Parameter K is called the *reuse factor* of the frequency reuse scheme. The scheme is referred to as narrowband FDMA. A narrowband frequency channel can be divided in time so as to be shared among multiple users, in which case the scheme is called FDMA/TDMA. Figure 6.1 shows an example of frequency reuse pattern for $K = 7$.

Signals on the same channel used in different cells interfere with each other. Interference between different channels is usually negligible. Clearly, the greater the frequency reuse factor, the better the protection against inter-cell interference but at the cost of less bandwidth per cell. The primary traffic in the early generations of the cellular systems is

---

[1] In Chapter 10, we will explore beyond the conventional cellular framework and show that those basic tools are not always effective in the new paradigm where small base stations are deployed in an unplanned manner and a user may not be allowed to connect to any base station (restricted association).

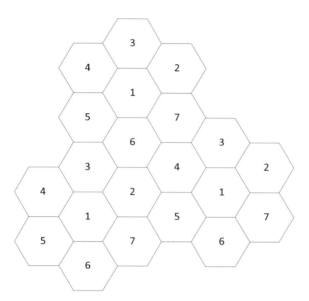

Figure 6.1 Frequency reuse pattern in narrowband system with reuse factor $K = 7$.

circuit-switched voice. Voice frames require quite stringent SINR to decode. For example, the full rate codec in GSM requires an SINR of 7 to 13 dB. Furthermore, narrowband FDMA does not achieve much averaging for inter-cell interference, because a user is interfered with only by one user in an interfering cell using the same channel. Thus, frequency reuse has to be designed conservatively to protect against the *worst-case* interference. The combination of the high SINR requirement of circuit-switched voice and the lack of interference averaging (and frequency diversity to mitigate channel fading) leads to the need for a large K. In GSM, typically $K = 4$ or 7.

From the experience of narrowband FDMA, it had been widely believed that $K > 1$ is necessary until CDMA with $K = 1$ was shown to achieve a higher capacity [168]. CDMA is a spread spectrum technique where every transmitter spreads its signal to the whole bandwidth using a spreading code. For this reason, CDMA is considered wideband, as opposed to narrowband FDMA. The whole bandwidth is reused at every base station (universal frequency reuse). Such aggressive frequency reuse leads to strong interference. A circuit-switched CDMA voice system typically operates at a very low (sample) SINR. Obviously the tradeoff is large bandwidth versus low SINR. A receiver despreads the signal plus interference according to the spreading code. After despreading, interference power is reduced by a factor equal to a spreading gain and SINR is thus boosted by the same factor. For example, using a spreading code of length 128, a symbol is spread into 128 samples and the spreading gain is 128. If a voice codec requires 7 dB (post-despreading) SINR, then the sample SINR can be as low as -14 dB. A benefit of using a wideband signal in CDMA is interference averaging and frequency diversity. Because interference comes from all users in an adjacent cell as opposed to only one in narrowband FDMA, the averaging effect reduces interference variance such that the system capacity is limited not by the worst-case interference but by the average. Besides, frequency diversity reduces channel variance due to multipath fading.

In summary, narrowband FDMA and wideband CDMA exhibit very different design principles for the SINR versus bandwidth tradeoff. Narrowband FDMA eliminates intra-cell interference, but uses a large frequency reuse factor to meet the SINR requirement in the worst-case interference scenario. CDMA enables universal frequency reuse thanks to interference averaging, but suffers from both intra-cell and inter-cell interference. As noted in Section 2.2, OFDMA is also a spread spectrum technique with interference averaging being achieved via tone hopping. Therefore, in principle, OFDMA can employ universal frequency reuse and is expected to achieve a higher capacity than CDMA because of its additional benefit of no intra-cell interference.

However, two important system design questions are yet to be answered in the data-centric mobile broadband context:

- Is interference averaging always desirable for inter-cell interference management? Does universal frequency reuse achieve the best SINR versus bandwidth tradeoff? The tradeoff is different in a data system as opposed to a voice system.
- Should OFDMA follow the same CDMA power control principle? How should the uplink transmit power be determined? Arbitrarily increasing transmit powers leads to excessive inter-cell interference but no appreciable SINR improvement. Uplink power control is directly related to SINR target setting. Clearly, not all SINR targets are achievable. And in a data system, SINR targets are usually not fixed beforehand.

In this chapter, we will attempt to answer these questions.

6.1 Analysis of SIR distributions

In this section, we study the impact of inter-cell interference on the SINR distribution in narrowband FDMA, CDMA, and OFDMA. Chapter 3 introduced the different techniques, such as frequency reuse, interference averaging, and frequency diversity, used in narrowband FDMA, CDMA, and OFDMA to manage inter-cell interference, and qualitatively discussed the differences. In this section, we will quantitatively analyze the effect of these techniques on the SINR distribution. Recall from Chapter 5 that the SINR of a user depends on power and bandwidth allocation within a cell, in addition to inter-cell interference. To facilitate the analysis, we first study the *Carrier-to-Interference* (C/I) distribution, which is mostly determined by cell geometry, and then derive the SINR distribution in the three cellular systems.

To obtain the C/I distribution, we utilize tools from *stochastic geometry* [6] to derive C/I expressions in the cellular network topology. We make the following simplifying assumptions for the analysis:

- Base stations are placed in regular hexagonal grids and each uses an omni-directional antenna (no sectorization)

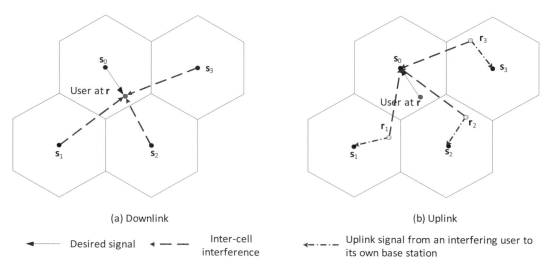

Figure 6.2 Desired signal and inter-cell interference.

- Users are uniformly distributed on a plane according to a *Poisson Point Process* (PPP)
- Thermal noise is ignored and *Signal-to-Interference-Ratio* (SIR) is used instead of SINR

Under these assumptions, we can get closed-form (integral) expressions for the C/I distribution, which can then be evaluated numerically.

6.1.1 Downlink SIR

First consider the downlink. Without loss of generality, focus on a single cell 0 and denote by $\mathcal{A}_0$ the cell region of interest, that is, users in $\mathcal{A}_0$ are served by the same base station in cell 0. All other cells are indexed as $n = 1, 2, \ldots$ Figure 6.2(a) shows the desired signal and the inter-cell interference for a given user i at $\mathbf{r}$ with universal reuse $K = 1$, where $\mathbf{s}_n, n = 0, 1, \ldots$ represent the positions of the base stations. We will first calculate the C/I and then derive the SIR for FDMA/CDMA/OFDMA.

C/I distribution and frequency reuse

Ignoring intra-cell interference, the C/I $\gamma_i'(\mathbf{r})$ is given as

$$\gamma_i'(\mathbf{r}) = \frac{G_{\mathbf{s}_0,\mathbf{r}} P_0}{\sum_{n>0} G_{\mathbf{s}_n,\mathbf{r}} P_n}, \qquad (6.1)$$

where $G_{\mathbf{r}_1,\mathbf{r}_2}$ represents the channel power gain from location $\mathbf{r}_1$ to location $\mathbf{r}_2$, and P_n the transmit power used at base station n. In the downlink, it is usually assumed that all the transmit powers are the same and equal to the maximum transmit power of a base station. Thus, we drop P_n out of the C/I Equation (6.1).

For the sake of simplicity, we ignore the effect of shadowing and assume that a user is connected with the closest base station. In this case, region $\mathcal{A}_0$ is the hexagonal cell

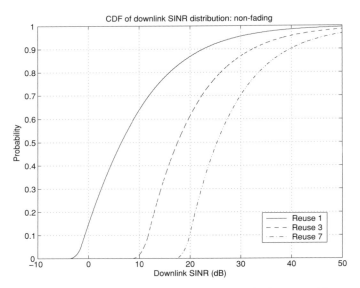

Figure 6.3 Downlink C/I distribution in hexagonal deployment with reuse factor $K = 1, 3, 7$ and a non-fading path loss channel model $r = 4$.

centered at the origin s_0 shown in Figure 6.2. Consider a path loss model (4.5) where $G(\mathbf{r}_1, \mathbf{r}_2) \propto |\mathbf{r}_1 - \mathbf{r}_2|^{-r}$, with r the path loss exponent. Equation (6.1) becomes

$$\gamma_i'(\mathbf{r}) = \frac{|\mathbf{s}_0 - \mathbf{r}|^{-r}}{\sum_{n>0} |\mathbf{s}_n - \mathbf{r}|^{-r}}, \tag{6.2}$$

which is fully determined by the geometrical position of the user $\mathbf{r}$.

The downlink C/I distribution is

$$\mathbb{F}_{\gamma'}(x) = \frac{1}{\Lambda(\mathcal{A}_0)} \int_{\mathcal{A}_0} \mathbb{P}(\gamma_i'(\mathbf{r}) < x) \Lambda(d\mathbf{r}) \tag{6.3}$$

$$= \frac{1}{|\mathcal{A}|} \left| \left\{ \mathbf{r} \in \mathcal{A}_0 : \frac{|\mathbf{s}_0 - \mathbf{r}|^{-r}}{\sum_{n>0} |\mathbf{s}_n - \mathbf{r}|^{-r}} < x \right\} \right|. \tag{6.4}$$

Here $\Lambda(\mathcal{S})$ represents the number of users in any area $\mathcal{S}$. We assume homogeneous PPP for user distribution with density λ, and thus $\Lambda(\mathcal{A}_0) = \lambda |\mathcal{A}_0|$, which is the average number of nodes within region $\mathcal{A}_0$. Assuming all cells have the same size, we drop the subscript 0 and denote it $|\mathcal{A}|$. $\mathbb{F}_{\gamma'}(x)$ in (6.4) is simply the fraction of the area within cell 0 such that the above inequality is met, and can be numerically calculated.

When the frequency reuse factor $K > 1$, the desired signal power from base station s_0 does not change. However, the summation in (6.2) and (6.4) only includes base stations reusing the same channels as base station 0. The interfering base stations are farther away, as the Euclidean distance between two nearby base stations with the same channels increases by a factor of $\sqrt{K}$. The difference in SIR distribution with different reuse factors is shown in Figure 6.3.

A few observations are worthy of note. First, with $K = 1$ the worst C/I is about -3 dB. This is because a user in the intersection of three cells sees two interferers as strong as its own base station. A significant fraction (about 15 percent) of users stay below 0 dB C/I. Second, from $K = 1$ to $K = 3$, the C/I distribution shifts to the right for about 10 dB. The C/I boost is more significant for low C/I users than for high C/I ones. Third, a similar C/I boost is seen from $K = 3$ to $K = 7$, although the amount is smaller.

Next, let us add multipath fading in the channel model. Assume that despite multipath fading, a user is still connected with the closest base station. This is because base station selection happens at a slow time scale during which multipath fading is averaged out. Assume that fading between any user and any base station is i.i.d. Rayleigh; that is, $G_{\mathbf{r}_1,\mathbf{r}_2} \propto |\mathbf{r}_1 - \mathbf{r}_2|^{-r} z_{\mathbf{r}_1,\mathbf{r}_2}$, where $z_{\mathbf{r}_1,\mathbf{r}_2}$ is an exponentially distributed random variable with mean 1.

Under the Rayleigh fading and PPP assumptions, we can get a closed form expression for $\mathbb{P}(\gamma_i'(\mathbf{r}) < x)$. Denote total inter-cell interference $I = \sum_{n>0} z_{\mathbf{s}_n,\mathbf{r}} |\mathbf{s}_n - \mathbf{r}|^{-r}$. It follows that

$$\mathbb{P}\left(\gamma_i'(\mathbf{r}) < x\right) = \mathbb{P}\left(z_{\mathbf{s}_0,\mathbf{r}} |\mathbf{s}_0 - \mathbf{r}|^{-r} < xI\right)$$

$$= \int_t \mathbb{P}\left(z_{\mathbf{s}_0,\mathbf{r}} |\mathbf{s}_0 - \mathbf{r}|^{-r} < xI | I = t\right) d\mathbb{F}_I(t)$$

$$= \int_t 1 - \exp\left(-\frac{x}{|\mathbf{s}_0 - \mathbf{r}|^{-r}} t\right) d\mathbb{F}_I(t)$$

$$= 1 - \mathbb{E}\left(\exp\left(-\frac{x}{|\mathbf{s}_0 - \mathbf{r}|^{-r}} I\right)\right)$$

$$= 1 - \prod_{n>0} \mathbb{E}\left(\exp\left(-\frac{x|\mathbf{s}_n - \mathbf{r}|^{-r}}{|\mathbf{s}_0 - \mathbf{r}|^{-r}} z_{\mathbf{s}_n,\mathbf{r}}\right)\right)$$

$$= 1 - \prod_{n>0}\left(1 + \frac{x|\mathbf{r} - \mathbf{s}_n|^{-r}}{|\mathbf{r} - \mathbf{s}_0|^{-r}}\right)^{-1}. \tag{6.5}$$

Now an integral form for the downlink C/I distribution is given as

$$\mathbb{F}_{\gamma'}(x) = 1 - \frac{1}{|\mathcal{A}|} \int_{\mathcal{A}_0} \prod_{n>0}\left(1 + \frac{x|\mathbf{r} - \mathbf{s}_n|^{-r}}{|\mathbf{r} - \mathbf{s}_0|^{-r}}\right)^{-1} d\mathbf{r}, \tag{6.6}$$

where the integral is taken over the hexagonal region of cell 0. Note that (6.6) is different from (6.4) in that there does not exist a clean closed geographic region with C/I exceeds x. Instead, channel uncertainty caused by fading makes it possible for every point in $\mathcal{A}_0$ to have a C/I better or worse than x.

It is straightforward to evaluate the expression (6.6) numerically and the C/I distribution with Rayleigh fading is shown in Figure 6.4. Comparison with the non-fading case in Figure 6.3 shows that with fading, more users are at the lower end of the SIR range. At $K = 1$, about 25 percent of users are below 0 dB C/I and the worst scenario (5th percentile) C/I is as bad as -8 dB. This is simply due to the fact that with fading, the signal at a given time instant can be weaker than the interference. Overall, cell edge users see more significant C/I degradation than interior users do.

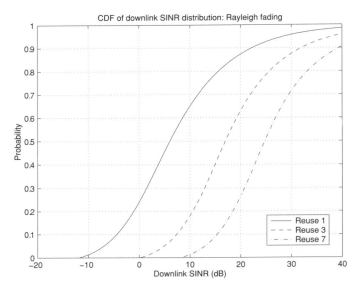

Figure 6.4 Downlink C/I distribution in hexagonal deployment with reuse factor $K = 1, 3, 7$ and a Rayleigh fading path loss channel model $r = 4$.

Discussion notes 6.1 An analysis of C/I distribution with randomly-placed base stations

We next analyze the scenario where base stations are placed in random locations, which represents an extreme case of an unplanned deployment often seen with femtocells. Femtocells are studied in detail in Section 10.1.2.

The process of placing base stations is modeled by a PPP with a different density than user dropping PPP. We still assume that the same user association rule applies to this case and each base station serves the users falling into the Vonoroi region centered at the base station, as shown in Figure 6.5.

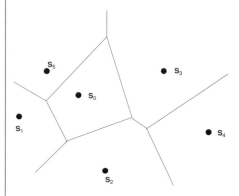

Figure 6.5 Cell boundaries with randomly placed base stations.

We place one base station s_0 at the origin and study the C/I distribution for nodes falling into the Vonoroi region $\mathcal{V}_0$ of base station s_0. Suppose that a user at location $\mathbf{r}$ is connected to s_0. If we draw a circle centered at $\mathbf{r}$ with radius $|\mathbf{r}|$, then there are no base stations within the circle. Using the stochastic geometry tools provided in [6], it can be shown [5] that with Rayleigh fading and $r = 4$ path loss,

$$\mathbb{F}_{\gamma'}(x) = \int_{\mathcal{R}^2} \mathbb{P}(\gamma_i'(\mathbf{r}) < x, \mathbf{r} \in \mathcal{V}_0) \Lambda(d\mathbf{r})$$

$$= \frac{\sqrt{x} \left(\frac{\pi}{2} - \arctan \frac{1}{\sqrt{x}} \right)}{1 + \sqrt{x} \left(\frac{\pi}{2} - \arctan \frac{1}{\sqrt{x}} \right)}. \tag{6.7}$$

Note that similar analytical results can be obtained with other values of r and fading distributions. Please refer to [5] for more details.

SIR comparisons

We next apply the above C/I analysis to determine the SIR in narrowband FDMA, CDMA, and OFDMA.

Since narrowband FDMA has negligible intra-cell interference, assuming each user is allocated an equal fraction of the total downlink transmit power and bandwidth, the SIR γ is indeed equal to the C/I γ' in (6.4) and (6.6), and the distribution is the same as what is shown in Figures 6.3 and 6.4. Because of the lack of frequency diversity, the fading case is particularly important. Frequency reuse is an important technique to ensure that a majority of the users, for example, 95 percent, have satisfactory performance. In Figure 6.4, note that even at $K = 3$, the 5th percentile users see a downlink SIR of only 5 dB. This justifies the need for a large frequency reuse factor K in a practical narrow FDMA system.

The situation in CDMA systems is quite different in several aspects:

- *Intra-cell interference.* In the absence of multipath, the CDMA downlink can use orthogonal spreading codes to eliminate intra-cell interference. However, signal paths with different delays, due to the wireless channel or transmit/receive filters, have non-zero cross correlation. Denote by ψ the fraction of the signal power that becomes intra-cell interference, where $0 \leq \psi \leq 1$.
- *Power allocation.* Base station 0 allocates a fraction β_i of its total transmit power to user i, with $\sum_i \beta_i = 1$ for all the users in the cell.
- *Spreading gain.* As noted before, a user projects the received signal plus interference to the subspace of its own spreading code, and boosts the post-despreading SINR by a factor of the spreading gain φ_i.
- *Frequency diversity.* If the channel consists of L taps each with i.i.d. Rayleigh fading, then a diversity order of L is achieved in CDMA. It can be shown that with maximum ratio combining of L taps, the SIR can be calculated as if $G_{\mathbf{r}_1,\mathbf{r}_2} \propto |\mathbf{r}_1 - \mathbf{r}_2|^{-r} \sum_{l=1}^{L} z_{\mathbf{r}_1,\mathbf{r}_2,l}$, where $z_{\mathbf{r}_1,\mathbf{r}_2,l}$'s are i.i.d. random variables with an exponential distribution. Figure 6.6 compares the SIR distribution with Rayleigh

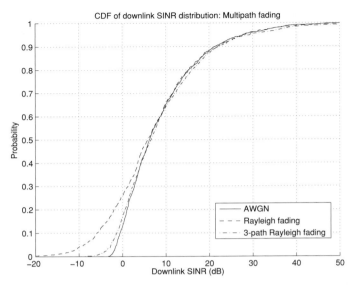

Figure 6.6 Downlink SIR distribution with no fading and L i.i.d. taps of Rayleigh fading, with $L = 1$ no frequency diversity and $L = 3$ frequency diversity.

fading with and without frequency diversity and no fading cases, and shows that even with only $L = 3$, the SIR tail probability is improved significantly. Therefore, we assume that fading is averaged out in CDMA and focus on the non-fading case.

Hence, the downlink SIR in CDMA of user i is given as

$$\gamma_i = \beta_i \varphi_i \frac{\gamma_i'}{\psi \gamma_i' + 1}, \qquad (6.8)$$

where γ_i' is defined in (6.2) with $K = 1$. Note that γ_i' entirely depends on the geometry. To control γ_i, rather than choosing frequency reuse factor K, one can keep $K = 1$ and choose β_i and φ_i. For example, one scheme is to split the transmit power evenly among all the CDMA orthogonal spreading codes $\beta_i = \frac{1}{K_c}$ and let $\varphi_i = L_c$, where K_c represents the number of the orthogonal spreading codes used in a cell and L_c represents the length of the orthogonal spreading codes. We refer to $\frac{L_c}{K_c}$ as the *code reuse factor*, which serves a similar purpose to the frequency reuse factor K in the sense that a large code or frequency reuse factor boosts SIR but reduces bandwidth available in each cell. In CDMA, because of frequency diversity and multipath fading being averaged out, the code reuse factor can be set smaller than the frequency reuse factor in narrowband FDMA.

Furthermore, power allocation β_i can be different for different users to shape the downlink SIR distribution, for example, to improve the SIR of cell edge users. If the intra-cell interference is negligible ($\psi \approx 0$), one can apply power and bandwidth optimization similar to the idea studied in Chapter 5 for OFDMA. In particular, allocate more power to cell edge users and more bandwidth to interior users – in CDMA, bandwidth means the number of orthogonal spreading codes. Note that the intra-cell interference ψ in

CDMA is similar to the cross interference discussed in Section 5.1.2. If ψ is significant, the performance gain from power and bandwidth optimization diminishes.

An important practical benefit of no frequency reuse planning is that it is easier to deploy a CDMA system. For example, adding a base station to an existing deployment is not hard in CDMA; however, in narrowband FDMA, one needs to change the frequency reuse pattern, which is a much more complex optimization task.

The downlink SIR in OFDMA can be calculated similarly, except that no intra-cell interference $\psi = 0$. Denote by α_i and β_i the fractions of bandwidth and power allocated to user i. The downlink SIR in OFDMA of user i is given as

$$\gamma_i = \frac{\beta_i}{\alpha_i} \gamma_i'. \qquad (6.9)$$

The C/I γ_i' here is the same as the nominal SIR defined in (5.1). Similar to CDMA, one can shape γ_i with parameters β_i and α_i in OFDMA. Comparison between (6.8) and (6.9) shows that spreading gain φ_i in CDMA and the inverse of bandwidth allocation $1/\alpha_i$ in OFDMA play a similar role.

6.1.2 Uplink SIR

In the downlink, the total transmit power is assumed to be the same in all the base stations. However, in the uplink, different schemes can be used to set the transmit power of the users. In particular, we consider two schemes, namely a constant transmit power scheme and a constant receive power scheme. In the latter scheme, the transmit power is inversely proportional to the channel power gain between a user and its own base station. Constant transmit power is often used in narrowband FDMA, while constant receive power is more popular in CDMA.

Let us continue to focus on cell 0. First, consider narrowband FDMA with constant transmit power. Figure 6.2(b) shows the desired signal and the inter-cell interference for a given user at $\mathbf{r}$ with universal reuse $K = 1$. A user in cell 0 is interfered with by exactly one user in each neighboring cell reusing the same channel. In the non-fading case, the SIR $\gamma_i(\mathbf{r})$ for a user i at $\mathbf{r}$ is given by

$$\gamma_i(\mathbf{r}) = \frac{|\mathbf{s}_0 - \mathbf{r}|^{-r}}{\sum_{n>0} |\mathbf{s}_0 - \mathbf{r}_n|^{-r}}, \qquad (6.10)$$

where $\mathbf{r}_n$ is the position of the interfering user in cell n. Compared with (6.1) in the downlink, the uplink SIR expression (6.10) involves more random variables, which are interferer locations $\mathbf{r}_n$ in addition to user location $\mathbf{r}$. We assume $\mathbf{r}_n$ is uniformly distributed in the hexagonal area $\mathcal{A}_n$.

Similar to (6.4), the uplink SIR distribution is given by

$$\mathbb{F}_\gamma(x) = \mathbb{E}_{\mathbf{r}_1, \mathbf{r}_2, \dots} \left(\frac{|\{\mathbf{r} \in \mathcal{A}_0 : \gamma_i(\mathbf{r}) < x\}|}{|\mathcal{A}|} \right). \qquad (6.11)$$

This expression has to be averaged over multiple regions for $\mathbf{r}_n$, $n = 1, 2, \dots$. With $K > 1$, the above summation only includes the cells reusing the same channel. Figure 6.7 plots the uplink SIR distribution. Comparing with the $K = 1$ downlink case in Figure 6.3,

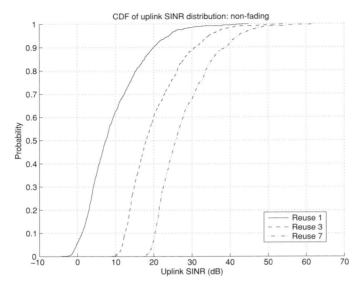

Figure 6.7 Uplink SIR distribution in a FDMA system with reuse factor $K = 1, 3, 7$ and a non-fading path loss channel model with $r = 4$.

we note that the uplink SIR distribution is slightly better. The reason is that at cell edges even the strongest interference can be much weaker than the desired signal.

In the Rayleigh fading case,

$$\mathbb{P}(\gamma_i(\mathbf{r}) < x) = \mathbb{P}\left(\frac{z_{\mathbf{s}_0, \mathbf{r}} |\mathbf{s}_0 - \mathbf{r}|^{-r}}{I} < x\right)$$

$$= 1 - \mathbb{E}\left(\exp\left(-\frac{x}{|\mathbf{s}_0 - \mathbf{r}|^{-r}} I\right)\right), \quad (6.12)$$

with total inter-cell interference $I = \sum_{n>0} z_{\mathbf{s}_0, \mathbf{r}_n} |\mathbf{s}_0 - \mathbf{r}_n|^{-r}$. The above expectation can be solved using the Laplace transform similar to (6.5) as the following:

$$\mathbb{F}_\gamma(x) = 1 - \frac{1}{|\mathcal{A}|}\int_{\mathcal{A}_0}\left(\Pi_{n>0}\frac{1}{|\mathcal{A}|}\int_{\mathcal{A}_n}\left(1 + x\frac{|\mathbf{s}_0 - \mathbf{r}_n|^{-r}}{|\mathbf{s}_0 - \mathbf{r}|^{-r}}\right)^{-1} d\mathbf{r}_n\right) d\mathbf{r}. \quad (6.13)$$

The above uplink SIR distribution can be numerically calculated and the results are shown in Figure 6.8. Similar to the downlink case, Rayleigh fading makes the SIR tail much worse. Again, in narrowband FDMA, a large frequency reuse factor is needed to ensure reasonable performance for the majority of the users.

Next consider CDMA with constant receive power. Normalizing the receive power to be equal to 1, the transmit power of the reference user at position $\mathbf{r}$ of cell 0 is $\frac{1}{|\mathbf{r}-\mathbf{s}_0|^{-r}}$. Similar to the downlink, the differences from the above narrowband FDMA scenario are intra-cell interference, spreading gain, and frequency diversity. What makes the CDMA uplink unique is that because of the wideband nature, inter-cell interference comes from not just one but all the users in a neighboring cell. As in the downlink, we assume that fading is averaged out because of frequency diversity. Let I include both intra-cell and

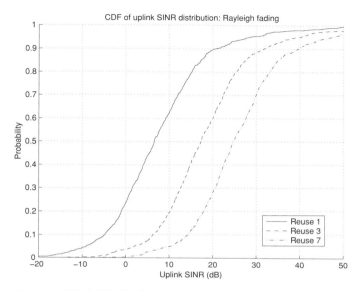

Figure 6.8 Uplink SIR distribution in a FDMA system with reuse factor $K = 1, 3, 7$ and a Rayleigh fading path loss channel model with $r = 4$.

inter-cell interference,

$$I = \sum_{n=0,1,\ldots} \int_{\mathcal{A}_n} \frac{|\mathbf{r}_n - \mathbf{s}_0|^{-r}}{|\mathbf{r}_n - \mathbf{s}_n|^{-r}} \Lambda(d\mathbf{r}_n), \qquad (6.14)$$

where the integral $\int_{\mathcal{A}_n}$ takes place over cell n, $n = 0, 1, \ldots$. Note that the above summation includes cell $n = 0$ to represent intra-cell interference. $\Lambda(d\mathbf{r}_n)$ represents the number of of users in a small area between $\mathbf{r}_n$ and $\mathbf{r}_n + d\mathbf{r}_n$ in a realization of the PPP. I is a random variable depending on the PPP realization. When the total number of interfering users in a cell is large, the Central Limit Theorem takes effect and I becomes approximately Gaussian. The mean of I is

$$\mathbb{E}(I) = K_u \sum_{n=0,1,\ldots} \frac{1}{|\mathcal{A}|} \int_{\mathcal{A}_n} \frac{|\mathbf{r}_n - \mathbf{s}_0|^{-r}}{|\mathbf{r}_n - \mathbf{s}_n|^{-r}} d\mathbf{r}_n, \qquad (6.15)$$

and the variance

$$\mathrm{var}(I) = K_u \sum_{n=0,1,\ldots} \frac{1}{|\mathcal{A}|} \int_{\mathcal{A}_n} \frac{|\mathbf{r}_n - \mathbf{s}_0|^{-2r}}{|\mathbf{r}_n - \mathbf{s}_n|^{-2r}} d\mathbf{r}_n, \qquad (6.16)$$

where K_u is the number of users in a cell and equal to $\lambda |\mathcal{A}|$. We can write $\mathbb{E}(I) = K_u c_1$ and $\mathrm{var}(I) = K_u c_2$, where c_1 and c_2 are constants determined by the geometry and can numerically calculated from (6.15) and (6.16). When $r = 4$, $c_1 = 1.41$ and $c_2 = 1.15$. Because of constant receive power, the intra-cell interference is equal to K_u from (6.14) if we assume that the receive power of all the users in cell 0 contributes to the intra-cell interference. In this case, on average, the ratio between the inter-cell and intra-cell interference is $c_1 - 1 = 0.41$. Clearly, intra-cell interference dominates the overall interference.

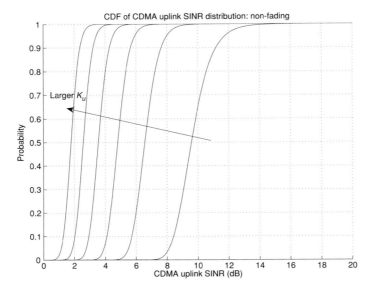

Figure 6.9 Uplink SIR distribution in a CDMA system with no fading. $r = 4$ and $\varphi_i = 128$. From right to left, K_u increases from 10 to 60.

Hence, the uplink SIR in CDMA is

$$\gamma_i = \frac{\varphi_i}{I}, \tag{6.17}$$

with φ_i the spreading gain, which is determined by the spreading code length L_c. Similar to the downlink, rather than adjusting frequency reuse factor K, CDMA keeps reuse factor $K = 1$ and controls γ_i with parameters K_u and φ_i. The SIR distribution is fully determined by the statistics of I:

$$\mathbb{F}_\gamma(x) = \mathbb{P}\left(\frac{\varphi_i}{I} < x\right) = \mathbb{P}\left(I > \frac{\varphi_i}{x}\right) \tag{6.18}$$

$$\approx \int_{\frac{\varphi_i}{x} - K_u c_1}^{\infty} \frac{1}{\sqrt{2\pi}} \exp\left(\frac{-t^2}{2}\right) dt. \tag{6.19}$$

The last step assumes the Gaussian approximation. Figure 6.9 plots the distribution (6.19) evaluated at different loading K_u. Clearly, due to interference averaging, the spread of the SIR is much smaller than that in Figure 6.8.

Now consider OFDMA. If we continue to target constant receive power, then the same analysis of inter-cell interference is applicable, except that cell 0 is excluded in the summation (6.14) thanks to the lack of intra-cell interference. The uplink SIR is now

$$\gamma_i = \frac{1}{\alpha_i I}, \tag{6.20}$$

similar to (6.17).

So far, we assume that the power control is such that all users have the same receive power. In principle, different users can be assigned different receive powers. However, in CDMA for circuit-switched voice, because the spreading gains are the same for all users, constant receive power is a natural choice to achieve the same SIR target. For OFDMA mobile broadband communications, on the other hand, it is not necessarily optimal to target constant receive power or SIR. In the next section, we will extend our study to more general ideas on power control and SINR assignment in OFDMA.

6.2 Uplink power control and SINR assignment in OFDMA

Power control has been critical in the CDMA uplink to deal with the near-far problem; a nearby user has to determine its transmit power carefully so as not to overwhelm the signal from a faraway user in the same cell. In OFDMA, the near-far problem is not as critical because of no intra-cell interference. A natural question is *whether power control is needed in OFDMA*. The answer is *yes*, and the reason is not intra-cell interference but inter-cell interference.

In the downlink, if the total transmit power is fixed, then the inter-cell interference from a base station to a user depends only on the total transmit power but not on how it is allocated in frequency among users within that cell.[2] As a result, power control becomes an intra-cell user multiplexing problem, which is studied in Chapter 5.

However, the situation is very different in the uplink, because the transmitters are distributed. Power control is no longer an intra-cell problem because of coupling among neighboring cells. To see this, consider the uplink SINR of user i,

$$\gamma_i = \frac{u_i}{\alpha_i(\sigma^2 + \sum_{j:c_j \neq c_i} S_{i,j} u_j)}, \tag{6.21}$$

where c_i represents the index of the serving base station of user i. For notational convenience (6.21) uses u_i to represent the *receive* power of user i at its serving base station

$$u_i = P_i G_{c_i, i}, \tag{6.22}$$

where P_i is the transmit power of user i in the uplink and c_i indicate the index of the serving base station of user i.

In this chapter, we assume zero intra-cell interference in OFDMA and that the interference all comes from users j in neighboring cells ($c_j \neq c_i$). Let α_i to be the fraction of bandwidth allocated to user i with $0 < \alpha_i \leq 1$. The desired signal from user i is received at its serving base station c_i with power u_i, and the signal from an interfering user j is received at its serving base station c_j with power u_j. Define $S_{i,j}$ the ratio of the channel gains from user j to base station c_i and to c_j as

$$S_{i,j} = \frac{G_{c_i, j}}{G_{c_j, j}}. \tag{6.23}$$

---

[2] Strictly speaking, while the total transmit power determines the average power of inter-cell interference, the power allocation affects its distribution, an effect similar to what is shown in Figure 2.9. We ignore that effect here.

$S_{i,j}$ is the inter-cell interference power received at base station c_i when the signal of the interfering user j arrives at its serving base station c_j with unit power.

As shown in (6.21), uplink power control of a user j in a neighboring cell affects the SINR of user i. The coupling of power control depends on $S_{i,j}$. As a result, in general uplink power control cannot be reduced to a single cell problem. For example, consider the following resource allocation problem:

$$\max_{\{\alpha_i\},\{u_i\}} \sum_i w_i R_i \qquad (6.24)$$

$$\text{s.t.} \quad R_i = \alpha_i \log(1 + \gamma_i), \qquad (6.25)$$

$$u_i \leq G_{c_i,i} P_m, \qquad (6.26)$$

where P_m is the maximum transmit power of a user. Note that for notational convenience, we use the receive power, instead of transmit power, as the optimization parameter. Let $u_{m,i} = G_{c_i,i} P_m$ to be the maximum receive power budget of user i. Clearly, the above problem is a joint optimization for all users across all base stations. Moreover, the problem is well known to be highly non-convex. A simple way to see this is to consider a scenario of two small cells where the thermal noise σ^2 is negligible and a base station serves only one user near the boundary between the two cells. Suppose $w_1 = w_2$. In this case, both power allocation schemes of $(u_1, u_2) = (u_{m,1}, 0)$ and $(u_1, u_2) = (0, u_{m,2})$ are optimal but any convex combination of the two schemes are not, indicating the non-convexity nature of the problem. Therefore, it is hard to directly solve (6.24). To tackle the non-convexity difficulty, we next decompose the general resource allocation problem to a few steps.

6.2.1 SINR feasibility region

Denote by $\boldsymbol{\gamma}, \boldsymbol{\alpha}, \boldsymbol{u}$ and $\boldsymbol{u}_m$ the vectors consisting of γ_i, α_i, u_i and $u_{m,i}$ for all i. We start with a simple question: given bandwidth assignment ($\boldsymbol{\alpha}$), how can we characterize the SINR feasibility region? A SINR vector $\hat{\boldsymbol{\gamma}}$ is feasible if a non-negative power vector $\boldsymbol{u}$ exists such that

$$\boldsymbol{\gamma} \geq \hat{\boldsymbol{\gamma}}, \qquad (6.27)$$

$$\boldsymbol{u} \leq \boldsymbol{u}_m, \qquad (6.28)$$

where γ_i is determined by $\boldsymbol{u}$ in (6.21). The SINR feasibility region $\Gamma(\boldsymbol{u}_m)$ is defined to be the set of all feasible SINR vectors. Furthermore, we use the notation Γ to represent the SINR feasibility region when the power constraint vanishes with $\boldsymbol{u}_m$ going to infinity.

The answer to this question has been extensively studied in the literature on power control in cellular networks. With no maximum power constraint, there is a clean characterization of the feasibility condition as follows. For notational convenience, define inter-cell interference plus noise power

$$q_i = \sum_{j:c_j \neq c_i} S_{i,j} u_j + \sigma^2, \qquad (6.29)$$

and rewrite everything in a vector form,

$$q = Su + \eta; \tag{6.30}$$

$$u = D(\alpha \cdot \gamma)q, \tag{6.31}$$

where η represents the thermal noise σ^2, operator $D(\alpha \cdot \gamma)$ indicates the diagonal matrix whose entries are from the components of vector $\alpha \cdot \gamma$, which is component-wise product of α and γ. Combining the above two equations, we get

$$u = D(\alpha \cdot \gamma)Su + D(\alpha \cdot \gamma)\eta \tag{6.32}$$

$$q = SD(\alpha \cdot \gamma)q + \eta. \tag{6.33}$$

Solving the above equations, it follows that

$$u^* = (I - D(\alpha \cdot \gamma)S)^{-1} D(\alpha \cdot \gamma)\eta \tag{6.34}$$

$$q^* = (I - SD(\alpha \cdot \gamma))^{-1} \eta \tag{6.35}$$

Hence, a necessary and sufficient condition for an SINR target vector $\hat{\gamma}$ to be feasible with no maximum power constraint is ([24])

$$\rho(D(\alpha \cdot \hat{\gamma})S) < 1, \tag{6.36}$$

where $\rho(D(\alpha \cdot \hat{\gamma})S)$ represents the spectral radius of matrix $D(\alpha \cdot \hat{\gamma})S$, which is the maximum of the absolute value of the eigenvalues. As ρ approaches 1, both the power and interference tend to infinity. In practice, it is desirable that ρ be set not too close to 1.

6.2.2 Distributed power control

Given a feasible SINR vector $\hat{\gamma}$, a distributed iterative algorithm is shown in [52] to converge to the solution u^* of (6.34), which satisfies (6.27) and requires a minimum power. For simplicity, consider a synchronous (slotted) system with time index $t = 1, 2, \ldots$. The algorithm works as follows. User i needs to know its own SINR target $\hat{\gamma}_i$ and measure inter-cell interference $q_i[t]$ seen in the current slot t, and then updates its power at time $t + 1$,

$$u_i[t+1] = \alpha_i \hat{\gamma}_i q_i[t]. \tag{6.37}$$

To show the convergence, we write (6.37) in a vector form,

$$u[t+1] = D(\alpha \cdot \hat{\gamma})q[t], \tag{6.38}$$

and check the $\mathbb{L}_2$ distance between $u[t]$ and u^*:

$$\|u[t+1] - u^*\|^2 = \|D(\alpha \cdot \hat{\gamma})q[t] - u^*\|^2$$
$$= \|D(\alpha \cdot \hat{\gamma})S(u[t] - u^*)\|^2$$
$$\leq \rho(D(\alpha \cdot \hat{\gamma})S)\|u[t] - u^*\|^2. \tag{6.39}$$

If $\rho(D(\alpha \cdot \hat{\gamma})S) < 1$, then $u[t]$ in algorithm (6.38) converges to u^* exponentially at a rate determined by the spectral radius.

With maximum power constraint, it is harder to obtain a simple necessary and sufficient characterization. Clearly, (6.36) remains a necessary condition, and $\Gamma(u_m) \subset \Gamma$ for a finite power constraint vector u_m. Intuitively, we can modify (6.37) to

$$u_i[t+1] = \min(\alpha_i \hat{\gamma}_i q_i[t], u_{m,i}). \tag{6.40}$$

The convergence of (6.40) and its asynchronous version is studied in [180]. It has been shown that the algorithm converges if $\hat{\gamma}$ is feasible. Otherwise, for users whose SINR targets are feasible, the power converges to a feasible solution, while other users keep on transmitting at maximum power, yet are unable to achieve the SINR target.

6.2.3 SINR assignment

Distributed power control converges to the optimal solution if an SINR target is feasible. In principle, whether an SINR target is feasible can be checked using (6.36). However, verifying the condition would require a centralized entity to know S and compute the spectral radius. It would be desirable to determine the feasibility of an SINR target in a distributed manner. Furthermore, to fully utilize the system capacity, it is desirable to assign an SINR target close to the Pareto-boundary of the SINR feasibility region Γ.

As in power control, we first consider the case without maximum power constraint. From the SINR feasibility condition given in (6.36), suppose that we set to operate at

$$\Gamma_\rho = \{\hat{\gamma} : \rho(D(\alpha \cdot \hat{\gamma})S) = \rho, \rho < 1\}. \tag{6.41}$$

A distributed approach is presented in [64] to check the feasibility without directly computing the spectral radius. Specifically, according to the Perron-Frobenius theorem, (6.41) is satisfied if and only if there exists a real non-negative left eigenvector s, corresponding to the eigenvalue of ρ, such that

$$s^T D(\alpha \cdot \hat{\gamma})S = \rho s^T. \tag{6.42}$$

Define another vector

$$l^T = s^T D(\alpha \cdot \hat{\gamma})/\rho. \tag{6.43}$$

(6.42) becomes

$$l^T S = s^T. \tag{6.44}$$

In a scalar form, for each user i, we have

$$l_i = s_i \alpha_i \hat{\gamma}_i / \rho; \tag{6.45}$$

$$s_i = \sum_{j:c_j \neq c_i} S_{j,i} l_j = \sum_{j:c_j \neq c_i} \frac{G_{c_j,i}}{G_{c_i,i}} l_j = \frac{1}{G_{c_i,i}} \sum_{b:b \neq c_i} G_{b,i} L_b, \tag{6.46}$$

with b being the base station index and

$$L_b = \sum_{j:c_j=b} l_j. \tag{6.47}$$

We call l_i the *load* of user i, as l_i is proportional to the amount of system resource allocated to user i (bandwidth fraction α_i and SINR target $\hat{\gamma}_i$). l_i also depends on s_i. For the same amount of allocated system resource, a different s_i results in a different load. From (6.46), s_i is a function of $S_{j,i}$s. Typically, a cell edge user has greater $S_{j,i}$s than a cell interior one. Therefore, the load would be higher to allocate the same amount of system resource to an edge user than to an interior one.

We call s_i the *spillage* of user i, because it mimics the total inter-cell interference, weighted by l_j, that user i causes to all other base stations. Clearly, s_i depends on not only the geometry of user i relative to neighboring base stations ($S_{j,i}$s), but also the load of other users (l_js). Given the geometry, the spillage of user i increases with l_js.

While the notions of load and spillage intuitively explain the dynamics of resource allocation and interference management, the challenge due to the coupling of power control in neighboring cells does not disappear and it is still unclear how to achieve SINR assignment at Pareto-boundary in a distributed manner. Fortunately, (6.47) shows that the load of users served by the same base station can be lumped together to become the load of the base station (L_b). The introduction of L_b allows us to decouple the interdependence among neighboring base stations with the following distributed SINR assignment algorithm.

We rewrite (6.45), (6.46), (6.47) to

$$L_b = \sum_{i:c_i=b} s_i \alpha_i \hat{\gamma}_i / \rho, \tag{6.48}$$

$$s_i = \sum_{b:b \neq c_i} S_{b,i} L_b, \tag{6.49}$$

with channel gain ratio $S_{b,i}$ defined as

$$S_{b,i} = \frac{G_{b,i}}{G_{c_i,i}}. \tag{6.50}$$

As shown in (6.49), to calculate the spillage factor s_i, user i only needs to know the loads of the neighboring base stations, rather than those of all individual users. This can be enabled by letting every base station b to broadcast its load L_b to all users consisting of not only the ones served by base station b but also those in neighboring cells. User i measures $S_{b,i}$ for all neighboring base stations b's with $b \neq c_i$, calculates the spillage s_i from (6.49) and reports to its serving base station c_i. Base station c_i collects the spillage reports from all the served users and assigns them the SINR target $\hat{\gamma}_i$ such that the total base station load (6.48) is satisfied for c_i. The required message exchange is shown in Figure 6.10.

Base station b can treat L_b as a form of inter-cell interference budget and assign $\hat{\gamma}_i$ to fully utilize the budget so as to operate at the Pareto-boundary of the SINR feasibility region. A base station should set the value of L_b as a function of its congestion level and

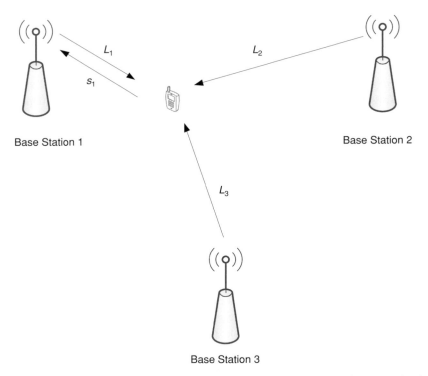

Figure 6.10 Message exchange of the distributed algorithm of SINR assignment using base station load L_b. The serving base station of the shown user is base station 1.

service priority. Note that only the relative value of L_b among neighboring base stations matters, since from (6.48) and (6.49), increasing all L_bs by the same factor does not change SINR assignment.

To understanding the setting of L_b, let us examine a scenario of two adjacent cells A and B. Let base station index b now equal to A or B. First, suppose that L_b is set the same in both cells ($L_A = L_B$). The SINR assignment constraint in cell A is reduced to

$$\sum_{i:c_i=A} S_{B,i} \alpha_i \hat{\gamma}_i = \rho. \quad (6.51)$$

As noted previously, a cell edge user with greater $S_{B,i}$ consumes more interference budget for the same amount of allocated system resource ($\alpha_i \hat{\gamma}_i$). The same situation occurs in cell B. Next, suppose that cell A carries more important traffic than cell B and needs to be prioritized. Set $L_A = \kappa L_B$ with $\kappa > 1$. The SINR assignment constraint becomes

$$\sum_{i:c_i=A} S_{B,i} \alpha_i \hat{\gamma}_i = \rho\kappa, \text{ for users in cell } A \quad (6.52)$$

$$\sum_{i:c_i=B} S_{A,i} \alpha_i \hat{\gamma}_i = \frac{\rho}{\kappa}, \text{ for users in cell } B. \quad (6.53)$$

Compared with (6.51), the budget of users in cell B becomes more stringent while the budget in cell A is relaxed.

Hence, if base station load L_b is used as an interference budget, then the uplink SINR assignment becomes an intra-cell problem. This is similar to the downlink where the constraint is instead the total transmit power budget. The above distributed algorithm obtains an SINR assignment on the Pareto-boundary of the feasibility region. Distributed power control is needed to further find the optimal power solution to achieve assigned SINR. It should be pointed out that as shown in (6.37) distributed power control requires iterations among neighboring cells due to coupling. In contrast, the downlink does not require any iterations to determine power allocation from SINR assignment.

It is not favorable to operate exactly on the Pareto-boundary 1. The reason is that as we have seen in (6.34), the receive power $\mathbf{u}$ would go to infinity if the spectral radius of the matrix $\rho(\mathbf{D}(\boldsymbol{\alpha} \cdot \hat{\boldsymbol{\gamma}})\mathbf{S})$ is 1. One way to control the operating point in the SINR feasibility region is to set $\rho < 1$. Another way is to impose an explicit maximum power constraint $\mathbf{u}_m$. It can be shown ([63]) that SINR assignment $\hat{\boldsymbol{\gamma}}$ is Pareto-optimal in $\Gamma(\mathbf{u}_m)$ if and only if it is achieved with power solution $\mathbf{u}$, with $\mathbf{u} \leq \mathbf{u}_m$ and $u_i = u_{m,i}$ for at least one i.

To obtain an SINR assignment on the Pareto-boundary of $\Gamma(\mathbf{u}_m)$, a simple scheme is to initially assign SINR target $\hat{\boldsymbol{\gamma}}$ as if maximum power constraint does not exist. Then, the maximum power constraint is used to determine power $\mathbf{u}$ in the power control algorithm (6.40). A nice property of the power control algorithm is that it will converge to a power vector $\mathbf{u}$ such that for a subset of users, the receive power is at maximum level $u_i = u_{m,i}$ with their achieved SINR still below the assigned SINR target, while the rest of users already achieve their SINR target with $u_i \leq u_{m,i}$.

Finally, we link the above load analysis with the inter-cell interference budget Q_m studied in Chapter 5. Suppose that the load is the same for all base stations. Consider base station b_0. Now (6.48) and (6.49) are reduced to

$$\sum_{i:c_i=b_0} \left(\sum_{b:b\neq b_0} S_{b,i} \right) \frac{G_{b_0,i} P_i}{q_i} = \rho, \quad (6.54)$$

where q_i is given in (6.29) and P_i the transmit power of user i. Note that q_i is the total inter-cell interference plus noise power received at base station b_0 and thus the same for all users in base station b_0. We rewrite q_i to q_{b_0} to reflect the dependency. Comparison of the Equation (6.54) with (5.17) shows that they are equivalent when we set

$$Q_m = \rho q_{b_0} \quad (6.55)$$

$$V_i = G_{b_0,i} \left(\sum_{b:b\neq b_0} S_{b,i} \right). \quad (6.56)$$

The case where the load is not the same for all base stations can be addressed similarly.

6.2.4 Joint bandwidth and SINR assignment

So far we have assumed bandwidth allocation $\boldsymbol{\alpha}$ is given. The base station can furthermore determine bandwidth and SINR assignment jointly. By fixing base station load L_b in (6.48) and (6.49), a multi-cell joint optimization problem (6.24) is reduced to the following single-cell bandwidth/SINR allocation problem, which can be solved independently. In base station b_0,

$$\max_{\{\alpha_i\},\{\hat{\gamma}_i\}} \sum_{i:c_i=b_0} w_i R_i \tag{6.57}$$

$$\text{s.t.} \quad R_i = \alpha_i \log(1+\hat{\gamma}_i), \tag{6.58}$$

$$\sum_{i:c_i=b_0} \alpha_i s_i \hat{\gamma}_i / \rho = L_b, \tag{6.59}$$

where s_i is given in (6.46) and fixed in the above optimization. With a slight change of variable, we rewrite the problem to be

$$\max_{\{\alpha_i\},\{l_i\}} \sum_{i:c_i=b_0} w_i R_i \tag{6.60}$$

$$\text{s.t.} \quad R_i = \alpha_i \log\left(1 + \frac{l_i \rho}{\alpha_i s_i}\right), \tag{6.61}$$

$$\sum_{i:c_i=b_0} l_i \leq L_{b_0}. \tag{6.62}$$

We have studied this type of problem in Chapter 5, and shown that it is a convex optimization problem in l and α under linear constraints. Thus, the base station can easily apply a gradient descendent type of algorithms to search for the optimal SINR and bandwidth allocation.

6.2.5 Utility maximization in SINR assignment

By fixing base station load $\{L_b\}$, we decompose the multi-cell resource allocation problem (6.24) to convex single cell bandwidth/SINR optimization problems (6.57). As a next step, we introduce a slow time scale adaptation algorithm to determine $\{L_b\}$ so as to further optimize performance across all the base stations. For simplicity, we fix bandwidth assignment α and consider the following utility maximization problem where the utility is defined upon SINR:

$$\max \sum_i U_i(\hat{\gamma}_i) \tag{6.63}$$

$$\text{s.t.} \quad \hat{\gamma} \in \Gamma_\rho. \tag{6.64}$$

Rather than fixing L_b, consider the following update rule of the load

$$l_i[t+1] = l_i[t] + \delta_l \left(\frac{U_i'(\hat{\gamma}_i[t])\hat{\gamma}_i[t]}{q_i[t]} - l_i[t] \right). \tag{6.65}$$

Every base station b starts with some initial load $l[0]$, calculates its aggregate load L_b with (6.47) and broadcasts it. The users listen to the broadcast messages from neighboring base stations, and calculate the spillage with (6.49) and report to their serving base stations. The base stations assign the SINR target $\hat{\gamma}_i[t]$ with (6.45) using the reported spillage. After distributed power control converges, every base station measures the resultant interference-plus-noise power q_i, and update the load with (6.65). It has been shown in [64] that for $\rho \to 1$ and sufficiently small step size δ_l, the update is along an ascent direction, and the above algorithm converges to the optimal solution of (6.63).

Now consider the utility maximization problem with maximum power constraint,

$$\max \sum_i U_i(\hat{\gamma}_i) \qquad (6.66)$$

$$s.t. \ \mathbf{u} \leq \mathbf{u}_m. \qquad (6.67)$$

It turns out ([64]) that the preceding algorithm is applicable here, except that the spillage is defined as

$$s_i = \sum_{b: b \neq c_i} S_{b,i} L_b + v_i, \qquad (6.68)$$

where v_i represents a price to pay if the power constraint is not satisfied. If $u_i > u_{m,i}$. $v_i[t]$ is updated as follows:

$$v_i[t+1] = \left(v_i[t] + \frac{\delta_v}{t} (u_i[t] - u_{i,m}) \right)^+, \qquad (6.69)$$

for a step size δ_v.

Practical example 6.1 Uplink power control in LTE

Uplink power control is used to deal with interference and improve link reliability against large scale channel fading and small scale multipath channel fading. In general, two types of power control, open-loop and closed-loop, are applied jointly. In open-loop power control, a user determines its transmit power on the basis of the measured receive power of the downlink signal. In closed-loop power, the user receives explicit power control commands from the base station.

In CDMA circuit-switched voice systems, the focus of power control is the near-far problem and intra-cell interference. Every user transmits continuously at the same data rate and thus requires the same SINR target. Power control is used to equalize the receive power of all users. Power control commands are sent frequently at a rate of around 1 kHz, as imperfect power control would cause loss in capacity.

On the other hand, OFDMA packet-switched data systems exhibit different characteristics. Because of the orthogonality, power control is mostly used to control inter-cell interference. A user usually does not transmit continuously, but only at scheduled time intervals. Different users can have different data rates and SINR targets. Thus the receive power does not have to be equalized. Closed-loop power control can be relatively slow at a no more than a few hundred Hz.

We next describe uplink power control in LTE. In LTE, power control of the PUCCH and PUSCH is quite different. The data rate of the PUCCH depends on the PUCCH formats and power control of the PUCCH is used to meet the corresponding

SINR targets. The PUSCH, on the other hand, has the flexibility of setting different SINR targets for different users depending on whether they are nearby or faraway from the base station.

Specifically, the transmit power of the PUCCH is given, in dB, by

$$P_{\text{PUCCH}} = \min(P_m, P_{0,\text{PUCCH}} + PL + \Delta_F + \delta_{\text{PUCCH}}). \quad (6.70)$$

P_m represents the maximum device transmit power. PL is the path loss estimated as the difference (in dB) between the base station transmit power and the receive power of the downlink cell-specific reference signal. Any discrepancy between downlink and uplink path loss is corrected by the closed-loop term δ_{PUCCH}. Δ_F is an offset to compensate for a different SINR required for a different PUCCH format. For example, $\Delta_F = 0$ dB for PUCCH format 1 (1-bit H-ARQ-ACK, BPSK), $\Delta_F = 3$ dB for PUCCH format 1 (2-bit H-ARQ-ACK, QPSK), and Δ_F is greater for PUCCH format 2 where more information bits are sent.

$P_{0,\text{PUCCH}}$ is used to control the average received SINR and can vary in response to overall interference fluctuation. $P_{0,\text{PUCCH}}$ consists of a cell-specific component for all users sent as part of the broadcast system information and a user specific offset configured via upper layer signaling, respectively. As a result, $P_{0,\text{PUCCH}}$ does not vary rapidly.

More rapid interference and channel fluctuations have to be overcome with the closed-loop term δ_{PUCCH}. For PUCCH, power control commands are accumulative: each received command increases or decreases δ_{PUCCH} by some amount specified by the command. Power control commands are provided in the PDCCH in two ways. First, a command can be included in each PDSCH assignment. The purpose is to adjust the transmit power of the H-ARQ-ACK corresponding to the assigned PDSCH. Second, a special PDCCH can provide commands for multiple users simultaneously. Those commands are usually sent periodically to control periodic PUCCH transmissions such as CQI/PMI/RI reports.

The transmit power, in dB, of the PUSCH is given by

$$P_{\text{PUSCH}} = \min(P_m, P_{0,\text{PUSCH}} + \alpha \cdot PL + \Delta_{MCS} + \delta_{\text{PUSCH}} + 10\log_{10} M_{\text{PUSCH}}). \quad (6.71)$$

$P_{0,\text{PUSCH}}$ and Δ_{MCS} serve similar purposes as $P_{0,\text{PUCCH}}$ and Δ_F. In particular, Δ_{MCS} depends on the chosen modulation and coding scheme (MCS). M_{PUSCH} represents the number of PRBs assigned to the PUSCH. Note that all the other terms in (6.71) control the receive power per PRB. If multiple PRBs are assigned, the user needs to linearly scale up the total transmit power P_{PUSCH} so as to keep the same receive power target. For PUSCH, power control commands are accumulative or absolute where a command directly specifies the value of δ_{PUSCH}. Power control commands are provided similar to the PUCCH case except that a command can be included in the PUSCH, not PDSCH, assignments.

The most important difference between (6.71) and (6.70) is that the PUSCH employs fractional power control with $0 \leq \alpha \leq 1$. Ignoring Δ_{MCS}, δ_{PUSCH}, M_{PUSCH}, the receive power is given by $P_{0,\text{PUSCH}} + (1-\alpha)PL$. When $\alpha = 1$, the receive power is constant. Power control fully compensates path loss difference between users. (6.71) becomes similar to (6.70). When $\alpha = 0$, the transmit power is the same for all users. The difference in the receive power is as much as the difference in the path loss. When $0 < \alpha < 1$, the receive power depends on the path loss and is higher for a nearby user than for a faraway user. Power control only partially compensates path loss difference between users. Therefore, a nearby user achieves a higher modulation and coding scheme and thus a higher data rate than a faraway user. This is an important property of LTE.

On the other hand, fractional power control (6.71) does not take into consideration the path loss to the neighboring base stations. Because of shadowing, it is possible that a user has small path loss to both the serving and neighboring base stations. In that case, targeting a large receive power causes excessive inter-cell interference. This drawback is because a user is connected to one base station at a time in LTE. If the user is connected to both base stations in an MBB manner as studied in Section 9.3, then the user can estimate not only the path loss to its serving base station but also the inter-cell interference to the neighboring base station.

Equations (6.70) and (6.71) assume that the PUCCH and PUSCH are transmitted separately in time so that they do not have to share the total power P_m. When they occur simultaneously, according to Example 2.2, the PUCCH resource is not used, and control and data are multiplexed in the PUSCH. In this case, the power budget P_m is first used for the control part, and whatever remains is used for the data part.

The transmit power of the SRS follows a similar scheme as (6.71):

$$P_{\text{SRS}} = \min(P_m, P_{0,\text{PUSCH}} + \alpha \cdot PL + \delta_{\text{PUSCH}} + 10\log_{10} M_{\text{SRS}} + \Delta P_{\text{SRS}}), \quad (6.72)$$

where M_{SRS} is the SRS bandwidth and ΔP_{SRS} a configurable offset.

6.3 Fractional frequency reuse

In [168], CDMA is shown to achieve a higher capacity than narrowband FDMA because of universal frequency reuse. Inter-cell interference averaging (through code spreading in CDMA or tone hopping in OFDMA) helps achieve universal frequency reuse. Does universal frequency reuse always outperform frequency reuse $K > 1$? Is interference averaging the optimal solution to manage inter-cell interference?

Let us consider a simple example of a two-cell downlink where each base station serves one user. Denote by $G_{n,i}$ the channel power gain from base station n to user i, and σ^2 noise power. Assume the scenario of the symmetric channel where $G_{1,1} = G_{2,2}$ and $G_{1,2} = G_{2,1}$. Focus on user 1. With universal frequency reuse, the downlink data rate is

$$R_1 = \log\left(1 + \frac{PG_{1,1}}{\sigma^2 + PG_{2,1}}\right). \quad (6.73)$$

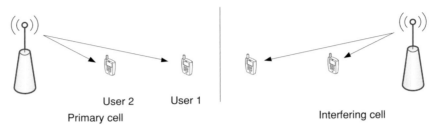

Figure 6.11 A symmetric two-cell two-subband two-user scenario.

With frequency reuse $K = 2$, the whole bandwidth is divided into two pieces of equal size and each base station operates only in one piece. The rate is

$$R_1 = \frac{1}{2} \log\left(1 + \frac{2PG_{1,1}}{\sigma^2}\right). \tag{6.74}$$

In a thermal limited case with $\sigma^2 \gg PG_{2,1}$, the inter-cell interference in universal reuse is much smaller than the thermal noise. Because of the concavity of the log function, universal reuse outperforms frequency reuse $K = 2$. However, in an interference limited case with $PG_{2,1} \gg \sigma^2$, reuse $K = 2$ eliminates the inter-cell interference and results in a greater SINR boost than what is needed to make up for the loss in bandwidth.

Generally in a system some users are interference limited while others thermal limited. Therefore, neither universal reuse or reuse $K > 1$ always outperforms the other. In this section, we introduce a new frequency reuse scheme, called *fractional frequency reuse* (FFR), to deal with such a mixed user scenario. Chapter 3 introduces the notion of subbands in OFDMA. As in FDMA/TDMA using narrowband channels, in OFDMA only users allocated to the same subband interfere with each other. The size of a subband is not important for the purpose of our discussion. Within a subband, we still employ tone hopping to achieve interference averaging.

The idea of FFR is that neighboring base stations use different transmit power patterns among the subbands in the downlink or different interference budget patterns in the uplink. As will be shown next, in order to obtain a capacity gain, the transmit power or interference budget patterns are designed so as to control or "shape" the SINR distribution of interference-limited and thermal-limited users. We will concentrate on the downlink and the ideas are applicable to the uplink as well.

6.3.1 A two-cell analysis

For simplicity, we first consider a symmetric two-cell scenario in the downlink, as shown in Figure 6.11. Similar to the model used to study intra-cell user multiplexing in Chapter 5, each base station serves two users, an inferior user and a superior user indexed as 1 and 2, respectively. The superior user is near the serving base station and assumed to be thermal limited, and the inferior one is close to cell edges and is interference limited. The channel conditions of the users in both cells are symmetric such that we focus on one cell, referred to as the primary cell, and call the other one the interfering cell.

6.3 Fractional frequency reuse

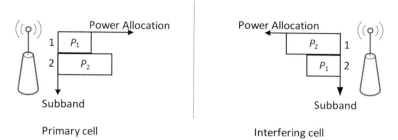

Figure 6.12 Two-subband symmetric cell power allocation. The primary cell uses power pattern (P_1, P_2) and the interfering cell uses (P_2, P_1), with $P_2 > P_1$.

Similar to frequency reuse $K = 2$, the whole bandwidth is divided into two pieces of equal size, called subbands. In subband i, the transmit power is P_i in the primary cell, and P_i' in the interfering cell. Suppose $P_2 > P_1$. Denote by (P_1, P_2) and (P_1', P_2') the power vectors. In the primary cell, a fraction $\alpha_{l,i}$ of the subband l is allocated to user i, $i = 1, 2$ and $l = 1, 2$. In this section, we fix the power vectors and vary the bandwidth allocation to study the capacity region

$$R_1 = \alpha_{1,1} \log\left(1 + \frac{G_{1,1}P_1}{\sigma^2 + G_{2,1}P_1'}\right) + \alpha_{2,1} \log\left(1 + \frac{G_{1,1}P_2}{\sigma^2 + G_{2,1}P_2'}\right), \quad (6.75)$$

$$R_2 = \alpha_{1,2} \log\left(1 + \frac{G_{1,2}P_1}{\sigma^2 + G_{2,2}P_1'}\right) + \alpha_{2,2} \log\left(1 + \frac{G_{1,2}P_2}{\sigma^2 + G_{2,2}P_2'}\right), \quad (6.76)$$

where $\alpha_{1,1} + \alpha_{1,2} = 1$ and $\alpha_{2,1} + \alpha_{2,2} = 1$.

Due to the channel symmetry, we restrict ourselves to the following two symmetric setting of the power vectors: $(P_1', P_2') = (P_1, P_2)$ or $(P_1', P_2') = (P_2, P_1)$. With the former choice, the same transmit power is used in each subband. It can be shown that this choice is always inferior to the latter choice, in which the two base stations use different power patterns as shown in Figure 6.12.

Let us focus on the choice of $(P_1', P_2') = (P_2, P_1)$. We can obtain a simple characterization of the capacity region of the two users. From (6.75) and (6.76), R_i is the sum of rates that user i obtained from the two subbands. Thus, the achievable rate region is the Minkowski sum of the rate regions of each subband; that is, if (x_i, y_i) is in the rate region of subband i, then $(x_1 + x_2, y_1 + y_2)$ is in the rate region of (6.75) and (6.76). Because bandwidth allocation is the only optimization variable, the Pareto-boundary of the rate region of each subband is simply a straight line, as shown in Figure 6.13. The rate region of subband 2 completely dominates that of subband 1, because not only it uses larger transmit power $P_2 > P_1$ but also sees smaller inter-cell interference $P_2' < P_1'$. In the primary cell, subband 2 is called the strong subband and subband 1 the weak one. One interesting observation here is that the slope of the Pareto-boundary of subband 2 is less steep than that of subband 1, for the following reason. Obviously, for either user, the SINR and rate will increase if it is allocated to subband 2 as opposed to subband 1. The

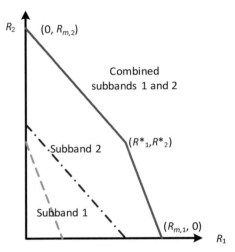

Figure 6.13 Rate regions of the primary cell in a two-subband symmetric downlink model of Figure 6.12.

increase is greater for the inferior user than for the superior one, because the inferior user is in a low SINR regime and the rate is more sensitive to a change in SINR.

The Pareto-boundaries of the capacity region are shown in Figure 6.13 and characterized as the polygon with vertexes given by $(0, 0)$, $(R_{m,1}, 0)$, $(0, R_{m,2})$ and (R_1^*, R_2^*) given by

$$R_{m,1} = \log\left(1 + \frac{G_{1,1}P_1}{\sigma^2 + G_{2,1}P_2}\right) + \log\left(1 + \frac{G_{1,1}P_2}{\sigma^2 + G_{2,1}P_1}\right), \qquad (6.77)$$

$$R_{m,2} = \log\left(1 + \frac{G_{1,2}P_1}{\sigma^2 + G_{2,2}P_2}\right) + \log\left(1 + \frac{G_{1,2}P_2}{\sigma^2 + G_{2,2}P_1}\right), \qquad (6.78)$$

$$R_1^* = \log\left(1 + \frac{G_{1,1}P_2}{\sigma^2 + G_{2,1}P_1}\right), \qquad (6.79)$$

$$R_2^* = \log\left(1 + \frac{G_{1,2}P_1}{\sigma^2 + G_{2,2}P_2}\right). \qquad (6.80)$$

$R_{m,i}$ ($i = 1, 2$) is the capacity of user i when it is allocated the entire two subbands. (R_1^*, R_2^*) is obtained when the superior user is allocated to the weak subband 1 only ($\alpha_{1,2} = 1$, $\alpha_{2,2} = 0$) and the inferior user to the strong subband 2 only ($\alpha_{1,1} = 0$, $\alpha_{2,1} = 1$). Furthermore, it can be shown that to obtain *any* point on the Pareto-boundary, either one of the two following conditions has to be satisfied:

- The superior user is *only* allocated to the weak subband.
- The inferior user is *only* allocated to the strong subband.

The same FFR principle applies in the interfering cell. Because of the symmetry of power vectors, (R_1^*, R_2^*) in the interfering cell corresponds to the bandwidth allocation

6.3 Fractional frequency reuse

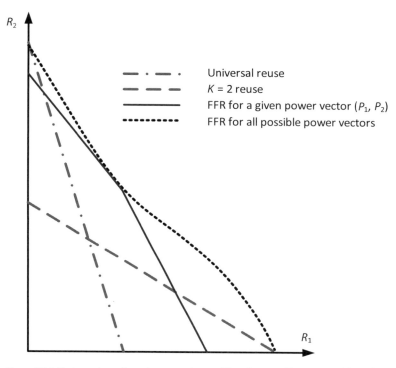

Figure 6.14 Rate regions for a two-user two-subband two-cell system with under universal reuse, reuse $K = 2$ and FFR with a given choice of power P_1, P_2, and the optimal FFR with all possible choices of P_1, P_2 assuming $P_1 + P_2 = P_m$ and $P_2 \geq P_1$.

where the superior user is allocated to the weak subband only and the inferior user to the strong subband.

After we introduce the idea of FFR, a natural question is how it compares with universal frequency reuse or reuse $K = 2$? Suppose that the constraint is the total transmit power (P_m). In FFR, P_m is split into $P_m = P_1 + P_2$ with $P_2 > P_1$ and the two neighboring base stations use symmetric power vectors (P_1, P_2) and (P_2, P_1) in two subbands. Universal reuse can be viewed as a special case of FFR where the same power is used in both subbands with $P_1 = P_2 = P_m/2$. Reuse $K = 2$ is another special case where $P_2 = P_m$, $P_1 = 0$.

Figure 6.14 qualitatively compares the rate regions of those three schemes. As expected, when all the bandwidth is allocated to the superior user, its rate is maximized with universal reuse. When all the bandwidth is allocated to the inferior user, its rate is maximized with reuse $K = 2$. The FFR scheme with a given choice of P_1, P_2 outperforms the other two schemes when the operating point is away from those two extreme cases and the system has to serve mixed superior and inferior users, which is a more realistic scenario.

Also shown in Figure 6.14 is the rate region when one can choose any P_1, P_2 as long as $P_1 + P_2 = P_m$ and $P_2 \geq P_1$. It is the union of the rate regions under all possible power

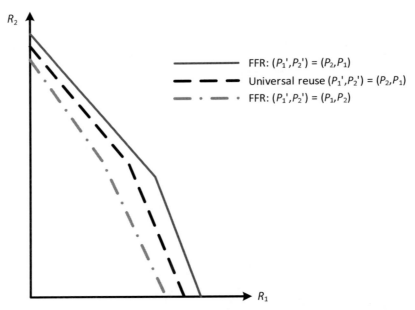

Figure 6.15 Rate regions for a two-user two-subband two-cell system under FFR and under universal frequency reuse. In FFR, inter-cell interference occurs between the two cells in subband 1 and in subband 2, but there is no inter-cell interference between subband 1 of one cell and subband 2 of the other cell. In universal frequency reuse, inter-cell interference is averaged over the two subbands. The outmost curve is the FFR rate region when the interfering cell uses $(P_1', P_2') = (P_2, P_1)$. The curve in the middle represents universal frequency reuse. The innermost curve is obtained with FFR when $(P_1', P_2') = (P_1, P_2)$. Recall the primary cell power pattern is (P_1, P_2).

allocation choices. We observe that either universal reuse or reuse $K = 2$ is the optimal solution only in an extreme case when R_2 or R_1 is to be maximized; otherwise we can always find a choice of P_1, P_2 with which FFR achieves a higher data. The performance gap can be potentially significant.

As a final note, we point out the similarity of FFR and joint power/bandwidth allocation studied in Chapter 5. The context is obviously very different in that FFR deals with inter-cell interference and spectrum reuse while Chapter 5 considers intra-cell user multiplexing. However, the resultant strategy is similar: transmit to inferior users at a high power in some part of the bandwidth and at a low power to superior users in the rest of the bandwidth. FFR furthermore requires that neighboring cells employ different power patterns in a coordinated manner.

Given the similarity, does the gain we observe in Figure 6.14 actually come from FFR or user multiplexing that we have observed in Chapter 5? To answer this question, note that a key ingredient of FFR is that the inter-cell interference is averaged only locally within each subband and not across the whole bandwidth as in universal reuse. It can be shown that using the same power vector (P_1, P_2) in the primary cell and $(P_1', P_2') = (P_2, P_1)$ in the interfering cell, the rate region with universal frequency reuse (inter-cell interference averaging across both subbands) is a subset of that with FFR.

The rate region comparison is shown in Figure 6.15. Clearly, user multiplexing within a cell alone cannot achieve the same gain as FFR. For any power vector (P_1, P_2), we can improve the rate region from universal frequency reuse by introducing subbands and local interference averaging. It is also shown in Figure 6.15 that if the interfering cell employs a wrong power pattern of $(P'_1, P'_2) = (P_1, P_2)$, FFR actually performs worse than universal frequency reuse.

> **Discussion notes 6.2** Motivation of fractional frequency reuse from a different angle
>
> We can motivate the idea of FFR from the perspective of *carrier diversity*. Specifically, suppose that the operator first deploys one channel in a geographic area. As traffic demand increases, the operator deploys a second channel. Usually, the second channel is deployed with the same transmit power setting as the first one. As a result, the cell boundary is the same in both the first and second channels, as shown in Figure 6.16(a). A user located in the cell boundary area experiences low SINR (≤ 0 dB) in either channel.
>
>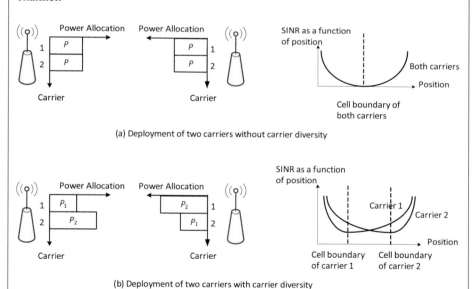
>
> **Figure 6.16** Illustration of carrier diversity.
>
> Alternatively, the operator can offset the transmit power used in the two carriers as in Figure 6.12. Then the cell boundary of the first carrier moves to the left and that of the second carrier moves to the right, as shown in Figure 6.16(b). Now any user can always find a carrier on which the user is not located at the cell boundary. For example, a user at the cell boundary of carrier 1 switches to carrier 2 and will then no longer be at the cell boundary of carrier 2. This way, the coordinated power allocation across neighboring cells leads a new form of diversity, namely carrier diversity, and eliminates the adverse effect of users being at low SINR cell boundary.

Reduced-power channel reuse [94] is yet another early example of FFR. Specifically, consider the following problem in a linear cellular system where the cells are indexed from left to right as $1, 2, \ldots$. The total bandwidth is divided into two distinct channels A and B. In each cell, one channel is transmitted at the full power and is called the primary channel, and the other channel is transmitted at a reduced power and is called the secondary channel. In cell $1, 3, 5, \ldots$, A is primary and B is secondary. In cell $2, 4, 6, \ldots$, A is secondary and B is primary.

Consider a simple channel selection scheme using a probabilistic policy; the user at location x selects to use the primary channel with probability $p(x)$ or the secondary channel with probability $1 - p(x)$. Let $g_P(x)$ and $g_S(x)$ be the traffic load densities at location x on the primary and secondary channels, respectively. Note that from $g_P(x), g_S(x)$, one can easily derive the channel selection decision probability

$$p(x) = \frac{g_P(x)}{g_P(x) + g_S(x)}. \qquad (6.81)$$

Denote by $P_P(x)$, $P_S(x)$ the capture probabilities of traffic being successfully delivered on the primary and secondary channels, respectively. Clearly, the capture probabilities depend on both the location where packet transmission occurs in the cell and the interference from adjacent cells. The throughout density at location x is given by

$$s(x) = g_P(x) P_P(x) + g_S(x) P_S(x). \qquad (6.82)$$

Let us consider a general channel selection problem of maximizing a utility function $f(s(\cdot))$ subject to fairness constraint $h(s(\cdot)) \geq 0$ for a wide range of functions f, h. It turns out that the optimal solution is in the following form of bang-bang control. Each cell is divided into two complementary regions. Users in one region use only the secondary channel ($g_P(x) = 0$), and those in the other region use only the primary channel ($g_S(x) = 0$). The cell partition is determined by comparing the ratio $P_P(x)/P_S(x)$ with a constant threshold. As a result, nearby users only use the secondary (reduced power) channel and faraway users the primary (full power) channel.

6.3.2 Static FFR in a multi-cell scenario

In this section, we extend the principle of FFR developed in the preceding two-cell analysis to a multi-cell scenario. Figure 6.17 illustrates a few power/subband configuration examples.

- In (a), the whole bandwidth is divided into four subbands, f_0, f_1, f_2, f_3. Subband f_0 is transmitted at a low power for serving superior users and reused in every cell. Subbands f_1, f_2, f_3 are transmitted at a high power for inferior users and reused with frequency reuse pattern $K = 3$.
- In (b), there are three subbands, f_1, f_2, f_3, which are reused with $K = 3$ at a high power for inferior users. However, rather than using a separate subband f_0 as in (a), the same subbands f_1, f_2, f_3 also serve superior users at a low power. For example, in

6.3 Fractional frequency reuse

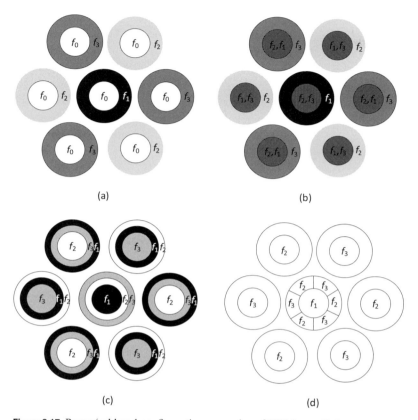

Figure 6.17 Power/subband configuration examples of FFR in a cellular system.

a cell with f_1 serving inferior users (the center cell shown in b), f_2, f_3 serve superior users. Other cells employ similar reuse patterns.

- The idea of (b) is refined in (c) where users are divided into three groups, namely interior, middle, and edge users. In a cell, the interior users are served at a lowest power, followed by the middle and then the edge users at a highest power. The subband pattern varies from one cell to another. For example, in the center cell shown in (c), the subbands are f_1, f_2, f_3 for interior, middle and edge users, respectively. In a neighboring cell, the pattern changes to f_2, f_3, f_1 or f_3, f_1, f_2.
- In (a) to (c), all inferior users in a cell use the same subband. However, note that the strongest interfering base station of a different inferior user may be different. A fine granular power coordination pattern can thus be employed as shown in (d). In the center cell, subband f_1 is transmitted at a low power for superior users, and subbands f_2 and f_3 are transmitted at a high power for inferior users. Inferior users are grouped depending on their strongest interfering base stations. In (d), three neighboring cells use f_2 at a low power. In the center cell, any inferior user close to one of those cells uses f_2. Similarly, three other neighboring cells use f_3 at a low power, and any inferior user in the center cell close to one of those cells uses f_3.

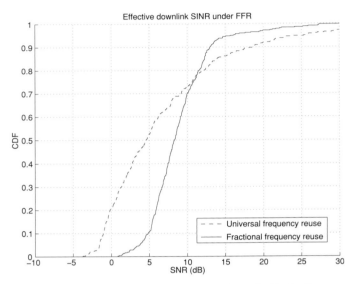

Figure 6.18 Comparison of effective downlink SIR distribution between universal frequency reuse and fractional frequency reuse. Power vector is $(P, P/4, P/16)$.

We next numerically analyze the performance of the FFR scheme in (c). Specifically, subbands f_1, f_2, f_3 are of the same size. Each base station employs a static power scaling, $P, P/\eta, P/\eta^2$, for the edge, middle and interior users, with some $\eta > 1$. A base station schedules users over time and subbands. For simplicity, we assume that only one user is scheduled at a subband at a given time slot. We do not consider joint power bandwidth optimization within a subband. Base stations are located at regular hexagonal grids, and users are uniformly distributed according to a Poisson point process. Channel gain is stationary and depends on path loss with exponent $r = 3.5$. Any user is connected with the strongest base station. Noise power is ignored.

We briefly describe the scheduling algorithm. Recall that the principle of FFR is that inferior users are allocated to a high power subband and superior users to a low power subband. In effect, the use of different power vectors at different base stations induces channel fluctuation over frequency, which from Section 4.3 can be effectively exploited with proportional fair scheduling. Specifically, the user scheduling algorithm (4.102) is slightly modified to work in an individual subband so as to maximize $\frac{R_{i,l}}{T_i}$, where $R_{i,l}$ is the data rate of user i in subband f_l. It is interesting to note that because rate is a log function of SIR, inferior users operating at a low SIR are more sensitive to power fluctuation over frequency than superior ones at a high SIR, and thus in proportional fair scheduling, is more likely scheduled to a high power subband than to a low power subband.

Figure 6.18 shows the distribution of the scheduled SIR, which is the SIR when a user is scheduled. The C/I of a user only depends on the geometry characterized by the channel gains with the base stations. In universal frequency reuse, the scheduled SIR is the same as the C/I, because without channel variation, the proportional fair scheduler

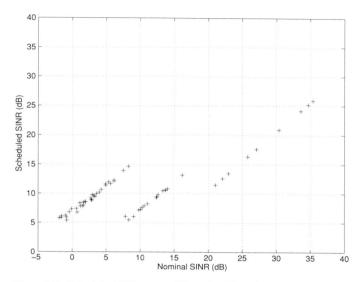

Figure 6.19 Scheduled SIR versus C/I under FFR. The left portion is for the subband using power P, the middle portion for the subband using power $P/4$, and the right portion for the subband using power $P/16$.

simply schedules the users in a round-robin fashion and all the users get scheduled for the same amount of time. On the other hand, in FFR, the inferior users are scheduled to a high power subband and experience low power inter-cell interference, thereby getting a significant SIR boost. Meanwhile, superior users are scheduled to a low power subband, and the scheduled SIR is lower than the C/I. Overall, the dynamic range of the scheduled SIR becomes significantly smaller than that of the original C/I.

The effect of shaping SIR can be seen in Figure 6.19, which compares the scheduled SIR and C/I in a given cell. For the users with poor C/I, the scheduled SIR is boosted by 6–9 dB. For the users with good C/I, the scheduled SIR is reduced by about 9 dB. In addition, we observe that the number of users scheduled to a high power subband is much greater than to a low power subband. This indicates that for the superior users, which are bandwidth limited, although the scheduled SIR is lower than the C/I, they are scheduled more often and get more bandwidth allocation. The inferior users, which are power limited, are scheduled less frequently but get a significant SIR boost when scheduled. As a result, all the users improve under FFR. This is verified in Figure 6.20, which shows that the throughput of almost all the users increases.

So far, we have studied FFR in a non-sectorized cell deployment with omni-directional antennas. Figure 6.21 illustrates two examples of configuring power patterns in a sectorized system. It is desirable to employ the same power/subband pattern among sectors of the same cell and different patterns for neighboring cells, as shown in Figure 6.21(a). In this case, the performance gain will be similar to that in a non-sectorized cellular system studied before. Recall that within a cell, the power/subband pattern in Figure 6.21(a) is similar to what is advocated from the perspective of user multiplexing in Sector 5.3.2.

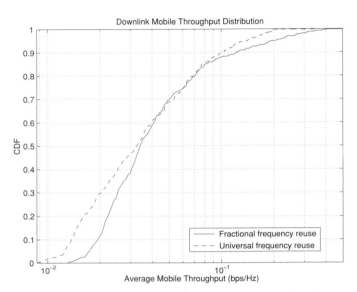

Figure 6.20 Comparison of throughput distribution between universal frequency reuse and fractional frequency reuse. Power vector is $(P, P/4, P/16)$.

From the perspective of keeping a large dynamic range in the presence of imperfect sectorization, this configuration works better than the one shown in Figure 6.21(b), where the sectors of a given cell employs different power/subband patterns and the same configuration is reused in all cells.

Finally, it should be pointed out that the benefit of FFR comes from channel variation in frequency coordinated among neighboring cells. In a practical system, time and frequency selective fading already exists and causes independent channel variations among users. A smart scheduler such as a proportional fair scheduler can harness multiuser diversity by opportunistically scheduling users at their relatively good channel conditions. Therefore, the additional gain brought by FFR will be less significant in a fading scenario than in a non-fading scenario. For example, as noted in Sector 4.3.1, channel fading boosts the scheduled SINR of a cell edge user due to multiuser diversity, even without FFR.

6.3.3 Breathing cells: FFR in the time domain

The FFR idea of introducing channel variation in frequency can be extended to the time domain. Suppose there is only one subband. Neighboring base stations can vary transmit power according to some coordinated pattern. We call such a scheme *breathing cells* and the time-varying power pattern *breathing pattern*, because the coverage of a cell changes over time. Figure 6.22 shows an example where neighboring cells choose different power variation curves, $P_1(t)$, $P_2(t)$, $P_3(t)$ to create time-varying SINR. Figure 6.23 illustrates one example of $P_1(t)$, $P_2(t)$, $P_3(t)$ where a cell varies its power between $0.5P_m$ and $1.5P_m$ in a period of 100 time slots. In the breathing cells scheme, the power variation

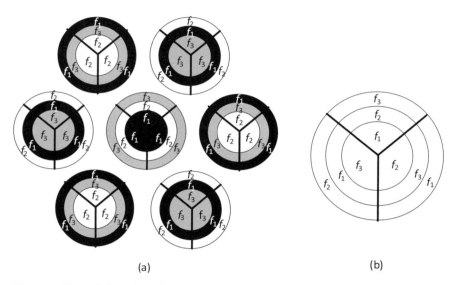

(a)　　　　　　　　　　　　　　　(b)

Figure 6.21 Power/subband configuration examples of FFR in a sectorized cellular system. In (b), the power/subband configuration is for three sectors of a given cell, and the same configuration is reused in all cells.

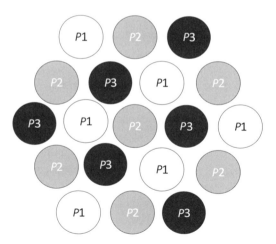

Figure 6.22 An exemplar power pattern in which a cell uses one of three power variation curves, denoted by $P1$, $P2$, $P3$, in a hexagonal deployment.

occurs at a much slower time scale than communications so as to allow the users to track the channel variation.

A regular breathing pattern as shown in Figure 6.22 requires cell planning. For example, the three curves in Figure 6.23 all have the same frequency but different phase offsets, which need to be planned. In addition, base stations need to be synchronized to the planned phases. Alternatively, each base station can employ a different frequency,

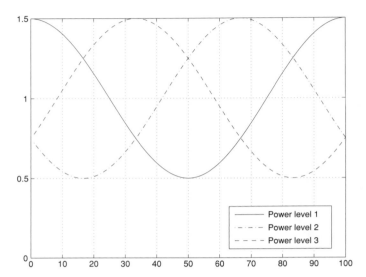

Figure 6.23 An example of power variation curves: power versus time. Each cell chooses one of three possible power level curves and varies its transmit power between 0.5 and 1.5 times its average power constraint P_m.

thereby eliminating the requirements of cell planning or base station synchronization. Simulations show that similar capacity improvement can be obtained as compared to FFR with multiple subbands.

One drawback of the breathing cells scheme is that because the power variation is slow, it leads to long outage period for edge users and delay sensitive traffic may suffer due to stringent latency budget. To address the issue, we combine power variation in time and in frequency. Specifically, consider again the scenario of FFR with three subbands. Instead of static power vectors, vary power across subbands. For example, vary the power of each subband according to one of the three curves shown in Figure 6.23. As a result, at any time, at least one subband has a decent SINR to serve delay sensitive traffic.

In essence, the breathing cells scheme induces channel fluctuation and a cellular system with smart scheduling benefits from multiuser diversity. Recall from Section 4.3.2 a similar form of induced multiuser diversity is opportunistic beamforming, where a base station uses multiple transmit antennas to form slowly varying beams sweeping across all users in the cell. The differences between those two schemes are worth pointing out:

- The breathing cells scheme does not require the use of multiple antennas.
- The gain of breathing cells can only be seen when each cell has users with different channel conditions. However, when all users are superior, breathing cells can actually lead to a capacity loss. In this case, opportunistic beamforming can still see a substantial gain as long as users have different angular directions from the base station.
- In breathing cells the scheduler does not schedule all users at their peaks. In particular, the scheduler tends to schedule inferior users at good channel conditions but superior

users at weak conditions. In opportunistic beamforming, all users are scheduled around their peaks assuming enough users are available in the cell for scheduling.

6.3.4 Adaptive FFR

So far we have assumed fixed power/subband patterns in FFR, which perform well in the presence of mixed superior and inferior users. However, as noted in the two-cell analysis, FFR is suboptimal in other scenarios. For example, if all the users are superior, introducing a disparity in transmit power in different subbands actually causes a capacity loss, because the Shannon capacity is a concave function of the transmit power, therefore it follows that

$$\sum_{l=1}^{L} \log\left(1 + \frac{GP_l}{\sigma^2}\right) \leq L \log\left(1 + \frac{G \sum_{l=1}^{L} P_l}{L\sigma^2}\right). \tag{6.83}$$

Ideally, we would like FFR to adapt to the scenario. Assuming fixed subband allocation in frequency, we consider adapting power vector and bandwidth allocation:

$$\max_{\boldsymbol{\alpha}, \boldsymbol{P}} \sum_{b} \sum_{i:c_i=b} U_i(R_i) \tag{6.84}$$

$$\text{s.t. } R_i = \sum_{l} \alpha_{l,i} \log\left(1 + \frac{G_{c_i,i} P_{c_i,l}}{\sigma_l^2 + \sum_{b \neq c_i} P_{b,l} G_{b,i}}\right), \tag{6.85}$$

$$\sum_{i:c_i=b} \alpha_{l,i} \leq 1, \quad \forall b, \tag{6.86}$$

$$\sum_{l} P_{b,l} \leq P_m, \quad \forall b, \tag{6.87}$$

$$\boldsymbol{\alpha} \geq 0; \boldsymbol{P} \geq 0. \tag{6.88}$$

where b is the base station index, l the subband index, and i the user index. σ_l^2 is the noise power in subband l. $\boldsymbol{\alpha}$ is the bandwidth allocation matrix, where $\alpha_{l,i}$ represents the steady state fraction of time or frequency that user i is allocated to subband l. $\boldsymbol{P}$ is the power matrix, where $P_{b,k}$ represents the power allocated to subband l at base station b. For simplicity we ignore the power bandwidth tradeoff for users sharing the same subband. As we have seen in previous sections, a smart scheduler will push users with similar channel and interference conditions to the same subband and thus there is not much additional gain to exploiting intra-subband power bandwidth optimization.

It is difficult to develop either analytical characterization or iterative/distributed algorithms for the above optimization. Actually, a much simplified version where each base station only serves one user is well studied in a different setup, the so-called DSL problem (see [184] and the references therein). The DSL problem is a well-known non-convex optimization and only centralized solutions are known in an asymptotic regime where the number of subbands goes to infinity. The problem of (6.84) introduces additional complexity of multiuser scheduling to the original DSL problem. In this section, we will borrow a few ideas from the DSL literature and propose adaptive FFR schemes to solve (6.84) iteratively.

The first scheme is called iterative water-filling. The idea is that each base station greedily and locally updates its power and bandwidth allocation with no explicit information exchange between base stations. In (6.84), let base station b only try to solve for $\boldsymbol{\alpha}_b$ and $\boldsymbol{P}_b$ to maximize its own utility,

$$\max_{\boldsymbol{\alpha}_b, \boldsymbol{P}_b} \sum_{i:c_i=b} U_i(R_i) \tag{6.89}$$

$$\text{s.t. } R_i = \sum_l \alpha_{l,i} \log\left(1 + \frac{G_{b,i} P_{b,l}}{\sigma_l^2 + I(l,i)}\right), \text{ for all } i \text{ such that } c_i = b, \tag{6.90}$$

$$\sum_{i:c_i=b} \alpha_{l,i} \leq 1, \tag{6.91}$$

$$\sum_l P_{b,l} \leq P_m, \tag{6.92}$$

$$\boldsymbol{\alpha}_b \geq 0; \boldsymbol{P}_b \geq 0, \tag{6.93}$$

where $I(l, i)$ represents the inter-cell interference seen at user i in subband l. Assuming fixed $I(l, i)$, the above objective function (6.89) is not jointly, but marginally concave in $\boldsymbol{\alpha}_b$ and $\boldsymbol{P}_b$. Specifically, given bandwidth allocation $\boldsymbol{\alpha}_b$, it is concave with respect to $\boldsymbol{P}_b$; and given power allocation $\boldsymbol{P}_b$, it is again concave with respect to $\boldsymbol{\alpha}_b$. Thus, a natural heuristic algorithm is to introduce iterations at two different time scales. A fast time scale inner loop determines the optimal bandwidth allocation $\boldsymbol{\alpha}_b$ given $\boldsymbol{P}_b$. An outer loop updates the power allocation $\boldsymbol{P}_b$ in a slow time scale once $\boldsymbol{\alpha}_b$ has converged. When the utility function is a log function of the rate, the base station runs a proportional fair scheduler for a number of time slots in the inner loop, and records the steady state bandwidth allocation fraction in each subband and the observed inter-cell interference as the input to the outer loop. In the outer loop, the base station calculates the optimal power allocation according to the water-filling principle. Obviously after the power allocation is updated in all the cells, the inter-cell interference $I(l, i)$ also changes, and the iteration will continue.

The performance of iterative water-filling has been studied extensively in DSL literature [184] and also in the context of spectrum sharing for unlicensed bands [46]. It is noted that the algorithm usually does not lead to the optimal solution when a base station only serves one user. The reason is that the algorithm may converge to a local Nash equilibrium point of equal power allocation in all the subbands, which is highly sub-optimal when the users are inferior. Interestingly, in a multiuser scenario with mixed superior and inferior users, iterative water-filling does not get stuck at equal power allocation. Consider the simple two-cell two-user two-subband example shown in Figure 6.12. Suppose that both cells start with equal power allocation in the two subbands. When each base station tries to optimize the bandwidth and power allocation over the subbands, it tends to apply a different power to a different subband and to schedule superior and inferior users accordingly, thanks to power bandwidth optimization in intra-cell user multiplexing. A neighboring base station then responds by allocating its power according to the water-filling principle; that is, it allocates more power to the subband of weak interference and less power the one of strong interference. This reinforces the convergence of the iterative algorithm towards unequal power allocation.

Iterative water-filling requires very low system overhead as no messaging exchange is required among base stations. Explicit collaboration among cells by exchanging control messages through backhaul will improve the algorithm performance and overcome local equilibria. We next introduce an algorithm proposed in [152], where base stations exchange the partial derivatives of the utility function with respect to power allocation at individual subbands.

Define *local marginal utility* the partial derivative of the utility function at base station b' with respect to the power choice in subband l at base station b,

$$D_l^{b,b'} = \frac{\partial U_{b'}}{\partial P_{b,l}}, \tag{6.94}$$

$$= \sum_{i:c_i=b'} \frac{\partial U_{b'}}{\partial R_i} \frac{\partial R_i}{\partial P_{b,l}}$$

$$\approx \sum_{i:c_i=b'} \frac{\partial U_{b'}}{\partial R_i} \alpha_{l,i} \frac{\partial R_{l,i}}{\partial P_{b,l}}, \tag{6.95}$$

where the approximation ignores the impact of $P_{b,l}$ on bandwidth allocation $\boldsymbol{\alpha}$. $R_{l,i}$ is the rate of user i at subband l

$$R_{l,i} = \log\left(1 + \frac{G_{b',i} P_{b',l}}{\sigma_l^2 + \sum_{b'' \neq b'} G_{b'',i} P_{b'',l}}\right). \tag{6.96}$$

To estimate $D_l^{b',b}$ in (6.95), base station b' needs the following input: R_i, $\alpha_{l,i}$, transmit power $P_{b'',l}$, and channel power gain $G_{b'',i}$. To obtain the first two, base station b' can run a fast time scale inner loop with a proportional fair scheduler for a number of time slots, similar to iterative water-filling. Base stations exchange the transmit power allocation information among themselves via backhaul. User i reports to its serving base station b' the channel gains with respect to neighboring base stations b''.

Suppose that base station b' estimates $D_l^{b,b'}$ for all b, l, and sends the information to neighboring base stations. To know the effect of changing any of its transmit power over any subband to the total system utility, base station b calculates

$$D_{b,l} = \sum_{b'} D_l^{b,b'} = \sum_{b'} \frac{\partial U_{b'}}{\partial P_{b,l}}, \tag{6.97}$$

The idea is to locally improve the total utility function by increasing (decreasing) the power in a subband that has a positive (negative) utility derivative. Base station b takes the following three steps to realize this idea. First, it finds subband l_1 that has the smallest $D_{b,l}$ for all l with $D_{b,l} < 0$ and $P_{b,l} > 0$, and reduces P_{b,l_1} by a step size Δ,

$$P_{b,l_1} = \max(P_{b,l_1} - \Delta, 0). \tag{6.98}$$

Second, it calculates the total transmit power $P_b = \sum_l P_{b,l}$. If $P_b < P_m$, there is unused power budget left, it increases the power of the subband l_2 that has the largest $D_{b,l}$ for all l with $D_{b,l} > 0$,

$$P_{b,l_2} = P_{b,l_2} + \min(\Delta, P_m - P_b). \tag{6.99}$$

Third, if $P_b = P_m$, it picks a pair of subbands l_3, l_4 where the corresponding D_{b,l_3} is the largest and D_{b,l_4} is the smallest among all l satisfying $P_{b,l} > 0$. It re-allocates the power as follows:

$$P_{b,l_3} = P_{b,l_3} + \min(\Delta, P_{b,l_4}), \tag{6.100}$$

$$P_{b,l_4} = \max(P_{b,l_4} - \Delta, 0). \tag{6.101}$$

Practical example 6.2 Inter-cell interference coordination in LTE

In addition to randomizing inter-cell interference through hopping and physical resource allocation as described in Example 2.1, LTE employs a more active mechanism, called inter-cell interference coordination (ICIC). The basic idea is similar to FFR; that is, to balance different reuse factors ($K = 1$ or $K > 1$) for superior and inferior users, to trade off SINR versus bandwidth, and to coordinate power patterns among neighboring cells. In particular, if one cell is using high transmit power in a time-frequency resource to serve an inferior user (either in the downlink or uplink), the neighboring cell should avoid using high power in the same resource.

Static power patterns studied in Section 6.3.2 are certainly applicable in LTE. Furthermore, to handle dynamic interference due to time-varying channel and traffic situations, ICIC defines various signaling messages to be exchanged between neighboring base stations over the backhaul, called X2 interface, to coordinate the use power and physical resource.

To coordinate the downlink interference, a base station sends to its neighboring base stations a Relative Narrowband Transmit Power (RNTP) indicator. An RNTP indicator is a bitmap where each bit represents a PRB and indicates whether the base station intends to transmit high power in that PRB. Upon the reception of an RNTP indicator, a neighboring base station should not schedule a cell edge user in that PRB to avoid strong interference.

The message similar to RNTP for coordinating the uplink interference is the High Interference Indicator (HII). An HII indicates the PRBs over which the base station intends to schedule cell edge users. Upon the reception of an HII, a neighboring base station should anticipate strong interference in that PRB and control its own SINR accordingly, for example, avoiding scheduling a cell edge user in that PRB. In addition to HII, the Overload Indicator (OI) is another message in which the base station indicates the measured interference level in each PRB. Upon the reception of a OI, a neighboring base station should reduce the transmit power scheduled in a heavily-loaded PRB to suppress the inter-cell inference to the base station that sends the OI.

> While all the three techniques aim at coordinating power and bandwidth usage among neighboring cells to achieve dynamic FFR, both RNTP and HII are proactive to forecast high interference to be generated beforehand while OI is reactive and responds to high interference that has already occurred.

6.4 Summary of key ideas

- Narrowband FDMA/TDMA, CDMA, and OFDMA are three important medium access technologies with different tradeoffs of spectrum reuse and intra-cell/inter-cell interference. In order to understand the tradeoffs, we can numerically study C/I and SINR distributions via simulation or analysis. The downlink C/I distribution only depends on the geometry. However, the uplink SINR distribution also depends on the transmit power scheme. Constant transmit or receive power is commonly used, but other power control ideas should be explored in a data-centric OFDMA system. SINR can then be derived from C/I.
- Power control is much more complex in the uplink than in the downlink because of the coupling effect among neighboring cells. The problem is non-convex and cannot be simply decomposed to a single-cell optimization. We divide uplink power control into a few simple steps: SINR assignment and feasibility check, power control, and joint bandwidth and SINR assignment. The load-spillage characterization allows SINR assignment to be decoupled among neighboring cells when the base station load is pre-assigned. Utility maximization can be done in a slow time scale to adjust the base station load.
- In an interference-limited scenario, a key parameter of uplink power control and SINR assignment is channel gain ratio, instead of just channel gain to its own base station. A large channel gain ratio means a user generates high inter-cell interference in order to deliver a unit signal power to its own base station. Everything else being equal, users with large channel gain ratio should be assigned with a low SINR target so as to reduce interference to neighboring cells.
- Universal frequency reuse and interference averaging are not always optimal. The idea of FFR is that the bandwidth is divided into subbands and inter-cell interference only occurs within a subband but not between different subbands. Neighboring base stations coordinate, either statically or dynamically, their power allocation to the subbands. FFR outperforms universal frequency reuse in the presence of mixed superior and inferior users.
- Power/bandwidth allocation in FFR is consistent with joint power/bandwidth optimization in intra-cell user multiplexing; that is, superior users are allocated lower power but more bandwidth, and inferior users are allocated higher power but less bandwidth. FFR in addition coordinates power patterns used in neighboring cells.
- FFR can be done in the frequency or time domain, and statically with fixed power/subband patterns or dynamically to adapt to user channel scenarios and to

maximize utility. Collaboration among neighboring cells by exchanging control messages such as marginal utility through backhaul helps improve FFR performance.
- We have focused on downlink FFR. The same ideas are applicable to the uplink by noticing the similarity between the transmit power in the downlink and the inter-cell interference budget in the uplink. In particular, introduce a different load in each subband in a cell, and neighboring cells employ different load/subband patterns.
- We have introduced the notion of FFR in an OFDMA system where the whole bandwidth divided into non-overlapping subbands. The same power/bandwidth allocation and coordination idea is applicable to other cases where subbands are not contiguous in a given frequency carrier or even belong to different frequency carriers, as shown in Figure 6.16 of carrier diversity. One such case is the carrier aggregation scenario discussed in Example 10.1.

7 Use of multiple antennas

In general, the use of multiple antennas provides three types of gains: diversity gain, power/SINR gain (including interference suppression), and multiplexing (bandwidth or degree of freedom) gain. In Chapter 4, we studied the first two and learned that

- Antenna diversity changes the channel statistics and reduces the probability of the channel being in a deep fade, thereby improving the link reliability. Receive diversity is relatively easy to obtain. With space-time coding, transmit diversity can be achieved without the transmitter knowing the channel.
- Beamforming achieves power/SINR gain by increasing signal power and/or reducing interference. Maximal ratio combining is commonly used for receive beamforming and is applicable for transmit beamforming if the channel is known at the transmitter. Without precise channel knowledge, transmit beamforming is not possible in a single user link, but can be opportunistically achieved in a multiuser system with multiple receivers and limited SINR feedback. Power/SINR gain is mostly beneficial in the low SINR regime.

In general, the three types of gains can be simultaneously obtained, but there is a tradeoff of how much of each type a given communication scheme can achieve. A well-known example is the fundamental tradeoff of diversity and multiplexing studied in [186]. For the sake of simplicity, we focus on multiplexing gain in this chapter, and do not address the tradeoff.

We first present a simple calculation to demonstrate the potential gain of spatial multiplexing. Denote by $\mathbf{H}$ a $K_r \times K_t$ complex channel matrix between a transmitter with K_t transmit antennas and a receiver with K_r receive antennas, where each entry $H_{m,n}$ represents the channel gain between transmit antenna n and receive antenna m. The Shannon capacity [50, 51, 157] of the MIMO channel is

$$C = \log \det \left(\mathbf{I} + \frac{\mathbf{H}\mathbf{H}^*}{K_t} \frac{P}{\sigma^2} \right)$$

$$= \sum_{i=1}^{K_r} \log \left(1 + \frac{\lambda_i}{K_t} \frac{P}{\sigma^2} \right), \quad (7.1)$$

where P is total transmit power of all the K_t antennas, σ^2 the noise power, and $\{\lambda_i\}$ the set of eigenvalues of the channel matrix product $\mathbf{H}\mathbf{H}^*$. Clearly, only non-zero eigenvalues make positive contributions to the capacity of the channel. In the high SNR regime, the capacity is significantly determined by the *rank* of the matrix $\mathbf{HH}^*$, which is equal to the number of non-zero eigenvalues. Since the matrix $\mathbf{HH}^*$'s rank is bounded by $\min(K_t, K_r)$, a maximum *multiplicative* gain is $\min(K_t, K_r)$. Hence, the Shannon capacity of the MIMO channel increases *linearly* with the number of antennas $\min(K_t, K_r)$. In effect, the MIMO channel adds a new dimension (space) to our previous notion of bandwidth (time and frequency) via spatial multiplexing. Thus the benefit is called spatial multiplexing gain or bandwidth gain.

As compared the diversity schemes, spatial multiplexing can contribute to a *pre-log* multiplicative factor in the channel capacity, as opposed to a power or diversity gain. However, we should keep in mind that there is a hardware cost associated with multiplexing. Specifically, every transmit or receive antenna is connected with a separate analog front-end that processes the signal between the antenna and the digital signal processor. The analog front-end typically includes filters for noise rejection, mixers for up or down frequency conversion, and power amplifiers at the transmitter or low noise amplifiers at the receiver. Obviously, the front-end adds costs of components and power consumption. To fully harness the multiplexing gain, multiple receiving chains and analog front-ends are required, as illustrated in Figure 7.1(a). As a comparison, it is feasible to have one analog front-end, which is connected with multiple antennas via a switch, as shown in Figure 7.1(b). The digital processor dynamically controls the switch depending on which antenna provides the best channel quality. For example, the transceiver quickly samples the channel quality of each antenna in a brief measurement interval every so often and stays with the best antenna in the remaining time until the next measurement interval. The transceiver in (b) achieves antenna selection diversity but no power gain. The biggest advantages are low component costs and low power consumption, making (b) attractive for low-cost power-sensitive mobile devices. Therefore, we should keep in mind that MIMO may not be desirable in all circumstances. The cost issue discussed here is more applicable to SU-MIMO than to MU-MIMO. As will be shown in Section 7.4, in MU-MIMO the complexity is handled at the base station and no costs are added to the users.

In this chapter, we will focus on the use of multiple antennas for improving system capacity, in particular in the high SINR regime where the benefit of power gain diminishes and spatial multiplexing gain is much desirable. We will investigate the practical constraints to achieve spatial multiplexing gain and the unique benefits of OFDMA in MIMO. To this end, we will start with MIMO channel modeling and then study a few use scenarios from single user, to multiuser, and to multi-cell/multi-sector MIMO.

7.1 MIMO channel modeling

In general, there are two widely considered methods in using multiple antennas: linear antenna arrays and polarized antennas. Between the two, linear antenna arrays are much better covered in literature.

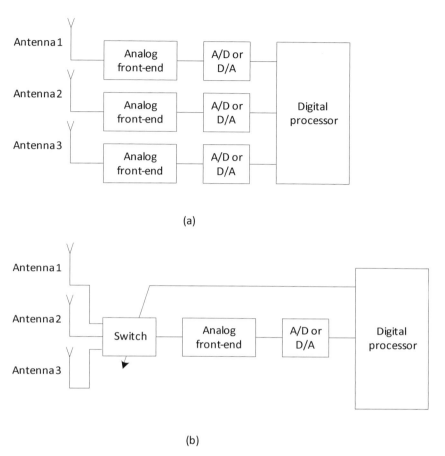

Figure 7.1 Comparison of transceivers using multiple antennas: (a) with multiple analog front-ends each connected with an antenna, (b) with a single analog front-end connected with all antennas via a switch.

7.1.1 Linear antenna arrays

In a linear antenna array, multiple antennas of the same type are placed on a straight line with preferably equal separation between any two adjacent elements. Figure 7.2 shows a typical MIMO channel with linear antenna arrays at both the transmitter and the receiver.

Denote by d_t and d_r the inter-element spacings at the transmitter and the receiver, respectively, and normalize them to

$$\Delta_t = d_t/\lambda_c, \; \Delta_r = d_r/\lambda_c, \quad (7.2)$$

where λ_c represents the wavelength of the carrier frequency. Denote by K_t, K_r the numbers of transmit and receive antennas. The normalized antenna size is defined to be

$$L_t = (K_t - 1)\Delta_t, \; L_r = (K_r - 1)\Delta_r. \quad (7.3)$$

Figure 7.2 MIMO channel with linear antenna arrays: single path scenario.

Denote by θ_t the angle between the direction of signal departure and the orientation of the transmit antenna array, and θ_r the angle between signal arrival and receive antenna array. Next, we study the MIMO channel matrix **H** under linear antenna arrays.

Single path scenario

First, consider the scenario of only one signal propagation path from the transmitter to receiver. Assume that the antenna physical sizes $L_t \lambda_c$ and $L_r \lambda_c$ are much smaller than the distance between the transmitter and receiver. A signal from transmit antenna n arrives at any receive antenna with approximately the same energy. The only difference between the signals at two receive antennas is a phase rotation determined by the difference of the signal propagation distances between transmit antenna n and the receive antennas. From Figure 7.2, the distance difference is approximated by $d_r \cos(\theta_r)$. Thus, a column vector in the channel matrix **H**, which characterizes the channel between transmit antenna n and all the receive antennas, can be written as $a_n \mathbf{e}_r(\Omega_r)$, where a_n is a scalar complex channel gain and $\mathbf{e}_r$ is given by

$$\mathbf{e}_r(\Omega_r) = \begin{pmatrix} 1 \\ \exp(-j2\pi \Delta_r \Omega_r) \\ \exp(-j2\pi 2\Delta_r \Omega_r) \\ \vdots \\ \exp(-j2\pi (K_r - 1)\Delta_r \Omega_r) \end{pmatrix}, \quad (7.4)$$

with $\Omega_r = \cos(\theta_r)$ referred to as receive directional cosine. We call $\mathbf{e_r}$ the receive angular vector.

Similarly, a row vector of the channel matrix **H** represents the channel between all the transmit antennas and receive antenna m, and can be written as $b_m \mathbf{e}_t(\Omega_t)^*$, with b_m a scalar complex channel gain and $\mathbf{e}_t$ given by

$$\mathbf{e}_t(\Omega_t) = \begin{pmatrix} 1 \\ \exp(-j2\pi \Delta_t \Omega_t) \\ \exp(-j2\pi 2\Delta_t \Omega_t) \\ \vdots \\ \exp(-j2\pi (K_t - 1)\Delta_r \Omega_t) \end{pmatrix}, \quad (7.5)$$

with transmit directional cosine $\Omega_t = \cos(\theta_t)$.

7.1 MIMO channel modeling

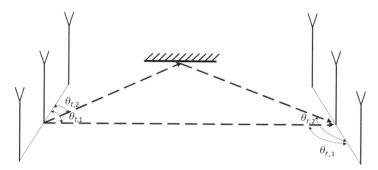

Figure 7.3 MIMO channel with linear antenna arrays: multipath scenario.

In summary, with one signal path, the channel matrix $\mathbf{H}$ has a simple structure: the elements in a given row or column only differ by a linear phase difference. $\mathbf{H}$ can thus be written as

$$\mathbf{H} = c\mathbf{e}_r(\Omega_r)\mathbf{e}_t(\Omega_t)^*, \tag{7.6}$$

with c a complex scalar equal to the channel gain between transmit antenna 1 and receive antenna 1. The rank of $\mathbf{H}$ is 1 and clearly there is no spatial multiplexing gain in this case.

Multipath scenario

Now suppose there are multiple paths from the transmitter to receiver, as shown in Figure 7.3. Denote by l index of the paths, $l = 1, \ldots, L$. Every path is characterized with complex channel gain c_l, delay τ_l, and transmit and receive directional cosines $\Omega_{t,l}, \Omega_{r,l}$. Normalize τ_l to $\tilde{\tau}_l = \tau_l \mathcal{W}$, where $\mathcal{W}$ the system bandwidth. Assume that the OFDMA cyclic prefix covers delay spread so that inter-symbol interference is ignored. As in the SISO case, delay spread causes frequency selectivity in the MIMO channel. Denote by N_c the total number of tones, or subcarriers, in the system. At tone k, the channel matrix is

$$\mathbf{H}[k] = \sum_{l=1}^{L} c_l e^{-j2\pi \frac{k\tilde{\tau}_l}{N_c}} \mathbf{e}_r(\Omega_{r,l})\mathbf{e}_t(\Omega_{t,l})^*. \tag{7.7}$$

Assume that the transmit and receive directional cosines are independent of frequency. Then the frequency selectivity of the MIMO channel solely comes from the scalar phase rotation $e^{-j2\pi \frac{k\tilde{\tau}_l}{N_c}}$. If $|\tilde{\tau}_l| \ll 1$ for all l, the frequency selectivity is negligible and the phase term can be absorbed into complex gain c_l, in which case we can omit the tone index k.

As noted before, the multiplexing gain of a MIMO channel depends on the rank of the channel matrix product $\mathbf{HH}^*$, which is the same as the rank of $\mathbf{H}$ itself. From the scenario of single path to the scenario of multiple paths, the rank of $\mathbf{H}$ does not necessarily increase. For example, suppose $L = 2$ and the two paths have the same transmit directional cosine, $\Omega_{t,1} = \Omega_{t,2}$. $\mathbf{H}$ is still of rank 1 no matter what the receive directional cosines are, because

$$\mathbf{H} = c_1 \mathbf{e}_r(\Omega_{r,1})\mathbf{e}_t(\Omega_{t,1})^* + c_2 \mathbf{e}_r(\Omega_{r,2})\mathbf{e}_t(\Omega_{t,2})^*$$
$$= (c_1 \mathbf{e}_r(\Omega_{r,1}) + c_2 \mathbf{e}_r(\Omega_{r,2})) \mathbf{e}_t(\Omega_{t,1})^*. \tag{7.8}$$

Use of multiple antennas

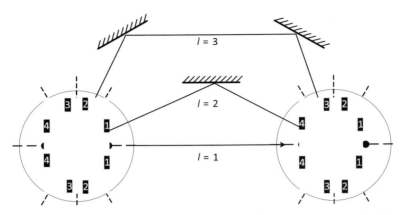

Figure 7.4 MIMO channel spatial resolvability with linear antenna arrays. In this example, $L_t = L_r = 2$ and there are four bins labeled from 1 to 4. There are three paths between the transmitter and the receiver. The first path and second path ($l = 1$ and $l = 2$) are not spatially resolvable at either the transmitter or receiver side. The third path ($l = 3$) is resolvable from the other two.

To make **H** rank 2, at least two paths must exist with substantially different transmit *and* receive directional cosines.

Angular domain representation of MIMO channels and signals

The above argument can be made more precise. It is shown in [159] that when the antennas are critically spaced ($\Delta_r = \Delta_t = 1/2$), if

$$|\Omega_{r,1} - \Omega_{r,2}| \ll \frac{1}{L_r}, \qquad (7.9)$$

two paths are not *resolvable* at the transmitter, and if

$$|\Omega_{t,1} - \Omega_{t,2}| \ll \frac{1}{L_t}, \qquad (7.10)$$

two paths are not resolvable at the receiver because the channel matrix is ill-conditioned. Hence, the antenna array size L_t and L_r define the *spatial resolvability* of different paths.

A simple way to model spatial resolvability is to separate the angular space around the antenna into small bins as shown in Figure 7.4, each angular bin containing paths whose transmit (or receive) directional cosines differ less than $1/L_t$ (or $1/L_r$). Since a directional cosine spans between $[-1, 1]$, $2L_t$ angular bins are formed on the transmitter side and $2L_r$ bins on the receiver side. Clearly, the larger antenna size, the more spatial bins and thus the finer precision in differentiating spatial paths, which further translates into possibly a higher rank of the channel matrix.

In this simple model, one can transform the signal to the angular domain. Define transmit orthonormal basis of K_t vectors to be

$$\mathcal{S}_t = \left\{ \frac{1}{\sqrt{K_t}} \mathbf{e}_t(0), \frac{1}{\sqrt{K_t}} \mathbf{e}_t\left(\frac{1}{L_t}\right), \ldots, \frac{1}{\sqrt{K_t}} \mathbf{e}_t\left(\frac{K_t - 1}{L_t}\right) \right\}, \qquad (7.11)$$

7.1 MIMO channel modeling

and receive basis of K_r vectors

$$S_r = \left\{ \frac{1}{\sqrt{K_r}} \mathbf{e}_r(0), \frac{1}{\sqrt{K_r}} \mathbf{e}_r\left(\frac{1}{L_r}\right), \ldots, \frac{1}{\sqrt{K_r}} \mathbf{e}_r\left(\frac{K_r-1}{L_r}\right) \right\}. \tag{7.12}$$

Represent transmit and receive signal column vectors $\mathbf{X}$, $\mathbf{Y}$ in the angular domain

$$\mathbf{X} = \mathbf{U}_t \mathbf{X}^a, \mathbf{Y} = \mathbf{U}_r \mathbf{Y}^a, \tag{7.13}$$

with $\mathbf{U}_t$, $\mathbf{U}_r$ the unitary matrices whose columns are the basis vectors of S_t, S_r.

Ignoring the noise term, transform the channel equation

$$\mathbf{Y} = \mathbf{H}\mathbf{X} \tag{7.14}$$

to the angular domain

$$\mathbf{Y}^a = \mathbf{H}^a \mathbf{X}^a, \tag{7.15}$$

where $\mathbf{H}^a$ is the angular domain representation of the channel matrix $\mathbf{H}$

$$\mathbf{H}^a = \mathbf{U}_r \mathbf{H} \mathbf{U}_t^*, \tag{7.16}$$

and (m, n) element of $\mathbf{H}_a$ is equal to

$$H^a_{m,n}[k] = \sum_l^L c_l e^{-j2\pi \frac{k\tilde{\tau}_l}{N_c}} \left(\mathbf{e}_r(m/L_r)^* \mathbf{e}_r(\Omega_{r,l})\right) \left(\mathbf{e}_t(\Omega_{t,l})^* \mathbf{e}_t(n/L_t)\right)$$

$$\approx \sum_l^L c_l e^{-j2\pi \frac{k\tilde{\tau}_l}{N_c}} \mathbf{1}_{\{|\Omega_{r,l}-m/L_r|<1/L_r\}} \mathbf{1}_{\{|\Omega_{t,l}-n/L_t|<1/L_t\}}. \tag{7.17}$$

Therefore, $H^a_{m,n}[k]$ is approximately the aggregate channel gain contributed by all paths that fall into transmit bin n and receive bin m, as shown in Figure 7.4.

Hence from (7.17) in order for $\mathbf{H}^a$ to be full rank, at least $\min(K_t, K_r)$ paths are needed. In addition, any path that falls into the same transmit (or receive) bin as another path does not contribute to the rank. As shown in Figure 7.5, when the angular spread of the paths is small at the transmitter (or receiver), they fall into the same transmit (or receive) bin. Spatial multiplexing requires wide angular spread at both the transmitter and receiver.

Frequency selectivity in MIMO channels

From (7.17), the L paths are divided into individual transmit/receive bins in the MIMO case. Thus the number of paths in each transmit/receive bin (m, n) tends to be smaller than L so that $H^a_{m,n}[k]$ exhibits a less degree of frequency selectivity. In the extreme scenario where each bin contains at most one path, there is no frequency selectivity in any bin. As a comparison, in the SISO case, the paths are all lumped together and the wireless channel becomes $\sum_{l=1}^L c_l e^{-j2\pi \frac{k\tilde{\tau}_l}{N_c}}$, thereby resulting in potentially substantial frequency selectivity.

When multiple paths fall into the same transmit/receive bin, the magnitude of $H^a_{m,n}[k]$ may still vary significantly with k. As a result, the channel matrix may be

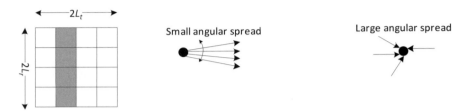

(a) Angular spread is small at transmitter and large at receiver: no spatial multiplexing

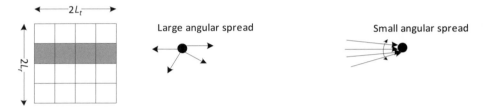

(b) Angular spread is small at receiver and large at transmitter: no spatial multiplexing

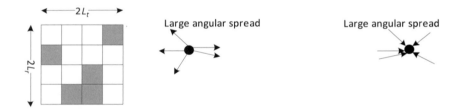

(c) Angular spread is large at receiver and transmitter: full spatial multiplexing

Figure 7.5 Angular spread, transmit/receive bins, and spatial multiplexing.

well-conditioned in some tones while ill-conditioned in others. In this case, it is desirable to allocate the user to tones or subbands in a frequency selective manner rather than spreading over the entire bandwidth.

The rank of the channel matrix does not vary over the frequency at all. In the scenario of Figure 3.9(a) or Figure 7.5(a) where the angle separation of the paths of the user transmitter is small, the channel matrix can be modeled similar to (7.8):

$$\mathbf{H}[k] = \left(\sum_l c_l \mathbf{e}_r(\Omega_{r,l}) \right) \mathbf{e}_t(\Omega_{t,1})^T. \tag{7.18}$$

In a frequency selective channel, the above sum term varies with k; however, since the signal direction $\mathbf{e}_t(\Omega_{t,1})$ is the same for all the paths, frequency selectivity does not improve the rank of the channel matrix.

7.1 MIMO channel modeling

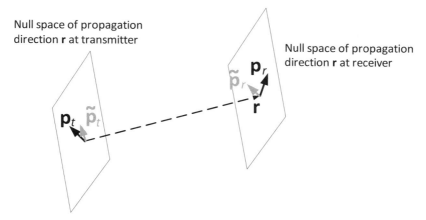

Figure 7.6 Wave propagation between one transmit dipole located at **0** and one receive dipole at **r**.

7.1.2 Polarized antennas

That study of linear antenna arrays shows that rich scattering is required in the wireless channel to obtain spatial multiplexing gain. This means a number of signal propagation paths with wide angular spread at both the transmitter and receiver. Multiplexing gain is limited when the angular spread is small at the transmitter or receiver. For example, when a single (LOS) path dominates, there is no multiplexing gain since the channel matrix is of rank 1. In this section, we introduce another form of multiple antenna arrays with different polarizations, which in contrast works well when channel scattering is not rich.

The polarization of an electromagnetic wave is defined to be the orientation of its electric field vector. The polarization of an antenna is defined to be the polarization of the electromagnetic wave radiated by the antenna. Most antennas used in wireless communication devices are electrical dipoles and generate linearly polarized electromagnetic wave whose electrical field is fixed in one direction. If propagated in free space, the polarization of the wave is the same as the direction of the electrical dipole. We next describe the channel characteristics between two electric dipole antennas in free space.

Free space channel model between a pair of electric dipoles

Let the transmit antenna to be located at the origin in Euclidean space. The receive antenna is located at $\mathbf{r}$. Denote by $\mathbf{p}_t$ the *dipole moment* of the transmit antenna. The norm of $\mathbf{p}_t$ indicates the strength of the input current and the direction coincides with the orientation of the electric dipole. The induced electric field at location $\mathbf{r}$ can be approximated as

$$\mathbf{E}(\mathbf{r}) = A(|\mathbf{r}|)\cos(\theta_t)|\mathbf{p}_t|\tilde{\mathbf{p}}_t, \tag{7.19}$$

where $A(|\mathbf{r}|)$ only depends on $|\mathbf{r}|$, the Euclidean distance between the transmit antenna and the receive antenna. The direction of the electric field is determined by the unit vector $\tilde{\mathbf{p}}_t$, the projection of the dipole moment $\mathbf{p}_t$ on the *null space* of vector $\mathbf{r}$ normalized to $|\tilde{\mathbf{p}}_t| = 1$, as shown in Figure 7.6. We refer to $\tilde{\mathbf{p}}_t$ as the *effective* dipole moment with respect to a receiver at location **r**. Only the electric component perpendicular to the propagation direction **r** has a non-zero contribution to the electric field at location **r**. Denote by θ_t the angle between the original dipole moment $\mathbf{p}_t$ and the unit projected effective dipole moment $\tilde{\mathbf{p}}_t$.

However, not all the electric field at location **r** can induce an electric current on the receive dipole. Denote by $\mathbf{p}_r$ the orientation of the receive dipole and $\tilde{\mathbf{p}}_r$ its unit projection at the null space of the propagation direction **r**. The channel response at the receive dipole can be written as

$$H = B(|\mathbf{r}|)\cos(\theta_t)\cos(\theta_r)\cos(\theta_{tr}), \qquad (7.20)$$

where θ_r is the angle between $\mathbf{p}_r$ and $\tilde{\mathbf{p}}_r$ and θ_{tr} is the angle between the two unit effective electric dipoles $\tilde{\mathbf{p}}_t$ and $\tilde{\mathbf{p}}_r$. $B(|\mathbf{r}|)$ depends on distance $|\mathbf{r}|$ but not the orientation or polarization direction of the antennas. Equation (7.20) shows that the antenna orientations have a significant impact on the channel. In particular, $H = 0$, if either the transmit or receive dipole has an orientation that coincides with the propagation direction **r** ($\cos(\theta_t) = 0$ or $\cos(\theta_r) = 0$), or if the two effective dipole moments are orthogonal ($\cos(\theta_{tr}) = 0$).

Free space channel model with multiple polarized antennas

To recover the energy loss due to mismatch of polarization orientations, one can introduce multiple dipoles with different orientations at either the transmitter or the receiver or both. For example, a *dual-polarized* antenna uses two electrical dipoles with different orientations.

In existing deployments, cross-polarized antennas are widely used at base stations for transmit polarization diversity. Assuming one antenna at the user receiver, the downlink channel vector is equal to

$$\mathbf{H} = B(|\mathbf{r}|)\left(\cos(\theta_{t_1})\cos(\theta_r)\cos(\theta_{t_1 r}), \cos(\theta_{t_2})\cos(\theta_r)\cos(\theta_{t_2 r})\right), \qquad (7.21)$$

where θ_{t_m} ($m = 1, 2$) represents the angle between the dipole moment of transmit antenna m and its effective dipole moment on the null space of the propagation direction **r**, and $\theta_{t_m r}$ the angle between the effective dipole moment of transmit antenna m and that of the receive antenna. The receiver sees a linear superposition of the two signals transmitted from the transmit antennas. Transmit polarization diversity is achieved when the same signal is transmitted from multiple antennas. Note that polarization diversity is helpful here even for non-fading channels since channel degradation is caused by a mismatch between the polarizations at the transmit and the receive antennas. Providing multiple transmit polarizations reduces the chance of such a mismatch. On the other hand, even with two transmit antennas, there is still a possibility that the receive signal has a bad signal strength. This happens, for example, when one of the effective transmit dipole moments is orthogonal to the null space, and the other one is orthogonal to the effective receive dipole moment. This possibility drops as the number of transmit antennas increases.

Similarly, suppose that the receiver uses multiple dipoles with different dipole moments, and the transmitter uses one antenna. The channel matrix now becomes

$$B(|\mathbf{r}|)\begin{pmatrix} \cos(\theta_t)\cos(\theta_{r_1})\cos(\theta_{tr_1}) \\ \cos(\theta_t)\cos(\theta_{r_2})\cos(\theta_{tr_2}) \end{pmatrix}. \quad (7.22)$$

The receiver can use maximal ratio combining (see Section 4.2.3) to achieve receive polarization diversity.

This idea can be extended to build a MIMO scenario. Let both transmit and receive antenna use two electric dipoles. Suppose the two transmit dipole antennas have orientations $\mathbf{p}_{t_1}$ and $\mathbf{p}_{t_2}$, and the two receive dipoles have orientations $\mathbf{p}_{r_1}$ and $\mathbf{p}_{r_2}$. The channel matrix is given by

$$\mathbf{H} = B(|\mathbf{r}|)\begin{pmatrix} \cos(\theta_{t_1})\cos(\theta_{r_1})\cos(\theta_{t_1 r_1}), & \cos(\theta_{t_2})\cos(\theta_{r_1})\cos(\theta_{t_2 r_1}) \\ \cos(\theta_{t_1})\cos(\theta_{r_2})\cos(\theta_{t_1 r_2}), & \cos(\theta_{t_2})\cos(\theta_{r_2})\cos(\theta_{t_2 r_2}) \end{pmatrix}. \quad (7.23)$$

Similar to the linear antenna array case, we want to check the rank of the matrix. To this end, we rewrite the matrix as a product of three matrices

$$\mathbf{H} = B(|\mathbf{r}|)\begin{pmatrix} \cos(\theta_{r_1}) & 0 \\ 0 & \cos(\theta_{r_2}) \end{pmatrix}\begin{pmatrix} \cos(\theta_{t_1 r_1}) & \cos(\theta_{t_2 r_1}) \\ \cos(\theta_{t_1 r_2}) & \cos(\theta_{t_2 r_2}) \end{pmatrix}\begin{pmatrix} \cos(\theta_{t_1}) & 0 \\ 0 & \cos(\theta_{t_2}) \end{pmatrix}. \quad (7.24)$$

A sufficient and necessary condition for the matrix $\mathbf{H}$ to be full rank is that all three matrices in (7.24) are full rank. Two of three matrices are diagonal and thus full rank as long as the diagonal elements are not equal to zero. Denote by $\tilde{\mathbf{p}}_{t_m}$ and $\tilde{\mathbf{p}}_{r_n}$ the unit effective dipole moments. It follows that

$$\cos(\theta_{t_m r_n}) = \tilde{\mathbf{p}}_{r_n}^T \tilde{\mathbf{p}}_{t_m}, \quad (7.25)$$

$$\begin{pmatrix} \cos(\theta_{t_1 r_1}), & \cos(\theta_{t_2 r_1}) \\ \cos(\theta_{t_1 r_2}), & \cos(\theta_{t_2 r_2}) \end{pmatrix} = \begin{pmatrix} \tilde{\mathbf{p}}_{r_1}^T \\ \tilde{\mathbf{p}}_{r_2}^T \end{pmatrix}\begin{pmatrix} \tilde{\mathbf{p}}_{t_1}, & \tilde{\mathbf{p}}_{t_2} \end{pmatrix}. \quad (7.26)$$

Thus, in order for H to be full rank, the above transmit and receive effective dipole matrices both have to be full rank. The effective dipole matrix fails to be full rank if one of the following events happen:

- One of the effective polarization moments is zero when it is perpendicular to the null space of the propagation vector $\mathbf{r}$.
- Both effective polarization moments have the same direction.

Assuming that the propagation vector and the antenna orientation are uncorrelated, these events only happen with a small probability. Therefore, in the free space propagation model, the channel matrix in (7.24) is full rank with high probability. In contrast, the rank of a MIMO channel with linear antenna arrays at the transmitter and receiver is 1 in free space.

Impact of reflections

With reflection, the transmitter and the receiver see different propagation vectors $\mathbf{r}_t$ and $\mathbf{r}_r$. Nevertheless, we can still use the notions of $\tilde{\mathbf{p}}_t$ and $\tilde{\mathbf{p}}_r$ the unit projected dipole moments to the null plane of the propagation vectors and model the channel as (7.20) with the same definition of θ_t and θ_r.

However, the definition of θ_{tr} is more involved due to reflection, because reflection may change the direction of the electric field as compared to the propagation direction. In particular, if the direction of the electric field vector of the incident wave is either completely parallel or perpendicular to the plane of incidence formed by the normal vector to the reflecting surface and the propagation vector of the incident wave, then the polarization of the incident wave is preserved upon reflection. This observation is important to study channel XPD later in this section. However, if the incident wave has both parallel and perpendicular components, the polarization is not preserved with respect to the propagation direction.

Most importantly, it can be shown that in the presence of reflections, the channel matrix with multiple polarized antennas is still full rank with high probability, similar to the free space case. A detailed study of modeling polarized MIMO channels can be found in [139].

Cross polarization discrimination (XPD)

So far, we have seen that the use of multiple dipoles with different polarized antennas can lead to a full rank channel matrix without relying on a rich scattering environment. Let us focus on the case of dual-polarized antennas. A popular choice to construct orthogonal polarization is vertical and horizontal polarization (V/H). As the name suggests, the polarization direction is perpendicular or parallel to the surface of the earth. Another choice is the $+45°/-45°$ (slant) polarizations, a mere rotation of the V/H axis by $45°$. We assume that both transmitter and receiver use the same polarization choice. In practice, a user has somewhat random orientation, thereby reducing the efficiency of polarized antennas (smaller XPD as discussed next).

An important characteristic of the dual-polarized channel is the cross-polar discrimination (XPD), which is a measure of *depolarization*. XPD is defined as the ratio of the average power of received signal on the co-polar branch to the average power of the received cross-polar signal. For example, in the V/H polarization, co-polar branches are V to V and H to H, and cross-polar ones are V to H and H to V. Define

$$\text{XPD} = \frac{\mathbb{E}(|H_{V,V}|^2) + \mathbb{E}(|H_{H,H}|^2)}{\mathbb{E}(|H_{V,H}|^2) + \mathbb{E}(|H_{H,V}|^2)} \qquad (7.27)$$

to reflect the relative leakage between different polarized antennas. The overall XPD depends on the antenna purity measured by antenna XPD, and depolarization in the wireless channel (e.g., reflections) measured by channel XPD. We are interested in channel XPD.

In general, channel XPD is higher in an LOS channel than in a non LOS channel, especially when the user is oriented in an upright (vertical) position. In an urban environment, most prominent reflectors are vertical surfaces like walls of buildings. Vertical and horizontal polarization, which are perpendicular or parallel to the plane of incidence, do not depolarize upon reflection from these surfaces. Thus a high channel XPD is maintained in spite of reflection. However, reflections from inclined surfaces, although infrequent, could lead to depolarization of V/H polarization. Also, a user is usually surrounded by many local scatterers and equipped with an antenna of much larger elevation beamwidth than the base station. As a result, the user probably picks up a number of reflected signals from these scatterers, whose electric field vectors are neither vertical nor horizontal. Thus, local scatterers around the user may reduce channel XPD. As another factor, the user orientation is usually not completely vertical or horizontal.

Measurements reported in the literature ([11]) suggest that XPD values vary in the range 7–15 dB and are typically much higher in suburban than in urban areas and tend to increase significantly in the presence of LOS. As the distance increases, the XPD has been reported to decrease because of more reflection.

7.2 SU-MIMO techniques

SU-MIMO techniques for spatial multiplexing have been extensively studied in the literature (see [159] and the references therein). In this section we briefly cite the major results for a single-tap channel

$$\mathbf{Y} = \mathbf{H}\mathbf{X} + \mathbf{W}, \qquad (7.28)$$

where column vector $\mathbf{X}$ consists of symbols transmitted in multiple antennas at a given time, and $\mathbf{Y}$ of symbols received in multiple antennas. $\mathbf{H}$ is a complex channel matrix. In OFDMA, when the channel is frequency selective, the above channel model applies separately to individual tones,

$$\mathbf{Y}[k] = \mathbf{H}[k]\mathbf{X}[k] + \mathbf{W}[k], \qquad (7.29)$$

with k the tone index. MIMO techniques developed for the single-tap channel can be employed individually for different tones, as shown in Section 3.5. If frequency selectivity is not severe and $\mathbf{H}[k]$ remains somewhat constant for a subband consisting of a number of tones, then MIMO implementation can be done for the entire subband as a group.

We consider two scenarios.

7.2.1 Channel state information at both transmitter and receiver

First, assume that the channel matrix is known to both transmitter and receiver. Perform singular value decomposition (SVD) on $\mathbf{H}$,

$$\mathbf{H} = \mathbf{U}\mathbf{\Lambda}\mathbf{V}^*, \qquad (7.30)$$

where $\mathbf{U}$ and $\mathbf{V}$ are unitary matrices and $\mathbf{\Lambda}$ a rectangular matrix with non-negative real diagonal elements $(\lambda_1, \lambda_2, \ldots, \lambda_{K_{\min}})$ and all-zero off-diagonal elements. $\lambda_1 \geq \lambda_2 \geq \cdots \geq \lambda_{K_{\min}}$ are the ordered singular values of $\mathbf{H}$, and $K_{\min} = \min(K_t, K_r)$. We have a channel equivalent to (7.28)

$$\tilde{\mathbf{Y}} = \mathbf{\Lambda}\tilde{\mathbf{X}} + \tilde{\mathbf{W}}, \qquad (7.31)$$

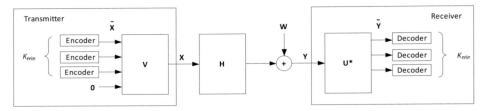

Figure 7.7 Communication architecture based on SVD to achieve full multiplexing gain.

where $\tilde{\mathbf{Y}} = \mathbf{U}^*\mathbf{Y}$, $\tilde{\mathbf{X}} = \mathbf{V}^*\mathbf{X}$ and $\tilde{\mathbf{W}} = \mathbf{U}^*\mathbf{W}$. Because of the structure of $\mathbf{\Lambda}$, (7.31) represents $K_{\min}$ parallel subchannels, called eigenchannels,

$$\tilde{Y}_i = \lambda_i \tilde{X}_i + \tilde{W}_i, i = 1, 2, \ldots, K_{\min}. \tag{7.32}$$

Eigenchannels do not interfere with each other.

Under a typical sum constraint of input power $\mathbb{E}[\|\mathbf{X}\|^2] \leq P_m$, the channel capacity is given by

$$C = \sum_{i=1}^{K_{\min}} \log\left(1 + \frac{|\lambda_i|^2 P_i^*}{\sigma^2}\right), \tag{7.33}$$

where P_i^* is the optimal power allocated on the eigenchannel i

$$P_i^* = \left(\mu - \frac{\sigma^2}{|\lambda_i|^2}\right)^+, \tag{7.34}$$

where μ is a positive constant such that $\sum_i P_i^* = P_m$.

Note that the rank of the channel matrix cannot exceed $K_{\min}$. Thus at most $K_{\min}$ parallel eigenchannels in (7.31) have non-zero power allocation. A MIMO channel creates at most $K_{\min}$ spatial degrees of freedom. In the high SNR regime, where SNR $\gamma = P_m/\sigma^2$, it can be shown that

$$\lim_{\gamma \to \infty} \frac{C}{\log(\gamma)} \leq K_{\min}, \tag{7.35}$$

where the equality holds if the matrix $\mathbf{H}$ is full rank. Comparing with SISO capacity $\log(\gamma)$, MIMO capacity increases at most by a factor of $K_{\min}$.

To achieve the capacity (7.33), with channel side information fully known at the transmitter, one can use an architecture shown in Figure 7.7. *Precoding* matrix $\mathbf{V}$ is used at the transmitter to transform the encoded bits to the signal sent over the transmit antennas. At the receiver, the signal from the receive antennas is transformed with matrix $\mathbf{U}^*$ before decoding. Since eigenchannels do not interfere with each other, separate encoding and decoding of an independent data stream can be done in each eigenchannel with the optimal power allocation determined by the water-filling principle (7.34).

7.2.2 Channel state information only at receiver

To obtain the channel state information at the transmitter can be quite expensive due to time-selective and frequency-selective fading. In particular in an FDD system where the

7.2 SU-MIMO techniques

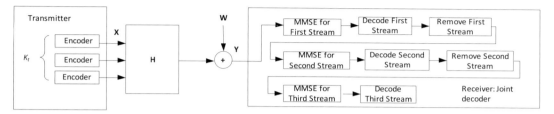

Figure 7.8 V-BLAST architecture where the receiver uses combined MMSE and SIC, which is capacity-achieving.

downlink and uplink carrier frequencies are widely separated, the receiver has to report the channel state information to the transmitter. Large signaling overhead is required for the transmitter to track the channel over time, frequency, and space. In addition, the delay in a feedback channel may cause transmitter side information to be obsolete. In the following we consider the scenario where the channel state information is only known to the receiver.

Without knowing the channel, the transmitter splits the total power P_m equally to every antenna. It can be shown[1] that the ergodic capacity of the channel is given by

$$C = \mathbb{E}\left(\log\det\left(\mathbf{I} + \frac{\gamma}{K_t}\mathbf{HH}^*\right)\right). \tag{7.36}$$

Suppose $\lambda_1, \lambda_2, \ldots, \lambda_{K_{\min}}$ are the ordered singular values of $\mathbf{H}$, which are random and not known at the transmitter. The above capacity is then equal to

$$C = \mathbb{E}\left(\sum_{i=1}^{K_{\min}} \log\left(1 + \frac{\gamma}{K_t}\lambda_i^2\right)\right). \tag{7.37}$$

Therefore, if the channel matrix is sufficiently random and well-conditioned, the same factor of $K_{\min}$ spatial multiplexing gain is obtained in the high SNR regime even with only receiver side channel knowledge and no joint encoding across the transmit antennas. This is a very encouraging information theoretic result.

The penalty of transmitter not knowing the channel is that the total power is equally allocated, which is not the optimum. In particular, when $K_t > K_r$, the optimal waterfilling scheme allocates the total power to only $K_{\min} = K_r$ eigenchannels. In the high SNR regime, the SNR of eigenchannel i is $\frac{\gamma}{K_r}\lambda_i^2$. Comparison with (7.37) indicates a factor of K_t/K_r loss in SNR, which is not significant in the high SNR regime.

One slightly surprising result in information theory is that joint encoding at the transmitter is *not* required to achieves the capacity in (7.36). Indeed, the V-BLAST architecture [60, 170] as shown in Figure 7.8 can achieve the capacity. Specifically, K_t independently encoded data streams are sent from the transmit antennas with no

---

[1] An important assumption for the capacity Equation (7.36) is that the channel matrix contains i.i.d. components. For the more general case where the channel matrix contains correlated components, the capacity is studied in [160, 164].

precoding matrix. Rewrite (7.28) to

$$\mathbf{Y} = \sum_{n=1}^{K_t} \mathbf{H}_n X_n + \mathbf{W}, \qquad (7.38)$$

where $\mathbf{H}_n$ is column n of channel matrix $\mathbf{H}$ and X_n the data stream sent from transmit antenna n. The MIMO channel (7.38) is very similar to a CDMA multi-access channel in the uplink where users use different *temporal signature sequences* (spreading codes) to send their independently-encoded bits. In the MIMO channel, X_n is received with a *spatial signature sequence* $\mathbf{H}_n$. Therefore, the ideas of multiuser detection ([165]) are applicable here. In particular, the combined MMSE and successive interference cancellation (SIC) receiver shown in Figure 7.8 works as follows. X_1 is first decoded using a linear MMSE receiver in the presence of interfering $X_2, \ldots, X_{K_t}$. The decoded data stream is subtracted from $\mathbf{Y}$. X_2 is then decoded using a linear MMSE receiver in the presence of $X_3, \ldots, X_{K_t}$ and subtracted. The process goes on until the last data stream is decoded.

7.2.3 Multiplexing with polarized antennas

Spatial multiplexing can create multiple spatial streams between a transmitter and a receiver, and the number of streams is limited by the *rank* of the channel matrix. As we have discussed before, linear antenna arrays may not lead to a high rank channel matrix without a rich scattering channel, such as in an LOS environment. However, polarized antennas work very well in such a channel. Thus, if the base station and the mobiles are equipped by polarized antennas, it is straightforward to introduce spatial multiplexing between the transmitter and receiver. Of course, the rank is limited to 2.

Thanks to the separation between the vertical and horizontal field (small XPD), the orthogonality between the two streams is largely preserved. In that case, receiver complexity can be further reduced. For example, an MMSE receiver or even a simple matched filter may suffice to suppress cross-stream interference without the need of a combined MMSE and SIC receiver.

7.3 Multiuser MIMO techniques

Linear antenna arrays provide maximum spatial multiplexing gain in a rich scattering environment with multiple spatial resolvable paths between the transmitter and receiver. Without rich scattering, spatial multiplex gain diminishes. In a cellular system, the base station is usually placed in a tower higher than the surroundings without many local reflectors or scatterers. The angular spread for a given user is thus small (see Figure 3.9a). Spatial multiplexing gain of a single user is very limited in many practical deployment scenarios. This limits the single-user MIMO benefit in a practical cellular system. In particular, the peak data rate of the user may not always see a factor of $K_{\min}$ boost. However, the system sum capacity can still be boosted by a factor of $K_{\min}$ by spatially

7.3 Multiuser MIMO techniques

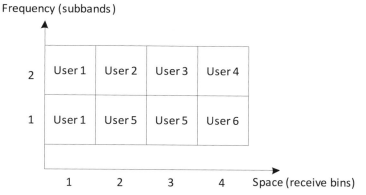

Figure 7.9 Illustration of frequency selective SDMA allocation in OFDMA.

multiplexing users that are geographically separated (Figure 3.9b). MU-MIMO is often referred to as *Spatial Division Multiple Access* (SDMA) in the uplink, and *beamforming* in the downlink.

Consider the uplink case. MU-MIMO in the downlink can be modeled similarly. For simplicity, assume that each user is equipped with one transmit antenna. The uplink model is the same as (7.14), except that the elements of column vector X represent the signals from different users instead of those from the antennas of the same user. A column vector in the channel matrix $\mathbf{H}$ at tone k characterizes the channel between user i and all the receive antennas and can be written as

$$\sum_{l=1}^{L_i} c_{i,l} e^{-j2\pi \frac{k\tilde{\tau}_{i,l}}{N_c}} \mathbf{e}_r(\Omega_{i,l}), \tag{7.39}$$

where $c_{i,l}, \tilde{\tau}_{i,l}, \Omega_{i,l}$ represent path l from user i. Note that a different user i may have a different number of paths L_i to the base station.

The rank of MU-MIMO channel $\mathbf{H}$ cannot exceed $\min(K_r, K_u)$ where K_u is the number of users being multiplexed. From the above angular representation, we know that if the paths from all users fall into the same receive bin, then the channel is ill-conditioned. To achieve spatial multiplexing, set $K_u = K_r$, and the path from a different user has to be in a different receive bin to achieve the full benefit of spatial multiplexing. Moreover, in the multipath scenario, suppose that all the paths from a given user fall into only one receive bin because of small angular spread at the base station. Due to frequency selectivity, the channel gain of a user may fade on some tones and thus make the channel matrix ill-conditioned. In OFDMA, MU-MIMO should avoid those tones and allocate the user to other tones in which the channel gain is strong.

Figure 7.9 illustrates frequency selective SDMA allocation. Users 1, 2, 3, 4 are resolvable in the angular domain and thus occupy the same subband with SDMA. User 1 is allocated to both subbands 1 and 2 because its channel is strong in both subbands; however users 2, 3, 4 take only subband 1 because their channel fade in subband 2, and they are replaced by users 5, 6 in subband 2. The paths from user 5 arrive at bins 2 and 3 to allow user 5 take both bins.

In the remainder of this chapter, we assume the wireless channel to be frequency flat, as a frequency selective channel can be addressed in the frequency domain as shown in Figure 7.9.

7.3.1 Uplink SDMA

We assume that the base station is equipped with K_r receive antennas, and for simplicity assume that each user is equipped with a single transmit antenna.

The fundamental constraint in the uplink is that the user transmitters cannot perform joint encoding. Joint decoding is possible at the base station receiver assuming sufficient processing capability. Recall that the V-BLAST architecture does not require joint encoding at the transmitter antennas either, yet still achieves the full spatial multiplexing gain ($K_{\min}$). In this case, independently encoded data streams are sent from the transmit antennas over the same time-frequency resource. The receiver separates the streams according to their *distinct* spatial signature sequences using combined MMSE and SIC. Hence, the V-BLAST architecture is readily applicable to SDMA by viewing uplink users as separate transmit antennas and can achieve a full spatial multiplexing gain of $\min(K_r, K_u)$, where K_u is the number of users being multiplexed, without the users performing joint encoding.

Clearly, the key to the success of MU-MIMO is to have distinct spatial signature sequences. Unlike SU-MIMO, rich scattering is not required to achieve full multiplexing gain in SDMA. Instead, the users need to be geographically separate. Specifically, denote by $H_{m,i}$ the channel response from user i to base station antenna m and column vector $\mathbf{H}_i = [H_{1,i}, \ldots, H_{K_r,i}]^T$ the channel response vector. To maximize spatial multiplexing, let $K_u = K_r$. Suppose K_r users, $1, \ldots, K_r$, are selected among the total M users in a cell to transmit their signals, $X_1, \ldots, X_{K_r}$, respectively, using the same time-frequency resource (tone-symbols in OFDMA). The received symbol vector at the base station is

$$\mathbf{Y} = [\mathbf{H}_1, \ldots, \mathbf{H}_{K_r}][X_1, \ldots, X_{K_r}]^T + \mathbf{W}, \qquad (7.40)$$

where each element of $\mathbf{W}$ is noise of power σ^2 and each transmit signal is subject to a maximum power constraint

$$\mathbb{E}(|X_i|^2) \leq P_m. \qquad (7.41)$$

Define the channel matrix $\mathbf{H} = [\mathbf{H}_1, \ldots, \mathbf{H}_{K_r}]$. If $\mathbf{H}$ is well-conditioned, $[X_1, \ldots, X_{K_r}]^t$ can be fully recovered from Y, thereby achieving a factor of K_r spatial multiplexing gain.

In the following, we would like to highlight two points.

The first point is that given K_r, the likelihood that one can find K_r users to make $\mathbf{H}$ well-conditioned increases with M, a form of multiuser diversity. The concept of multiuser diversity has been explored in Chapter 4 in the context of multiuser scheduling in a fading channel.

7.3 Multiuser MIMO techniques

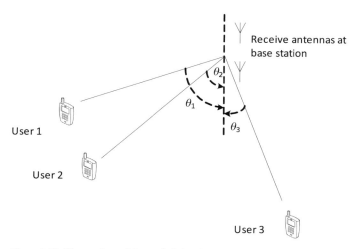

Figure 7.10 Illustration of line-of-sight channel model in SDMA.

To provide a concrete example, consider $K_r = 2$ and assume that each user transmits at the same power P_m. The SDMA capacity is a region given by

$$R_1 \leq \log\left(1 + \frac{\|\mathbf{H}_1\|^2 P_m}{\sigma^2}\right), \tag{7.42}$$

$$R_2 \leq \log\left(1 + \frac{\|\mathbf{H}_2\|^2 P_m}{\sigma^2}\right), \tag{7.43}$$

$$R_1 + R_2 \leq \log\det\left(\mathbf{I} + \frac{P_m}{\sigma^2}\mathbf{H}\mathbf{H}^*\right). \tag{7.44}$$

For the sake of simplicity, assume $\mathbb{E}(\|\mathbf{H}_i\|^2) = K_r$ for all i. Thus the limiting constraint is the sum rate (7.44), denoted by $(R_1 + R_2)_{\max}$, which depends on whether $\mathbf{H}$ is well-conditioned.

Let us consider two special channel models.

First, assume a single line-of-sight path from user i to the receive antennas at the base station, as illustrated in Figure 7.10 where θ_i represents the angle between the signal arrival path and the receive antenna array. Thus $\mathbf{H}_i = c_i \mathbf{e}_r(\Omega_i)$ where c_i is a complex number with $|c_i| = 1$ and $\mathbf{e}_r(\Omega_i)$ defined in (7.4) with $\Omega_i = \cos(\theta_i)$. The sum rate constraint (7.44) is thus given by

$$(R_1 + R_2)_{\max} = \log\left(1 + 4\gamma_0 + 2\gamma_0^2\left(1 - \cos\left(2\pi\Delta_r(\Omega_1 - \Omega_2)\right)\right)\right), \tag{7.45}$$

where $\gamma_0 = P_m/\sigma^2$ is per antenna SNR. Apparently, when $|\Omega_1 - \Omega_2| \approx 0$, the sum rate $(R_1 + R_2)_{\max}$ drops to a minimum. Figure 7.10 illustrates a three-user scenario in which the SDMA capacity is smaller when users 1 and 2 are selected than when users 1 and 3 are selected.

Suppose that the channel $\mathbf{H}_i$ is fixed for all i. The base station first randomly picks a user and then selects from among the remaining $M - 1$ users the one that maximizes

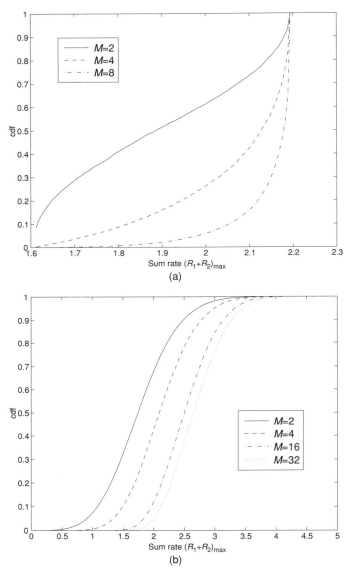

Figure 7.11 Cumulative distribution functions of the sum rate constraint $(R_1 + R_2)_{\max}$ in the line-of-sight model (a) and in the i.i.d. Gaussian model (b). $\Delta_r = 1/2$.

the sum rate $(R_1 + R_2)_{\max}$. Figure 7.11(a) plots the cdf of the sum rate assuming $\gamma_0 = 0$ dB and θ_i is uniformly distributed in $[0, 2\pi]$. We observe that as M increases from 2 to 8, the sum rate at the 10 percent probability tail increases roughly from 1.6 to 2.1, which is already close to the maximum value of $(R_1 + R_2)_{\max}$ in (7.45) indicating that multiuser diversity starts to saturate even when M is modest.

The second channel model assumes $H_{m,i}$ is fixed i.i.d. complex Gaussian with zero mean and unit variance. The base station follows the same scheduling policy.

Figure 7.11(b) plots the cdf of the sum rate. Compared with the line-of-sight case, the sum rate exhibits a larger dynamic range, and the multiuser diversity benefit only saturates when M is relatively large. The reason is that in the Gaussian model, both amplitude and phase vary in $H_{m,i}$, thereby increasing the potential of multiuser diversity.

The second point to be highlighted is that although a combined MMSE and SIC receiver is much simpler than a general maximum-likelihood (ML) receiver, its complexity is still nontrivial. Moreover, as pointed out in Section 5.2.2, the effectiveness of SIC strongly depends on the quality of channel estimation. In a mobile environment, channel estimation is inherently subject to errors. Therefore applying SIC in practice should be cautious, especially if users with a large dynamic range are multiplexed. On the other hand, since the base station selects geographically separate users to be multiplexed in SDMA, an MMSE only receiver may suffice without the need of SIC if user selection is done carefully.

To elaborate, from the received signal $\mathbf{Y}$ (7.40), the MMSE receiver is a linear receiver that maximizes the SINR. To detect X_i for user i, project $\mathbf{Y}$ to a scalar Y_i

$$Y_i = \mathbf{u}_i^* \mathbf{Y}, \qquad (7.46)$$

where MMSE receive beamforming vector $\mathbf{u}_i^*$ is given by

$$\mathbf{u}_i^* = \mathbf{H}_i^* \left(\gamma_0 \sum_{i':i' \neq i} \mathbf{H}_{i'} \mathbf{H}_{i'}^* + \mathbf{I} \right)^{-1}. \qquad (7.47)$$

The SINR of user i is thus equal to

$$\gamma_i = \gamma_0 \mathbf{H}_i^* \left(\gamma_0 \sum_{i':i' \neq i} \mathbf{H}_{i'} \mathbf{H}_{i'}^* + \mathbf{I} \right)^{-1} \mathbf{H}_i. \qquad (7.48)$$

We assume the rate to be $R_i = \log(1 + \gamma_i)$.

Continue the above example of $K_r = 2$ with the single line-of-sight path model. Figure 7.12 plots $R_1 + R_2$ as receive angle difference $|\theta_1 - \theta_2|$ increases from $0°$ to $360°$. The baseline scheme is FDM in which each user concentrates its power P_m in half of the bandwidth. Notice that the SDMA scheme with the MMSE only receiver does not always outperform the FDM scheme. When two users with poor spatial separation are scheduled to the same time-frequency resource, the MMSE only receiver is inferior. However, as γ_0 increases, the MMSE only receiver outperforms the FDM scheme almost everywhere except for a small region where the angle difference is close to $0°$ (or $360°$). With either a combined MMSE and SIC receiver or an MMSE only receiver, the performance of SDMA peaks at $|\theta_1 - \theta_2| = 180°$. In that region, the scheduled users are spatially well separated, and the performance difference between the two receivers is negligible.

This observation suggests that in a *scheduled* SDMA system, the MMSE only receiver can closely approach the SDMA capacity with only linear complexity. Figure 7.13 corroborates this conjecture by plotting the sum data rate $R_1 + R_2$ versus γ_0. We observe that as the number of users M increases from 2 to 4 and then 8, the loss due to the absence of SIC becomes very small.

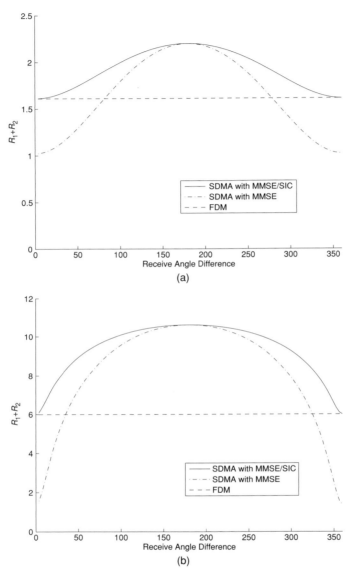

Figure 7.12 Plot of sum rate $R_1 + R_2$ versus receive angle difference (in degrees). Comparison of a combined MMSE and SIC receiver and an MMSE only receiver. $\gamma_0 = 0$ dB in (a) and 20 dB in (b). Note the transmit power per degree of freedom in the FDM scheme is twice as much as that in the SDMA scheme to maintain the same total power for any given user.

So far we have assumed fixed power for the scheduled users. With the MMSE receiver, the base station can apply power control jointly with SDMA. Denote by P_i the transmit power of user i, with $P_i \leq P_m$. The SINR of the MMSE receiver (7.48) now becomes

$$\gamma_i = P_i \mathbf{H}_i^* \left(\sum_{i':i' \neq i} P_{i'} \mathbf{H}_{i'} \mathbf{H}_{i'}^* + \sigma^2 \mathbf{I} \right)^{-1} \mathbf{H}_i. \quad (7.49)$$

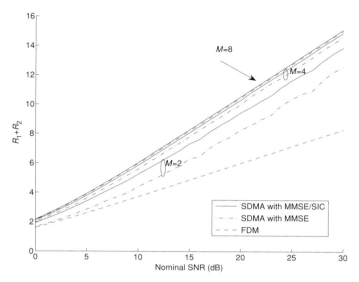

Figure 7.13 Comparison of sum data rate $R_1 + R_2$ with a combined MMSE and SIC receiver and an MMSE only receiver in a scheduled SDMA system.

We employ an iterative search algorithm similar to Section 6.2 to solve joint SDMA and power control. As an example, suppose the goal is to component-wise minimize P_i with a given SINR target $\hat{\gamma}_i$. The search algorithm starts with some initial $P_i[0]$. At slot t, calculate the MMSE beamforming vector of user i similar to (7.46)

$$\mathbf{u}_i^* = \mathbf{H}_i^* \left(\sum_{i':i' \neq i} P_{i'}[t] \mathbf{H}_{i'} \mathbf{H}_{i'}^* + \sigma^2 \mathbf{I} \right)^{-1}, i' \neq i. \quad (7.50)$$

Calculate the SINR γ_i from (7.49) and update the transmit power

$$P_i[t+1] = \min\left(P_m, P_i[t] \frac{\hat{\gamma}_i}{\gamma_i} \right). \quad (7.51)$$

It can be shown that as long as the SINR target $\hat{\gamma}_i$ is feasible, the above simple iterative algorithm converges to the optimal power control and beamforming vector.

7.3.2 Downlink beamforming

To study downlink MU-MIMO, in this section we will compare and contrast it with uplink SDMA. It turns out that downlink MU-MIMO (beamforming) exhibits interesting similarity, more precisely duality, to uplink SDMA; yet it faces unique challenges because of structural difference between the downlink and uplink.

We assume that the base station is equipped with K_t transmit antennas, and each user with one receive antenna. The channel model of a user i is

$$Y_i = \mathbf{H}_i \mathbf{X} + W_i, \quad (7.52)$$

where channel matrix $\mathbf{H}_i$ is simply a row vector and represents the channel response from K_t transmit antennas to the user i. Column vector $\mathbf{X}$ represents the signals sent from the antennas.

A key difference from (7.40) is that the power constraint is not on individual user but on the total $\mathbb{E}(|\mathbf{X}|^2) \leq P_m$. In addition, noise W_i power may be different for a different user i. More importantly, unlike the uplink, joint decoding is not possible as the user receivers do not cooperate. Therefore, one cannot apply the V-BLAST architecture in a straightforward way. In this case, the base station takes the burden of joint encoding and multiplexes the signals to the users in such a way that each user can easily separate its desired signal from the rest without any multiuser signal processing technique. As will be shown below, the challenge of joint encoding is that it requires channel state information at the transmitter. We have noted that in a TDD system, by harnessing the channel symmetry between the downlink and uplink, the base station learns the downlink channel $\mathbf{H}_i$ to any user i from its uplink receiver. In an FDD system, the base station needs explicit reports of channel state information from the user, thereby incurring signaling overhead, errors (such as quantization errors and communication errors), and delay, which could be significant in a severe time-frequency selective channel (high mobility, large delay spread) with a large K_t. As a comparison, for joint decoding, channel state information is only needed at the receiver. We will address the signaling overhead issue later in this section.

Suppose that there is a total of M users with $K_t \leq M$. It can be shown that similar to uplink SDMA, downlink beamforming can achieve a spatial multiplexing gain of $\min(K_t, M)$ without the users performing joint decoding. The base station selects $K_u = K_t$ geographically separate users, and transmits to them over the same time-frequency resource.

Let us first restrict to linear transmit beamforming. The transmitted signal is constructed by

$$\mathbf{X} = \sum_{i=1}^{K_t} X_i \mathbf{u}_i, \qquad (7.53)$$

where $\mathbf{u}_i$ is the transmit beamforming vector with unit power ($\|\mathbf{u}_i\| = 1$ for all i) and X_i complex scalar representing the desired information signal targeted to user i with $P_i = |X_i|^2$ being the power allocated to user i. The total power budget is $\sum_{i=1}^{K_t} P_i \leq P_m$. Joint encoding needs to solve power allocation (P_i) and beamforming vector ($\mathbf{u}_i$).

Equation (7.52) now becomes

$$Y_i = \mathbf{H}_i \mathbf{u}_i X_i + \sum_{i' \neq i} \mathbf{H}_i \mathbf{u}_{i'} X_{i'} + W_i. \qquad (7.54)$$

The SINR of user i is equal to

$$\gamma_i = \frac{P_i |\mathbf{H}_i \mathbf{u}_i|^2}{\sigma_i^2 + \sum_{i' \neq i} P_{i'} |\mathbf{H}_i \mathbf{u}_{i'}|^2}, \qquad (7.55)$$

where σ_i^2 is the power of noise W_i.

Recall that in uplink SDMA, receive beamforming vector $\mathbf{u}_i$ only affects the SINR of user i and therefore can be solved separately in the MMSE sense for i. An important difference in downlink beamforming is that transmit beamforming vector $\mathbf{u}_i$ affects not only the SINR of user i but also the SINRs of all other users, and it becomes very difficult to decompose the problem into individual MMSE problems at each receiver. Fortunately, this problem can solved as a consequence of the following duality between the downlink and uplink MU-MIMO.

Rewrite (7.52) in a vector form to include the received signals of all users

$$\mathbf{Y}_{dl} = \begin{bmatrix} \mathbf{H}_{1,dl} \\ \vdots \\ \mathbf{H}_{K_t,dl} \end{bmatrix} \mathbf{X}_{dl} + \mathbf{W}_{dl}. \tag{7.56}$$

Rewrite (7.40) here

$$\mathbf{Y}_{ul} = \begin{bmatrix} \mathbf{H}_{1,ul} \cdots \mathbf{H}_{K_r,ul} \end{bmatrix} \mathbf{X}_{ul} + \mathbf{W}_{ul}. \tag{7.57}$$

Subscripts dl, ul are added to differentiate the downlink and uplink cases. The downlink (7.60) is a *dual* of the uplink (7.61) if the following three conditions all hold:

- Channel symmetry:

$$\begin{bmatrix} \mathbf{H}_{1,dl} \\ \vdots \\ \mathbf{H}_{K_t,dl} \end{bmatrix} = \begin{bmatrix} \mathbf{H}_{1,ul} \cdots \mathbf{H}_{K_r,ul} \end{bmatrix}^*; \tag{7.58}$$

Thus, $\mathbf{H}_{i,dl} = \mathbf{H}_{i,ul}^*$, for all i;
- Same noise statistics of $\mathbf{W}_{dl}$ and $\mathbf{W}_{ul}$;
- Same *total* transmit power:

$$|\mathbf{X}_{dl}|^2 = |\mathbf{X}_{ul}|^2. \tag{7.59}$$

Every component power does not have to be the same.

An important result shown in [72] is that the dual downlink and uplink achieve the same capacity region using the same optimal transmission strategy. Note that the power constraint in the uplink is usually on individual user (7.41). The duality definition assumes the individual power constraints are pooled (7.59) to a total power budget, which is shared among all the user transmitters. Thus, the uplink capacity region is the union of the capacity regions of all individual power constraints whose total is the same as the power constraint of the downlink dual.

In the special case of linear transmit beamforming (7.53) and receive beamforming (7.46), (7.56), and (7.57) becomes

$$\begin{bmatrix} Y_{1,dl} \\ \vdots \\ Y_{K_t,dl} \end{bmatrix} = \begin{bmatrix} \mathbf{H}_{1,dl}^* \\ \vdots \\ \mathbf{H}_{K_t,dl}^* \end{bmatrix} \begin{bmatrix} \mathbf{u}_{1,dl} \cdots \mathbf{u}_{K_t,dl} \end{bmatrix} \begin{bmatrix} X_{1,dl} \\ \vdots \\ X_{K_t,dl} \end{bmatrix} + \begin{bmatrix} W_{1,dl} \\ \vdots \\ W_{K_t,dl} \end{bmatrix}, \tag{7.60}$$

and

$$\begin{bmatrix} Y_{1,ul} \\ \vdots \\ Y_{K_r,ul} \end{bmatrix} = \begin{bmatrix} \mathbf{u}_{1,ul}^* \\ \vdots \\ \mathbf{u}_{K_r,ul}^* \end{bmatrix} \begin{bmatrix} \mathbf{H}_{1,ul} \cdots \mathbf{H}_{K_r,ul} \end{bmatrix} \begin{bmatrix} X_{1,ul} \\ \vdots \\ X_{K_r,ul} \end{bmatrix} + \begin{bmatrix} W_{1,ul} \\ \vdots \\ W_{K_r,ul} \end{bmatrix}. \quad (7.61)$$

Suppose the beamforming vectors are symmetric,

$$\mathbf{u}_{i,ul} = \mathbf{u}_{i,dl} \quad (7.62)$$

for all i. It is shown in [159] that the dual downlink and uplink achieve the same SINR for any user i.

From the above duality result, we now solve the joint power allocation and transmit beamforming problem in downlink MU-MIMO. Recall that in the uplink the MMSE receive beamforming vector $\mathbf{u}_{i,dl}^*$ is chosen to maximize the SINR of user i. In the downlink, we cannot optimize the transmit beamforming vector $\mathbf{u}_{i,dl}$ just for the SINR of user i, because the SINR also depends on the other transmit beamforming vectors. Likewise, power allocation has to be solved jointly for all users. Therefore, similar to the joint SDMA and power control problem in the uplink, we rely on an iterative search algorithm to jointly determine P_i and $\mathbf{u}_i$.

Specifically, suppose that the goal is to minimize the total downlink transmit power $\sum_i P_i$ with a given set of SINR targets $\hat{\gamma}_i$s. The first step is to solve the joint SDMA and power control problem of the uplink dual, as shown in Section 7.3.1. Then, the transmit beamforming vector $\mathbf{u}_{i,dl}$ of the downlink problem is set equal to the receive beamforming vector $\mathbf{u}_{i,dl}$ of the uplink dual (with proper normalization such that $\|\mathbf{u}_{i,dl}\| = 1$ for all i). The total transmit power of the downlink problem is equal to that of the uplink dual; however, the individual transmit powers of the downlink problem are not necessarily equal to those of the uplink dual and therefore need to be solved in the second step[2]: given $\mathbf{u}_{i,dl}$, iteratively update P_is to satisfy the target $\hat{\gamma}_i$s. The iteration starts with some P_is, calculates the SINR γ_is from (7.55), and updates the transmit power similar to (7.51)

$$P_i[t+1] = P_i[t] \frac{\hat{\gamma}_i}{\gamma_i}, \quad (7.63)$$

and then normalizes $P_i[t+1]$ such that the total is equal to what is obtained from the uplink dual.

Similar to the uplink, linear transmit beamforming alone is not capacity-achieving. In the uplink, the capacity-achieving strategy is a combined linear MMSE receive beamforming and SIC receiver. In the downlink, the counterpart is a combined linear transmit beamforming and *dirty-paper precoding*.

Dirty paper precoding is similar to the SIC strategy but used at the transmitter. Consider the following communication channel similar to AWGN except for an additional interference s,

$$y = x + s + z. \quad (7.64)$$

---

[2] An alternative scheme is provided in (7.71) to (7.73) to a similar problem of power allocation in multi-cell downlink MIMO.

7.3 Multiuser MIMO techniques

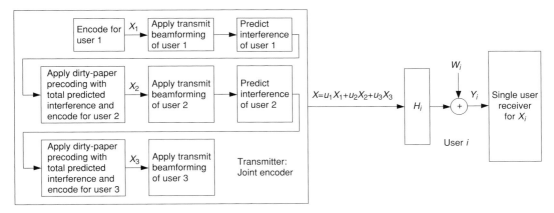

Figure 7.14 Transmitter architecture with combined linear transmit beamforming and dirty-paper precoding, which is capacity-achieving.

Suppose that s is known to the transmitter but not to the receiver. A somewhat surprising and remarkable result shown in [29] is that with dirty-paper (Costa) precoding the above channel has the same capacity as the AWGN channel where s does not exist. In dirty-paper precoding, the complexity resides in the transmitter while the receiver operates the same way as in the AWGN channel. This property makes it very applicable to downlink MU-MIMO.

Figure 7.14 illustrates a transmitter architecture where linear transmit beamforming and dirty-paper precoding are combined to achieve the capacity. Note its similarity with the V-BLAST receiver architecture shown in Figure 7.8. Rewrite (7.54) to

$$Y_i = \mathbf{H}_i^* \mathbf{u}_i X_i + \sum_{i'<i} \mathbf{H}_i^* \mathbf{u}_{i'} X_{i'} + \sum_{i'>i} \mathbf{H}_i^* \mathbf{u}_{i'} X_{i'} + W_i. \tag{7.65}$$

Recall that in linear transmit beamforming, X_i represents the desired scalar signal to user i and X_i's are independently encoded. Now, apply dirty-paper precoding to determine X_i by treating $\sum_{i'<i} \mathbf{H}_i^* \mathbf{u}_{i'} X_{i'}$ as known interference and $\sum_{i'>i} \mathbf{H}_i^* \mathbf{u}_{i'} X_{i'}$ as unknown Gaussian noise. The process goes on from $i = 1$ to $i = K_t$. Details on dirty-paper precoding and its application to downlink MU-MIMO can be found in [159].

So far in downlink MU-MIMO (linear transmit beamforming with or without dirty-paper precoding) we have assumed that channel state information of all the users is available at the base station transmitter. On the other hand, as noted previously, it may be costly or even impractical to obtain *perfect* channel state information through a feedback channel. How much spatial multiplexing gain can be achieved with only *partial* channel state information at the base station?

Recall that in Section 4.3.2 it is demonstrated that with only SINR feedback, opportunistic beamforming achieves power gain in a multiuser downlink. As proposed in [142], a similar idea can be applied to obtain spatial multiplexing gain. Specifically, at a given time, let the base station to choose K_t random orthonormal vectors $\mathbf{u}_k, k = 1, \ldots, K_t$ to

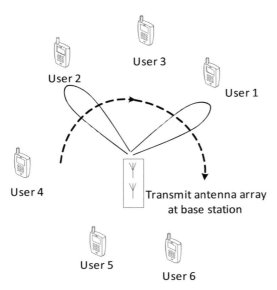

Figure 7.15 Opportunistic spatial multiplexing. In this example, two beams, orthogonal with each other, are formed by transmit antenna arrays and sweep over the space. Users in the beamforming directions are opportunistically multiplexed.

represent separate beams as shown in Figure 7.15. The base station transmits a known pilot signal A_k in beam k. $\mathbf{u}_k A_k$ are sent in orthogonal time-frequency resources.

Every user i measures the SINR of each beam ($k = 1, \ldots, K_t$) using the received pilot signals

$$\gamma_{i,k} = \frac{|\mathbf{H}_i \mathbf{u}_k A_k|^2}{\sigma^2 + \sum_{k' \neq k} |\mathbf{H}_i \mathbf{u}_{k'} A_{k'}|^2}. \tag{7.66}$$

and reports to the base station the best SINR and the corresponding beam index. For each beam, the base station schedules the user that reports the best SINR. When the number of users M is much greater than K_t, the base station is likely able to find users close to the present directions of all the K_t beams. Therefore, the base station transmits to separate data streams X_i simultaneously in orthogonal beam directions to K_t scheduled users,

$$\mathbf{X} = \sum_{i=1}^{K_t} \mathbf{u}_i X_i, \tag{7.67}$$

thereby achieving a spatial multiplexing gain of K_t. The orthonormal vectors $\mathbf{u}_i$ vary over time so that the beams sweep through all the users.

Note the similarity of the above opportunistic spatial multiplexing scheme and the opportunistic beamforming scheme enhanced with subband selection (Figure 4.20). In both schemes, multiple beams are formed in a given time. In the opportunistic spatial multiplexing scheme, multiple beams are made orthogonal with each other in space by using orthonormal beamforming vectors. In the enhanced opportunistic beamforming scheme, multiple beams are made orthogonal in frequency by using different subbands.

Separate pilots are needed for individual beams, which vary with time. Therefore, in addition to spatial multiplexing gain, the above opportunistic spatial multiplexing scheme obtains the same benefit of reducing scheduling latency as the enhanced opportunistic beamforming scheme does.

Opportunistic spatial multiplexing may not perform well when the number of users is not significantly greater than the number of transmit antennas; that is, $M \approx K_t$. The reason is that at a given time the formed beams may not be well aligned with any user's channel. One way is to for a user to report a quantized version of channel state information [55]. In general, the amount of feedback overhead is a function of the frequency of feedback, the number of parameters to be quantized, and the quantization resolution. The frequency of feedback depends on the rate of wireless channel fluctuation, while the number of parameters and the quantization resolution have been addressed using limited feedback precoding [107]. The key idea is to quantize the precoding vectors rather than the channel itself. In an SU-MIMO scenario, denote by $\mathbf{H}$ the channel matrix, and $\mathbf{U}$ the precoding matrix, which is a function of $\mathbf{H}$. Quantize the space of $\mathbf{U}$ into $\{\mathbf{U}_1, \mathbf{U}_2, \ldots, \mathbf{U}_{N_B}\}$, where $N_B = 2^B$ for an integer B. Therefore, the choice of a particular precoding matrix can be reported from the receiver using B bits of feedback.

A similar idea can be applied to a multiuser downlink MIMO scenario, where each user equipped with a single receive antenna uses a random quantization codebook consisting of 2^B quantization vectors chosen from the isotropic distribution on the K_t-dimensional unit sphere [73]. An interesting observation is that the feedback requirements have to scale linearly with SINR (in dB) in order to achieve the full spatial multiplexing gain if the transmitter employs zero-forcing precoding on the basis of feedback. In contrast, in SU-MIMO, feedback does not need to scale with SINR. This indicates that compared with SU-MIMO, multiuser downlink MIMO is generally more sensitive to the imperfection of channel state information and it is more important to provide accurate channel feedback. The reason is that joint decoding is permitted in SU-MIMO but not in multiuser downlink MIMO. Recall that in SU-MIMO with channel state information at the transmitter, power can be optimally allocated; however, even without the transmitter side channel state information, the transmitter equally allocates power to the antennas and the V-BLAST receiver can still achieve the full spatial multiplexing gain at some power penalty. On the other hand, in multiuser downlink MIMO, imperfect channel state information leads to an error-floor caused by residual inter-user interference. As SINR grows, the feedback quantization error has to drop accordingly in order to balance noise and inter-user interference.

7.4 Multi-cell MIMO techniques

In this section, we extend our study of multiple antenna techniques to a multi-cell system. Cellular spectrum reuse is a form of spatial multiplexing in which transmitters and receivers in different cells use the same bandwidth and the total system capacity increases as spectrum reuse becomes denser. Inter-cell interference is a main constraint for spectrum reuse. Recall that without the use of multiple antennas, we use frequency

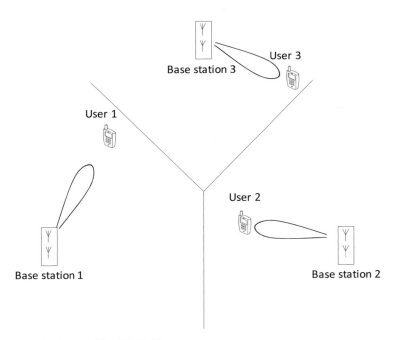

Figure 7.16 A multi-cell MIMO system.

reuse factor $K > 1$ to handle inter-cell interference, interference averaging to achieve universal frequency reuse $K = 1$, and also base station coordination of power and bandwidth usage for fractional frequency reuse.

Figure 7.16 depicts a multi-cell MIMO system. In a cell, K_u users use the same time-frequency resource, and from the previous section, MU-MIMO can handle intra-cell interference among those K_u users. Consider universal frequency reuse where users in different cells use the same resource as well. How would multiple antennas mitigate inter-cell interference?

7.4.1 Coordinated beamforming

As before, we assume that a base station is equipped with multiple antennas (K_t at the transmitter or K_r at the receiver) and a user with a single antenna. From Figure 7.16, multi-cell MIMO is similar to MU-MIMO, except that now base stations are not co-located and joint encoding/decoding becomes difficult to implement. Inter-cell interference mitigation with multi-cell MIMO depends on the level of coordination among base stations.

The simplest form assumes no coordination at all. In this case, each cell employs MU-MIMO and treats inter-cell interference as background noise. In the uplink, the base station employs the MMSE receive beamforming scheme (7.46) to maximize the SINR. Obviously the noise power now depends on the location of the interfering users in adjacent cells. If an interfering user happens to be in the same direction as a desired user, then it would significantly damage the SINR even thought the receiver is already optimal from a single cell perspective. On the other hand, if an interfering user is in an orthogonal direction, then it would easily be nulled out in MMSE receive beamforming without degrading the desired signal strength. Clearly, some level of coordinated scheduling to select users in appropriate spatial locations can boost the achieved SINR. The same observation can be made in the downlink.

The other extreme is networked MIMO where antennas of multiple base stations form a single antenna array. Conceptually the multi-cell problem now becomes a multiuser one and the techniques in the previous section can be applied straightforwardly. The challenge, however, is to support base station cooperation, which involves the following:

- *Backhaul overhead.* A base station needs to know the channel state information not only of itself but also of other base stations. The base stations needs to exchange raw data information for joint encoding/decoding.
- *Processing complexity.* Joint encoding/decoding requires tremendous processing capability.
- *Synchronization requirement.* Transmissions in different cells need to be tightly synchronized, on the order of μs, to ensure the accuracy of joint processing.

We will study networked MIMO in the context of cooperative communications in Section 10.2.2.

Compared with no coordination, networked MIMO achieves great performance gains, but at substantial costs of cooperation at the signal level. To be cost effective and practical in the real world, a multi-cell MIMO scheme has to strike a good balance between performance and cost. Coordination at the beamforming and scheduling level seems appealing as it requires much less overhead and is easier to implement. In the uplink, the single cell MMSE receiver is optimal, assuming users have been selected via coordinated scheduling. In the downlink, neighboring cells need to coordinate transmit beamforming, because a transmit beamforming vector used in one cell affects the SINR in adjacent cells. In the following, we assume that appropriate coordinated scheduling has already been employed to select K_u users in every cell and focus on the transmit beamforming aspect.

Consider linear beamforming. Extending (7.53), let $X_{b,i}$ be a complex scalar representing the desired information signal to user i in cell b and $\mathbf{u}_{b,i}$ the corresponding transmit beamforming vector with unit power. Denote by $\mathbf{H}_{b',b,i}$ the channel vector from base station of cell b' to user i of cell b. Then similar to (7.54), the received signal $Y_{b,i}$ includes the desired signal as well as intra-cell and inter-cell interference:

$$Y_{b,i} = \mathbf{H}^*_{b,b,i}\mathbf{u}_{b,i}X_{b,i} + \sum_{i' \neq i}\mathbf{H}^*_{b,b,i}\mathbf{u}_{b,i'}X_{b,i'} + \sum_{b' \neq b, i'}\mathbf{H}^*_{b',b,i}\mathbf{u}_{b',i'}X_{b',i'} + W_{b,i}, \quad (7.68)$$

where $W_{b,i}$ is a circularly symmetric complex AWGN with $\sigma^2/2$ on each of the real and imaginary components. The SINR is given by

$$\gamma_{b,i} = \frac{P_{b,i}|\mathbf{H}^*_{b,b,i}\mathbf{u}_{b,i}|^2}{\sigma^2 + \sum_{i' \neq i} P_{b,i'}|\mathbf{H}^*_{b,b,i}\mathbf{u}_{b,i'}|^2 + \sum_{b' \neq b, i'} P_{b',i'}|\mathbf{H}^*_{b',b,i}\mathbf{u}_{b',i'}|^2}, \quad (7.69)$$

with $P_{b,i} = |X_{b,i}|^2$.

Suppose the goal is to minimize the total downlink transmit power $\sum_{b,i} P_{b,i}$ such that $\gamma_{b,i} \geq \hat{\gamma}_{b,i}$ for some feasible SINR target $\hat{\gamma}_{b,i}$. Recall that a similar problem is solved in the MU-MIMO context and the tool we use is the uplink-downlink duality, which is established in the single cell scenario. A natural question is whether the duality can be generalized to the multi-cell scenario. The answer is yes, as shown in [33]. The generalized result is as follows.

We can furthermore generalize the preceding objective function a little, and minimize the weighted total downlink power $\sum_{b,i} w_b P_{b,i,dl}$, where w_b is a weight for base station b. The variables $\gamma_{b,i}, P_{b,i}, \mathbf{u}_{b,i}$ in (7.69) should be added with subscript dl. The uplink dual is used to minimize the total uplink power $\sum_{b,i} \sigma^2 P_{b,i,ul}$, subject to the same SINR target $\gamma_{b,i,ul} \geq \hat{\gamma}_{b,i}$, where the uplink SINR is given by

$$\gamma_{b,i,ul} = \frac{P_{b,i,ul} |\mathbf{H}^*_{b,b,i} \mathbf{u}_{b,i,ul}|^2}{\sum_{(b',i') \neq (b,i)} P_{b',i',ul} |\mathbf{H}^*_{b,b',i'} \mathbf{u}_{b,i,ul}|^2 + w_b \|\mathbf{u}_{b,i,ul}\|^2}. \quad (7.70)$$

Here $P_{b,i,ul}$ is the uplink transmit power of user i in cell b, $w_b \|\mathbf{u}_{b,i,ul}\|^2$ is the noise power at base station b, and $\mathbf{u}_{b,i,ul}$ is the uplink receive beamforming vector.

We can use an iterative search algorithm similar to (7.50) and (7.51) to determine power control ($P_{b,i,ul}$) and beamforming ($\mathbf{u}_{b,i,ul}$). $\mathbf{u}_{b,i,dl}$ is a normalized version of $\mathbf{u}_{b,i,\mu l}$ such that $\|\mathbf{u}_{b,i,dl}\| = 1$ and downlink power allocation is such that

$$\sqrt{P_{b,i,dl}} \mathbf{u}_{b,i,dl} = \sqrt{\delta_{b,i}} \mathbf{u}_{b,i,ul}. \quad (7.71)$$

The scaling factor $\delta_{b,i}$ can be solved with the following system of linear equations (7.73) by noting that the optimal solution is such that the SINR target is met with equality ([33]):

$$\gamma_{b,i,dl} = \hat{\gamma}_{b,i}. \quad (7.72)$$

Rearranging (7.69), it follows that

$$\frac{1}{\hat{\gamma}_{b,i}} |\mathbf{H}^*_{b,b,i} \mathbf{u}_{b,i,ul}|^2 \delta_{b,i} - \sum_{i' \neq i} |\mathbf{H}^*_{b,b,i} \mathbf{u}_{b,i',ul}|^2 \delta_{b,i'} - \sum_{b' \neq b,i'} |\mathbf{H}^*_{b',b,i} \mathbf{u}_{b',i',ul}|^2 \delta_{b',i'} = \sigma^2. \quad (7.73)$$

The essence of downlink multi-cell transmit beamforming is to align the beamforming direction to the desired user and meanwhile the null directions to other users in adjacent cells. Recall that in the single cell multiuser case, the number of multiplexed users K_u, selected from M in total, is set to K_t in order to maximize the spatial multiplexing gain. In an ideal world, the channel vectors of those K_t users are orthogonal with each other and downlink beamforming completely separates them with no intra-cell interference. However, the choice of $K_u = K_t$ with the single cell approach of linear beamforming leaves no room to spatially null inter-cell interference. In the multi-cell case, we may set $K_u < K_t$ so as to improve inter-cell interference suppression. Clearly there is a tradeoff between intra-cell spatial multiplexing and inter-cell interference suppression.

7.4 Multi-cell MIMO techniques

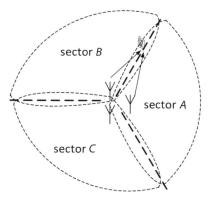

Figure 7.17 Gain patterns of directional antennas for sectorization and beamforming at sector boundary.

7.4.2 Inter-sector beamforming

As noted in the preceding section, a key practical constraint for base station coordination or cooperation comes from the fact that base stations are not co-located and any joint processing requires backhaul exchange among base stations. An important exception is the multi-sector scenario, in which case local coordination is much easier to accomplish among sectors in the same cell.

Directional antennas are designed to provide high gain across a range of signal arrival/departure angles. Use of directional antennas can create multiple sectors (usually 3) in a cell, thereby increasing spatial reuse. Figure 7.17 illustrates the gain patterns of directional antennas used to partition a cell into three 120° sectors.

In a typical cellular sectorization deployment, directional antennas are used to well cover the interior area of a sector. At the sector boundary, the antenna gain does not drop suddenly. As a result, a user on the boundary between two sectors shown in Figure 7.17 receives signals from both with similar strength and thus ends up with a low SINR (0 dB or less), even if it is close to the base station and far from other base stations.

Section 5.3 introduces the concept of *synchronized sectors* by coordinating inter-sector interference such that inferior users of one sector only interfere with inferior ones of another sector and superior users of one sector only interfere with superior ones of another sector. Coordination of inter-sector interference helps preserve large dynamic range of power and bandwidth allocation when multiplexing users in OFDMA. We next show that sector synchronization also facilitates coordinated MIMO between sectors.

One strategy is to introduce multi-sector beamforming to serve sector boundary users to boost their SINRs. In particular, two antennas in two adjacent sectors are used to transmit simultaneously to a sector boundary user. As shown in Figure 7.17, both transmit antennas in sector A and sector B are transmitting to a user in sector A. Meanwhile, the antenna in sector B is not serving any user in its own sector in the same time-frequency resource. This becomes possible because the sectors are synchronized

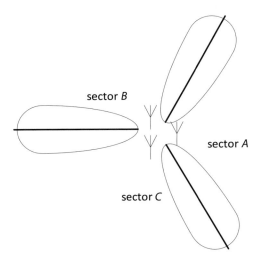

Figure 7.18 Sector antennas form multiple beams to cover all sector boundaries simultaneously.

and now the adjacent sectors can coordinate the transmit strategy in a given fraction of the time-frequency resource, as pointed out in Section 5.3.3.

The above scheme obviously suffers from bandwidth loss, because sector B's antenna helps the sector boundary user of sector A at the cost of not being able to transmit to a user in its own sector. However, we can do better than that; the three antennas jointly form three independent beams to cover the sector boundaries so that three users at the different sector boundaries are scheduled *simultaneously* in the same fraction of the time-frequency resource. Specifically, denote by $X_1, X2, X3$ scalar symbols to be sent to users 1, 2, 3, respectively, in the synchronized sectors. User 1 is located on the boundary between sectors A and B, user 2 between A and C, and user 3 between B and C. Let the three antennas transmit the following signals: antenna in sector A transmits $(X_1 + X_2 - X_3)/2$, antenna in sector B transmits $(X_1 - X_2 + X_3)/2$ and antenna in sector C transmits $(-X_1 + X_2 + X_3)/2$. Generate the transmit signal by precoding X_1, X_2, X_3 as follows:

$$\mathbf{X} = \begin{bmatrix} \frac{1}{2} & \frac{1}{2} & -\frac{1}{2} \\ \frac{1}{2} & -\frac{1}{2} & \frac{1}{2} \\ -\frac{1}{2} & \frac{1}{2} & \frac{1}{2} \end{bmatrix} \begin{bmatrix} X_1 \\ X_2 \\ X_3 \end{bmatrix}. \quad (7.74)$$

We use the above precoding matrix assuming that each user at the boundary of two sectors sees the same complex channel from those two sectors. In general, the precoding matrix to be used depends on the complex channel gains. The effective gain patterns created by the above precoding is shown in Figure 7.18. By eliminating inter-sector interference, we effectively eliminates the low SINR effect at the sector boundaries of the original sectorization. In summary, with sector synchronization, we apply the precoding matrix in a fraction of the time-frequency resource to serve the sector-boundary users, and the normal antenna patterns for the sector-interior users in the remaining resource.

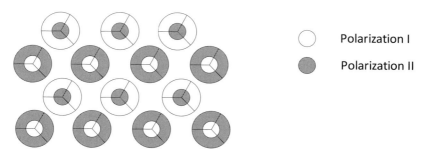

Figure 7.19 Fractional reuse pattern with polarized antennas.

7.4.3 Inter-cell interference avoidance with polarized antennas

Polarized antennas can be used to mitigate inter-cell interference. For example, suppose that a base station uses a vertically or horizontally polarized antenna. If an adjacent base station uses a different polarization than the serving base station does, then the interference is at least XPD below the desired signal. High XPD gain protects the SINR of the cell boundary users. However, since there are only two type of polarizations available but a cell is adjacent to more than two others, it is simply not possible that any pair of adjacent cells use different polarizations. In a hexagonal cellular deployment, each cell has six adjacent cells. As shown in Figure 7.19, the polarization pattern can be arranged such that each cell has only two (out of six) adjacent cells with the same polarization. The use of polarized antennas in effect leads to a factor of 2 polarization reuse to suppress inter-cell interference. In this scheme, a user needs to be equipped with both polarizations to match its antenna polarization with a given serving base station.

Furthermore, we can combine the FFR idea with polarization reuse. In particular, each base station uses both polarizations, but puts more power on one of its polarized antennas (the preferred antenna) and less on the other. The choice of preferred polarized antenna follows a coordinated pattern among adjacent base stations. One example is shown in Figure 7.19. The FFR techniques studied in Section 6.3 can be used to schedule inferior and superior users to maximize the system capacity.

Practical example 7.1 Multiple antenna techniques in LTE

The use of multiple antennas has been a key technical feature of LTE. Because of cost, size, power, and other considerations, it is easier to equip multiple transmit antennas at an eNB than at a UE. LTE specifies a variety of *transmit* multiple antenna techniques in the downlink but provides only limited use in the uplink. Specifically, in addition to the SISO mode, LTE defines transmit diversity, spatial multiplexing, and beamforming in the downlink assuming multiple transmit antennas at an eNB. A UE needs to be equipped with multiple receive antennas only in the spatial multiplexing mode with rank greater than one. On the other hand, only transmit antenna selection and MU-MIMO are defined in the uplink. In the former case, a UE employs a low-cost transceiver architecture similar to Figure 7.1(b), which includes multiple transmit antennas with only one transmit chain. In the latter case, no multiple transmit antennas are required at a UE. Many of the downlink techniques are later extended to the uplink in LTE Advanced, as will be described in Example 10.1. In the following, we describe the multiple antenna techniques in LTE, in both the downlink and uplink.

It should be pointed out that *receive* multiple antenna techniques are out of the scope of LTE specifications and left to the implementation. We will not cover receiver techniques.

Downlink

As we have learned in this chapter, channel information is essential to exploit multiple antennas. To assist channel estimation, an eNB sends a reference signal, which is a pilot known to all UEs. An *antenna port* is a logical entity associated with a reference signal. An antenna port does not necessarily correspond to a physical antenna. For example, if an eNB sends a reference signal simultaneously from two physical antennas, then the combination of those antennas is an antenna port; if an eNB sends two separate reference signals, one from each antenna, then those antennas are two separate antenna ports. In the former case, two signals from the antennas are combined over the air and received at a UE with an overall channel response. To demodulate a data stream sent via an antenna port, the receiver uses the corresponding reference signal for channel estimation. In LTE, antenna ports 0–3 are cell-specific, port 4 MBSFN specific (not discussed here), and port 5 UE specific.

Figure 7.20 shows the physical resource of the reference signal CS-RS in the case where an eNB is equipped with one, two or four antennas. A UE estimates the channel separately for each antenna port using the corresponding reference signal.

Although Figure 7.20 shows the resource only in a PRB, the CS-RS span the entire downlink bandwidth to assist channel estimation. The CS-RS are used to coherently demodulate all the downlink signals except for the beamforming technique using antenna port 5. In the case of multiple antenna ports, to avoid interference between the reference signals of different antenna ports, when an antenna port sends its reference signal, other antenna ports keep silent in the corresponding physical resource elements labeled as null signal. As a result, the resource overhead of the CS-RS increases with the number of antenna ports. To reduce the overhead, in the case of four antenna ports, antenna ports 0, 1 have twice the physical resources of antenna ports 2, 3. The justification is that four-antenna spatial multiplexing is likely used in a low mobility scenario and thus the CS-RS does not have to be as dense. The set of reference signal physical resource elements shown in Figure 7.20 can be cyclically shifted in frequency to form another non-overlapping set. There are six frequency shifts in total. Neighboring cells should use different frequency shifts to avoid interference between their reference signals. The modulation symbols mapped to those physical resource elements are derived from one of 504 reference signal sequences depending on the physical layer cell identifier, which a UE has acquired during cell search (see Example 9.3).

7.4 Multi-cell MIMO techniques

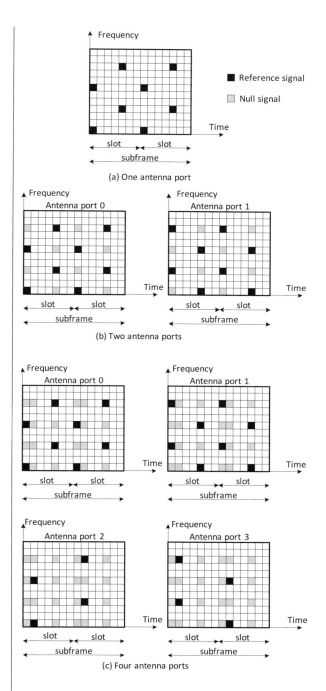

Figure 7.20 An example of the CS-RS in a PRB: (a) one antenna port, (b) two antenna ports, and (c) four antenna ports. Normal CP.

The reference signal UE-RS is associated with antenna port 5. Unlike the CS-RS, the UE-RS is sent only in the PDSCH physical resource assigned to a UE and used for the UE to demodulate that PDSCH. Clearly it is not meant for all UEs. Figure 7.21 shows the physical resource of the UE-RS. The above cyclic frequency shift idea is applicable here to form a different UE-RS. Note that the CS-RS and UE-RS do not overlap in time, thereby no collision between the two. Besides, from Figure 5.7, the control channels only reside in the first third OFDM symbols and therefore are not affected by any UE-RS.

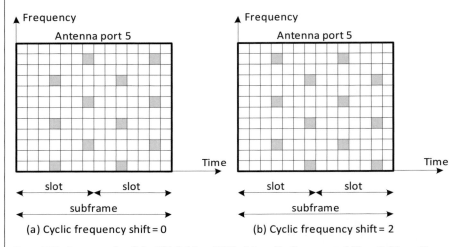

Figure 7.21 An example of the UE-RS in a PRB: (a) cyclic frequency shift = 0 (b) cyclic frequency shift = 2. Normal CP.

We next describe the transmit diversity, spatial multiplexing and beamforming techniques in LTE.

Transmit diversity. Transmit diversity is achieved when an eNB is equipped with two or four transmit antennas. In the case of two antenna ports, the technique is Space Frequency Block Code (SFBC), which applies the Alamouti scheme studied in Section 4.2.3 to the frequency domain. Specifically, a pair of modulation symbols $\{X_1, X_2\}$ are transmitted in adjacent tones from the first antenna port, and then $\{-X_2^*, X_1^*\}$ are transmitted in the same tones from the second antenna port:

$$\begin{bmatrix} X_1 & X_2 \\ -X_2^* & X_1^* \end{bmatrix}, \tag{7.75}$$

where the row index is antenna port and the column index represents tones. SFBC achieves a diversity order of 2. In the case of four antenna ports, the technique is a combination of SFBC and Frequency Switched Transmit Diversity (FSTD). Specifically, four modulation symbols $\{X_1, X_2, X_3, X_4\}$ are transmitted in four adjacent tones simultaneously:

$$\begin{bmatrix} X_1 & X_2 & 0 & 0 \\ 0 & 0 & X_3 & X_4 \\ -X_2^* & X_1^* & 0 & 0 \\ 0 & 0 & -X_4^* & X_3^* \end{bmatrix}. \qquad (7.76)$$

Note that the above choice is superior to an alternative choice

$$\begin{bmatrix} X_1 & X_2 & 0 & 0 \\ -X_2^* & X_1^* & 0 & 0 \\ 0 & 0 & X_3 & X_4 \\ 0 & 0 & -X_4^* & X_3^* \end{bmatrix}. \qquad (7.77)$$

The difference is that in the latter choice, X_1, X_2 are sent from antenna ports 0, 1, and X_3, X_4 are sent from antenna ports 2, 3, while in the former choice, X_1, X_2, X_3, X_4 are from all four antenna ports. Note that antenna ports 0, 1 have a higher CS-RS density than antenna ports 2, 3 and thus achieve better channel estimation quality. Thus, the former interlaced choice leads to more balanced demodulation performance across all symbols.

Spatial multiplexing. Let us first define a few terms. A *codeword* is the output of the modulation and coding module from the input of a single transport block. In the case of spatial multiplexing in a MIMO channel, a *layer* corresponds to a data stream to be sent over a spatial multiplexing subchannel. The number of layers, N_L, is called *transmission rank* and is no greater than the rank of the MIMO channel, which in turn is no greater than the number of antenna ports N_A. The number of codewords is no greater than the number of layers such that a codeword is mapped to one or more layers.

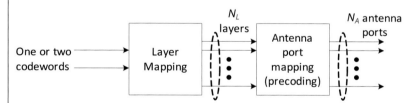

Figure 7.22 Basic transmit signal processing structure of spatial multiplexing in LTE.

Figure 7.22 depicts the basic transmit signal processing structure of spatial multiplexing in LTE. In LTE, the maximum number of codewords to be multiplexed is 2, and $N_A, N_L \leq 4$.

One or two codewords are first mapped to N_L layers as shown in Figure 7.23. The number of modulation symbols, N, is the same on each layer. After layer mapping, a block of N_L modulation symbols, each from one layer, are linearly processed with a $N_A \times N_L$ precoding matrix to generate N_A symbols, which are then mapped to N_A antenna ports, respectively, and sent in a given resource element.

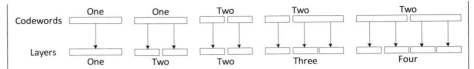

Figure 7.23 Layer mapping from codewords to layers in LTE.

There are two forms of spatial multiplexing: open-loop or closed-loop. The basic idea of closed-loop spatial multiplexing comes from the SVD scheme studied in Section 7.2.1. A UE explicitly recommends an N_L (Rank Indicator, RI) and a precoding matrix (Precoding Matrix Indicator, PMI) in the uplink control channel, and reports channel SNR (Channel Quality Indicator, CQI) on each layer. Roughly speaking, PMI and RI jointly indicate the spatial directions of the MIMO channel and CQI tells the channel SNR quality of the corresponding directions. The UE estimates the MIMO channel using the CS-RS and derives the recommended RI and PMI to maximize the capacity given the estimated MIMO channel. To reduce the control overhead in the uplink, the precoding matrix is selected from a predefined set, called *codebook*, and the PMI is the index to that set. For example, in the case of $N_L = 2$, $N_A = 2$, the predefined codebook consists of three precoding matrices with index equal to 0, 1, 2, respectively:

$$\frac{1}{\sqrt{2}}\begin{bmatrix} 1 & 0 \\ 0 & 1 \end{bmatrix}, \frac{1}{\sqrt{2}}\begin{bmatrix} 1 & 1 \\ 1 & -1 \end{bmatrix}, \frac{1}{\sqrt{2}}\begin{bmatrix} 1 & 1 \\ j & -j \end{bmatrix}. \tag{7.78}$$

The special case of $N_L = 1$ is referred to closed-loop rank-1 precoding and in essence beamforming for power gain. The predefined codebook is

$$\frac{1}{\sqrt{2}}\begin{bmatrix} 1 \\ 1 \end{bmatrix}, \frac{1}{\sqrt{2}}\begin{bmatrix} 1 \\ -1 \end{bmatrix}, \frac{1}{\sqrt{2}}\begin{bmatrix} 1 \\ j \end{bmatrix}, \frac{1}{\sqrt{2}}\begin{bmatrix} 1 \\ -j \end{bmatrix}. \tag{7.79}$$

The basic idea of open-loop spatial multiplexing is the V-BLAST scheme studied in Section 7.2.2. Open-loop spatial multiplexing is desirable, for example, in a high mobility environment where the MIMO channel changes rapidly. The UE does not recommend any PMI. The precoding matrix is equal to the multiplication of three matrices $W(i)D(i)U$, where i represents the index of modulation symbols in each layer, $i = 0, \ldots, N-1$. $W(i)$ is a $N_A \times N_L$ matrix selected from a subset of the same predefined codebook as in closed-loop spatial multiplexing except that the precoding matrices are selected cyclically in turn without UE feedback. For example, for $N_L = 2$, $N_A = 2$, $W(i)$ is always equal to

$$\frac{1}{\sqrt{2}}\begin{bmatrix} 1 & 0 \\ 0 & 1 \end{bmatrix}. \tag{7.80}$$

In some other cases such as when $N_A = 4$, $W(i)$ varies with i. $D(i)U$ is to make different layers to experience the same average channel quality. Specifically, $D(i)$ is a $N_L \times N_L$ matrix to support large-delay Cyclic Delay Diversity (CDD) by inducing channel quality variation in frequency. U is a $N_L \times N_L$ unitary DFT-based precoding matrix. For $N_L = 2$,

$$D(i) = \begin{bmatrix} 1 & 0 \\ 0 & e^{-j2\pi i/2} \end{bmatrix}, U = \frac{1}{\sqrt{2}} \begin{bmatrix} 1 & 1 \\ 1 & e^{-j2\pi/2} \end{bmatrix}. \qquad (7.81)$$

Beamforming. As noted before, closed-loop spatial multiplexing with $N_L = 1$ is a form of beamforming. In this case, the beamforming vector is selected from a predefined codebook. In the uplink control channel the UE recommends a preferred beamforming vector index. The eNB may or may not follow the recommendation; if not, then the eNB explicitly inform the UE of the index of the beamforming vector actually used so that the UE can estimate the beamforming channel correctly:

$$H = \sum_{k=1}^{N_A} w_k H_k, \qquad (7.82)$$

where H_k is the channel estimate of antenna port k from the corresponding CS-RS, and w_k is the weight in the beamforming vector.

LTE specifies another mode of beamforming in which the eNB is not limited to a predefined codebook but can apply any beamforming vector. Indeed the eNB does not even have to inform the UE of the beamforming vector actually used. To assist channel estimation, the eNB sends the UE-RS using the same beamforming vector. This *non-codebook-based* transmission is said to be from antenna port 5 to be distinguished from antenna ports 0–3 studied before.

MU-MIMO. In principle, the eNB can apply the above closed-loop rank-1 precoding technique to enable MU-MIMO. Specifically, the eNB transmits to two UEs simultaneously, and uses two rank-1 precoding vectors, one for each UE. Ideally, the eNB scheduling algorithm selects the two UEs that are spatially orthogonal, and moreover the direction of each UE is aligned with a rank-1 precoding vector in the predefined codebook such that inter-user interference is suppressed without the need for interference cancellation. However, such an idealistic scenario may not always happen in reality, in particular because of coarse quantization of the predefined precoding vector codebook. In this case, a UE would need to know the precoding vectors of the other UE in order to suppress inter-user interference (e.g., linear MMSE or cancellation), thereby complicating the receiver design and SNR calculation for selecting the modulation and coding scheme.

In addition, an important challenge is the accuracy of channel state information. As indicated in the end of Section 7.3.2, MU-MIMO requires more accurate channel state information at the transmitter than SU-MIMO does. In the single user case, the loss is an SNR degradation whose impact is not significant in the high SNR regime. However, in the multiuser case, imperfect channel state information leads to residual inter-user interference and a loss in spatial multiplexing gain, which is more significant to the capacity.

Overall, the support of MU-MIMO is quite limited in LTE. This limitation is greatly overcome in LTE-A as described in Example 10.1.

> **Uplink**
> Simultaneous transmissions from multiple transmit antennas of a given UE is not supported. Instead, LTE supports the following two multiple antenna techniques in the uplink.
>
> **Closed-loop transmit antenna selection**. In this case, a UE is equipped with multiple transmit antennas but only one power amplifier. Recall from Example 4.2 that the UE sends the SRS to allow the eNB measures the quality of the uplink channel. Now the UE alternates between the multiple transmit antennas to send the SRS in successive SRS symbols. After measuring the received SRS, the eNB instructs which antenna to be used for the PUSCH by coding the information in the uplink grant control message.
>
> **MU-MIMO**. Two UEs, each with one transmit antenna active, send the PUSCH on the same physical resource. The eNB separates their codewords with multiple receive antennas. To differentiate the signals, orthogonal DM-RS are assigned to the two UEs. Specifically, Figure 2.26 shows the PUSCH signal structure where two OFDM symbols are used per subframe for DM-RS, which is generated from a cyclic shift of some base sequence. The eNB assigns the UEs different cyclic shifts of the same base sequence so as to generate orthogonal DM-RS, which are used for channel estimation with minimum inter-user interference. In LTE, the maximum number of cyclic shifts is eight and therefore up to eight UEs can be multiplexed in MU-MIMO.

7.5 Summary of key ideas

- There are three main types of gains using multiple antennas: diversity gain that alters channel statistics, power gain that boosts SNR or SINR, and multiplexing gain that increases bandwidth. The focus of this chapter is multiplexing gain. Multiplexing gain is important in the high SNR regime.
- Multiplexing gain depends on the rank of the MIMO channel matrix. The maximum gain is achieved when the channel matrix is full rank in which case the channel capacity increases by a factor of $\min(K_r, K_t)$ from the SISO capacity.
- Linear antenna arrays are most commonly used in practice so far. To be full rank, the channel needs rich scattering at both the transmitter and the receiver. When a single path dominates, there is no multiplexing gain. In general, multiplexing gain depends on the number of non-zero rows and the number of non-zero columns of the angular domain representation of the channel matrix. In a multipath scenario, paths are divided into transmit/receive bins such that the number of paths in any bin is generally less than in the SISO case and the corresponding channel response exhibits a lesser degree of frequency selectivity. However, when multiple paths fall into the same bin, the channel response still varies with frequency and frequency selective scheduling is important.
- Polarized antennas can lead to a full rank channel matrix without rich scattering. XPD measures the power leakage between different polarized antennas and depends on

7.5 Summary of key ideas

antenna design and wireless channel. In general, channel XPD is higher in an LOS channel than in a non-LOS channel.

- In a single user case, when the channel matrix is known to the transmitter, the singular value decomposition scheme achieves the MIMO channel capacity. When the channel matrix is only known to the receiver, there is no loss of multiplexing gain, only a power penalty where the transmitter equally splits its total power among all the antennas as opposed to the optimal water-filling power allocation along eigenchannels. At the high SNR regime, the loss is minor. The V-BLAST receiver using combined linear MMSE and SIC is capacity-achieving.
- SU-MIMO capacity depends on the channel between the base station and the user, which in many practical scenarios does not provide rich scattering due to small angular spread at the base station. However, MU-MIMO does not suffer from this problem, as long as geographically separate users are scheduled. In either downlink or uplink MU-MIMO, users do not cooperate and the complexity burden resides at the base station. In the uplink, the base station receiver employs the V-BLAST architecture to achieve the capacity. The counterpart in the downlink is linear beamforming and dirty-paper precoding. Uplink-downlink duality helps address the downlink problem using the uplink solution.
- To achieve downlink MU-MIMO capacity, channel state information is needed at the base station. In contrast, in the SU-MIMO case, the same multiplexing gain can be achieved with or without channel state information at the base station. In an FDD system, the base station obtains channel state information from users' feedback reports at the cost of signaling overhead. With only partial channel state information, opportunistic spatial multiplexing works in a similar fashion to the opportunistic beamforming studied in Section 4.3.2 and the benefit increases with the number of users.
- Inter-cell interference is often a capacity limiting factor in a cellular system. A large percentage of users operate in a low SINR regime (5 dB or below). Their SINR would be high if inter-cell interference were removed from the equation. Ignoring inter-cell interference in MIMO design will fail to achieve meaningful MIMO gain in practice. One relatively simple idea of multi-cell MIMO is coordinated beamforming where neighboring base stations employ linear transmit/receive beamforming techniques to suppress inter-cell interference. An even simpler application is inter-sector beamforming where sectors of the same cell performs only local coordination.
- It should be pointed out that the study in this chapter applies to the high SINR regime. At the low SINR regime, power and diversity gains are much more important. Indeed, transmitting in all spatial dimensions can decrease the capacity because of increased dimensionality of channel estimation. A better strategy is to concentrate power on the direction of the greatest eigenvalue of the channel matrix.

8 Scheduling

In general scheduling addresses two problems:

- *User selection*: which users to transmit in the uplink or receive in the downlink.
- *Resource allocation*: what time-frequency bandwidth to be allocated to the selected users and what transmit power to be used.

Good scheduling strives to achieve two goals, namely quality of service (QoS) on the user level, measured by data rate, delay, loss, and fairness among users, and efficiency on the system level, measured by the total amount of traffic supported by the system.

A key feature in an OFDMA mobile broadband cellular system is scheduling by which a base station dynamically selects users and allocates time-frequency-power resource to them. In contrast, when a user is admitted to the system in a circuit-switched voice system, it is statically assigned a piece of bandwidth resource (time, frequency, or code) over which the voice traffic is transported without explicit dynamic scheduling. In a CDMA voice system, for example, the only dynamic resource allocation job is power allocation or power control [168]. The situation is quite different in mobile broadband, because data traffic is bursty and has different QoS requirements. Static resource allocation cannot simultaneously meet the QoS requirements and achieve high system efficiency. Scheduling becomes a necessity to dynamically match user selection and resource allocation with traffic needs and wireless channel conditions. On the other hand, scheduling has been well studied in wireline broadband networks (see [13, 137]), and many of those design and analysis ideas are applicable to the wireless counterpart. However, a unique challenge in mobile broadband is that the link capacity is not fixed or completely predictable. In OFDMA, power and bandwidth allocation introduces a new dimension to optimize.

Scheduling in a cellular system has been a very active research area, especially since the introduction of EV-DO [12]. In this chapter, we will review a few important scheduling algorithms designed for different traffic types.

- *Infinitely backlogged traffic*: this is a simplified model and is also known as the full-buffer model. The key scheduling concept here is *channel-awareness*: the scheduler selects a user when its channel is favorable. This property is also referred to as multiuser diversity as discussed in Chapter 4.

- *Elastic traffic*: the arrival rate is flexible and is regulated by a congestion control protocol, such as TCP. The key challenge here is the cross-layer coupling between scheduling and congestion control.
- *Inelastic traffic*: the arrival rate is fixed. One example is real-time voice or video. For real time traffic, queueing dynamics are very important because of stringent delay constraints. Thus, in addition to being channel-aware, scheduling has also to be *queue-aware*. In addition to scheduling, admission control and policing are also needed to ensure QoS.
- *Flow level traffic*: this type consists of a large number of short data sessions, each of which only consists of a small amount of data. Scheduling should consider flow level dynamics instead of queue dynamics.

In addition to the scheduling algorithms, we will study the signaling mechanisms to support the scheduling function. The overall system design has to take into account the tradeoff of scheduling benefits and signaling overheads. For this reason, we will introduce semi-persistent scheduling for VoIP traffic and MAC state scheduling to provide QoS to a large number of users by exploiting large time scale data traffic burstiness.

8.1 Scheduling for infinitely backlogged traffic

Infinitely backlogged traffic is a simplified model where the queue length of every user is much longer than what can be served in a scheduling cycle, which is typically on the order of milliseconds. In this case, the QoS is measured by fairness among different users. Fairness can be quantitatively defined using a utility function as follows.

8.1.1 Fairness based on utility functions

A utility function indicates the level of user satisfaction after receiving service. A utility function can be defined on data rate, delay, loss, or a combination of them. For infinitely backlogged traffic, a utility function is defined on data rate only: $U(R_i)$ where R_i is the throughput of user i. The optimization problem as illustrated in Figure 8.1 can be formulated as

$$\max_{\mathbf{R}} \sum_{i}^{M} U(R_i), \quad (8.1)$$

$$\mathbf{R} = [R_1, R_2, \ldots]^T \in \mathcal{R}, \quad (8.2)$$

where $\mathbf{R}$ is the throughput vector, M is the total number of users, and $\mathcal{R}$ is the rate region. Here we first assume that the rate region $\mathcal{R}$ is time-invariant and known to the scheduler. We will relax this constraint in the next subsection.

A few common properties of function $U(R)$ are that it is non-decreasing, differentiable, and concave. The non-decreasing property is obvious. Differentiability is mainly needed for mathematical convenience. Concavity assumes that when the throughput is

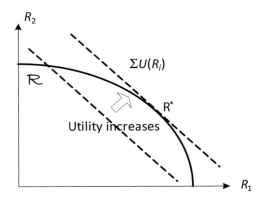

Figure 8.1 Illustration of sum utility maximization. A dotted contour represents the set of rate vectors that have the same value of total utility. In this example, $a = 1$. $\mathbf{R}^*$ represents the optimal throughput vector given the rate region $\mathcal{R}$.

already large, any further increase in the throughput leads to only marginal return in the level of user satisfaction. The following family of utility functions, usually referred to as the a-utility functions, satisfies these properties and provides a wide range of fairness:

$$U(R) = \begin{cases} \frac{R^a}{a}, & a \leq 1, a \neq 0 \\ \log(R), & a = 0. \end{cases} \quad (8.3)$$

Note that $\log(R) = \lim_{a \to 0} \frac{R^a - 1}{a}$. Thus the log utility is a constant (equal to $\frac{1}{a}$) away from $\frac{R^a}{a}$ as $a \to 0$ and either utility function at this point ($a = 0$) leads to the same optimal solution to the utility maximization problem (Equation 8.1).

Figure 8.2 shows some examples of the a-utility functions. The smaller the parameter a, the more fair in the sense that the a-utility function intends to protect low data rate users to a greater extent. In particular, the log utility function is also referred to as the *proportional fair* utility function. The name comes from the observation that the optimal throughput vector $\mathbf{R}^* = [R_1^*, R_2^*, \ldots]^T$ satisfies the property that for any small perturbation $\delta \mathbf{R}$ from $\mathbf{R}^*$,

$$\sum_i \frac{\delta R_i}{R_i^*} \leq 0. \quad (8.4)$$

In other words, from $\mathbf{R}^*$, one cannot increase the proportion of one user's rate without decreasing the proportion from another. In the two user case, for example, to increase user 1's rate by 1 percent, the scheduler has to decrease user 2's rate by at least 1 percent.

Next we look at a simple example of downlink scheduling with different utility functions. Consider a slotted TDMA downlink with stationary channels where only one user can be scheduled in a time slot. Denote by γ_i the SINR of user i, and α_i the fraction of slots allocated to user i subject to $\sum_i \alpha_i = 1$. Thus, the throughput of a user is $R_i = \alpha_i \log(1 + \gamma_i)$. For an a-utility function, the optimal scheduling policy can be

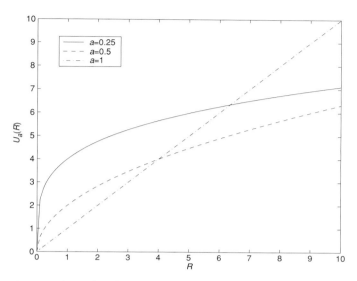

Figure 8.2 Examples of the a-utility functions. The smaller the a is, the more concave the utility function becomes.

computed as

$$\alpha_i^* = \frac{(\log(1+\gamma_i))^{\frac{a}{1-a}}}{\sum_{j=1}^{M}(\log(1+\gamma_j))^{\frac{a}{1-a}}}. \tag{8.5}$$

Now we look at the effect of setting a in the utility function:

- $a = 1$: *max rate scheduler*. In this case, the objective is to maximize the sum throughput of the system. As $\frac{a}{1-a} \to \infty$, the user with the best SINR dominates the sum in the denominator of (8.5) and the optimal policy is to schedule that user in all the time slots. The total system throughput is maximized and equal to $\max_i \log(1+\gamma_i)$.
- $a = 0$: *proportional fair scheduler*. In this case, $\alpha_i^* = \frac{1}{M}$, for all users. The proportional fair scheduler assigns an equal number of the time slots to every user, a policy also referred to as *resource fair* scheduling. The total system throughput is equal to the arithmetic mean of the capacity of all the users, $\frac{1}{M}\sum_i \log(1+\gamma_i)$.
- $a = -\infty$: *equal throughput scheduler*. In this case, $\alpha_i^* = \frac{(\log(1+\gamma_i))^{-1}}{\sum_j (\log(1+\gamma_j))^{-1}}$. All the users achieves the same throughput, a policy also referred to as *rate fair* scheduling. The total system throughput is equal to the harmonic mean of the capacity of all the users, $M\left(\sum_i (\log(1+\gamma_i))^{-1}\right)^{-1}$.

In summary, $a = 1$ results in the most aggressive scheduler, which targets maximization of the total system throughput but sacrifices fairness the most. The most conservative scheduler, with $a = -\infty$, respects the fairness the most but sacrifices the total system throughput. The proportional fair scheduler with $a = 0$ performs somewhere between those two extreme choices. This observation is consistent with the previous interpretation of a in (8.3). In general, by choosing a different value of a, we trade off between maximizing the total system throughput and maximizing the fairness.

8.1.2 Gradient-based scheduling schemes

In the previous section, we solve the utility maximization assuming a time-invariant rate region. In practice, because of channel fading, the rate region is not fixed. A key insight of scheduling in mobile broadband is to harness channel fading for multiuser diversity. We next assume the rate region to be time-varying but known to the scheduler.

Specifically, assume that the channel conditions are captured with a stochastic channel state $s[t] \in S$ at time t, where S is the channel state space. Any channel state $s \in S$ corresponds to a rate region $\mathcal{R}_s$, which is the convex hull of all rates $\mathbf{R}_s$ supportable with any user selection and resource allocation when the channel state is s. Furthermore, assume that the channel variation due to fading is ergodic and there exists a steady state distribution of the channel state probabilities π_s, with $\sum_s \pi_s = 1$. Hence, the steady state capacity region $\overline{\mathcal{R}}$ can be defined as

$$\overline{\mathcal{R}} = \left\{ \overline{\mathbf{R}} = [\overline{R}_1, \overline{R}_2, \cdots, \overline{R}_M] : \overline{R}_i = \sum_s \pi_s R_{s,i}, \mathbf{R}_s \in \mathcal{R}_s \right\}, \quad (8.6)$$

where $\mathbf{R}_s = [R_{s,1}, \ldots, R_{s,M}]^T$ and $R_{s,i}$ is user i's rate at the channel state s. $\overline{\mathcal{R}}$ is the set of all achievable steady state long-term empirical throughput vectors. Given that $\mathcal{R}_s$ is convex and compact for any s, it can be shown that $\overline{\mathcal{R}}$ is also convex and compact.

The utility maximization problem becomes

$$\overline{\mathbf{R}}^* = \arg\max_{\overline{\mathbf{R}} \in \overline{\mathcal{R}}} \sum_{i=1}^{M} U(\overline{R}_i). \quad (8.7)$$

At a given time when the channel state is s, the scheduler is to determine $\mathbf{R}_s^* \in \mathcal{R}_s$ so as to achieve

$$\overline{\mathbf{R}}^* = \sum_s \pi_s \mathbf{R}_s^*. \quad (8.8)$$

At time t, one may consider selecting the rate as follows:

$$\mathbf{R}_{s[t]}[t] = \arg\max_{\mathbf{R} \in \mathcal{R}_{s[t]}} \sum_{i=1}^{M} U(R_i). \quad (8.9)$$

In general, the above choice is suboptimal, because maximizing $\sum_{i=1}^{M} U(R_i)$ at each t does not necessarily maximize overall $\sum_{i=1}^{M} U(\overline{R}_i)$.

At first it may appear that in order to solve the optimal scheduling $\mathbf{R}_s^*$, one needs to know the channel state probability distribution π_s, for all s. However, it is shown in [2] the following gradient opportunistic scheduling algorithm using only online channel state information converges to the optimal solution $\overline{\mathbf{R}}^*$.

Denote by $\mathbf{T}[t]$ the *measured* empirical throughput vector at time t. Suppose that the rate at time t is $\mathbf{R}_{s[t]}[t]$. In practical systems, a widely adopted scheme to measure

8.1 Scheduling for infinitely backlogged traffic

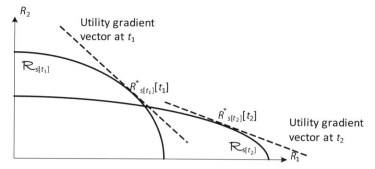

Figure 8.3 Illustration of online gradient opportunistic scheduling. The rate region varies with time because of channel fading. The marginal utility also varies as the users are scheduled and their throughput evolves.

throughput is to use *exponential averaging* as in

$$\mathbf{T}[t+1] = \mathbf{T}[t] + \epsilon \left(\mathbf{R}_{s[t]}[t] - \mathbf{T}[t]\right), \tag{8.10}$$

where ϵ is a small scheduling step size. Initially, $\mathbf{T}[0] = 0$. We are interested in maximizing $U(\mathbf{T}[t]) = \sum_{i=1}^{M} U(T_i[t])$ as $t \to \infty$. At time t, all the past scheduling decisions $\mathbf{R}_{s[0]}[0], \ldots, \mathbf{R}_{s[t-1]}[t-1]$ are given, and the scheduling decision is to maximize

$$U(\mathbf{T}[t+1]) - U(\mathbf{T}[t]) = U(\mathbf{T}[t] + \epsilon \left(\mathbf{R}_{s[t]}[t] - \mathbf{T}[t]\right)) - U(\mathbf{T}[t])$$
$$\approx \epsilon \nabla U(\mathbf{T}[t])^T \cdot \left(\mathbf{R}_{s[t]}[t] - \mathbf{T}[t]\right), \tag{8.11}$$

where the gradient of the utility function $\nabla U(\mathbf{T}[t])^T = [U'(T_1[t]), \ldots, U'(T_M[t])]$, with $U'(x) = \frac{dU(x)}{dx}$. The optimal solution is thus given by

$$\mathbf{R}^*_{s[t]}[t] = \arg\max_{\mathbf{R} \in \mathcal{R}_{s[t]}} \nabla U(\mathbf{T}[t])^T \cdot \mathbf{R}. \tag{8.12}$$

The optimal scheduling within the present rate region is to maximize the projection onto the gradient of the utility. Note the difference between the above gradient scheme (8.12) and (8.9), which is suboptimal in general.

We can view $\nabla U(\mathbf{T}[t])^T$ as a weight vector. Hence, (8.12) is a utility maximization problem with a linear utility function $\sum_i w_i R_i$, where $w_i = U'(T_i[t])$ represents the marginal utility of user i. The weight vector is known to the scheduler and varies with time as the users are scheduled and the exponential-average throughput evolves. As noted before, the rate region also varies with time. Figure 8.3 illustrates the time-varying nature of the linear utility optimization.

Now let us apply the gradient scheme (8.12) to a TDMA system where the scheduler selects only one user at a time and allocates all the power bandwidth resource to the scheduled user. We have seen such a system in Example 4.1. Consider the log utility function. The marginal utility is $U'(x) = 1/x$. Hence, the optimal scheduling is to select the user $i^*[t]$ such that

$$i^*[t] = \arg\max_i \frac{R_i[t]}{T_i[t]}. \tag{8.13}$$

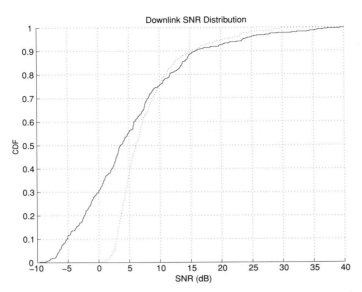

Figure 8.4 Comparison of the distributions of instantaneous SINR (solid curve) provided by the wireless channel and scheduled SINR (dotted curve) with the proportional fair scheduler at $M = 20$ in a cellular downlink as in Figure 6.3 with universal reuse and Rayleigh fading.

This is the proportional fair scheduling algorithm studied in (4.102) where the user with the highest ratio of its instantaneous rate and exponential-average throughput is scheduled. With channel fading, the multiuser diversity benefit is that such a scheduling policy boosts up the scheduled SINR, that is, the SINR of the user that satisfies (8.13) and is scheduled. The improvement roughly scales with $\log(M)$. Figure 8.4 shows the improvement in the distribution of scheduled SINR versus instantaneous SINR provided by the wireless channel. In typical cellular deployment, a significant fraction of the mobiles have instantaneous SINR below 0 dB due to inter-cell interference and fading. In principle, a proportional fair scheduler attempts to schedule all users at their relatively favorable channel conditions. For users operating in the low SINR regime, the same amount of channel improvement can lead to a larger capacity gain and thus a better metric as determined by (8.13), since the capacity here is linear to the channel improvement, as opposed to logarithmic for users operating in the high SINR regime. Thus, a slight subtlety here is that a proportional fair scheduler favors channel improvement for low SINR users. This is also reflected in Figure 8.4, where we can observe a significant improvement of the scheduled SINR at the low-end tail.

Next we consider an OFDMA downlink. In principle, the scheduling problem is an integer programming since a user cannot be allocated with a fraction of a tone. For simplicity, assuming infinite granularity of resource allocation, at each time slot t, the scheduling problem becomes the following power bandwidth allocation problem:

$$\max_{\boldsymbol{\alpha},\boldsymbol{\beta}} \sum_i w_i \alpha_i \log\left(1 + \frac{\beta_i \gamma_i}{\alpha_i}\right) \tag{8.14}$$

$$\text{subject to } \sum_i \alpha_i = 1, \sum_i \beta_i = 1, \boldsymbol{\alpha}, \boldsymbol{\beta} \geq 0,$$

where α_i and β_i are the fractions of total bandwidth and total power assigned to user i, respectively. The weight w_i is the marginal utility. For example, w_i is equal to $\frac{1}{T_i[t]}$ in the proportional fair scheduling. For notion brevity, we omit the time dependency of the variables in this formulation. Chapter 5 has characterized the rate region of the OFDMA downlink and considered the linear utility function in (5.8). A detailed optimization algorithm can be found in [68].

As noted in Chapter 5, an important advantage of joint power bandwidth allocation in OFDMA as opposed to TDMA is that bandwidth-limited superior users can be scheduled simultaneously with power-limited inferior users so as to fully utilize both power and bandwidth resource. The formulation in (8.14) allows the scheduler to take into account both opportunistic scheduling and power bandwidth allocation. For example, consider a scenario where only an inferior user would be selected in a TDMA system as in (8.13). In an OFDMA system, the scheduler will probably select another superior user to pair with the inferior user to better use the system resources. Specifically, the inferior users occupy a small fraction of the bandwidth but a large fraction of the power while a superior user occupies a small fraction of the power but a large fraction of the bandwidth. On the other hand, it should be noted that the power bandwidth optimization in OFDMA has *less* gain over the TDMA scheduling when multiuser diversity is fully exploited than when multiuser diversity is not exploited. This is simply because with multiuser diversity, the scheduled SINR is improved significantly for the low-SINR users as shown in Figure 8.4 and there is less value to optimizing power and bandwidth allocation among users.

In (8.14) we have not taken into account channel frequency selectivity, as the SINR γ_i is assumed to be the same in any tone. In a wideband system, we should further harness frequency selectivity. As noted in Chapter 4, the total bandwidth is often divided into K_S subbands. The channel condition is assumed to be the same within a subband and to vary across different subbands. In this case, (8.14) needs to expand and cover individual subbands:

$$\max_{\boldsymbol{\alpha},\boldsymbol{\beta}} \sum_i w_i \left(\sum_{k=1}^{K_S} \alpha_{i,k} \log\left(1 + \frac{\beta_{i,k}\gamma_{i,k}}{\alpha_{i,k}}\right) \right) \quad (8.15)$$

$$\text{subject to } \sum_i \sum_k \alpha_{i,k} = 1, \sum_i \sum_k \beta_{i,k} = 1, \boldsymbol{\alpha}, \boldsymbol{\beta} \geq 0,$$

where $\gamma_{i,k}$ is the SINR of user i in subband k. Fortunately, the utility function in (8.15) remains joint convex in $\boldsymbol{\alpha}$ and $\boldsymbol{\beta}$ and we can use a gradient algorithm to search for the optimal power bandwidth allocation over multiple subbands. Finally, it should be noted that power allocation across subbands in neighboring cells results in frequency selective inter-cell interference. Therefore, the scheduler has to take into account inter-cell interference. Some of the optimization and heuristic ideas are studied in Chapter 6.

8.2 Scheduling for elastic traffic

The preceding infinitely backlogged traffic model is a simplification of real systems. In practice, traffic queues are always of finite lengths. Indeed an important QoS requirement

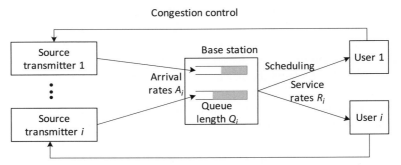

Figure 8.5 Illustration of scheduling in conjunction with congestion control for elastic traffic. Here it is assumed that user i receives traffic from only one source transmitter i. In practice, multiple transmitters may send traffic to the same user simultaneously and each of them performs congestion control by itself.

is to keep queue lengths short because long queue lengths imply long queueing delays. In this section, we consider elastic traffic where traffic sources adjust arrival rates so as to avoid long queues, and at the same time prevent queues from being empty since otherwise the channel capacity would be under-utilized. The mechanism in which arrival rates are adjusted is called *congestion control*. Another motivation of congestion control is that the capacity region $\overline{\mathcal{R}}$ defined in (8.6) is not known beforehand because of channel dynamics. Traffic sources adjust arrival rates to probe the channel capacity.

Consider the downlink case. Figure 8.5 shows an end-to-end system with congestion control and scheduling. While scheduling happens at the base station for the wireless link, congestion control is usually done between individual source/destination pairs.

8.2.1 Congestion control and scheduling

Congestion control is used to determine the arrival rates A_i, and scheduling is used to determine the service rates R_i within the capacity region $\mathcal{R}_s$ at the present state s. Consider the following optimization problem:

$$\max_{A_i} \sum_i U(A_i) \tag{8.16}$$

subject to $A_i \leq \overline{R}_i, [\overline{R}_1, \overline{R}_2, \cdots] \in \overline{\mathcal{R}}$.

The above optimization could be solved in principle if $\overline{\mathcal{R}}$ is known beforehand and all the sources coordinate their arrival rates according to the optimal solution. However, as pointed out before, in practice $\overline{\mathcal{R}}$ is not precisely known, and congestion control is usually not coordinated among different sources. In the following, we will present a joint congestion control and scheduling scheme, following [43, 99, 100, 116, 150] in the framework of stochastic network optimization, to achieve fair and efficient sharing of system resource defined in (8.16) and to reduce queueing delays.

Denote by $Q_i[t]$ the queue length of user i at time t. The arrival rate is determined by

$$A_i[t] = \arg\max_{A_i} \left(U(A_i) - \eta Q_i[t] A_i \right), \quad (8.17)$$

with η a tuning parameter. Increasing A_i obviously leads to higher utility $U(A_i)$. However, this has to be balanced by the cost of queue length increase, which is represented by the second term in (8.17). In fact, the cost is set to be proportional to queue length $Q_i[t]$. Thus the arrival rate can be determined by solving the equation

$$U'(A_i[t]) = \eta Q_i[t]. \quad (8.18)$$

The above congestion control scheme states that each traffic source reacts to the queueing congestion by selecting its arrival rate such that the marginal utility is equal to the congestion cost. For example, for the utility function (8.3), it follows that

$$A_i[t] = \frac{1}{(\eta Q_i[t])^{1-a}}. \quad (8.19)$$

Scheduling is determined by taking into account both the channel state $s[t]$ and queue length,

$$\mathbf{R}^*_{s[t]}[t] = \arg\max_{\mathbf{R} \in \mathcal{R}_{s[t]}} \mathbf{Q[t]}^T \cdot \mathbf{R}, \quad (8.20)$$

with queue length vector $\mathbf{Q[t]}^T = [Q_1[t], \ldots, Q_M[t]]$. Note the difference between (8.12) and (8.20). (8.19) and (8.20) show the joint mechanism of reducing queueing delays. Specifically, as the queue length of user i increases, the scheduler prioritizes user i when the system resource is allocated, and meanwhile the congestion controller reduces its arrival rate.

With arrival rate $A_i[t]$ and service rate $R_i[t]$, the queue length is updated as follows:

$$Q_i[t+1] = (Q_i[t] + A_i[t] - R_i[t])^+. \quad (8.21)$$

It is shown in [43] that by choosing η very small, the above congestion control and scheduling schemes converge to a solution that can be arbitrarily close to the optimal solution of (8.16). However, the drawback of using a small η is large queueing delays because the cost of the queue length is not sufficiently weighted in the optimization objective in (8.17).

In practice, the queue length information may not be precisely known at any time t. For example, as described in Section 8.6.1, to reduce the signaling overhead in the uplink scheduling case, the base station scheduler often gets delayed reports from the users about the queue length information. In this case, the scheduling algorithm can be the same as (8.20) with $\mathbf{Q[t]}$ being replaced by the latest update of the queue length vector [183]. Furthermore, [43] extends the deterministic congestion control scheme of (8.19) to the following stochastic policy where $A_i[t]$ is a random variable such that

$$\mathbb{E}(A_i[t]|Q_i[t]) = \frac{1}{(\eta Q_i[t])^{1-a}}, \quad (8.22)$$

and $\mathbb{E}(A_i[t]|Q_i[t]) \leq M_1$ and $\mathbb{E}(A_i^2[t]|Q_i[t]) \leq M_2$ for some constants M_1, M_2. The reason for making $A_i[t]$ a random variable is to allow for various sources of randomness in actual implementation, such as window-based implementations of congestion control.

In the above framework of joint congestion control and scheduling, the scheduler does not directly control the fairness among users, and it is up to the individual sources to determine the arrival rate on the basis of marginal utility (8.18). In practice, the individual traffic sources are usually not controlled by the operator. It would be difficult to enforce every source to implement the same utility function. Thus, it is desirable that the scheduler is the one that controls fair sharing of the system resource, because the scheduler is part of the base station implementation controlled by the operator.

Below, we briefly describe an alternative framework in which scheduling and congestion control are very much decoupled. Specifically, the base station maintains a queueing buffer for every user i and the buffer size is equal to the round-trip delay between the source i and the base station multiplied by the maximum possible rate[1] of user i. The maximum possible rate depends on the long term channel condition (path loss) as well as short term fading characteristics (Rayleigh versus Ricean fading), and may vary as the user moves from one place to another. For example, when the user moves close to the base station, its maximum possible rate increases and so does the buffer size.

At the end of each scheduling slot, if user i is scheduled and some number of bytes are successfully served, the base station informs the source i to backfill the same number of bytes so as to keep the buffer size. In this case, the sources are not in control of fair sharing of the system resource. Indeed, the sources are even not aware of the utility function employed by the base station scheduler. Meanwhile, since the buffer sizes are sufficiently large, as far as the scheduler is concerned it works as if it deals with infinitely backlogged queues. It can apply any fairness criterion deemed to be appropriate, and maximize the utility as shown in Section 8.1.

As far as the mean rates are concerned, the above decoupled approach should perform the same as the joint congestion control and scheduling approach, as numerically verified in [44]. It would be interesting to compare the queueing delay performance of the two approaches under a variety of parameter settings (η, buffer size) and network assumptions (end-to-end delays, wireless channel characteristics).

Discussion notes 8.1 TCP performance over wireless

TCP is a transport protocol aiming at guaranteeing end-to-end reliable delivery of data packets over wired networks. In TCP, the sender uses a transmission window for flow control. The transmission window size is the minimum of a congestion window and a receiver window. The congestion window indicates the maximum number of bytes that can be outstanding at any time. The size of the congestion window needs to be tuned carefully. If the congestion window is too small, then the sender does not transmit enough traffic to adequately utilize the channel capacity. If the congestion

---

[1] To reduce the buffer size and thus the queueing delay, we can use, say, 95th percentile maximum rate in practice.

window is too large, then the sender injects too much traffic and overloads the network. TCP congestion control uses an additive increase/multiplicative decrease approach to adjust the congestion window. The sender detects the loss of a packet either by the reception of duplicate cumulative acknowledgments or the absence of an acknowledgment within a timeout interval (time-out). Because the link error rates are very low in wired networks, most packet losses are caused by network congestion. Therefore, the sender reacts to packet losses by drastically dropping its congestion window size to reduce the load.

Wireless channels exhibits very different characteristics from wired networks. As a result, TCP over wireless suffers from significant performance degradation. The performance issues are mostly related to the congestion control mechanisms used in TCP.

- *Channel errors.* In wireless systems, packet losses are often not because of network congestion but because of link errors. TCP congestion control results in an unnecessary reduction in end-to-end throughput. The vast majority of work on TCP over wireless [8] has aimed at reducing the adverse impact of mistreating channel errors as congestion losses. There are two basic ideas [9]. First, hide non-congestion related losses from the TCP sender. In particular, the link layer ARQ or H-ARQ in mobile broadband communications can reduce the channel errors to well below 1 percent, thereby minimizing the impact on the TCP performance. The latency caused by ARQ or H-ARQ has to be kept small to avoid TCP time-out. Second, make the sender aware of the existence of wireless channels and realize that some packet losses are not due to congestion; however, this requires changes to the TCP implementations.
- *Handoff.* Handoff may cause temporary disconnections resulting in packet losses and delays. To address this problem, the work in [59] proposes to freeze the connection and timers at the TCP sender, thus avoiding time-outs, by setting the receiver window size to zero until the connection is reestablished.
- *Rate variation and scheduling jitter.* Opportunistic scheduling maximizes link layer capacity but also results in significant variations in the throughput seen by TCP. Large delay variations lead to delay spikes and cause spurious time-outs where the sender assumes a packet is lost while it is only delayed. Various studies have examined the impact on the TCP performance and proposed solutions [22, 181].

8.3 Scheduling for inelastic traffic

By definition, the arrival rates of inelastic traffic are fixed and therefore no congestion control is considered. The scheduler aims at serving arrival traffic as much as possible. An important notion of scheduling inelastic traffic is *stability*: given arrival rates and channel conditions, can any scheduling policy make the system stable, meaning that none of the queues grow to infinity? Inelastic traffic is often associated with real-time

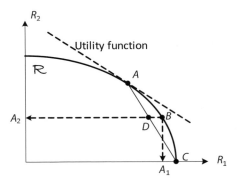

Figure 8.6 $[A_1, A_2]$ is the arrival rate vector. Point B is the throughput optimal. Without knowing the arrival vector, A is the utility optimal operating point. Point C is where the scheduler only serves user 1. D is obtained by operating between A and C in a TDM manner. At point D, the service rate for user 2 is equal to the arrival rate A_2, but the service rate for user 1 is less than A_1.

traffic[2] such as voice or video. Therefore, in addition to stability, we are also interested in delay performance.

8.3.1 Throughput optimal scheduling

As before, assume the arrival process to be ergodic. The stability region consists of all the arrival rates A_i under which a scheduling policy exists to make the system stable. Given the steady state capacity region $\overline{\mathcal{R}}$, arrival rate vector $[A_1, \ldots, A_M]$ belongs to the stability region if there exists a steady state throughput vector $[\overline{R}_1, \overline{R}_2, \cdots] \in \overline{\mathcal{R}}$ such that for all i,

$$A_i \leq \overline{R}_i. \tag{8.23}$$

The rate stability region is thus the same as the capacity region $\overline{\mathcal{R}}$.

Unlike the congestion/scheduling problem of elastic traffic where the arrival rate vector can be controlled, the scheduling problem here for inelastic traffic assumes the arrival rate vector is given and not flexible. The scheduling policy that stabilizes any arrival rate vector in the stability region is called *throughput optimal*. Obviously if the arrival rate vector is precisely known beforehand, then the scheduler simply picks the corresponding operating point within the capacity region. In Figure 8.6, suppose $[A_1, A_2]$ is the arrival rate vector. Then B is the throughput optimal operating point. However, this offline approach is not robust in practice because the capacity region and the arrival rates are not known precisely beforehand and usually vary over time.

An important and somewhat surprising result from the seminal work [156] is that a scheduler can be throughput optimal even without knowing the arrival rate vector or the capacity region. In particular, the scheduling policy in (8.20) is shown to be throughput

---

[2] It should be pointed out that real-time traffic is not necessarily inelastic, because a traffic source can adjust the source coding rate in response to the fluctuation of the available capacity. In this case, the receiver will experience variable quality of voice or video traffic.

optimal. The insight of (8.20) is that in order to overcome the difficulty of not knowing the arrival rate vector or capacity region, a throughput optimal scheduler should in general be both channel-aware and queue-aware. In other words, a throughput optimal scheduler cannot always schedule users at their best channel conditions. When its queue is becoming too long, a user has to be scheduled even if its channel condition is not favorable.

We have studied the importance of channel-awareness in the context of opportunistic scheduling for enhancing system efficiency. To see the importance of queue-awareness, assume that the scheduler is not queue-aware but is designed to maximize some fairness utility function as studied in Section 8.1. As shown in Figure 8.6, the utility-optimal operating point (A) differs from the arrival rate vector (B). Since neither of the two points dominates the other, without loss of generality we assume that $\overline{R}_2$ of point A is greater than $\overline{R}_2$ of point B. In other words, for user 2, the service rate is greater than the arrival rate. Thus, user 2's queue becomes empty from time to time. When that occurs, the scheduler serves user 1 exclusively (C in Figure 8.6). Overall, the scheduler in effect operates between A and C in a TDM manner. The overall data rate is represented by D that is on the line segment AC and whose corresponding $\overline{R}_2$ is equal to A_2. Because of the convexity of the capacity region, the effective operating point D is dominated by the throughput optimal point B. While user 2 achieves the throughput equal to the arrival rate A_2 and sees a stable queue, the queue of user 1 grows to infinite as the service rate at point D is below the arrival rate A_1.

Is (8.20) the only throughput optimal scheduler? The answer is no. Indeed, it turns out that many schedulers are actually throughput optimal as long as they are both queue-aware and channel-aware. More precisely, the following family of schedulers are proved to be throughput optimal [45]:

$$\mathbf{R}^*_{s[t]}[t] = \arg \max_{\mathbf{R} \in \mathcal{R}_{s[t]}} f(\mathbf{Q}[t])^T \cdot \mathbf{R}, \quad (8.24)$$

where $f(\cdot)$ is a positive function satisfying the following conditions:

- $f(\cdot)$ is a nondecreasing, continuous function with $\lim_{x \to \infty} f(x) = \infty$
- Given any $M_1, M_2 > 0$ and $0 < \epsilon < 1$, there exists $\chi < \infty$, such that for any $x > \chi$

$$(1 - \epsilon) f(x) \leq f(x - M_1) \leq f(x + M_2) \leq (1 + \epsilon) f(x). \quad (8.25)$$

The constraint (8.25) is visualized in Figure 8.7. It should be pointed out that (8.25) is not particularly stringent. For example, the constraint holds for any monotonically increasing concave function. Specifically, for given M_1, M_2, ϵ, the difference between $|f(x + M_2) - f(x - M_1)|$ decreases as x increases. Meanwhile, the size of the band equal to $2\epsilon f(x)$ increases with x. Thus, there must exist a constant χ such that (8.25) is satisfied. Furthermore, all polynomial functions in the form of $f(x) = x^\alpha$ with $\alpha \geq 0$ satisfy this constraint as well. One example that does not satisfy (8.25) is an exponential function $e^{\alpha x}$.

The study of throughput optimality only focuses on how to make queues stable but does not explicitly take into account queueing dynamics. Among all throughput optimal schedulers, even though the achieved arrival rates are the same, the delay

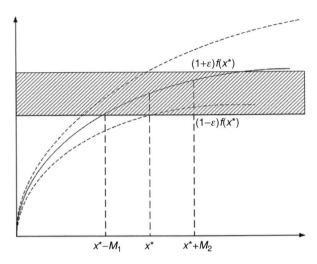

Figure 8.7 Illustration of a constraint for function $f(\cdot)$ to be used to construct a throughput optimal scheduler.

characteristics can be very different. Since throughput optimal schedulers are not unique, in the following we we characterize the performance tradeoff among a few popular throughput optimal schedulers, and study the parameters of the tradeoff.

8.3.2 Tradeoff between queue-awareness and channel-awareness

For the sake of simplicity, we consider a TDMA system where the scheduler selects a single user to serve in every scheduling instant. The throughput optimal scheduler (8.24) is reduced to user selection

$$i^*[t] = \arg \max_{i=1,\ldots,M} f(Q_i[t])R_i[t]. \quad (8.26)$$

To simplify the notation, we will drop $[t]$ in the following. With some effort, the single user schedulers can be extended to OFDMA multiuser ones with power and bandwidth allocation [129].

We consider the following three schedulers, namely Max Weight (MW), Exponential (EXP) and Log (LOG), which are all throughput-optimal:

$$\text{MW:} \quad i^*_{MW} = \arg \max_{i=1,\ldots,M} b_i Q_i^\beta R_i, \quad (8.27)$$

$$\text{EXP:} \quad i^*_{EXP} = \arg \max_{i=1,\ldots,M} b_i \exp\left(\frac{a_i Q_i}{c + (\frac{1}{M}\sum_j a_j Q_j)^\eta}\right) R_i, \quad (8.28)$$

$$\text{LOG:} \quad i^*_{LOG} = \arg \max_{i=1,\ldots,M} b_i \log(c + a_i Q_i) R_i, \quad (8.29)$$

for fixed positive $1/b_i, a_i, c, \beta$ and $0 \leq \eta < 1$. From Little's law, the average queue length is indicative of the average delay given the arrival rate. Moreover, queue length

Q_i can be replaced with head-of-line delay to explicitly incorporate the delay target of real-time traffic. For example, in [129] it is suggested to set $1/b_i$ to user i's expected rate and $a_i = 5/d_i$ in the LOG scheduler and $a_i \in [5/d_i, 10/d_i]$ in the EXP scheduler, where d_i is the 99th percentile delay target of user i's traffic.

A key design choice for a channel-aware and queue-aware scheduler is how to make the tradeoff between maximizing current transmission rate (channel-aware) and balancing unequal queues/delays (queue-aware). Let us examine how those three schedulers make this tradeoff.

Suppose we set $b_i = 1/T_i$, where T_i is the empirical throughput of user i up to the current time slot. All the three schedulers become proportional fair (PF) when the queue lengths Q_i are the same. When the queue lengths become unequal, in principle all the schedulers try to schedule users with longer queues even when the channel conditions are not the most favorable. The schedulers differ in how much weight they put on the queue length ($f(Q_i)$) as opposed to the channel condition (R_i). Comparison of (8.27) to (8.29) shows that as the queue length of one user grows big, the EXP rule reacts most strongly by assigning a disproportionally large weight, while the LOG rule saturates.

To visually see the effect, consider two users $M = 2$. Denote by $T_{i,PF}$, $T_{i,MW}(\mathbf{Q})$, $T_{i,EXP}(\mathbf{Q})$, $T_{i,LOG}(\mathbf{Q})$ the steady state throughput of user i achieved under the PF, MW, EXP, and LOG schedulers, respectively, when the queue lengths are equal to vector $\mathbf{Q} = [Q_1, Q_2]$. $T_{i,PF}$ assumes infinitely backlogged queues and does not depend on $\mathbf{Q}$. The queue length state space $[Q_1, Q_2]$ can be partitioned into a few regions, as shown in Figure 8.8.

$\mathcal{S}_{MW}^{PF}$ consists of $\mathbf{Q}$ at which $T_{i,MW}(\mathbf{Q}) = T_{i,PF}$ for all i. Clearly, the area along the 45° line where $Q_1 \approx Q_2$ is part of $\mathcal{S}_{MW}^{PF}$. As Q_1 increases or Q_2 decreases from that area, the scheduler becomes more biased in favor of user 1 over user 2 to balance the unequal queues. As a result, the MW scheduler moves away from being PF, and eventually schedules only user 1. $\mathcal{S}_{MW}^{i}$ consists of $\mathbf{Q}$ at which only user i is served. $\mathcal{S}_{EXP}^{PF}, \mathcal{S}_{EXP}^{i}, \mathcal{S}_{LOG}^{PF}, \mathcal{S}_{LOG}^{i}$ can be defined similarly for the EXP and LOG schedulers, respectively. While the exact shape of each region depends on the scheduler parameters and the distribution of rate R_i, it is shown in [132] that the asymptotic shapes only depend on the type of the scheduler. In particular, with the 45° line as the axis, $\mathcal{S}_{MW}^{PF}$ is a cone, $\mathcal{S}_{EXP}^{PF}$ a cylinder with gradually increasing radius, and $\mathcal{S}_{LOG}^{PF}$ a French horn.

The distinct asymptotic shapes of the queue length state space partitions of the three schedulers indicates very different behaviors when the queue lengths grow. For any queue vector $\mathbf{Q}$ and scalar $s > 0$, it is shown in [130] that

- $\sum_i b_i T_{i,MW}(s\mathbf{Q})$ is constant in s.
- $\sum_i b_i T_{i,EXP}(s\mathbf{Q})$ decreases with s, and in the limit where $s \to \infty$, only the longest queue is scheduled.
- $\sum_i b_i T_{i,LOG}(s\mathbf{Q})$ increases with s, and in the limit where $s \to \infty$, $T_{i,LOG}(s\mathbf{Q})$ converges to $T_{i,PF}$ for all i when $b_i = 1/T_i$.

Thus, as the queues grow linearly, the LOG scheduler de-emphasizes queue balancing in favor of increasing the weighted service rates. As $\mathcal{S}_{LOG}^{PF}$ becomes increasingly dominant,

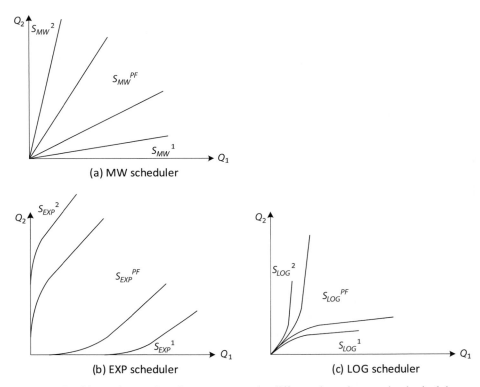

Figure 8.8 Partitions of queue length state space under different throughput-optimal schedulers.

the LOG scheduler behaves similar to the PF scheduler for the most part. On the other hand, the EXP scheduler tries to balance the queues at the cost of the service rates, as indicated by the observation that S_{EXP}^i grows rapidly with Q_i. Indeed, it can be shown that the EXP scheduler minimizes the asymptotic overflow probability of the longest queue $\max_i a_i Q_i$ [151], and meanwhile the log rule minimizes the asymptotic overflow probability of the sum queue $\sum_i b_i Q_i$ [132]. As a side note, [130] studies delay optimal scheduling policies that minimize the overall weighted average delay, and finds that those policies exhibit a common characteristic that we have seen in the LOG scheduler: as the queues grow proportionally, the LOG scheduler de-emphasizes queue balancing in favor of increasing the weighted service rates. However, the LOG scheduler is not provably delay optimal.

In summary, the EXP scheduler minimizes the delay tail distribution of the worst user, and the LOG scheduler achieves better mean delay. The difference between the two schedulers is particularly noteworthy when the system is congested. Note that, so far we have assumed that the arrival rates are within the stability region. However, that is very difficult to guarantee because of the uncertainty and time varying nature of the wireless channel and user mobility (moving from one cell to another and causing cell overloading). When the system is congested due to temporary overloading, it is desirable that the system performance *gracefully* degrades. In this case, the LOG scheduler sacrifices the performance of (hopefully) a small number of users while attempting to meet the QoS requirements of the remaining ones. As a result, there could be a wide disparity in the performance experienced by different users. The EXP scheduler, on the other hand, attempts to help the worst user and therefore runs the risk of not meeting the QoS requirements of any user. Hence, the choice of the EXP versus LOG scheduler depends on the desired behavior of graceful performance degradation.

8.3.3 Admission control

From the previous discussion, we have learned that when the system is overloaded, not all the users can meet their QoS requirements. The goal of admission control [118, 158] is to maintain the delivered QoS to users at a target level by limiting the number of users with inelastic traffic flows to be admitted in the system. When a new user arrives, admission control is used to ensure that admitting the new user does not violate the QoS agreements of already admitted users and that the QoS requirement of the new user can be met.

A certain amount of system resource is utilized in order to meet the QoS requirements of a user, either statically or statistically. In general, admission control policy is based on some threshold comparison. A new user is admitted if

$$I + \Delta I \leq I_0, \tag{8.30}$$

where I represents the total amount of resource required to serve the existing users, ΔI is the additional resource needed to serve the new user, and I_0 is some resource threshold. Obviously the goal of efficient admission control is to maximize the resource utilization while satisfying the QoS requirements. The required resource is usually predicted on the basis of wireless channel models, user mobility models, and traffic models. In particular:

- Wireless channel models describe channel shadowing and fading as well as the path loss effect that the user moves away from the base station to cell edge.
- User mobility models describe the handoff pattern of users moving from one cell to another.
- Traffic models describe the packet level statistics of arrival processes.

For example, consider user mobility [171]. If the user moves to a new cell, the new cell may not have sufficient resource to meet the user's QoS requirements even though the present cell has adequate resource. Even if the user stays in the present cell, when other users move into the cell, the cell may still become congested. Therefore, in order to guarantee the QoS agreements with the user, the system has to be conservative in predicting the required resource. For example, admission control should check whether sufficient resource exists not only in the present cell but also in the neighboring cells.

None of the above models is very accurate. Thus, the resource utilization may be compromised because of the inherent uncertainty and because of QoS guarantees. On the other hand, when dealing with a large number of users, admission control should take into account statistical multiplexing to improve efficiency.

In practice, to improve the overall resource utilization, we can admit elastic traffic to mix with inelastic traffic. Inelastic traffic – in particular real time traffic – is handled with QoS guarantees because the operator usually charges a premium for QoS guaranteed traffic. Elastic traffic expands to fully utilize the resource when it is under-utilized by inelastic traffic, but shrinks when demands for resource increases from inelastic traffic. In essence, elastic traffic is treated as best-effort traffic.

The mixture of elastic and inelastic traffic naturally leads to the task of multi-class scheduling.

8.4 Multi-class scheduling

Our basic assumption in the discussion that follows is that in multi-class scheduling, different classes are of different priorities. Typically, real time traffic is of higher priority than non real time traffic. One approach is to enforce *strict* priority in scheduling. First schedule high priority traffic and then schedule low priority traffic only if there is leftover resource. Strict priority scheduling is simple, attempting to satisfy the QoS requirements of high priority traffic to the maximum possible extent. The drawback, however, is the loss of efficiency. For example, at one time instant, suppose that the user with high priority traffic experiences an unfavorable channel condition. Clearly, scheduling that user immediately is not efficient, especially if the deadline of its traffic is still far from the present time. Hence, strict priority scheduling could be a decent choice if high priority traffic is much lighter than low priority traffic, perhaps consisting of only VoIP packets and some critical signaling packets such as TCP acknowledgments and SIP.

In general, we would like to treat priority in an opportunistic sense rather than a strict one and to take into account both traffic priority and channel-awareness in scheduling. For example, in [92], the scheduler takes a heuristic approach and gives equal priority to all packets until the deadlines of some packets are approaching, in which case the scheduler increases their priorities. To address multi-class scheduling in a more general way, we next expand the utility maximization framework studied in Section 8.1.

Define a generic utility $U_i(R_i)$ of user i at the present time instant if it is scheduled with rate R_i. The rate vector $[R_1, \ldots, R_M] = \mathbf{R} \in \mathcal{R}_{s[t]}$ belongs to the feasible set within the capacity region. The optimal solution maximizes the marginal utility return, and similar to (8.12), is given by

$$\mathbf{R}^*_{s[t]}[t] = \arg \max_{\mathbf{R} \in \mathcal{R}_{s[t]}} \sum_i U'_i(R_i) R_i. \tag{8.31}$$

Different utility functions and thus utility gradients can be used to represent users of different traffic classes. For best effort traffic, we already know the utility gradient

$U'_i(R_i)$. For example, for the log utility (proportional fairness),

$$U'_i(R_i) = \frac{1}{T_i}, \quad (8.32)$$

where T_i is the empirical throughput. The utility gradient of real time traffic can be obtained from the queue-aware and channel-aware throughput optimal schedulers (8.27) to (8.29). For example, let

$$U'_i(R_i) = b_i Q_i^\beta \quad (8.33)$$

for the Max Weight scheduling rule,

$$U'_i(R_i) = b_i \exp\left(\frac{a_i Q_i}{c + (\frac{1}{M} \sum_j a_j Q_j)^\eta}\right) \quad (8.34)$$

for the EXP rule, and

$$U'_i(R_i) = b_i \log(c + a_i Q_i) \quad (8.35)$$

for the LOG rule.

As far as scheduling is concerned, only the utility gradient $U'_i(R_i)$ matters. Of course, we can derive the utility function $U_i(R_i)$ from $U'_i(R_i)$. For example, noting the relationship in (8.10), the log utility function for best effort traffic is

$$U_i(R_i) = \log((1-\alpha)T_i + \alpha R_i), \quad (8.36)$$

where α is the filtering parameter for estimating the empirical throughput (4.103). Note that $Q_i[t+1] = Q_i[t] - R_i[t]$ ignoring any new arrival traffic. It is shown in [129] that the utility function of the LOG rule for real time traffic is

$$U_i(R_i) = -b_i\left(\left(Q_i - R_i + \frac{c}{a_i}\right)\log(c + a_i(Q_i - R_i)) - (Q_i - R_i)\right). \quad (8.37)$$

8.5 Flow level scheduling

So far we have assumed the number of users is fixed and traffic sources keep on injecting traffic into the system as if users would stay in the system forever. This model works well in use cases such as large file transfer and long voice/video streaming because of the *separation of timescales*. Specifically, the scheduler operates at the packet level, that is, on the order of milliseconds, while user level dynamics (user arrival or service completion) occur on the order of tens of seconds or longer. Thus, a service cycle at the user level takes many packet level schedules to complete. Hence, whenever the number of users in the system changes, the data rates of the users are assumed to be adjusted instantaneously to the optimal and fair rate allocation. Under this assumption, it has been shown[3] that such rate allocation policies can achieve the largest possible stability region.

---

[3] The work in [98] removes this assumption and shows that the largest possible stability region can still be achieved by a large class of control algorithms.

On the other hand, it is well known [22] that Internet traffic consists of a small number of long-lived flows and a large number of short-lived flows. A long-lived flow is in the system for a very long time and continually generates bits for transmission. A short-lived flow has only a finite number of bits to transmit, and the user will leave the system once the bits are transmitted. A typical use case of short-lived flows is for smartphone applications to retrieve a small amount of data through a mobile broadband communications system. One popular example is the Siri application on the Apple iPhone 4S™, which sends short packets to the cloud for analysis and retrieves interpretation back within a few seconds. Unlike users of long-lived flows, users of short-lived flows get in and out of the system quickly.

Our study so far has focused on scheduling long-lived flows. The questions we would like to address next are:

- Should the same scheduling policy be applied to short-lived flows?
- How should a mixture of long-lived and short-lived flows be scheduled?

Let us consider the first question. In particular, we are interested in throughput optimal scheduling policies. As far as flow level scheduling is concerned, a throughput optimal scheduling policy is defined as the one that can support any set of traffic flows that are supportable by any other scheduling policies. It is shown in [163] that the Max Weight scheduler, which is throughput optimal for long-lived flows, is not throughput optimal for short-lived flows. This is not surprising, because if a long-lived flow is not served for some period of time, its queue builds up, increasing its priority metrics in (8.27) and eventually getting the flow served. The situation is quite different for a short-lived flow, however, since the queue never builds up. Thus, the flow may stay in the system forever and never be served. Hence, a different form of scheduling policies are needed for short-lived flows.

Consider the following system model. Assume that the set of long-lived flows indexed by l is fixed. Short-lived flows arrive and depart and are indexed by i. The size of a short-lived flow B_i is bounded by a constant B, $B_i \leq B$, for all i. At time t, $Q_i[t]$ represents the number of bits remaining to be transmit of short-lived flow i, and $Q_l[t]$ the number of bits in the queue of long-lived flow l. Assume a TDMA system where the scheduler selects only one user at a time. Denote by $R_i[t]$, $R_l[t]$ the link rate if flow i or l is scheduled, respectively. Furthermore, assume that $R_i[t]$, $R_l[t]$ are discrete random variables with largest values equal to $R_{i,\max}$, $R_{l,\max}$. For short-lived flows, assume $R_i[t]$ can indeed achieve $R_{i,\max}$ at least for some positive fraction of time: $\mathbb{P}(R_i[t] = R_{i,\max}) \geq p$ for some positive constant p. $R_{i,\max}$ may be different for different i.

In the above model, if the scheduler knows $R_{i,\max}$ for all i, then it should schedule flow i when its channel condition achieves its maximum rate $R_i[t] = R_{i,\max}$. This way, the amount of time to serve flow i is minimized to $B_i/R_{i,\max}$, thereby allowing more flows to be served overall. This scheduling idea can also be expected from the sufficient and necessary stability condition[4] derived in [163].

---

[4] The system model studied in [163] is slightly different from the one in this section. Specifically, the short-lived flows belong to K types. A type-k flow has an i.i.d. random size of expectation $\mathbb{E}(B_k)$ and an i.i.d.

Clearly, the above "Max Rate" scheduling policy condition is very different from the queue-aware and channel-aware scheduling policy. A practical problem is how to estimate $R_{i,\max}$. A heuristic learning algorithm is to base on the observation of the channel rate over some period of time T:

$$\hat{R}_{i,\max}[t] = \max_{s \in (t-T, t)} R_i[s]. \tag{8.38}$$

Choosing an appropriate T is not easy, as it involves a tradeoff between throughput and delay. By setting T very large, one can get an accurate estimate assuming ergodic $R_i[t]$. However, it is possible that a flow may have to wait for a very long time if the probability $\mathbb{P}(R_i[t] = R_{i,\max})$ is tiny. Clearly the tradeoff depends on the stochastic process of channel fading.

Is it possible to develop an efficient scheduler without estimating $R_{i,\max}$? Recall that the Max Weight scheduler (8.27) is queue-driven and the queue length does not keep on growing because of the finite size of a short-lived flow. A delay-driven MW scheduler does not suffer from the same problem. Specifically, revise (8.27) to

$$\text{MW:}\ i^*_{MW} = \arg \max_{i=1,\ldots,M} b_i D_i^\beta R_i, \tag{8.39}$$

where D_i represents the head-of-line delay of flow i. If flow i has not been served for a while, D_i will keep on increasing, eventually forcing flow i to be scheduled. In fact, it is shown in [131] that this kind of delay-driven MW scheduler is throughput optimal for short-lived flows. As a comparison, for long-lived flows, using queue length or delay does not matter as far as the throughput optimality is concerned. The reason is that from Little's law, in the presence of a fixed number of flows, the head-of-line packet delay is proportional to the queue length of a flow.

However, as pointed out in [105], the above delay-driven MW scheduler works only when $R_{i,\max}$ is identical for all i. Indeed, the simulation results in [105] show that the scheduler does not solve the stability problem when there are two types of links with different $R_{i,\max}$. It appears that the following scheduler works better when the absolute rate R_i is replaced by a relative measure $R_i / R_{i,\max}$:

$$\text{Max Weight:}\ i^*_{MW} = \arg \max_{i=1,\ldots,M} b_i D_i^\beta \frac{R_i}{\hat{R}_{i,\max}}. \tag{8.40}$$

Unfortunately, we again need to estimate $R_{i,\max}$ in order to determine the relative measure.

---

random channel rate of maximum value $R_{k,\max}$. Denote by α_k the arrival rate of type-k flows. Then, the sufficient and necessary stability condition is that the total load ρ is less than 1, where

$$\rho = \sum_{k=1}^{K} \rho_k, \tag{8.41}$$

with the load of type-k flows given by

$$\rho_k = \alpha_k \mathbb{E}\left(\frac{B_k}{R_{k,\max}}\right). \tag{8.42}$$

In effect, $B_k / R_{k,\max}$ is the service interval of a type-k flow.

Finally, let us consider the mixture of long-lived and short-lived flows. The performance goals for those two types of flows are quite different. The goal is to maximize the throughput for long-lived flows but to minimize the average delay for short-lived flows. The heuristic algorithm proposed in [105] turns out to be throughput optimal. Specifically, the scheduler has to balance the needs of long-lived and short-lived flows. For a short-lived flow, the scheduling needs can be measured with workload (backlog) defined to be the amount of time required to transmit the remaining bytes of the flow $Q_i[t]/\hat{R}_{i,\max}$. For a long-lived flow, the scheduling needs are represented with a combination of workload (queue length) and opportunity (channel rate) $Q_l[t]R_l[t]$. Therefore, the scheduler compares the following two entities:

$$\Psi_S = \sum_i \frac{Q_i[t]}{\hat{R}_{i,\max}} \tag{8.43}$$

$$\Psi_L = \max_l Q_l[t]R_l[t]. \tag{8.44}$$

If Ψ_S/Ψ_L exceeds some threshold ψ, then the short-lived flows are scheduled. In that case, among the short-lived flows, the one is scheduled if it experiences its best channel ($R_i[t] = \hat{R}_{i,\max}$) or the remaining bytes are few enough to be completely transmitted within the present slot. If no such flow exists, then schedule any flow. Otherwise $\Psi_S/\Psi_L \leq \psi$, the long-lived flows are scheduled. In that case, among the long-lived flows, schedule the one that maximizes the metrics in the MW rule (8.27) assuming $\beta = 1$. Parameter ψ can be used to balance the performance between long-lived and short-lived flows.

8.6 Signaling for scheduling

8.6.1 Dynamic packet scheduling

A complete scheduling system consists of not only the scheduler (scheduling policies and algorithms) studied so far but also control signaling that supports the scheduler. In a cellular system, the scheduler is implemented as part of the base station software. The job of the supporting control signaling is to collect information to be used by the scheduler and to convey the scheduling decision to the users. Figure 8.9 illustrates downlink and uplink scheduling systems.

In the downlink scheduling system, all the users send to the base station scheduler the downlink channel related information including channel quality indicator and MIMO channel matrix information. Since the traffic to be transmitted in the downlink is stored in the base station, the base station collects the traffic related information within its internal modules with no over-the-air signaling, and feeds that information to the scheduler. The base station sends the downlink scheduling decision, including what users are scheduled, over what time-frequency resources, and with what coding and modulation scheme. The downlink scheduling decision is meant for data transmitted over corresponding

8.6 Signaling for scheduling

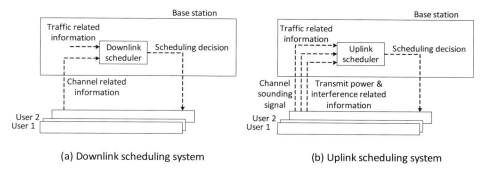

(a) Downlink scheduling system (b) Uplink scheduling system

Figure 8.9 Illustration of downlink and uplink scheduling systems.

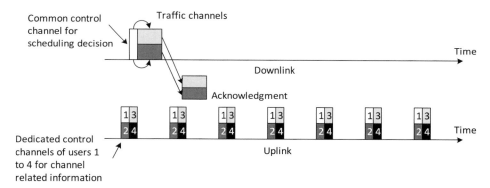

Figure 8.10 Illustration of resources of the control signaling that supports downlink scheduling. The common control channel conveys the downlink scheduling decision corresponding to the traffic channels shown. Also shown are the control channels for the corresponding acknowledgments.

traffic channel resources. Upon the reception of the downlink scheduling decision, the scheduled users attempt to decode the data and send positive or negative acknowledgment to inform the base station of whether decoding succeeds or not.

To avoid collision, the downlink channel related information is conveyed over uplink control channel resources dedicated to a given user. The user can be assigned in a semi-static manner with a set of resources to periodically report the information. The base station can assign additional resources on an on-demand basis to aperiodically solicit specific information. On the other hand, the downlink scheduling decision for given traffic channels is conveyed over downlink common control channel resources, since all the users attempt to decode it to see whether they are scheduled. The uplink resources for sending the corresponding acknowledgments are dedicated to whatever users are scheduled. Figure 8.10 depicts the uplink and downlink resources of the control signaling that supports downlink scheduling.

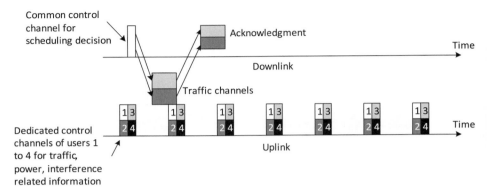

Figure 8.11 Illustration of resources of the control signaling that supports uplink scheduling. The common control channel conveys the uplink scheduling decision corresponding to the traffic channels shown. Also shown are the control channels for the corresponding acknowledgments.

In the uplink scheduling system, all the users send the uplink sounding signals to allow the base station to measure the uplink channel quality. The users may additionally report inter-cell interference information. Moreover, each user also reports how much transmit power headroom it still has so as to prevent the base station from over-scheduling the user. Since the traffic to be transmitted in the uplink is stored in the corresponding user, the user sends traffic related information to inform the base station scheduler of its scheduling needs. The base station sends the uplink scheduling decision, including what users are scheduled, over what time-frequency resources, and with what coding and modulation scheme. The uplink scheduling decision is meant for data transmitted over corresponding traffic channel resources. Upon the reception of the uplink scheduling decision, the scheduled users transmit the data over the scheduled traffic channel resources. The base station attempts to decode the data and send positive or negative acknowledgment to inform the user of whether decoding succeeds or not.

To avoid collision, the channel resources of the uplink sounding signal, interference, power or traffic related information are dedicated to a given user. Similar to the case of downlink channel related information, the channel resources can be assigned in a semi-static manner or on a on-demand basis. Moreover, the uplink scheduling decision is conveyed over downlink common control channel resources, since all the users attempt to decode it to see whether they are scheduled. The downlink resources for sending the corresponding acknowledgments are dedicated to whatever users are scheduled. Figure 8.11 depicts the uplink and downlink resources of the control signaling that supports uplink scheduling.

The effectiveness and efficiency of the scheduler depend on how timely and accurately the control information can be reported to the base station and how flexibly the base station can select users and allocate traffic resources. Clearly, there is a tradeoff between the signaling overhead and the scheduler performance. In the following practical example, we describe the control signaling design in LTE to illustrate the tradeoff.

8.6 Signaling for scheduling

Practical example 8.1 Signaling for scheduling in LTE

The following table summarizes the control signaling that supports scheduling in LTE.

Control information	Physical Channel	Scheduler
Downlink channel state report (CQI/PMI/RI)	PUCCH for periodic and PUSCH for aperiodic reporting	Downlink
Downlink scheduling assignment	PDCCH	Downlink
Acknowledgment (H-ARQ-ACK) for PDSCH	PUCCH	Downlink
Uplink channel sounding	SRS	Uplink
Uplink scheduling request (SR)	PUCCH	Uplink
Uplink buffer status report	PUSCH	Uplink
Uplink power headroom report	PUSCH	Uplink
Uplink scheduling grant	PDCCH	Uplink
Acknowledgment (H-ARQ-ACK) for PUSCH	PHICH	Uplink

The channel structures of SRS, PUSCH, and PUCCH are described in Example 2.2, and those of PDCCH and PHICH in Example 5.1. We next elaborate on the control information for downlink and uplink scheduling.

Downlink channel state report. A channel state report in general consists of CQI, RI, and PMI. CQI represents the highest coding and modulation scheme to be used with the expected block error probability not exceeding 10 percent. RI and PMI are reported only when a UE is in a spatial multiplexing mode. RI is the number of layers to be used and PMI is used to indicate the corresponding precoding matrices. Note that a channel state report is just a recommendation and the eNB can determine transmission parameters different from the recommended numbers. The actual parameters are specified in the downlink scheduling assignment.

There are two types of channel state reports: periodic and aperiodic. A UE is configured in a semi-static manner with a dedicated periodic subset of the PUCCH channel resources to send periodic reports as often as once every 2 ms. In addition, the eNB can explicitly request the UE to send aperiodic reports over dynamically assigned PUSCH resources. A typical use scenario of those two types of reports is that the UE sends lightweight control information with periodic reports so as to minimize the signaling overhead. On an on-demand basis, for example when a large amount of data has arrived to be sent to the UE, the eNB requests the UE to send an aperiodic report to provide more detailed channel state information to improve scheduling accuracy and efficiency. The potential gain of making PDSCH more efficient justifies the additional overhead of a large aperiodic report. Depending on the transmission mode of the UE, the channel state reports can use one of a variety of reporting modes to include wideband or subband CQI/PMI for frequency diversity or selective scheduling, respectively.

Downlink scheduling assignment. A downlink scheduling assignment is used to provide the assignment information of a corresponding PDSCH in the same subframe. There are several formats representing a tradeoff between the signaling overhead and the scheduling flexibility. The choice of the format also depends on the transmission mode of the assigned UE.

An assignment in general consists of five major components.

1. Identity of the assigned UE, which is 16 bits. Note that the identity is not explicitly sent as part of payload but is used to calculate the CRC so that another UE with a different identity will fail the CRC check and ignore the assignment.
2. PRB allocation, which indicates the resource blocks on which the assigned UE is to receive the PDSCH. As described in Example 2.1, there are three types of resource allocation that support contiguous and non-contiguous resource block allocation with the tradeoffs of frequency diversity versus selectivity and of overhead versus flexibility.
3. The modulation and coding scheme used (5 bits). The modulation order ranges from QPSK, 16-QAM to 64-QAM, and the coding rate is represented by a spectral efficiency of roughly 0.1 to 5.9 bits per second per symbol.
4. Hybrid-ARQ process number, which is used by the UE to determine soft combining.
5. The information about the use of multiple antennas, such as precoding information.

Acknowledgment for PDSCH. For FDD, the acknowledgment of a PDSCH transmission in subframe n is sent in a PUCCH in subframe $n+4$. If the acknowledgment is negative, then the eNB may decide to retransmit. Presumably, the scheduler treats retransmissions with higher priority than new transmissions so as to reduce ARQ latency. In the downlink, retransmissions can be scheduled at any time and frequency, just like new transmissions.

Uplink channel sounding. A UE is assigned with a dedicated periodic subset of the SRS channel resources to send the SRS signal, which allows the eNB to estimate the uplink channel quality from the UE. The bandwidth, frequency location, and periodicity of the assigned SRS channel resources dedicated to the UE are configurable.

Uplink scheduling request and buffer status report. A UE is assigned with a dedicated periodic subset of the PUCCH channel resources to send the SR to inform the eNB that the UE has traffic to send. To minimize the signaling overhead, the SR signal is ON or OFF, conveying no further information about the traffic queue (priority, amount, etc.). Once the SR triggers the eNB to assign some PUSCH to the UE, the UE can then provide a more detailed buffer status report in the assigned PUSCH. The detailed buffer status information of all the UEs allows the uplink scheduler to take into account traffic priority and queue lengths.

8.6 Signaling for scheduling

Uplink power headroom report. The power headroom is the difference in dB between the maximum output power and the transmit power of the present PUSCH. Similarly to a buffer status report, a power headroom report is sent when a UE is assigned with a PUSCH. The eNB uses power headroom reports to ensure that it does not assign a PUSCH that exceeds the transmit power capability of the UE, meaning that the assigned coding and modulation scheme is not too high or the size of the assigned resource blocks is not too large.

Uplink scheduling grant. An uplink scheduling grant is used to provide the assignment information of a corresponding PUSCH in a subsequent subframe. For FDD, an uplink scheduling grant in subframe n corresponds to a PUSCH in subframe $n + 4$. Compared with downlink scheduling assignments, uplink scheduling grants use much fewer formats. Indeed only one format is specified in LTE Release 8 (DCI format 0).

A grant in general consists of five major components.

1. Identity of the assigned UE, which is 16 bits. Similar to the case of downlink assignments, the identity is not explicitly sent as part of payload but is used to calculate the CRC.
2. PRB allocation, which indicates the resource blocks on which the assigned UE is to transmit the PUSCH. As described in Example 2.1, there are two types of resource allocation that support contiguous resource block allocation for SC-FDMA.
3. The modulation and coding scheme including redundancy version for hybrid-ARQ.
4. DM-RS phase information used to support multiuser MIMO. Recall from Example 7.1 that UEs that share the same PUSCH use orthogonal DM-RS so as to minimize inter-user interference in channel estimation.
5. Request for aperiodic channel state reports and PUSCH power control.

Acknowledgment for PUSCH. For FDD, the acknowledgment of a PUSCH transmission in subframe n is sent in a PHICH in subframe $n + 4$. If the acknowledgment is negative, then the UE will retransmit. To reduce the signaling overhead of another uplink scheduling grant, the UE is automatically assigned to the same set of resource blocks in subframe $n + 8$ for retransmission of the same hybrid-ARQ process. This is called synchronous non-adaptive hybrid-ARQ operation as opposed to asynchronous adaptive operation in the downlink. In addition, synchronous adaptive operation is possible by the eNB sending a new uplink scheduling grant. The retransmission still occurs in subframe $n + 8$ but can be moved to a different frequency location so as to avoid spectrum fragmentation, something particularly important to support contiguous resource block allocation in the uplink.

So far we have concentrated on dynamic packet scheduling where every packet is dynamically scheduled by control signaling for maximum flexibility and efficiency in the traffic channels. From Example 8.1, we note that there are two types of signaling overhead to support scheduling:

Scheduling

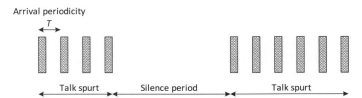

Figure 8.12 Illustration of VoIP traffic pattern. Talk spurts and silence periods alternate. No traffic arrives in a silence period. Traffic with similar sized payload arrives periodically in a talk spurt.

- Various uplink reports related to channel, traffic, and power to support scheduling decision making. Most reports are sent periodically – often frequently – to reduce signaling latency.
- Assignments and acknowledgments that are directly related to dynamic scheduling once scheduling decisions have already been made.

To reduce signaling overhead without sacrificing too much the system performance, we need to exploit some specific traffic characteristics. Below we show two examples that address those two types of signaling overhead.

8.6.2 Semi-persistent scheduling

Some traffic, such as VoIP, is characterized by regular arrival patterns with small payload sizes and very stringent delay and jitter requirements. Figure 8.12 illustrates a traffic pattern of VoIP.

Using the above dynamic packet scheduling to schedule VoIP leads to relatively significant overhead of assignments compared with the payload amount. Meanwhile, the benefit of dynamic packet scheduling becomes less significant because of stringent scheduling delay and small payload sizes. The number of simultaneous VoIP users is potentially large if all the circuit-switched voice users are converted to VoIP. As a result, the assignment channel resource may be the bottleneck, instead of the traffic channel capacity.

Persistent scheduling is a way of reducing assignment overhead. Specifically, a sequence of periodic traffic channel resources is allocated to a user beforehand with a fixed coding and modulation scheme until a new allocation is made. To support hybrid-ARQ, a packet is sent a fixed number of times and acknowledgment is not needed. The drawback, however, is the loss of flexibility, in particular due to the mismatch between resource that is allocated and that which is actually needed in a time-varying fading environment.

The idea of semi-persistent scheduling is to have persistent scheduling for initial transmissions and dynamic scheduling for retransmissions or for overriding persistent scheduling. Suppose the traffic arrives once every T time interval (in a typical voice codec, $T = 20$ ms). The user is persistently allocated a fixed set of traffic channel resources every T. If the first transmission of a packet succeeds, then the user simply waits until the next period. Otherwise, the scheduler dynamically assigns a traffic channel resource for retransmissions. As a result, the hybrid-ARQ process is adjusted according to the actual channel conditions. In addition, the scheduler may dynamically send a new assignment to override the persistent assignment, for example, when the payload size increases temporarily.

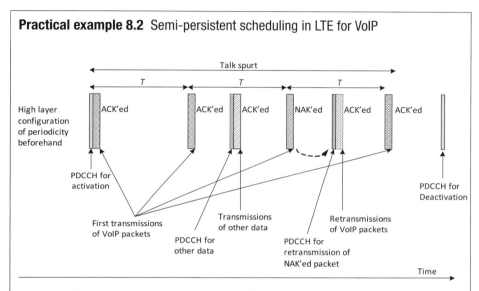

Practical example 8.2 Semi-persistent scheduling in LTE for VoIP

Figure 8.13 Illustration of semi-persistent scheduling in LTE for VoIP.

Figure 8.13 depicts an example of semi-persistent scheduling in LTE. The resource periodicity of semi-persistent scheduling is configured beforehand between the eNB and the UE via high layer messages. Semi-persistent scheduling can be activated when a talk spurt starts and deactivated when it ends, without changing the configuration. The eNB uses the PDCCH to toggle activation/deactivation with a UE identity different from the UE's normal identity of dynamical scheduling.

After being configured with semi-persistent scheduling, the UE continues to monitor the PDCCH for that identity. Once activated, the UE expects to transmit/receive traffic on the persistently assigned traffic resources. In addition, the UE keeps on monitoring the PDCCH for dynamic scheduling assignments (with the UE's normal identity as described in Example 8.1). This way, it is flexible to mix VoIP of semi-persistent scheduling and other data of dynamic scheduling. When a dynamic assignment is received in the same subframe as the persistent one, the dynamic one takes precedence.

8.6.3 MAC state scheduling

While the traffic pattern shown in Figure 8.12 is for VoIP, the overall structure of alternating active and inactive periods applies to other data traffic models such as web

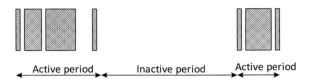

Figure 8.14 Illustration of VoIP traffic pattern. Talk spurts and silence periods alternate. No traffic arrives in a silence period. Traffic with similar sized payload arrives periodically in a talk spurt.

browsing traffic illustrated in Figure 8.14. A key difference is that in general data traffic there is no periodic arrival pattern during an active period, and the payload sizes can vary a lot. Moreover, the length of an active period is usually much shorter than the length of a talk spurt, which can last for a few seconds or longer. More importantly, the ratio of active versus inactive periods is much smaller for general data traffic than for VoIP.

To exploit the alternating active/inactive traffic pattern, we define two corresponding MAC states, ACTIVE and INACTIVE, and the MAC state scheduler puts the user into the ACTIVE (or INACTIVE) state when the traffic is in an active (or inactive) period. In the ACTIVE state, the user maintains all the signaling overhead and is under the control of the dynamic packet scheduler. In the INACTIVE state, the user gets rid of most of the signaling overhead at the cost of not being able to get scheduled in a fully flexible manner. The benefit, in addition to minimizing the overhead, is to save battery power.

The job of the MAC state scheduler is to move users between the ACTIVE and INACTIVE MAC states. Because of significant overhead required to stay in the ACTIVE state, the number of ACTIVE users at a given time is much smaller than the number of INACTIVE users. Therefore, in addition to saving power, the MAC state scheduler aims at best utilizing the MAC resource in the case of MAC resource congestion, for example, to determine what users should be in ACTIVE when more users intend to stay in ACTIVE than the system can support. The MAC state scheduler operates at a much slower time scale than the dynamic packet scheduler does, and therefore should base its scheduling decision more on long-term factors, such as user priority, than on short-term ones, such as channel fading.

> **Practical example 8.3** LTE DRX mode and Flash-OFDM HOLD state
>
> LTE supports a feature called discontinuous reception (DRX) for UE battery power saving. The basic idea is to configure a DRX cycle so that the UE monitors the downlink control signaling only in one subframe out of a DRX cycle. If the UE is not assigned in either the downlink PDSCH or the uplink PUSCH, then the UE goes to sleep and does not transmit or receive control signaling in the remaining subframes. However, if the UE is scheduled, it will continue to monitor the downlink control signaling for some configurable number of subframes just in case there are additional assignments. If there is no activity for some period of time, then the UE goes to sleep and will not wake up until the next DRX cycle.

8.7 Summary of key ideas 313

> Flash-OFDM further partitions the INACTIVE state into the SLEEP state and HOLD state [88]. The UEs in the HOLD state are timing controlled so that they are synchronized to the eNB receiver. However, they may not need to be power controlled to reduce overhead and save power. To support timing control operations, a HOLD state UE transmits a wideband signal at a very low periodicity (on the order of once per second). The eNB MAC state scheduler can move the HOLD state UE back to the ACTIVE state by paging it. Alternatively, the HOLD state UE can request a transition to the ACTIVE state by sending the request on a dedicated contention-free channel resource. Contention-free operation is made possible because the HOLD state UEs are all synchronized. The MAC state scheduling is QoS aware. In the case of MAC resource congestion, when a premium UE requests a transition from HOLD to ACTIVE, the MAC state scheduler can move a best effort ACTIVE UE to HOLD, free up the MAC resource and bring the premium UE to the ACTIVE state. Hence, operating at two very different time scales, the MAC state and dynamic packet schedulers can collectively provide QoS to a large number of UE population, not limited to only those in the ACTIVE state. The SLEEP state is the main power-saving mode where the UEs are not timing controlled. To transmit data, a SLEEP state UE has to go through random access to get timing synchronization.

8.7   Summary of key ideas

- In OFDMA mobile broadband, scheduling plays an essential role of satisfying the QoS needs of the users and achieving high system efficiency. In general, scheduling addresses user selection and resource allocation. The basic idea of an efficient scheduler is to be channel-aware and queue-aware.
- A simplified traffic model is infinitely backlogged traffic model. A variety of utility functions can be defined to maximize the overall efficiency subject to some fairness constraint. Different forms of the utility function can lead to different schedulers such as the max rate scheduler, the proportional fair scheduler, and the equal throughput scheduler, which are commonly used in the literature and in practice.
- Given a utility function, the utility maximization problem can be solved with a gradient-based approach. At a given time, the scheduler is used to maximize a linear utility where the weights represent the marginal utility of the present state. This leads to on-line scheduling algorithms.
- For elastic traffic, scheduling is often addressed together with congestion control, which is to determine the arrival rates in an end-to-end manner at individual traffic sources.
- For inelastic traffic, throughput optimal scheduling is shown to need queue-awareness in addition to channel-awareness. It is found that a large family of schedulers are throughput optimal including Max Weight, EXP, and LOG schedulers. Those schedulers exhibit distinct delay characteristics because of different tradeoffs between queue-awareness and channel-awareness. In particular, in the case of traffic overload,

the EXP scheduler minimizes the delay tail distribution of the worst user while the LOG scheduler achieves better mean delay. In addition, admission control is needed to minimize the probability of traffic overload.
- Flow level scheduling addresses short-lived flows. The scheduling policies turn out to be quite different from those developed for long-lived flows. In particular, it seems important to estimate and utilize the maximum channel rate $R_{i,\max}$ of individual users in order to develop an efficient flow level scheduler. This implies an inherent tradeoff between scheduling delay and efficiency.
- To achieve full scheduling flexibility, dynamic packet scheduling requires a potentially significant amount of control signaling. To balance the overhead cost and the scheduling benefits for further optimization, we need to exploit traffic characteristics. Two examples studied are semi-persistent scheduling for VoIP traffic and MAC state scheduling for general bursty data traffic.

9 Handoff in IP-based network architecture

The central design idea of mobile broadband is to adapt wireless to the Internet, not vice-versa. Compared with its wireline counterpart, mobile broadband faces two major technical challenges: *fading* and *interference*, which make the wireless link less reliable, and *mobility*, which requires handoff from one cell to another as a user moves. The previous chapters describe the physical and MAC layer approaches of dealing with fading and interference and improving link reliability and system capacity. In this chapter, we will expand our scope to view the airlink as part of an end-to-end network system and address the handoff issue.

Network architecture describes the necessary functions of the network system, partitions them to a set of logical nodes, and defines the interfaces between the nodes. An end-to-end network system is usually quite complex. To handle the complexity in a scalable manner, a good design practice is to adopt a layered structure. For example, the famous open system interconnection (OSI) model defines a networking framework of implementing protocols in seven layers, namely the application, presentation, session, transport, networking, data link, and physical layers. When two nodes communicate with each other, control is passed down from a higher layer to a lower one in one node, all the way to the bottom physical layer, then over the physical channel to the other node, and finally moving up the hierarchy in that node. The TCP/IP model of the Internet simplifies the layering model to four layers, namely the application, transport, Internet, and network access layers. Figure 9.1 provides a rough comparison of these two models. Since this book focuses on the IP-based network architecture, Figure 9.2 illustrates an internetworking scenario where two Internet hosts are connected via two routers.

In a wireless wide area network (W-WAN) system, a user communicates with a remote user or server via a base station. The airlink is a subsystem between the user and the base station, and usually includes only the network access layer, which can be further divided into the link, MAC, and physical layers. The base station implements IP as the network (Internet) layer that is separate from the network access layer associated with the W-WAN interface. In the TCP/IP layering model, the base station simply routes the packets between the user and another node in the network. Mobile wireless, as compared with fixed wireline, exhibits unique challenges:

- *Link reliability*. The wireless channel is inherently less reliable than the wireline counterpart because of channel fading. The problem could be especially severe when

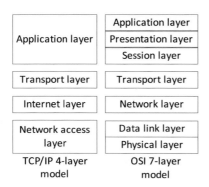

Figure 9.1 Comparison of OSI 7-layer model and TCP/IP 4-layer model.

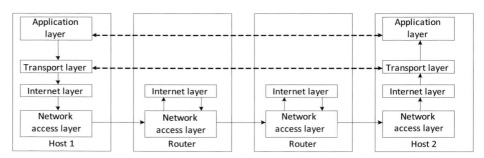

Figure 9.2 Data flow from host 1 to host 2 via two routers using the TCP/IP layering model. The link between a host and a router can be Ethernet, fiber, wireless, or other access technology.

the user is at the edge between two base stations, where sometimes the link with the serving base station becomes weaker than that with an interfering base station. It is desirable to improve the link reliability in such a scenario.
- *User mobility*. At a vehicular speed, the user may quickly migrate from one cell to its neighbor. The channel with the old base station rapidly deteriorates, and fast handoff is required such that the user continues to be served with the new base station. It is desirable to ensure seamless handoff and minimize service disruption including loss, delay, jitter, re-ordering.

Strictly speaking, according to the principles of the layered networking protocol model, the specifics of the network access layer of the airlink subsystem should be contained only between the user and the base station, and isolated from the rest of the network architecture. However, to overcome the above airlink challenges, some network architectures, such as in early CDMA networks, chose to deviate from the principle for good reasons. We will examine the rationales of such a design choice. More importantly, we will see whether the same rationales are applicable to OFDMA mobile broadband and argue about what the "right" network architecture should be.

9.1 IP-based cellular network architecture

9.1.1 Motivation for IP-based cellular network architecture

To a great extent, the success of IP-based wireline Internet is attributable to the architectural principle of pushing intelligence and complexity to the edge of the network, that is, hosts and servers, and keeping the core of the packet network simple and straightforward. Internet is very different from the traditional telephony network in which the core network is sophisticated and controls everything while the end nodes can be dumb and cheap. In Internet architecture, the function of the core is dumb: stateless (connectionless) packet forwarding. This enables the freedom of innovation at the edge. Specifically, innovative applications and services can be introduced to the end user without requiring the network infrastructure to be changed. Network infrastructure evolution usually takes place at a much slower pace and is controlled by a small number of large standards organizations and business entities. With open Internet architecture, the possibility of trying out new application or service ideas is, in effect, open to everyone.

As the cellular systems evolve from the voice centric 2G to the data centric 3G/4G, an IP-based network architecture is clearly the desired path towards mobile Internet. From the perspective of the end users, the IP-based network architecture seamlessly extends the broadband access experience that the users have in the wireline world. Existing applications and protocols run natively across the wired and wireless links without modification. Indeed, they need not be mobility-aware. Those applications include traditional server applications such as web browsing and peer-to-peer applications such as gaming, messaging, VoIP, and push-to-talk that have become increasingly popular. TCP and UDP, predominant transport layer protocols in the IP world, run end-to-end without wireless-specific protocol translation in the middle.

The IP-based network architecture benefits the operators by reducing the capital and operating expenses. Specifically, it enables fully converged networks to provide voice and data, fixed and wireless services using standard off-the-shelf IP core technology and equipment, such as AAA servers. Network deployment and management gets simpler and more flexible as the base stations become autonomous and can be deployed in a plug-and-play manner. Since the base stations are IP aware and thus more intelligent than just dumb transceivers, the operators can provide QoS-based services and ultimately maximize revenue per Hertz instead of bits per Hertz.

9.1.2 Description of IP-based cellular networks

Figure 9.3 shows a simplified diagram of an IP-based cellular network architecture. The architecture consists of a radio access network (RAN) and a core network. The RAN resides at the network edge and consists of the base stations and the users. A base station is in essence an IP access router with one or more backhaul interfaces connecting to the core network and another radio interface to serve users through the wireless channel. The service coverage area of a base station is a cell. Cells partially overlap with each other to allow a user to hand off seamlessly from one base station to another when it

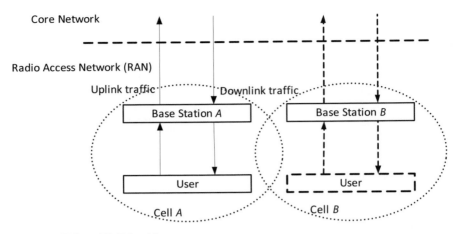

Figure 9.3 IP-based RAN architecture.

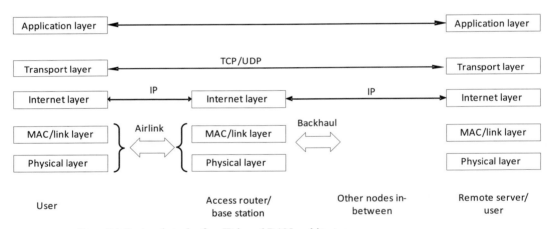

Figure 9.4 Protocol stack of an IP-based RAN architecture.

moves. When the user moves from cell A to cell B, it switches to a new base station to continue the downlink and uplink traffic.

Most of the functions covered in this book, such as coding, modulation, synchronization, power control, and scheduling, reside in the RAN. One exception is the mobility management function, which is split between the RAN and the core network, as will be studied in this chapter.

Figure 9.4 shows the layered protocol stack used by the user and the base station. Here, the airlink represents just another network access layer over which IP operates. This allows the IP protocol suite to use the network access layer capabilities and to control its resources. Moreover, IP traffic goes directly to the base station, thereby allowing the base station to exploit all existing IP features, for example, to be QoS aware.

The protocol layering shown in Figure 9.4 optimizes for IP-based data delivery and networking. Specifically, the network access layer, including physical, MAC, and link

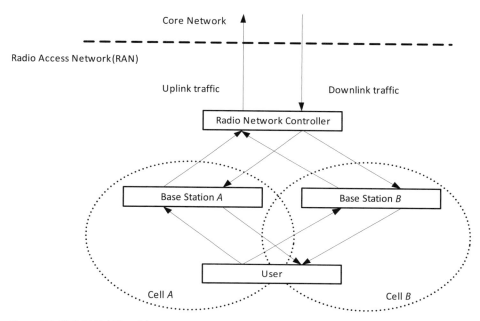

Figure 9.5 CDMA RAN architecture with soft handoff.

layers, are vertically integrated. Cross-layer design and joint optimization of the airlink helps deliver the ideal network access layer service to the IP protocols. The ideal network access layer is to adapt wireless to the Internet, for example, achieving high data throughput, high reliability, and low latency, so that the IP protocols, most of which were originally designed to operate in the wireline environment, run transparently over the wireless channel. At the IP layer and above, traditional horizontal layering is respected, thereby cohesively integrating the wireless connectivity into the greater Internet.

9.2 Soft handoff in CDMA

The IP-based network architecture has been widely recognized as the trend of 4G mobile broadband. However, early cellular systems were not based on that architecture. One important technical reason was that the dominant traffic then was circuit-switched voice and the network architectures were designed for the characteristics and requirements of that type of traffic:

- Low and somewhat constant data rate (as compared with bursty broadband data).
- Fixed MAC channel resource assignment (no packet-switched dynamic scheduling).
- Stringent latency requirement (overly delayed voice frames are useless and lost).
- High reliability (a call drops if a large number of voice frames are lost due to degraded link reliability).

As an example, we next examine the CDMA RAN architecture illustrated in Figure 9.5. The most noticeable difference from the IP-based RAN architecture Figure 9.3 is that

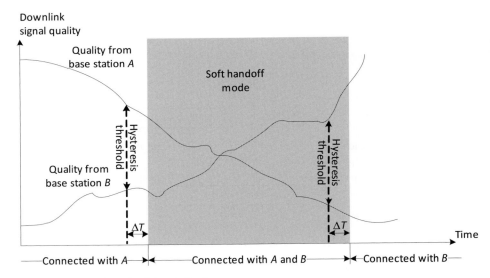

Figure 9.6 Illustration of soft handoff scheme.

a node called *Radio Network Controller* (RNC) sits on the data path between the base stations and the core network.

Why does the CDMA RAN architecture work well with circuit-switched voice traffic? The reason is *macro-diversity*. Macro-diversity exploits the property that the wireless channel, in particular multipath fading, is independent with different base stations. To obtain macro-diversity, a user is put in a *soft handoff* mode, in which it is connected simultaneously with multiple base stations as shown in Figure 9.5. This is different from hard handoff in the narrowband FDMA system in which a user is connected with at most one base station at a time. As a wideband transmission scheme with universal frequency reuse, CDMA makes it easy for the user to be connected with multiple neighboring base stations with a single RF front-end.

Soft handoff can be initiated by the user. The user keeps on monitoring the quality of the downlink channel[1] from the serving base station as well as from other potentially strong base stations. Suppose that the user moves away from the serving base station (A) and gets close to another base station (B), which becomes the handoff target. The quality from base station A deteriorates, while that from base station B improves. As shown in Figure 9.6, when the quality of base station B gets close enough to that of base station A for some period of time ΔT, the user enters the soft handoff mode by connecting to base station B. Then, as the user moves further, base station B becomes superior to base station A. Finally, when the quality of base station A drops beyond

---

[1] In CDMA, the downlink physical layer channel to be monitored is called pilot. Pilots from different base stations are constructed with different code sequences and can thus be separated at the user receiver. The quality is usually measured with the SINR or the signal strength of the pilot.

a certain hysteresis threshold for a period of time ΔT, the user exits from the soft handoff mode by disconnecting from base station A and is connected only with base station B.

During the soft handoff mode, both base stations simultaneously send the same signal[2] to the user in the downlink and receive a single signal from the user in the uplink. The function of the RNC is to facilitate the soft handoff operations. Specifically, in the downlink path it forwards a voice frame from the core network to both base stations and in the uplink path combines the voice frames from the two base stations. In addition, the RNC handles the control procedure of connecting the user to base station B and later disconnecting it from base station A.

Let us first examine the uplink case. If we treat the base stations as multiple receive antennas, then soft handoff in the uplink is the similar to receive diversity studied in Section 4.2.3. Specifically, the optimal receiver scheme is maximal ratio combining, where the base stations both send the soft values of demodulation symbols (after despreading in CDMA) to the RNC and the RNC combines the symbols and performs decoding. This requires a significant amount of raw data to be sent from the base stations to the RNC. To reduce the backhaul costs, in practice, selection combining is performed instead of maximal ratio combining. Each base station decodes its own demodulation symbols and sends the decoded information bits, together with the indication of decoding quality, to the RNC. The RNC simply selects one of the two branches with the higher quality.

Suppose that the data rate is fixed for circuited switched voice. A fixed SINR, γ_0, is required to successfully receive a voice frame. With selection combining, the user transmit power P_u is determined by

$$\max\left(\frac{P_u G_A}{\sigma_A^2}, \frac{P_u G_B}{\sigma_B^2}\right) = \gamma_0, \qquad (9.1)$$

where σ_A^2, σ_B^2 represent the total interference and noise powers at base stations A and B, respectively, and G_A, G_B the corresponding uplink channel power gains. Soft handoff reduces the transmit power P_u and therefore increases the system capacity.

Soft handoff becomes *softer* when handoff occurs between two sectors of the same cell. In this case, the base station is co-located with the antennas of different sectors and performs maximal ratio combining, instead of selection diversity, without incurring the backhaul costs.

Soft handoff in the uplink provides power and diversity gains with virtually no interference costs, because base station B would be receiving the signal energy from the user whether it participates in soft handoff or not. However, just like receive versus transmit diversity, the situation in the downlink soft handoff is quite different in that now base station B has to transmit to the user the signal energy, which could otherwise be allocated to another user if base station B were not in soft handoff. Hence, downlink soft handoff

---

[2] More precisely, the two base stations spread the same signal with different spreading codes such that when the user despreads the signal from base station A, the signal from base station B appears like noise.

does cost system resource, namely power,[3] which has to be taken into account when we study the performance benefit.

Next we view the base stations as multiple transmit antennas, and compare soft handoff in the downlink with transmit diversity. The signals from base stations A and B appear to the user receiver like multipath copies of one signal. The user uses the RAKE receiver to collect the signal energy with maximal ratio combining [168]. Suppose that each base station allocates β fraction of the total transmit power to the user. As in the uplink, power allocation β should meet the SINR requirement

$$\frac{\beta P_d G_A}{\psi P_d G_A + P_d G_B + \sigma_0^2} + \frac{\beta P_d G_B}{\psi P_d G_B + P_d G_A + \sigma_0^2} = \gamma_0, \qquad (9.2)$$

where P_d is the total transmit power of one base station and G_A, G_B the corresponding downlink channel power gains. Here we follow the traditional assumptions of CDMA RAKE receivers. Specifically, when the user receives the signal from one base station, the signal from other base stations is treated as interference. In addition, because the CDMA downlink is not orthogonal in the multipath channel, parameter $0 \leq \psi \leq 1$ represents the fraction of the total power from the same base station to be treated as self interference. σ_0^2 represents the total interference power from other base stations and the noise power.

Although soft handoff reduces the required power allocation β, the same signal power is sent in both base stations. After the total consumed power is normalized, the effective channel gain roughly becomes $(G_A + G_B)/2$. Hence, downlink soft handoff provides only a diversity gain, and no power gain. The benefit is similar to what is observed with transmit diversity in Section 4.2.3.

We note that the above scheme of using the same transmit power at the two base stations may not be cost effective. During soft handoff, the instantaneous channel quality of the two base stations may differ remarkably depending on the hysteresis threshold. For example, in Figure 9.6, before the two channel quality curves cross over, sending the signal from base station B is not as efficient as from base station A. This suggests an improved downlink soft handoff scheme in which the user sends a 1-bit feedback indicating which base station is preferred for an upcoming voice frame and only the preferred base station is sending the voice frame. The scheme works as long as the prediction is correct with high probability and the feedback bits can be reliably received in the uplink. Obviously, the delay in the feedback control loop makes it difficult for the prediction to be accurate. What makes it differ from hard handoff in narrowband FDMA is that both base stations are in the soft handoff mode so that switching from one base station to the other can be as rapid as for every voice frame. In contrast, to avoid the "ping-pong" effect where a user near the cell boundary is handed off back and

---

[3] The CDMA downlink uses a fixed number of orthogonal spreading codes to separate different users. A user in soft handoff occupies orthogonal codes with both base stations and doubles the number of orthogonal codes (bandwidth resource) it consumes. However, since the capacity of the CDMA downlink is limited by interference not by the number of spreading codes, we only consider power as the system resource cost in soft handoff.

Table 9.1. Design tradeoffs of soft handoff for circuit-switched voice and for broadband data.

	Circuit-switched voice	Broadband data
Link reliability	Soft handoff improves link reliability via macro-diversity and reduces the error rate of voice frames	Link reliability can be achieved via ARQ. For most traffic, the latency requirement is not as stringent as for circuit-switched voice, and temporary link outage is tolerable
Scheduling	No scheduling is used in a frame-by-frame manner	Multiuser diversity-based opportunistic scheduling is crucial. It is difficult and often counterproductive to synchronize multiuser scheduling across multiple base stations
Backhaul overhead	The data rate is low as compared with the backhaul capacity. It is not burdensome for the RNC to send/receive the same voice frames to/from multiple base stations	The data rate is high. Duplicating broadband data traffic to/from multiple base stations results in significant backhaul costs

forth several times from one base station to the other, in practice hard handoff occurs only after base station B is sufficiently superior to base station A. As a result, the user is connected with an inferior base station (A) for an extended period of time.

In summary, the diversity gain reduces the power margin required to combat fading, together with the power gain in the uplink, increases the system capacity. The capacity in CDMA is interference-limited and the reduction in transmit power directly translates into the reduction in interference and therefore the capacity. Hence, we conclude that soft handoff is the right solution for circuit-switched voice and list the reasons in Table 9.1.

9.3 Make-before-break handoff in OFDMA

Is soft handoff the right solution for broadband data as well? Table 9.1 highlights different characteristics and requirements of circuit-switched voice versus broadband data. From the table, it seems an alternative solution is desirable.

Another motivation for an alternative solution comes from viewing soft handoff from the perspective of RAN architecture. Suppose that we extend the CDMA soft handoff RAN architecture to transport IP data traffic beyond circuit-switched voice. Figure 9.7 shows a hypothetical layered protocol stack similar to the one presented in [12].

Comparing with Figure 9.4, we note that the difference is that the lower three layers of protocol functions are divided between the base station and the RNC. In particular, the base station only handles the physical layer protocol and is in essence a modem, whereas the RNC terminates the IP protocol and handles the link layer. As a result, the benefits of the IP-based network architecture mentioned in Section 9.1.1 are compromised.

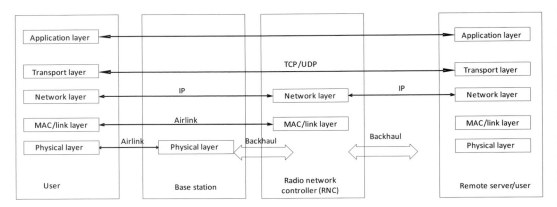

Figure 9.7 Protocol stack of a CDMA IP-based RAN architecture.

One solution to achieve macro-diversity and yet retain the IP-based RAN architecture, without soft handoff is *IP-based make-before-break* (MBB) handoff, in which the user acts as if it were multiple co-located users *independently* connected with base stations simultaneously. This means that each link runs separate physical layer control loops (frequency, time, and power) and transports its own control and data packets.

9.3.1 Parallel independent links to multiple base stations

Figure 9.8 illustrates the block diagrams of the user in the MBB handoff mode with two base stations that operate in the same bandwidth (universal reuse).

At the transmitter, the user has two parallel chains of digital signal processing modules, each of which processes the uplink signal targeted to one base station and is identical to the one shown in Figure 2.5 when the user is not in the MBB handoff mode. The inputs to the chains represent the information bits of control and data packets to be sent to the base stations and therefore are different for the two chains.

Recall from Section 2.2 that in OFDMA the user transmitter performs time and frequency synchronization. In particular, the user adjusts its transmit symbol time to compensate for the round-trip propagation delay with the base station. As a result, the transmit symbol time may be different for the two base stations, unless the user happens to be in the exact midway point between them. Figure 9.9 shows that the transmitter generates two sequences of digital OFDMA samples $x_1(n), x_2(n)$ to the two base stations where $x_1(n), x_2(n)$ are of different symbol times. The two sequences are added before the D/A convertor so that only one analog RF transmit chain is needed subsequently. Similarly, the transmitter may need different amounts of frequency compensation for the two base stations, for example, because of different Doppler shifts. In this case, the transmitter needs to digitally perform frequency compensation in at least one of the transmit chains, because the analog RF transmit chain is shared between the two chains. Digital frequency compensation means that each digital sample is to be rotated by $e^{j2\pi\delta_f n}$, where n is the sample index and δ_f is the amount of frequency to

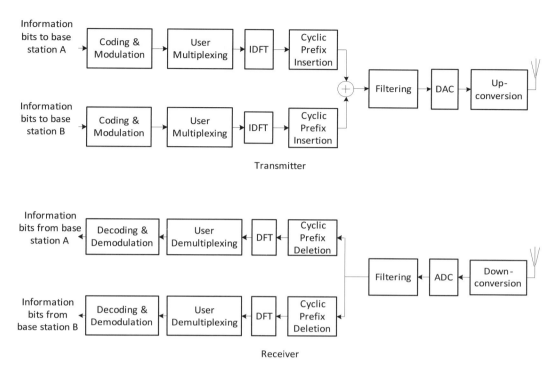

Figure 9.8 Simplified block diagram of OFDMA signal processing modules at a user in the MBB handoff mode. The user is connected simultaneously with two base stations. At either the transmitter or the receiver, there are two parallel digital processing chains, one for each base station. Assuming the two base stations reuse the same spectrum, only one analog RF chain is required.

be compensated. In addition, the transmit powers of $x_1(n)$, $x_2(n)$ are determined in the corresponding links and can be different because the user is power controlled by the two base stations independently.

Similarly, at the receiver, the user has two parallel chains of digital signal processing modules, each of which processes the downlink signal from one base station. Only one analog RF receive chain is used. After the A/D convertor, the digital samples $y(n)$ are duplicated and provided as the input to both receive chains, which are used to recover the information bits of control and data packets sent from the corresponding base stations. Each receive chain performs separate time and frequency synchronization and takes a corresponding portion of the sample sequence as the input to the DFT module for further processing.

In addition to time and frequency synchronization, the user also needs to synchronize its sampling rates to multiple base stations. The user can directly change the sampling rate of the A/D and D/A modules by adjusting its local oscillator. However, if the base stations have different sampling rates themselves, then that approach does not work because the same A/D and D/A modules are shared among parallel chains. Alternatively,

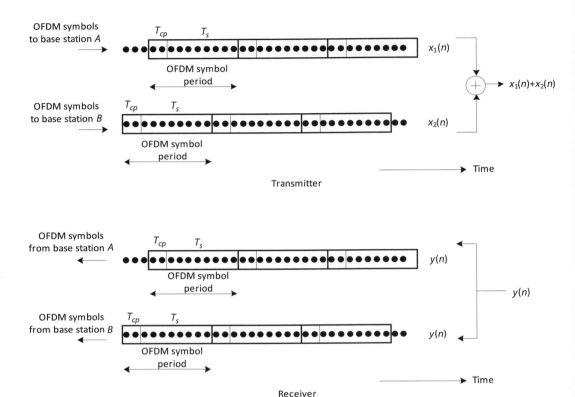

Figure 9.9 Illustration of transmit/receive symbol time synchronization of a user in the MBB handoff mode. The user performs separate symbol time synchronization independently with respect to each of the base stations it is connected with. At the transmitter, each little dot represents a digital sample of either $x_1(n)$ or $x_2(n)$, the OFDM signal to be sent to base stations A or B, respectively. As shown in the figure, the symbol times are different for $x_1(n)$ and $x_2(n)$. The two sequences of digital samples are added before the D/A convertor. Similarly, at the receiver, the sequence of digital samples is processed with the receive symbol times of the corresponding links. If the symbol times are different, as shown in the figure, different portions of the same sample sequence are used in the two receive chains to recover OFDM symbols from different base stations.

the user can periodically add or drop digital samples so as to keep the *average* sampling rate the same as the base station. For example, suppose that the user's sampling rate is 0.001 percent slower than base station's sampling rate. Then, the duration of every OFDM symbol measured at the user is slightly longer than that at the base station. Suppose that $N_c = 512$. After approximately 200 OFDM symbols, the accumulated discrepancy is equal to one sample, at which point the user drops one sample in the cyclic prefix insertion/deletion modules of its front-end sample rate signal processing paths so as to align itself with the base station sampling rate. The user carries out this operation periodically. Adding/dropping samples occurs independently in parallel chains.

9.3 Make-before-break handoff in OFDMA

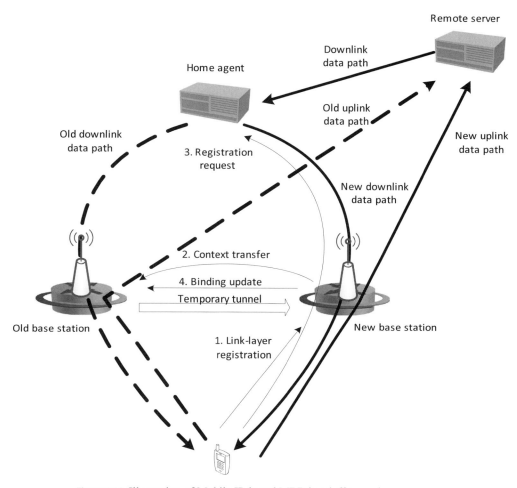

Figure 9.10 Illustration of Mobile IP-based MBB handoff procedure.

9.3.2 Mobile IP-based MBB handoff procedure

The user in the MBB handoff mode can independently send/receive different IP packets to/from the two base stations. The base stations are not necessarily aware that the user is in the MBB handoff mode. Indeed, the RAN architecture is completely IP-based, as shown in Figure 9.3, and does not require an RNC to control the base stations.

Exploiting the capability of independent connections with two base stations, a handoff procedure based on the Mobile IP protocol [123] is shown in Figure 9.10.

Suppose the user is communicating with a remote server. With Mobile IP, the user has a home agent, which is a fixed server in the core network. The user is initially connected only with one base station (old base station) and registers to the home agent the address of the old base station as its care-of address. The remote server sends IP packets to the old base station through the home agent.[4] However, for the uplink packets sent by the user, standard IP routing can be used to deliver the packets to the remote server without going through the home agent.

Subsequent to handoff initiation, the user brings up a new link with a new base station, while it continues receiving/sending packets from/to the old base station. The user sends a *link layer registration* message to the new base station, indicating the IP address of the old base station. The message triggers *context transfer*, a process that transfers from the old to new base station the user related state such as security keys, which would otherwise have to be recreated at the new base station. Note that in principle, all the context could simply be created at the new base station and then handoff performed via Mobile IP. Here context transfer is an optimization to minimize handoff delays and signaling traffic in the network and over the air. The handoff process also establishes a temporary tunnel for the user from the old to new base station.

Next, the user registers its new care-of address (the IP address of the new base station) to the home agent. Upon the reception of this *registration request* message, the home agent redirects the downlink traffic to the new base station. While link-layer registration, context transfer, and Mobile IP registration take place at the new base station, the user is still connected with the old base station, thereby avoiding delay or packet loss.

For some period of time, the old base station continues sending packets previously received from the home agent, while the new base station starts to send new packets. During this time, the arrived packets from the two base stations may be out of order. The user eventually severs the link with the old base station when no packets remain in flight from the home agent to the old base station. Furthermore, a binding update can be sent from the new base station to the old base station at the same time that the Mobile IP registration request is sent to the home agent. The packets in the old base station can be re-directed to the new base station via the temporary tunnel upon on the arrival of the binding update, or when the link with the old base station degrades/drops.

So far the study is focused on handoff between two base stations. The procedure for handoff between two sectors of the same base station is simpler. In particular, context transfer is not required, since all the authentication, security and configuration information is already known at the base station. More importantly, there is no need to send the Mobile IP registration request message. Inter-sector handoff can be made completely transparent to the home agent. After packets arrive at the base station, they are provided to the particular sector currently serving the user.

9.3.3 Uplink macro-diversity

While the user is keeping links with the two base stations, it has the freedom of sending the uplink packets to either base station. Because IP routing is connectionless and each IP packet carries its destination IP address, the base station knows where to forward the

---

[4] Route optimization is possible to eliminate triangle routing so that the packets from the remote server do not have to route through the home agent. Our discussion ignores route optimization for simplicity.

9.3 Make-before-break handoff in OFDMA

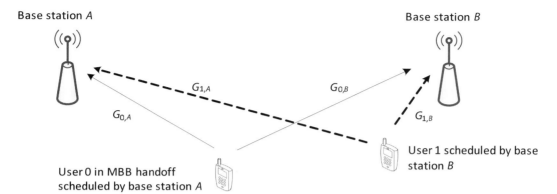

Figure 9.11 Calculation of uplink SINR of a user in MBB handoff at two base stations when they both receive the signal from the user.

packet upon the completion of packet assembly. Hence, in the MBB handoff mode, the user obtains uplink macro-diversity by dynamic link selection.

How does MBB handoff compare with soft handoff from the perspective of uplink macro-diversity? We next address this question in the following aspects.

Combining power gain

In soft handoff, the reception of the signal from the user at both base stations leads to a combining gain. In OFDMA mobile broadband, suppose the user is scheduled by one base station. The additional benefit for the other base station to also receive the signal is only marginal. The reason is that the airlink features of OFDMA mobile broadband, in particular, opportunistic link scheduling and zero intra-cell interference, make it very different from circuit-switched CDMA.

Note that in MBB handoff, the two links are independently scheduled by the corresponding base stations. When the user is scheduled, only the scheduling base station tries to recover the signal from the user. We will show next that conditional on that the user is scheduled by one base station, the SINR at the scheduling base station is usually much greater than the instantaneous SINR at the other base station.

Consider a simple scenario of only two cells served, respectively, by base stations A and B shown in Figure 9.11, and assume that the same spectrum is reused in both base stations. Suppose that a user (0) in MBB handoff is scheduled by base station A. Meanwhile, base station B schedules another user (1) to reuse the same spectrum.

The uplink SINR of user 0 at base station A is given in (4.108) and re-written as

$$\gamma_A = \frac{P_0 G_{0,A}}{\sigma_0^2 + \sigma_I^2}, \tag{9.3}$$

where σ_0^2 is the background noise power and σ_I^2 is the interference power from the users transmitting to base station B. P_0 and $G_{0,A}$ are the transmit power of the user and its channel power gain to base station A. Inter-cell interference σ_I^2 depends on power control. Assume a simple power control scheme of (4.111), where the transmit power of any scheduled user is set such that its interference to the other base station is equal to a target value of inter-cell interference budget $\bar{\sigma}_I^2$ such that

$$P_0 = \frac{\bar{\sigma}_I^2}{G_{0,B}}, \tag{9.4}$$

with $G_{0,B}$ the channel power gain to base station B. It follows that

$$\gamma_A = \frac{G_{0,A}}{G_{0,B}} \frac{\bar{\sigma}_I^2}{\sigma_0^2 + \bar{\sigma}_I^2}. \tag{9.5}$$

Denote by $G_{1,B}$, $G_{1,A}$ the channel power gains from user 1 to base stations B and A, respectively. The transmit power of user 1 is set with the same principle[5]:

$$P_1 = \frac{\bar{\sigma}_I^2}{G_{1,A}}. \tag{9.6}$$

Now let us see what the additional benefit is for base station B to also receive the signal from user 0. For the signal from user 0, the signal of user 1 acts as interference at the base station B receiver, as we assume no multiuser detection. The SINR of user 0 at base station B is thus

$$\gamma_B = \frac{\bar{\sigma}_I^2}{\sigma_0^2 + \frac{G_{1,B}}{G_{1,A}}\bar{\sigma}_I^2}. \tag{9.7}$$

In order to be worthwhile for base station B to also receive the signal from user 0, γ_B should be significant compared with γ_A. Let us examine γ_A/γ_B. It follows that when inter-cell interference dominates noise $\bar{\sigma}_I^2 \gg \sigma_0^2$,

$$\frac{\gamma_A}{\gamma_B} \approx \frac{G_{0,A}}{G_{0,B}} \frac{G_{1,B}}{G_{1,A}} \gg 1. \tag{9.8}$$

The inequality is for two reasons. First, given the fact that the two users are scheduled by the corresponding base stations, the above two instantaneous channel power gain ratios, $\frac{G_{0,A}}{G_{0,B}}$, $\frac{G_{1,B}}{G_{1,A}}$, are likely much greater than one, because of multiuser diversity scheduling – see (4.113). Second, since user 0 is in MBB handoff, $G_{0,A}$ and $G_{0,B}$ have roughly equal mean. However, if user 1 is not in MBB handoff, then the mean of $G_{1,B}$ is noticeably greater than that of $G_{1,A}$.

Hence, we reason that γ_A tends to be significantly greater than γ_B and therefore not much combining gain is obtained for base station B, in addition to base station A, to receive the signal of user 0.

Control of link selection

Links are selected by the transmitter (user) in MBB handoff but by the receiver (RNC) in soft handoff.

In soft handoff, both base stations attempt to recover every voice frame from the user, and the RNC selects the one that is more reliably decoded. However, in MBB handoff,

---

[5] The target value may be different in base station A or B. But that will not change the final conclusion in (9.8).

only one base station receives the link layer frames of a packet, and the user is the one responsible for selecting the better link. How does the user, as the transmitter, figure out which link is better?

One possibility is that the user measures and compares the signal strength of the downlink pilots of both base stations, and selects the one with a stronger downlink pilot, assuming channel symmetry between the downlink and uplink. The drawback of this open-loop approach is, however, that in an FDD system the difference in carrier frequencies results in different, almost independent, multipath fading in the downlink and uplink, and therefore sometimes the channel power gains are very different between the downlink and uplink. Making the uplink selection on the basis of the downlink measurement may not always lead to the right decision.

A better approach is to exploit closed-loop power control. In MBB handoff, the transmit powers on the two links are independently controlled by the corresponding base stations. In contrast, in soft handoff, the user has only one transmit power to be controlled. When the user receives closed-loop power control commands from both base stations, it only increases transmit power if *both* commands are "up."

When the user is connected with a base station, in general there are two types of uplink traffic. The first type is certain control traffic, such as reports of downlink channel quality, that the user periodically sends to the base station at a fixed data rate. The second type includes data traffic and other control traffic that occur aperiodically and usually at a variable data rate. The details can be found in Appendix A.3. Closed-loop power control means that the base station sends power control commands to adjust the transmit power of the first type of traffic. The transmit power of the second type of traffic is determined as an offset from that of the first type, and the offset depends on the variable data rate. Because the data rate of the first type is fixed, so is the received SINR target. Therefore, the receive power target only depends on the total interference plus noise power at the base station receiver. Assuming the total interference plus noise power is the same at the two base stations, the link that requires less transmit power is the one having greater uplink channel gain and should be selected for macro-diversity, as shown in Figure 9.12.

Hence, this approach of link selection is considered closed-loop, and is expected to have similar performance to the RNC selection approach in soft handoff as long as closed-loop power control is able to keep prompt track of uplink channel power gains.

Stickiness of link selection

Link selection occurs on a per IP packet basis in MBB handoff and per voice frame basis in soft handoff.

To follow the architectural principles that a base station is an IP access router forwarding packets from the radio interface to backhaul interface and that the airlink is just another network access layer over which IP operates, in MBB handoff, the user does not send a portion of an IP packet to one base station and the remaining portion to the other base station and expect the two base stations to combine them and reconstruct a complete packet. Consider sending an IP packet in Figure 9.4. At the transmitter, the packet is passed down to the link layer where it is segmented into link layer frames, each of which is sent separately. At the receiver, the frames are reassembled. If some frames

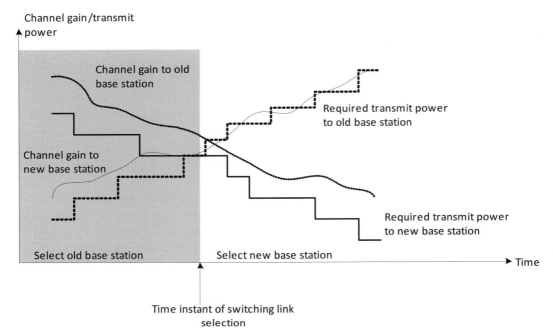

Figure 9.12 Illustration of link selection based on closed-loop power control to achieve uplink macro-diversity. In MBB handoff, the transmit powers of individual links are independently controlled by the closed-loop power control commands from the corresponding base stations. One can figure out which link exhibits better channel gain from closed-loop power control.

are lost, then reassembly fails. The packet cannot be reconstructed. Even those frames that have been successfully received become useless.

Suppose that the user has started to send some frames of the packet on one link. When handoff occurs, the user can either continue finishing the remaining frames, or switch to the other link in which case it gives up those unfinished frames and starts from scratch with the packet. In principle, it is possible that the pieces received at both base stations are reassembled together to reconstruct a complete IP packet. For example, one base station can tunnel all the received frames to the other. However, architecturally, it is not as clean as the one shown in Figure 9.3.

Hence, in MBB handoff, the link selection in the uplink is done on a per packet basis, as illustrated in Figure 9.13. As a comparison, in soft handoff, the RNC selects every voice frame from the better link, and therefore, the link selection is not sticky.

The above constraint of sticky link selection leads to the following two implications.

- First, the user may be stuck with an inferior link for some period of time. Clearly, the macro-diversity gain would be greater if the user were allowed to transmit on the better link for every frame. Note that the frame size depends on the physical layer data rate and is usually small at cell edges where handoff occurs; as a result, a packet may be segmented into many frames. Thus, the stickiness of the per packet link selection could force the user to transmit over the inferior link for an extended period of time.

9.3 Make-before-break handoff in OFDMA

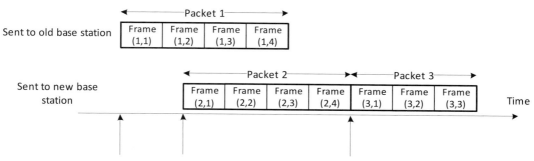

Figure 9.13 Illustration of per packet link selection in uplink macro-diversity. Each of packets 1, 2, 3 is segmented into several link layer frames. Frames of packet 1 are sent to an old base station. While frame (1, 2) is being sent, the user selects a new base station as the new link and starts to send frames of packets 2, 3. The remaining frames of packet 1, that is, frame (1, 3) and frame (1, 4), continue to be sent to the old base station.

- Second, suppose that if the old link deteriorates, the user starts to send new packets on the new link while continuing on the old link with existing packets. Since the two base stations do not coordinate their uplink scheduling, it is possible that the user is scheduled to transmit on both links simultaneously. In that case, the user may not have sufficient transmit power for both links, especially at cell edges where the user is likely constrained by its maximum transmit power. In addition, if SC-FDMA is used, the user is required to transmit on contiguous or equally spaced tones (see Section 2.3). When the user is transmitting on both links simultaneously, such a requirement is not easily satisfied unless tone allocation is carefully coordinated at both base stations.

It should be pointed out that handoff between two sectors of the same base station does not have to suffer from the constraint of sticky link selection. In particular, two sectors share a common frame reassembly unit that puts together frames received from different sectors to reconstruct complete IP packets. Now, the user can send some frames of a packet to one sector and the remaining to the other sector. A similar idea is applicable to downlink macro-diversity.

9.3.4 Downlink macro-diversity

Macro-diversity in the downlink is obtained by the user switching the packet forwarding route between the old and new base stations. The user sends a Mobile IP registration request message to the home agent to change the route.

Ideally, the base station that sends downlink packets is always the one with a superior channel gain. As pointed out in Section 9.2, this ideal selection macro-diversity outperforms downlink soft handoff because it never spends transmit power in an inferior link. Moreover, similar to the above discussion on the uplink, with multiuser diversity scheduling, when the user is scheduled by one base station, the SINR tends to be higher from the scheduling base station than from the other one. The SINR combining gain of receiving from both base stations, as opposed to just receiving from the scheduling base station, becomes insignificant. Therefore, it is not beneficial[6] for the inferior link to transmit simultaneously as in soft handoff.

In practice, however, the base station with a superior channel gain may not be the one that has downlink packets to send, because link selection is sticky. Indeed, link selection stickiness is more severe in the downlink than in the uplink, because link selection is usually purposely made less sensitive to channel fluctuation and takes place in a slower time scale than multipath fading, for the following reasons.

- First, to avoid a ping-pong effect, the user employs some hysteresis threshold to decide whether to send the Mobile IP registration request message. The threshold is compared with average, rather than instantaneous, channel gains. Thus, link selection may not reflect fast channel fluctuation.
- Second, after the user sends the message to initiate link switching, downlink packets do not immediately move from the old to new base stations. The reasons are that the forwarding route does not change instantaneously due to the latency of signaling through the home agent, that packets remain in flight from the home agent to the old base station, and that the old base station cannot instantaneously forward pending packets to the new base station. Therefore, for some transition time period, both the old and new base stations send packets to the user.

Because of sticky link selection, the base station with an inferior link may have pending packets to send to the user, thereby resulting in suboptimal performance. When the wireless channel fluctuates rapidly, multiuser diversity scheduling helps alleviate the suboptimality by only transmitting to the user when the channel temporarily changes to favorable. Therefore, having distinct packets pending in the old and new base stations is beneficial as it allows them to both exploit channel fluctuation.

One performance limiting factor due to sticky link selection is latency, especially for delay sensitive traffic. To address the latency issue, we next present a special downlink macro-diversity scheme different from the general Mobile IP-based MBB handoff protocol presented in Section 9.3.2. While the scheme is described in the context of supporting voice over IP (VoIP) traffic, the idea can be extended to other real time traffic.

Recall that in Figure 9.10, any downlink packet arrives at either the old or new base station, but not both. In the new scheme, every VoIP packet is made available at both base stations. For example, the primary serving base station that receives packets from the home agent keeps a list of other base stations that also serve the user in MBB handoff. After one base station receives a VoIP packet from the home agent, it forwards the packet to the other base stations on the MBB handoff list. The user compares the quality of the downlink channels from the base stations and selects one as the preferred downlink in a time interval of every 20 ms. Here the scheme assumes that a new VoIP packet arrives

---

[6] Strictly speaking, this statement is not entirely true. It should be pointed out that in the framework of cooperative communications, acting like multiple antenna transmitters, the two base stations transmit simultaneously to achieve a performance gain beyond the traditional SINR combining gain. OFDMA facilitates this new framework in the multiuser scenario. Chapter 10 will discuss cooperative communications in more detail.

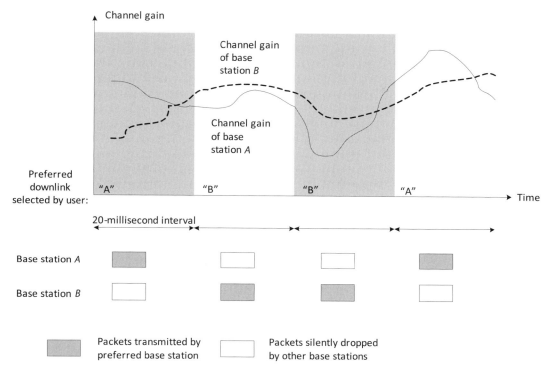

Figure 9.14 Illustration of downlink macro-diversity scheme for VoIP traffic. The MBB handoff list consists of base stations A and B.

from the source roughly every 20 ms, so that the selection of the preferred downlink is valid for one VoIP packet and can change from one packet to another. The user informs all the base stations on the MBB handoff list of its link selection, for example, sending a one-bit ("preferred" or "not preferred") indicator to each base station every 20 ms. Then, only the base station of the preferred downlink manages to transmit the VoIP packet within the current 20 ms interval, and the other base stations silently drop the packet. The downlink macro-diversity scheme for VoIP is illustrated in Figure 9.14.

9.3.5 MBB handoff in an FFR or multi-carrier scenario

So far we have studied MBB handoff where a user maintains simultaneous connections with multiple base stations on the same carrier. Figure 9.15 shows a special example in an FFR deployment where the total bandwidth is divided into two subbands F_1 and F_2. At base station A, F_2 is transmitted at high power and F_1 at low power. Base station B uses the reverse power pattern. When the user moves from base station A to B, it is initially connected with A on both F_1 and F_2, and then only on F_2. The user is then connected with B on F_1, and finally on both F_1 and F_2 when it is close to B. When the user is at the cell boundary between A and B, it is connected with both A (on F_2) and B (on F_1). As studied previously in Figure 9.8, only one RF front-end is required at the

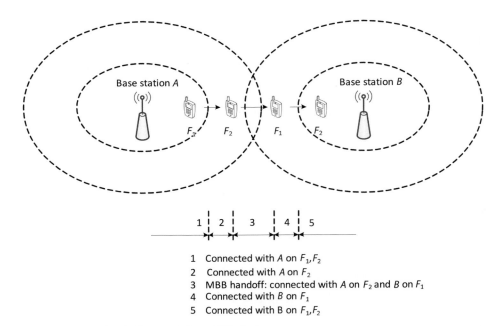

Figure 9.15 MBB handoff scenario in an FFR deployment.

1. Connected with A on F_1, F_2
2. Connected with A on F_2
3. MBB handoff: connected with A on F_2 and B on F_1
4. Connected with B on F_1
5. Connected with B on F_1, F_2

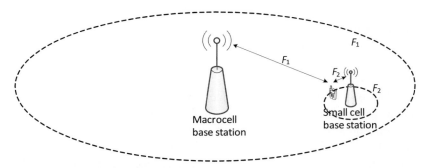

Figure 9.16 MBB handoff scenario in a multi-carrier deployment.

user implementation, as the RF front-end processes the whole bandwidth including F_1 and F_2.

MBB handoff is also possible in a multi-carrier scenario. Figure 9.16 illustrates a multi-carrier deployment example where one carrier F_1 is used to provide wide coverage and another carrier F_2 is to provide capacity enhancement in a hot spot area. Presumably, F_1 is sent from a macrocell base station at high power and F_2 is sent from a small cell (e.g., picocell in Section 10.1.2) base station at low power. The carrier frequency of F_1 may be lower than that of F_2 and thus supports a longer propagation range. In the example of Figure 9.16, the user is equipped with two RF front-end processing chains, and is connected with the macrocell base station on F_1 and simultaneously with the small cell base station on F_2.

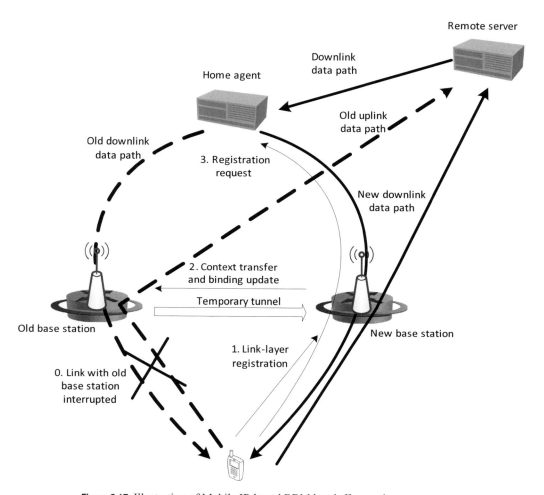

Figure 9.17 Illustration of Mobile IP-based BBM handoff procedure.

9.4 Break-before-make handoff in OFDMA

In some scenarios, MBB handoff may not be possible. For example, because of drastic deterioration of the wireless channel, the existing link with the old base station is interrupted before the connection with the new base station can be established and the protocol steps in Figure 9.10 completed. In this case, the user has to rely on break-before-make handoff, as shown in Figure 9.17.

First, if the old link is interrupted before a temporary tunnel is established, the old base station will immediately start to buffer all the downlink packets for that user. At the same time, the user will attempt to regain connectivity at the same base station or a different one. The user will send the same link layer registration message as in Figure 9.10, indicating the address of the old base station. If the user reestablishes connectivity with the same (old) base station, the base station will start to deliver the buffered packets immediately.

On the other hand, if the user establishes connectivity with a different (new) base station, the user will register its new care-of address with the home agent and redirect the traffic. Meanwhile, the new base station will establish a temporary tunnel between the two base stations. Before the home agent changes the route to the new base station, the packets in flight to the old base station as well as the buffered packets in the old base station can now traverse through the temporary tunnel to be delivered to the user. The tunnel can be established via the binding update, as described in Figure 9.10. As a result, BBM handoff completes with minimum or no packet loss, but at the cost of delay. The length of delay depends on how long it takes for the user to reconnect, and at what stage of handoff the connectivity to the old base station was interrupted. After the new base station starts to receive packets from the home agent and all the packets are tunneled from the old base station (with some time-out window), the temporary tunnel will be torn down.

9.4.1 BBM handoff in an FFR or multi-carrier scenario

In addition to the above example of sudden deterioration of the old link, other scenarios also require BBM handoff. Two obvious examples are (1) the user cannot be connected with more than one base station simultaneously due to hardware or other implementation constraints, and (2) the old and new base stations operate at different carriers and the user hard switches from one carrier to another during handoff (inter-carrier handoff). BBM handoff also occurs in an FFR deployment. Recall from Chapter 6 that in FFR the total bandwidth is divided into subbands, which are reused across cells according to an FFR pattern. As the user moves between cells or even within a cell, it switches from one subband to another, as shown in Figure 9.18.

In base station A, F_1 is used for nearby users and F_2 for faraway ones. The reverse applies in base station B. Initially, the user is close to base station A and uses F_1. As it moves away from base station A and gets close to the new one, it switches to F_2 but still with base station A, and then to F_1 with base station B and finally back to F_2. If the user operates in only one subband at a time, then all the three handoff events are BBM.

Handoff in FFR does not have to be BBM; as shown in Figure 9.15, MBB handoff is possible if the user is able to communicate in two subbands simultaneously during handoff. However, suppose that for the sake of reducing implementation costs, the user communicates in only one subband at a given time. Then handoff from one subband to another becomes BBM, even if it is with the same base station.

Similarly, if the user is equipped with only one RF front-end, then in the multi-carrier deployment of Figure 9.16, the user has to perform BBM handoff between the macrocell base station on F_1 and the small cell base station on F_2.

9.4 Break-before-make handoff in OFDMA

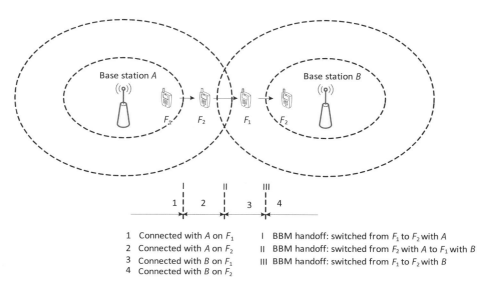

Figure 9.18 BBM handoff scenario in an FFR deployment.

9.4.2 Expedited BBM handoff

The above discussion shows that delay is a critical issue for BBM handoff. Clearly, it is desirable to reduce the delay in BBM handoff. Let us first examine the logical sequence of typical BBM handoff, as shown in Figure 9.19, to see what causes the delay.

The sequence starts when the link with the old base station drops. The user attempts to regain connectivity by searching for a new base station. After the new base station is identified, the user synchronizes to it and then invokes a random access procedure in which the user sends to the new base station a random access probe signal. Upon the reception of the signal, the new base station grants access by assigning airlink resource, such as MAC identifier and dedicated control channel. The user then carries out authentication, security, and configuration procedures via context transfer. The user requests the new base station to set up a temporary tunnel with the old base station to receive the buffered packets forwarded from the old base station. Meanwhile a Mobile IP registration request is sent to the home agent so that the user starts to receive new packets directly from the home agent. Note that binding update and registration request are sent in parallel, so timing between getting tunneled packets from the old base station and packets directly from the home agent is not necessarily as shown in Figure 9.19.

To reduce the delay in BBM handoff, one has to shorten the control signaling steps shown in Figure 9.19. The basic idea of proactive expedited BBM handoff is for the user to anticipate BBM handoff before the old link completely drops, so that the user can carry out the control signaling steps, to the extent possible, over the old link. In Figure 9.20, we illustrate the procedure of expedited BBM handoff. The key areas in which the delay is eliminated or minimized are highlighted in the following:

Handoff in IP-based network architecture

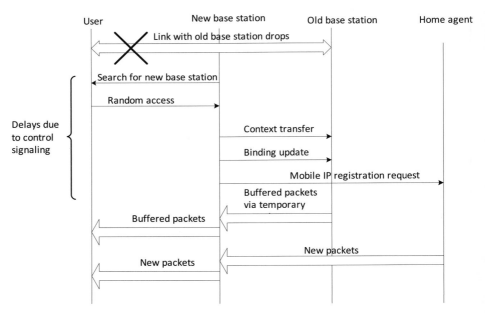

Figure 9.19 Delay analysis of BBM handoff procedure. Only the downlink traffic is illustrated. The user starts to send the uplink traffic after authentication, security, and configuration with the new base station.

- While the link with the current serving (old) base station is still active, the user keeps on monitoring broadcast signals from potential handoff candidate base stations. The user compares the signal strength from a candidate base station with that from the serving base station and initiates handoff if a certain hysteresis test has passed. The details of the broadcast signal design and processing aspect of handoff initiation will be studied in Section 9.5. This eliminates the delay due to searching for new base station in Figure 9.19.
- After the link with the old base station drops, the user gets connected with the new base station through reserved access, as opposed to random access in Figure 9.19. Random access is subject to contention from other users attempting to access the same base station simultaneously. When collision occurs, the user has to backoff in time. Even without collision, additional control signaling, for example, exchanging messages between the user and the new base station, is always required for the base station to check whether collision occurs. Reserved access means that the user is assigned with dedicated airlink resource to connect with the new base station, thereby eliminating those delay components altogether.
- Control signaling above the physical and MAC layers, such as authentication and security, is done between the two base stations via context transfer before the link with the old base station drops. Then, once the user has connected to the new base station with the reserved access, the user skips all those signaling steps with the new base station.

9.4 Break-before-make handoff in OFDMA

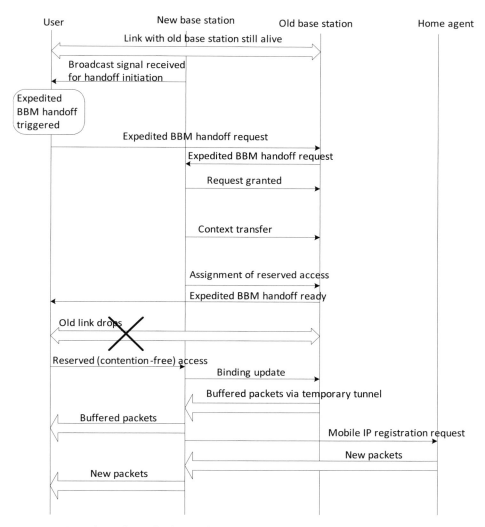

Figure 9.20 Procedure of expedited BBM handoff. Only the downlink traffic is illustrated. The user starts to send the uplink traffic after reserved access with the new base station.

- In Figure 9.20, packet forwarding and route updating are triggered after the user has secured its link with the new base station. Alternatively, the user could even send the Mobile IP registration request over the old link and ask the old base station to start to forward the buffered packets to the new base station, just before it is ready to access the new base station.

It should be pointed out that the user may establish an expedited BBM handoff relationship (context transfer, temporary tunnel) with more than one candidate base station, but seeks access and hands off to only one of them. For this reason, it might be desirable to delay the use of the temporary tunnel until the new link has been established.

The context information and tunnels in the other base stations will expire after some time-out window.

9.5 Handoff initiation

How is handoff triggered? In general, when handoff occurs within the same technology, the handoff decision is based on two factors: wireless channel quality and network loading condition. Not covered in this book, inter-technology handoff may also occur between two different systems, such as cellular in licensed spectrum and Wi-Fi in unlicensed spectrum. In that case, handoff decision is often policy-based.

In this section, we consider handoff initiation due to wireless channel quality. The basic idea is that the user measures broadcast channels from the serving base station and a candidate one, compares their downlink channel quality and initiates handoff once certain hysteresis tests have passed such as the one shown in Figure 9.6. We will ignore the detailed algorithms of hysteresis tests, but rather focus on the design of the broadcast signals to facilitate handoff initiation.

In the following we will first study the universal frequency reuse case and then extend to non-universal frequency reuse including FFR.

9.5.1 The universal frequency reuse case

In the universal frequency reuse case, the downlink signals from both the serving and any candidate base stations are in the same frequency bandwidth. The user receives signals not only from the serving base station, but also from other base stations. The latter signals are so far treated as inter-cell interference in most of the other chapters of this book. However, for the purpose of handoff initiation, they are also desired signals to be detected. In particular, the user processes the received signal to accomplish the following:

- *Cell search*: detect the presence of a candidate base station, and estimate the downlink channel quality.
- *Synchronization*: recover the downlink symbol timing, frequency and symbol clock as well.
- *System information*: decode additional system information, such as the identification of the detected base station.

Since handoff initiation occurs before any connection has been established with the candidate base station, the above tasks have to be based on detection of broadcast signals. One possible choice of broadcast signals is a wideband signal generated by modulating a sequence of complex symbols on many tones in an OFDM symbol. The broadcast OFDM symbol is time multiplexed with other control and traffic OFDM symbols, as shown in Figure 9.21. Different base stations use distinct sequences of complex symbols so that they can be distinguished in cell search.

9.5 Handoff initiation

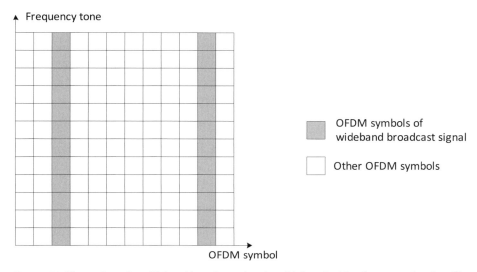

Figure 9.21 Illustration of a wideband broadcast signal multiplexed with other control and traffic signals in a TDM manner.

The design principle of the wideband broadcast signal is to make it easy to detect the presence of the signal, to determine the physical layer identifier (index of the specific broadcast sequence) of the base station, and to estimate the synchronization information, such as symbol timing. Specifically, assume a flat fading channel model

$$y(t) = h \cdot x(t) + w(t), \tag{9.9}$$

where y is the received signal, x the signal from a candidate base station, h the complex channel coefficient and w the noise plus interference. When the signal is the wideband broadcast signal, $x(t) = s_m(t)$ with m the signal sequence index of the corresponding base station. A base station chooses a locally unique m.

Without any prior knowledge of the candidate base station, the user has to detect m non-coherently. Assuming Gaussian noise and unknown but constant h in a duration of one OFDM symbol, it can be shown that the maximum likelihood detection algorithm of finding sequence index m^* and time synchronization parameter l_0^* is to calculate the sample cross-correlation between the received signal and all possible broadcast signal sequences m with any time offset l_0

$$[l_0^*, m^*] = \arg\max_{l_0, m} \left| \sum_{l=0}^{N-1} y_{l+l_0} s_{m,l}^* \right|, \tag{9.10}$$

where the subscript of y represents the index of the time domain samples and N the length of the wideband broadcast signal equal to the number of samples in an OFDM symbol. To control the probability of false alarm, the user only declares the candidate base station with sequence index m^* to be found if the maximum value of the above cross-correlation exceeds some threshold Θ.

Wideband broadcast signal based on Zadoff-Chu sequences

From (9.10), it is desirable that the wideband broadcast signal exhibits low autocorrelation at a non-zero time shift and low cross-correlation between any two sequences. One possible choice is to construct the signal similar to a pseudo-random (PN) sequence used in CDMA. A better choice is to use Zadoff-Chu (ZC) sequences ([185], [27]). Assuming N is a prime, the ZC sequence of prime length N is given by

$$z_u(k) = e^{-j\frac{\pi u k(k+1)}{N}} e^{-j2\pi l \frac{k}{N}}, k = 0, \ldots, N-1, \tag{9.11}$$

where $u \in \{1, \ldots, N\}$ is the root index and l is any integer representing a frequency shift. For simplicity, let $l = 0$. From the perspective of cell search, two important properties of the ZC sequences (see [124]) are

- Cyclic autocorrelation is a delta function, namely

$$\frac{1}{N} \sum_{k=0}^{N-1} z_u(k) z_u^*((k+k_0) \mod N) = \begin{cases} 1, & k_0 = 0 \\ 0, & k_0 \neq 0 \end{cases}. \tag{9.12}$$

- Cyclic cross-correlation is constant and given by

$$\frac{1}{N} \left| \sum_{k=0}^{N-1} z_{u_1}(k) z_{u_2}^*((k+k_0) \mod N) \right| = \frac{1}{\sqrt{N}}, \text{ for } u_1 \neq u_2 \text{ and any } k_0. \tag{9.13}$$

As a comparison, the autocorrelation (at $k_0 \neq 0$) and cross-correlation of a PN sequence have an average absolute value of $\frac{1}{\sqrt{N}}$, but show significant peaks due to random fluctuation.

The ZC sequence is modulated in the frequency domain[7] in an OFDM symbol to form the wideband broadcast signal

$$X_k = z_u(k), k = 0, \ldots, N-1. \tag{9.14}$$

Different base stations can use distinct root indices so that their wideband broadcast signals have small cross-correlation. Moreover, in a synchronous network deployment, neighboring base stations can use the same ZC sequence with distinct cyclic shifts, as long as the cyclic shifts are dimensioned such that the corresponding number of samples exceeds the maximum possible propagation difference of the downlink signals from

---

[7] The careful reader may notice a discrepancy. From the detection algorithm (9.10), the desired autocorrelation and cross-correlation properties are defined in the time domain. However, the ZC sequence is modulated in the frequency domain to form the wideband broadcast signal. How do its properties benefit the detection performance? From [14], it turns out that the DFT of a ZC sequence z_u, denoted by Z_u, is a time-scaled conjugate of the ZC sequence, multiplied by a constant factor

$$Z_u(k) = z_u^*(u^{-1}k) Z_u(0), \tag{9.15}$$

where $Z_u(0) = \sum_{k=0}^{N-1} z_u(k)$ and u^{-1} is the modular multiplicative inverse of u where $u \cdot u^{-1} = 1 \mod N$. Furthermore, $z_u^*(u^{-1}k)$ is itself another ZC sequence of root index u^{-1} with a frequency shift $\frac{1-u^{-1}}{2}$,

$$z_u^*(u^{-1}k) = z_{u^{-1}}^*(k) e^{j2\pi \frac{1-u^{-1}}{2} \frac{k}{N}}. \tag{9.16}$$

Therefore, the wideband broadcast signal constructed by modulating a ZC sequence in the frequency domain possesses the same desired auto-correlation and cross-correlation properties in the time domain.

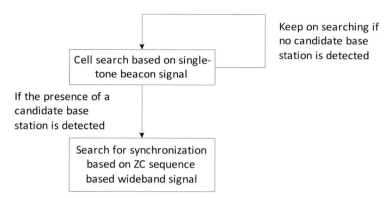

Figure 9.22 Block diagram of two-step search: cell search using a single-tone beacon signal and synchronization using a ZC sequence-based wideband signal.

neighboring base stations arriving at the user receiver. The zero autocorrelation property ensures that the wideband broadcast signals from those base stations are mutually orthogonal. Hence, the sequence index m is completely defined with root index u and cyclic shift.

Single-tone beacon signal

The cross-correlation metric (9.10) is commonly used in cell search and synchronization. One example is the RAKE receiver in CDMA. However, a main drawback is the computational complexity, as the algorithm requires sample level or even subsample level search [168]. The autocorrelation function peaks at zero time shift and drops rapidly with a time shift of only one sample, a highly desired property to estimate symbol timing but in the meantime forcing the receiver to examine a large number of possible time shifts.

One idea for reducing the computation burden is to separate cell search from synchronization. Specifically, a base station sends two separate broadcast signals, one for cell search and the other for synchronization. The user keeps on searching for the first signal in an attempt to detect the presence of a candidate base station. Once the presence is detected, a fine-grained search is then invoked to obtain synchronization based on the second signal. The two-step approach is illustrated in Figure 9.22. The advantage is that since synchronization does not run all the time, its complexity becomes less burdensome to the overall implementation.

Apparently the ZC sequence-based wideband signal is ideal for synchronization. What is a suitable broadcast signal for cell search? We would suggest the following two requirements:

- *Detectability at low SINR*. In either MBB or expedited BBM handoff, it is desirable to detect a candidate base station earlier rather than later in case the wireless channel fluctuates drastically and causes the existing link to drop suddenly. For example, in expedited BBM handoff, detecting a candidate base station earlier allows additional time budget to make a request to the detected base station, transfer context, and set

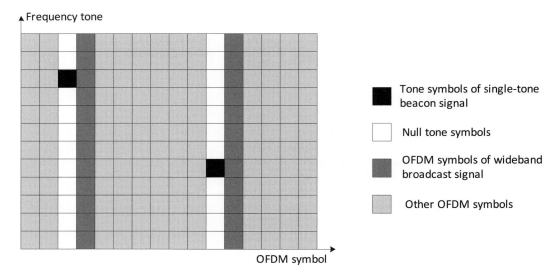

Figure 9.23 Illustration of a single-tone beacon signal and a wideband broadcast signal time multiplexed with other control and traffic channels. The beacon signal is for cell search and the wideband signal for subsequent synchronization in the two-step search approach.

up a temporary tunnel. This requires the capability of detecting at low SINR. Note that when the user tries to detect the presence of a candidate base station, the signal from the serving base station acts as interference, so the SINR of such detection can be quite low.

- *Low processing complexity.* Since the user does not know beforehand when a candidate base station will show up, it has to keep on searching for potential candidates in the background while processing the signal from the serving base station in the foreground. Therefore, the complexity of detection algorithms has significant impact on implementation costs.

In summary, a broadcast signal should penetrate deeply into neighboring cells and be easy to detect. The *single-tone beacon* signal presented next satisfies the above requirements.

Unlike a general OFDM signal we have seen so far, which distributes the transmit power among many tones, the single-tone beacon is a special OFDM signal where the power is concentrated on only one tone and all the other tones, called null tones, are left with no power

$$X_k = \begin{cases} A, & k = k_0 \\ 0, & \text{otherwise} \end{cases}, \tag{9.17}$$

where k_0 is the index of the beacon tone and $|A|^2$ is the beacon power. The beacon signal is designed to be the first signal to be detected from a candidate base station in cell search. The detection of the beacon signal is non-coherent, and the phase of A is of no interest here. The single-tone beacon is illustrated in Figure 9.23.

The index m of the base station is signaled with the beacon tone indices. In the simplest form, a base station takes a fixed and locally unique frequency for its beacon tone (constant k_0). However, such a scheme lacks frequency diversity and moreover leads to detection errors if the user has a large frequency offset from a candidate base station. A more sophisticated scheme is to let the beacon tone hop over frequency in different OFDM symbols, as shown in Figure 9.23. Over time, the beacon tones form a beacon sequence: $k_{0,n}, k_{0,n+1}, k_{0,n+2}, \ldots$, where $n, n+1, n+2, \ldots$ are time indices of the beacon signal and each element of the sequence belongs to the set of $\{0, \ldots, N-1\}$. Neighboring base stations use distinct beacon sequences. To detect the presence of a candidate base station and to further determine its index, the user does not rely on a single beacon tone but attempts to decode the beacon sequence over time. The design of beacon sequences is in essence a special coding problem. Unlike traditional coding, the beacon sequence does not have a specific starting point, and the ending point varies depending on when the user has successfully decode the sequence. With appropriate coding design, the use of beacon sequences overcomes the drawbacks of this simple scheme using a fixed beacon tone frequency.

The single-tone beacon signal is very different from a general wideband OFDM signal. In particular, the autocorrelation of the beacon signal decays only linearly as time shift increases. Therefore a fine-grained search using *sample-level* correlation (9.10), which checks all possible time shifts $l_0, l_0 + 1, \ldots$, is not very effective. Instead, we can adopt a much coarser *symbol-level search*, where the receiver only checks the correlation at time shift equal to $l_0, l_0 + N, l_0 + 2N, \ldots$, for some l_0. Figure 9.24 compares the correlation windows of sample-level and symbol-level search. Here, a correlation window is of the length of the prototype beacon signal, which is of one OFDM symbol. Clearly, compared with sample-level search, symbol-level search reduces the computation by a factor of N.

Moreover, to obtain the correlation with single-tone signals at all the possible frequencies, an FFT operation is far more efficient when N is large. Specifically, while the user is taking FFT and receiving OFDM symbols from the serving base station, the FFT output at tone k is treated as the correlation metrics of the received signal and the prototype beacon signal with $k_0 = k$. Therefore, no additional FFT operation is required. Figure 9.24(b) shows that the correlation metrics in symbol-level search are obtained with FFT in successive OFDM symbols. The correlation metrics are then fed into the decoding module of the beacon sequences to determine whether any candidate is present, and if so, its index.

Note that in symbol-level cell search, because the user receiver has not yet been synchronized with the downlink symbol time of the candidate base station, it is likely that any correlation window only partially overlaps with the beacon signal. To catch a whole OFDM symbol while keeping the simplicity of symbol-level search, one can cyclically extend the beacon signal to two OFDM symbols, as shown in Figure 9.25.

So far we show that the use of single-tone beacon signals reduces processing complexity in cell search. We next examine the advantages of beacon signals in terms of detectability at low SINR.

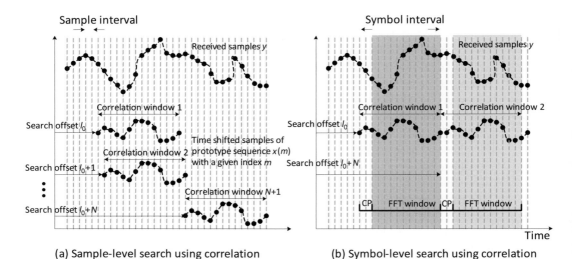

(a) Sample-level search using correlation (b) Symbol-level search using correlation

Figure 9.24 Comparison of correlation windows in sample-level and symbol-level search. A black dot represents a sample. The prototype samples represent the signal waveform to be searched for. Sample-level search is suitable for a wideband broadcast signal (9.14), and symbol-level search for a beacon signal (9.17).

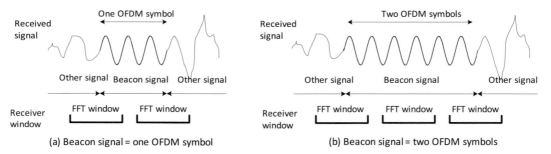

(a) Beacon signal = one OFDM symbol (b) Beacon signal = two OFDM symbols

Figure 9.25 Comparison of beacon signals with duration of (a) one OFDM symbol and (b) two OFDM symbols. In the absence of symbol synchronization, no single FFT window captures a whole OFDM symbol of the beacon signal in (a), but in (b) at least one of successive FFT windows captures a whole OFDM symbol.

First, as a single-tone signal, the beacon signal has a very low PAPR and therefore can boost its transmit power for a given power amplifier. Recall from Section 2.3, a signal with high PAPR requires a large power backoff, meaning that the average transmit power is backed off from the peak power allowed by the maximum linearity of the power amplifier. The PAPR of a general OFDM signal is known to be high, and a 10 dB power backoff is not unusual in practice. On the other hand, the beacon signal can be transmitted with a boosted power because of its low PAPR advantage. Specifically, denote by P_{OFDM}, P_{beacon} the transmission powers of a general OFDM signal, such as the wideband broadcast signal, and the beacon signal

$$P_{\text{OFDM}} = \sum_{k=0}^{N-1} |X_k|^2, \qquad (9.18)$$

$$P_{\text{beacon}} = |A|^2, \qquad (9.19)$$

where A is given in (9.17). We define power boost P_{boost}

$$P_{\text{boost}} = \frac{P_{\text{beacon}}}{P_{\text{OFDM}}}. \qquad (9.20)$$

Depending on the characteristics of the power amplifier, P_{boost} can be a few dB without increasing the out-of-band emission [121]. The power boost is called *PAPR gain*. Of course, the wideband broadcast signal can be designed, for example, using ZC sequences, to have a lower PAPR than a general OFDMA signal. Nonetheless, a single-tone signal has the ideal optimal PAPR.

Second, because sample-level search checks N times as many time-shifted correlations as symbol-level search, a given false alarm probability would require a relatively higher detection threshold. As a result, a higher SINR is needed to obtain the same missed detection probability.

To be specific, consider the model in (9.9). Assume that the channel h is constant and flat, and noise plus interference w Gaussian with power $\sigma_0^2 = \mathbb{E}[|w|^2]$. The SINR is defined to be

$$\gamma = \frac{|h|^2 \mathbb{E}[|x|^2]}{\sigma_0^2}. \qquad (9.21)$$

The detection problem is to distinguish two hypotheses about a particular candidate base station: $\mathcal{H}_0$ it is absent and $\mathcal{H}_1$ present. At any time shift l_0 to be checked, the detection algorithm declares

$$\begin{cases} \mathcal{H}_1, & \text{if } \frac{1}{\sqrt{N}} \left| \sum_{l=0}^{N-1} y_{l+l_0} s_l^* \right| > \Theta, \\ \mathcal{H}_0, & \text{otherwise} \end{cases} \qquad (9.22)$$

with Θ the detection threshold. Here, for simplicity, we assume that the target sequence m is known and thus drop the subscript m in prototype signal samples $s_{m,l}^*$ to become s_l^*; in reality, the user has to try out many possible sequence indices.

In the sample-level search, when the time shift l_0 is not aligned between prototype sequence s_l^* and the wideband broadcast signal in the received signal y, the correlation between $h \cdot x$ and s^* is approximately complex Gaussian with zero mean and power $\sigma_0^2 \gamma$. Thus the metric $\frac{1}{\sqrt{N}} \left| \sum_{l=0}^{N-1} y_{l+l_0} s_l^* \right|$ can be modeled as a Rayleigh random variable with pdf

$$p_0(v) = \frac{v}{\sigma_0'^2/2} e^{-\frac{v^2}{\sigma_0'^2}}, \qquad (9.23)$$

with $\sigma_0'^2 = \sigma_0^2(1+\gamma)$. When aligned, all the signal power is coherently combined in the metric such that the metric can be modeled as a Ricean random variable with pdf

$$p_1(v) = \frac{v}{\sigma_0'^2/2} e^{-\frac{v^2+S^2}{\sigma_0'^2}} I_0\left(\frac{Sv}{\sigma_0'^2/2}\right), \qquad (9.24)$$

where $I_0(\cdot)$ is the zeroth order modified Bessel function of the first kind, and $S^2 = N\sigma_0^2\gamma$ representing the coherently combined energy of the wideband broadcast signal. The detection threshold Θ_N is set such that the false alarm probability in a time interval of one symbol,[8] which consists of N time shifts to be checked in (9.22), does not exceed a target value P_F. Assuming the metrics of those N tests are i.i.d., Θ_N is solved as follows:

$$1 - \left[\int_0^{\Theta_N} p_0(v)\,dv\right]^N \approx Ne^{-\frac{\Theta_N^2}{\sigma_0'^2}} = P_F \qquad (9.25)$$

To simplify the analysis of the symbol-level search with the single-tone beacon signal, we assume that the beacon signal consists of two OFDM-symbols as shown in Figure 9.25(b) and ignore the FFT windows that only overlap partially with the received beacon signal.[9] Thus, the statistical model of the metrics when the beacon signal is absent or present in the FFT window is the same as (9.23) and (9.24), respectively. The detection threshold Θ_1 can be solved to obtain the same false alarm probability target P_F

$$1 - \int_0^{\Theta_N} p_0(v)\,dv = e^{-\frac{\Theta_1^2}{\sigma_0'^2}} = P_F. \qquad (9.26)$$

It is easy to see that $\Theta_1 < \Theta_N$ and

$$\Theta_N^2 - \Theta_1^2 = \log(N)\sigma_0'^2. \qquad (9.27)$$

Figure 9.26(a) plots Θ_1/σ_0', Θ_N/σ_0' as P_F decreases. Furthermore, given Θ_1, Θ_N, Figure 9.26(b) compares the required γ with respect to the target missed detection probability, given by $\int_0^{\Theta} p_1(v)\,dv$, with $\Theta = \Theta_1$ or Θ_N, respectively. We observe that for the same false alarm and missed detection probabilities, the symbol-level search requires roughly 2 dB fewer SINR.

In summary, combining the advantages of single-tone signals and symbol-level search, a beacon signal allows cell detection at a lower SINR than a general OFDM wideband signal, thereby facilitating handoff initiation as the user moves from one cell to another.

---

[8] To further reduce the false alarm probability, the user can exploit the channel structure that the wideband broadcast signal is sent repetitively with a fixed period and the beacon signal follows certain hopping patterns. For a simple back-of-envelope analysis, here we ignore those properties.

[9] With sophisticated signal processing algorithms, those partially overlapping FFT windows can be used to reduce the missed detection probability.

9.5 Handoff initiation

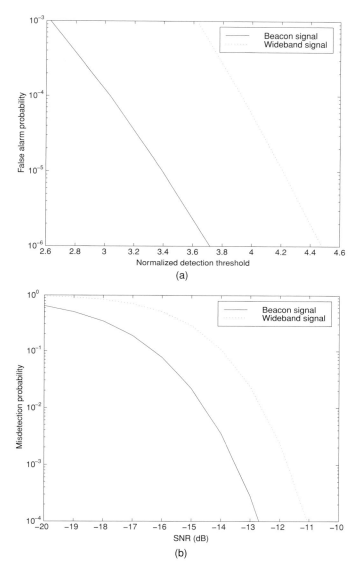

Figure 9.26 Comparison of false alarm probability versus required normalized detection threshold in (a) and missed detection probability versus required SINR given false alarm probability equal to 10^{-3} in (b). $N = 512$. Normalized detection thresholds are defined to be Θ_1/σ'_0, Θ_N/σ'_0, respectively.

Practical example 9.1 Flash signaling in Flash-OFDM

The on-off keying (9.17) is the simplest form of *flash signaling* studied in [167]. An input signal x is defined to be flash signaling if for all $\nu > 0$,

$$\lim_{\gamma \to 0} \frac{\mathbb{E}(|x|^2 \mathbf{1}_{\{|x|>\nu\}})}{\mathbb{E}(|x|^2)} = 1, \qquad (9.28)$$

where γ is the SNR. From the above definition, flash signaling exhibits a very special probability distribution where almost all its mass is concentrated at $x = 0$ and a tiny mass at $|x|^2 \to \infty$ and that tiny mass drops to zero asymptotically so that the power $\mathbb{E}(|x|^2)$ satisfies the SNR constraint γ. It is shown in [167] that for a general class of channels, flash signaling is optimal in the wideband low SNR regime if the channel is not fully known at the receiver. The optimality is defined in the sense of achieving $(E_b/N_0)_{\min}$ given by

$$\left(\frac{E_b}{N_0}\right)_{\min} = \lim_{\gamma \to 0} \frac{\gamma}{C(\gamma)}, \qquad (9.29)$$

where $C(\gamma)$ represents the channel capacity at SNR γ. In other words, flash signaling achieves the minimum received signal energy per information bit in the non-coherent regime.

The above property makes flash signaling suitable for use in a broadcast channel to facilitate initial cell search. As a special form of flash signaling, a beacon signal similar to what is shown in Figure 9.23 is employed in Flash-OFDM. Note that the beacon signal occupies a large bandwidth (the whole OFDM symbol in this case), although the signal energy is concentrated on only one tone. In particular, assuming orthogonal signaling, the amount of bandwidth grows exponentially with the number of bits representing the index of a base station. However, the remaining tones in that OFDM symbol do not have to go unused, as shown next.

Recall the idea of superposition-by-position coding in Section 5.2.3. Here we treat the beacon signal as X_1 intended to inferior users and in addition transmit another signal in the same OFDM symbol as X_2 for superior users. In Flash-OFDM, the two OFDM symbols labeled for "single-tone beacon signal" and "wideband broadcast signal" in Figure 9.23 are superposed on each other in a single OFDM symbol. The total transmit power is split into two, one part for the beacon signal and the other for the wideband signal. A user first searches for the beacon signal, and then processes the other tones once the beacon signal is located.

This idea of combining flash signaling and superposition-by-position coding is used in other downlink channels in Flash-OFDM. As one example, the assignment channel is used to transmit two assignment messages, a first one (8 bits) for users everywhere in a cell and the second one (28 bits) only for users close to the base station. Figure 9.27 shows the physical resource block, which consists of 64 tone-symbols divided into eight groups of eight tone-symbols each. The second message is coded using a convolutional code at rate $\approx 1/4.5$. The coded bits are converted to 64 QPSK modulation symbols, which are mapped to the entire physical resource block. The first message is coded using flash signaling where 3 bits determine which group to spend the transmit power, and the other 5 bits determine the phases of the selected 8 tone-symbols. There are totally 32 phase sequences constructed by multiplying 8 length-8 orthogonal Walsh codes with $1, -1, j$, and $-j$, respectively. The selected 8 tone-symbols finally puncture into the physical resource block to mix with the convolutional codeword.

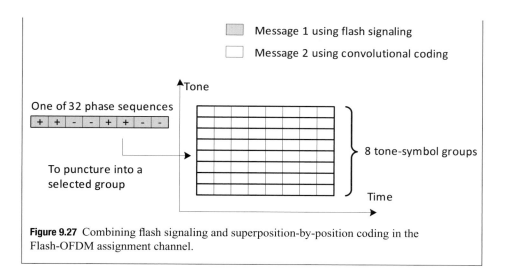

Figure 9.27 Combining flash signaling and superposition-by-position coding in the Flash-OFDM assignment channel.

Practical example 9.2 Handoff in a railway Flash-OFDM network

We next describe the handoff design of a railway wireless communication system using Flash-OFDM. A train is equipped with a Wi-Fi access point to provide wireless connectivity to passengers and uses a Flash-OFDM user modem to communicate with base stations as the backhaul channel. The base stations are deployed along the railway line, as shown in Figure 9.28(a). As the train moves quickly from one base station to another, handoff has to be fast enough to be seamless.

Unlike other cellular deployments, here the number of base stations exceeds the number of "users" (trains). A typical scenario is that at a given time a base station serves only one train while there is no train in the neighboring cells at all. Those cells are kept inactive. In order to minimize the interference from those inactive cells, the corresponding base stations do not transmit any conventional broadcast channel or even pilot except for a very infrequent single-tone beacon. The duty cycle of the beacon signal is about 0.2 percent, as a beacon is transmitted once every about 100 ms with a duration of roughly 200 μs (two OFDM-symbols). Moreover, a beacon concentrates its energy in one out of some 100 tons. Thus the inter-cell interference from the beacon signal is negligible.

Yet, since the beacon signal penetrates deeply into neighboring cells, a user can detect the presence of a handoff candidate base station well before it crosses the cell boundary. If the user has decided to hand off to a neighboring base station, the user sends a random access signal to the base station with the symbol timing slaved to the detected beacon signal, as shown in Figure 9.28(b). The base station always monitors the access channel in the uplink after the beacon signal. Once a random access signal is detected, the base station immediately turns on the broadcast channel and pilot in the downlink to allow the user to complete the handoff procedure and stay connected.

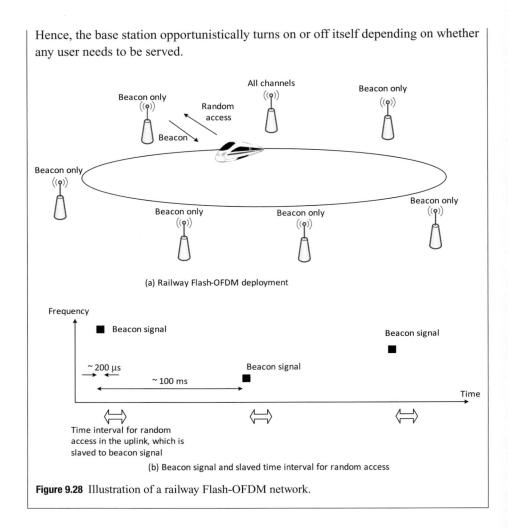

Figure 9.28 Illustration of a railway Flash-OFDM network.

9.5.2 The non-universal frequency reuse cases

Now let us extend our study to the non-universal frequency reuse cases. We focus on FFR and the discussion is applicable to the reuse factor $K > 1$ cases.

In FFR where every base station transmits in all the tones but at different power levels in subbands, there are a few possible ways of sending the beacon or wideband broadcast signals:

- Send the broadcast signals in a fixed subband, for example, the one in the middle of the bandwidth. The transmit power is set to a fixed value. The broadcast signals from neighboring cells interfere with each other and there is no FFR power coordination benefit.

9.5 Handoff initiation

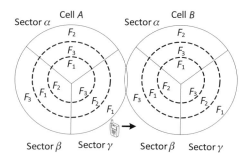

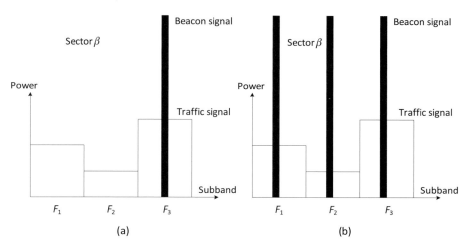

Figure 9.29 Illustration of beacon signals in FFR. The power profile of sector β is shown in (a) and (b). In (a), the beacon signals are sent only in the primary subband and in (b) beacon signals are sent in every subband. The power profile of sectors α and γ can be obtained similarly from Figure 6.21(b).

- Send the beacon and wideband broadcast signals in the subband in which the largest transmit power is applied according to the FFR power scaling rule. This is illustrated in Figure 9.29(a). In this case, neighboring base stations send their beacon and wideband broadcast signals in different subbands, thereby avoiding interfering with each other.

These schemes work as long as the user employs a wideband receiver so that it receives signals in all the subbands from all the neighboring base stations. However, it should be pointed out that the idea of FFR makes it possible to accommodate a narrowband user, which only transmits and receives in one of the subbands at any given time and can switch from one subband to another as needed. The obvious advantages are low hardware/software implementation costs and low battery power consumption. A narrowband user would not be able to detect the broadcast signals of a candidate base station unless the signals happen to be sent in the same subband in which the user is currently operating.

Figure 9.29 illustrates a scenario where the narrowband user moves from cell A to cell B. When the user is close to cell edges, it operates in subband F_1. In (a), the beacon and wideband broadcast signals of the candidate base station in cell B are only sent in subband F_3, and as a result, the user will miss the signals and fail to trigger handoff when it crosses the cell boundary. In order to detect the signals, the user may have to periodically suspend its operation in F_1 and switch to the other subbands (F_2, F_3) to detect the presence of any handoff candidate. How frequently the user checks the other subbands depends on the tradeoff between the length of service disruption and the delay in detecting a handoff candidate.

Figure 9.29(b) shows an alternative scheme, *beacon insertion*. The idea is to send a beacon sequence in every subband at the maximum transmit power. Now, while operating in F_1, the user is able to detect the beacon signal from cell B and therefore the presence of the candidate base station. The user can then initiate the expedited BBM handoff procedure. By comparing the strengths of the beacon signals from the two base stations, the user figures out an appropriate time to switch to the new base station in the new subband. In essence, with beacon insertion, the beacon signals are sent with universal frequency reuse, while the traffic signals follow the FFR power scaling rule. The beacon signals from neighboring cells are sent at different frequency tones so as to avoid inter-cell interference.

9.6 Mobile-controlled versus network-controlled handoff

To conclude this chapter, we briefly compare two basic handoff control approaches: mobile-controlled or network-controlled.

In mobile-controlled handoff, the user decides when handoff is to take place and to which base station, with possible assistance from the base stations. In contrast, in network-controlled handoff, the user measures the relative signal strengths from neighboring base stations and reports the measurement results to the serving base station or a network controller that controls a number of base stations, and it is up to the base station or network controller to make handoff decision.

The handoff procedures we have presented so far are fundamentally mobile-controlled, although it is possible to modify them to be network-controlled. On a very high level, mobile-controlled handoff is consistent with the Internet architectural principle of pushing intelligence to the edge. The user possesses a rich set of intelligence to make handoff decision. In particular, the user has the most updated and complete picture of channel and interference. Transporting the information to the network results in signaling overhead and delay. Moreover, the user has become increasingly sophisticated. In particular, the "multi-mode" feature with multiple airlink technologies, such as cellular and Wi-Fi, is increasingly common among smart phones. Inter-technology handoff is generally policy-based, and the user is the desired entity that has the whole picture for inter-technology handoff. The network of one technology usually has no control over another different technology.

9.6 Mobile-controlled vs network-controlled

A major advantage of network-controlled handoff is that the network can better balance traffic loadings in adjacent cells. However, if the base stations broadcast their current loadings and the user at the cell boundary makes handoff decision using the loading information, then mobile-controlled handoff can also balance traffic loadings to some extent.

> **Practical example 9.3** Cell search and random access in LTE handoff
>
> In LTE, there are two types of handoff, namely idle handoff when the UE is in the IDLE state and active handoff when the UE is in the CONNECTED state. Idle handoff is UE controlled while active handoff is network controlled. In the following, we focus on active handoff.
>
> Appendix A.5 describes the general procedure of handoff. In LTE, cell search is controlled by the eNB. Specifically, if the eNB decides that the UE needs to perform cell search, it will provide the UE with monitoring gaps during which the communications with the serving eNB are suspended and the UE performs intra-frequency monitoring (measurements on neighboring LTE cells using the same carrier frequency), inter-frequency monitoring (measurements on neighboring LTE cells using a different carrier frequency) and inter-RAT monitoring (measurements on another Radio Access Technology such as GSM or UMTS). The duration of a monitoring gap is 6 ms, which is slightly longer than the 5 ms interval when the PSS and SSS repeat. Taking into account switching overhead, the UE is able to capture a PSS/SSS of neighboring LTE cells in a monitoring gap. A monitoring gap occurs every several 10 ms frames and the gap periodicity is configurable to trade off between cell search performance and monitoring overhead. The network makes a handoff decision and instructs the UE to access the new eNB. Once the UE is connected with the new eNB, upper layer mobility management messages are exchanged over the traffic channel to complete handoff. In the following, we briefly describe cell search and random access in LTE handoff.
>
> **Cell search**
> A UE carries out cell search not only for handoff but also at power up. Cell search is used to acquire frequency, symbol, and frame synchronization with a candidate eNB, as well as the physical layer cell identifier and other system information. Appendix A.1 describes the general procedure of cell search.
>
> In LTE, there are 504 distinct physical layer cell identifiers, which are divided into 168 cell-identifier groups each consisting of three cell identifers. Denote by N_{ID} a physical layer cell identifier. Let $N_{ID} = 3N_{ID}^{(1)} + N_{ID}^{(2)}$, where $N_{ID}^{(1)} = 0, 1, \ldots, 167$, $N_{ID}^{(2)} = 0, 1, 2$. Every eNB uses a locally unique physical layer cell identifier, which is used in generating various cell-specific physical layer signals. The cell identifier of an eNB is conveyed in two downlink broadcast signals; the Primary Synchronization Signal (PSS) for $N_{ID}^{(2)}$ and Secondary Synchronization Signal (SSS) for $N_{ID}^{(1)}$.

Figure 9.30 shows the physical resource of the PSS and SSS. In a 10 ms frame, the PSS is sent in the last OFDM symbol of the first slot of subframes 0 and 5, and the SSS is sent one OFDM symbol prior to the PSS. Both the PSS and SSS occupy the middle 6 PRBs in the frequency, irrespective of the total system bandwidth. 6 PRBs represent the usable bandwidth of a 1.4 MHz system, which corresponds to the minimum bandwidth configuration in LTE. When the system bandwidth is greater than 1.4 MHz, other PRBs in those two OFDM symbols are used by the PDSCH channel.

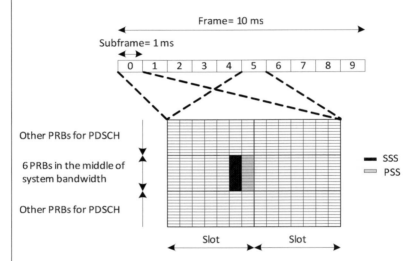

Figure 9.30 Illustration of the PSS and SSS in LTE. FDD frame structure 1 and normal CP.

The modulation symbols mapped to the tone-symbols of the PSS physical resource are generated from a Zadoff-Chu sequence using one of three distinct root sequence indices corresponding to $N_{ID}^{(2)} = 0, 1, 2$, respectively. As noted in Section 9.5, a Zadoff-Chu sequence exhibits low PAPR thereby improving coverage, and good auto-correlation and cross-correlation properties thereby improving receiver performance. The modulation symbols of the SSS are constructed by interleaving and concatenating a pair of distinct cyclic shifts of a m-sequence. Similar to Zadoff-Chu sequences, m-sequences also exhibit good frequency domain properties (spectrally flat) and time domain properties (low cross-correlation). The pair of cyclic shifts (s_1, s_2) uniquely represents $N_{ID}^{(1)}$.

The first step of cell search is to acquire $N_{ID}^{(2)}$ and symbol and frequency synchronization from the PSS. Without any prior knowledge of the candidate eNB, detection of the PSS is non-coherent. After the UE has locked to the PSS, the second step is to acquire $N_{ID}^{(1)}$ and frame synchronization from the SSS. Acquisition of the SSS is done coherently since the acquired PSS can be used for channel estimation.

Note that within a 10 ms frame, the two PSS are identical. Once a UE locates the position of a PSS, it gets to know the 5 ms (half) frame structure but cannot tell whether the PSS is in subframe 0 or 5. To acquire the 10 ms frame structure, the UE relies on the SSS because the pair of cyclic shifts is different between subframes 0 and 5. In particular, cyclic shift pairs (s_1, s_2) and (s_2, s_1) is used to construct the SSS in subframes 0 and 5, respectively. The SSS sequence design ensures that if the combination of (s_1, s_2) in subframe 0 and (s_2, s_1) in subframe 5 is valid, then the combination of (s_2, s_1) in subframe 0 and (s_1, s_2) in subframe 5 is invalid. Therefore, the structure of distinct pairs of cyclic shifts allows the UE to synchronize with the 10 ms frame of the downlink.

After the acquisition of the PSS and SSS, the UE proceeds to receive other system information such as the downlink and uplink system bandwidth, number of transmit antennas, random access channel configuration. The system information is conveyed in Physical Broadcast Channel (PBCH) or PDSCH.

Random access

After the UE acquires synchronization and system information from the eNB in cell search, the UE proceeds to the step of random access.

Random access is used in a general scenario where a UE intends to connect to an eNB when the UE has not been connected, has lost its connection, or has been out of uplink synchronization. Appendix A.2 describes the general procedure of random access. In most cases, random access is contention-based, and in LTE includes four steps shown in Figure 9.31(a).

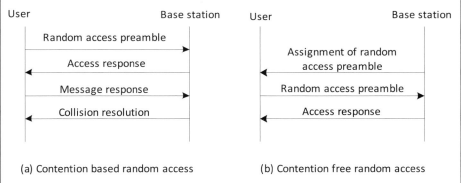

Figure 9.31 Four steps of random access.

- First, the UE transmits a random access preamble using Physical Random Access Channel (PRACH), which allows the eNB to detect the presence of the access request and to estimate the signal arrival time from the UE. In contention-based random access, multiple UEs may send random access preambles simultaneously, resulting in potential collision.
- Second, upon the detection of a random access preamble, the eNB transmits an access response, which includes a timing advance command to control the transmit

symbol time of the UE. In addition, the eNB assigns uplink channel resource (temporary identifier and PUSCH resource grant) for the UE to further exchange access information. In the case of a collision, the eNB may detect the occurrence of the collision and instruct the UEs to back off before any subsequent retry. However, the eNB may only detect the strongest preamble signal and not be aware of the occurrence of collision.

- Third, after transmitting a random access preamble, the UE keeps on monitoring the downlink to receive an access response corresponding to its random access preamble. If a response is not received within a time window, the UE transmits another random access preamble and the above steps repeat. Otherwise, the UE adjusts its transmit symbol time to be uplink synchronized, and in the assigned PUSCH resource transmits a message response including the UE identity and other information depending on the random access scenario.
- Fourth, after receiving the message response, the eNB transmits a collision resolution message, which echoes the UE identity. The purpose of this step is to resolve contention caused by multiple UEs sending the same random access preamble in the first step. The UE declares the random access succeeds if the UE identity in the message matches its own.

In some cases such as expedited handoff, random access can be contention free, shown in Figure 9.31(b), because a random access preamble has been assigned to the UE beforehand and will not be used by other UEs. Without the concern of collision, only the first two steps are needed and the last two steps can be skipped.

The design of the PRACH is different from any other uplink channels such as PUCCH or PUSCH in LTE. The reason is that the PRACH channel is used by UEs that have not been uplink synchronized yet. All the other channels are used by UEs already synchronized and therefore orthogonal with each other. The relationship between OFDMA orthogonality and uplink synchronization is discussed in Section 2.2. We next briefly describe the PRACH.

After cell search, the UE is synchronized to the downlink of the eNB and determines its transmit time according to open-loop timing control described in Section 2.2. Figure 2.17 illustrates that the uplink signal from the UE arrives at the eNB subject to the round trip delay. Thus, the PRACH signal arrives earlier if the UE is nearby than if the UE is faraway, with the maximum timing difference equal to the maximum round trip delay. How should the PRACH signal be designed to facilitate detection at the eNB given the uncertainty of arrival time?

One solution is to make the PRACH a single OFDM symbol whose cyclic prefix is long enough to cover the maximum round trip delay plus maximum delay spread. This way, the eNB receiver can use a fixed FFT window to capture the PRACH signal irrespective of where the UE is, as illustrated in Figure 9.32. A guard time is added to avoid inter-symbol interference (ISI) spilling over subsequent OFDM symbols.

Note that as a comparison, the cyclic prefix used in the other channels is only intended to cover the maximum delay spread. Since the maximum round trip delay is usually much greater than the maximum delay spread, the PRACH uses an excessive cyclic prefix overhead, which is the cost of non-orthogonal communications prior to uplink synchronization. For this reason, the number of information bits to be conveyed in the PRACH should be kept to a minimum. Indeed in LTE the UE identity is not sent in the PRACH but in the PUSCH after the UE is uplink synchronized.

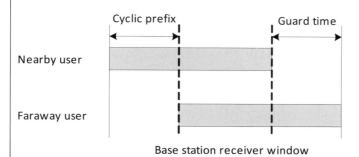

Figure 9.32 eNB window that captures the PRACH signal irrespective of propagation delay.

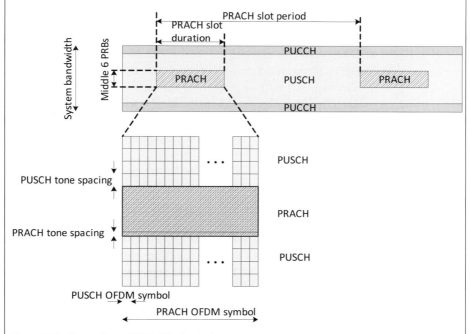

Figure 9.33 Illustration of PRACH physical resource.

Figure 9.33 shows the physical resource of the PRACH that implements the above design idea. It occupies the middle 6 PRBs in the system bandwidth over a PRACH slot duration in every PRACH slot period, which is configurable. In a PRACH slot, the PRACH signal consists of a special long OFDM symbol followed by a guard time, and the parameters depend on the PRACH formats summarized in the following table. The tone spacing and cyclic prefix of the special PRACH OFDM symbol are very different from the counterparts of a normal OFDM symbol of the other uplink channels. Specifically, the tone spacing of a normal OFDM symbol is equal to $\Delta_f = 15$ kHz. The tone spacing of the special PRACH OFDM symbol is given by $\Delta'_f = \Delta_f/12$ or $\Delta_f/24$, where the FFT window is doubled from 0.8 ms to 1.6 ms in the latter choice so as to boost the signal energy captured in the receiver window and increase the PRACH link budget. The cyclic prefix is roughly equal to the guard time, which depends on the cell size and is equal to the maximum round trip delay to be supported. The length of a PRACH slot duration is equal to 1, 2, or 3 subframes to support the various configurations of the OFDM symbol length.

Format	No. of Subframes	Tone spacing	Cyclic prefix (ms)	Guard time (ms)	Cell size (km)
0	1	$\Delta_f/12$	0.1	0.1	15
1	2	$\Delta_f/12$	0.68	0.52	77
2	2	$\Delta_f/24$	0.2	0.2	30
3	3	$\Delta_f/24$	0.68	0.72	100

The modulation symbols to be mapped to those PRACH tones, referred to as a PRACH sequence, are constructed from a Zadoff-Chu sequence because of its low PAPR and good auto-correlation and cross-correlation properties. It is desirable that the number of the PRACH tones (N_{ZC}) is a prime number. In the case of $\Delta'_f = \Delta_f/12$, the bandwidth of 6 PRBs provides $N_{ZC} \leq 864$. In LTE, N_{ZC} is selected to be 839. The remaining $(864-839)\Delta'_f = 31.25$ kHz is used as the guard bands (15.625 kHz on each side) between the PRACH and the adjacent PUSCH to reduce inter-carrier interference (ICI).

In a cell, there are 64 distinct PRACH sequences for the UE to select from to generate the PRACH signal in the first step in Figure 9.31. In a given PRACH slot, if two UEs select the same PRACH sequence, then collision occurs; otherwise, it is possible that the eNB detects both PRACH signals, therefore no collision. Clearly, it is desirable that those sequences be of low cross-correlation. Since orthogonal sequences can be obtained by cyclically shifting the same base sequence, it is desirable to generate 64 sequences of a given cell from a single base sequence as much as possible and to use a different base sequence only if the number of cyclic shifts is insufficient. What limits the number of cyclic shifts is the requirement that in order to ensure the orthogonality between the sequences, a cyclic shift should be equal to the number of samples whose time duration covers at least the maximum round trip delay plus maximum delay spread.

Handoff procedure summary

Finally we summarize the handoff procedure in Figure 9.34. Since the UE is connected to only one eNB at a time in LTE, the handoff procedure follows the general idea of expedited BBM handoff in Section 9.4.2. It is interesting to compare Figure 9.34 and Figure 9.20. The former exemplifies a network-controlled procedure and the latter reflects a UE controlled version, although they are both expedited BBM handoff.

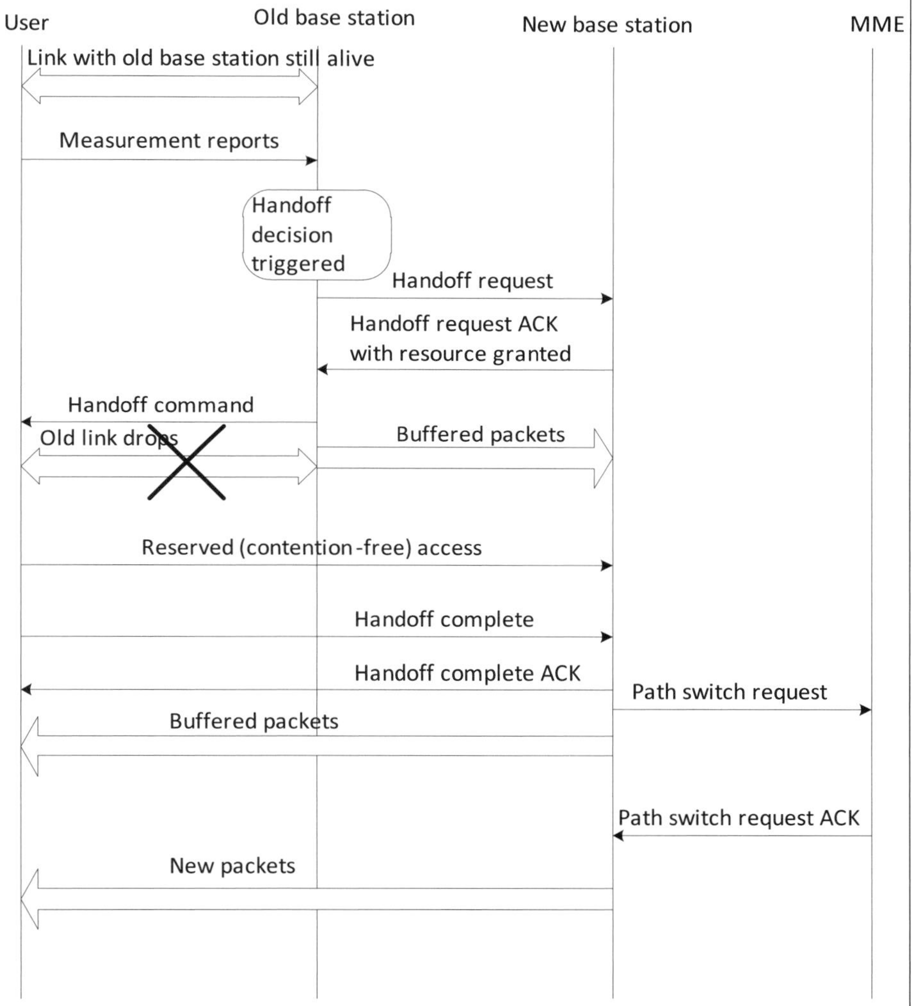

Figure 9.34 Procedure of LTE handoff. Only the downlink traffic is illustrated. The UE starts to send the uplink traffic after handoff complete ACK with the new eNB. MME (Mobility Management Entity) is a control node that manages mobility aspects.

9.7 Summary of key ideas

- Airlink design has to be viewed in the context of a greater end-to-end network architecture. Broadband data networks have converged to be IP-based Internet. Therefore, it makes sense to adapt wireless to the Internet, not vice-versa. This means that

airlink design has to handle link reliability and user mobility in a seamless manner so that Internet protocols, services and applications run transparently over the wireless channel.
- Though an effective way to obtain macro-diversity in circuit-switched CDMA voice, soft handoff is not the right solution for broadband OFDMA data because of network architecture and airlink scheduling considerations.
- MBB handoff is an alternative way to obtain macro-diversity. In MBB handoff, a user keeps independent links with multiple base stations. Each link maintains its own synchronization and power control and is used to send separate data and control traffic. When the base stations operate in the same bandwidth, the links can share the same RF front-end and use separate digital processing chains.
- In MBB handoff, the user chooses an appropriate link to send uplink traffic on a per IP packet basis for uplink macro-diversity. The user can fast select a link to receive real-time downlink traffic for downlink macro-diversity. The link for uplink traffic can be different from that for downlink traffic.
- If MBB handoff is not possible, the fallback is BBM handoff. The drawback is service disruption in the break period. Expedited BBM handoff anticipates BBM handoff before it actually happens and proactively performs control signaling so as to minimize the break period.
- To initiate handoff, the user needs to search for the presence of a candidate base station in the background while communicating with the serving base station. Two important requirements are detectability at low SINR and low processing complexity. A single-tone beacon signal seems a good choice for cell search and can be augmented with a wideband broadcast signal for subsequent synchronization.
- Overall, we advocate a mobile controlled handoff approach as opposed to network controlled handoff. The reason is that the user has better knowledge than the network of the channel environment and device capability and gets increasingly sophisticated to make intelligent handoff decisions.

10 Beyond conventional cellular frameworks

So far we have studied the system design principles of OFDMA-based mobile broadband under a conventional cellular network framework. The basic premises of the framework are:

- The base stations use high transmit power, and are placed at carefully chosen locations, ideally at the vertices of regular hexagons.
- A user is connected to the "best" base station. The best base station is usually the closest one that has the greatest downlink signal strength received at the user.
- A base station is open to all the users within a cell by providing "unrestricted" access service.
- Both the downlink and uplink communications are one-hop between the base station and the users. The users do not communicate directly even if they are nearby to each other.
- The time-frequency resource is reused spatially. Among cells reusing the same resource, a signal transmitted in one cell is treated as interference/noise in another cell.
- The spectrum to be used in a cell is fixed and known to both the base station and the users.

In this chapter, we explore several ideas that go beyond the conventional cellular framework in pursuit of the next performance leap.

The first such idea is *heterogenous network topology*. In wireless, moving the transmitter and receiver close to each other increases signal strength, reduces required transmit power and thus interference to other transmissions, and allows dense spectrum reuse. The conventional cellular topology is for *macrocell* deployment, where a cell radius of several miles is common. The macrocell deployment is optimized for coverage, resulting in homogeneous networks where cells are of similar sizes and planned with regular patterns. To reduce the communication distance, one can instead adopt a heterogeneous network topology; that is, deploy small cells (microcells, picocells and even femtocells) served by low power base stations; deploy intermediate nodes between the base station and the users (relays); and allow the communication to go direct between two nearby users rather than through the base station (device-to-device, or D2D, communications).

The second idea is *cooperative communications*. The wireless channel is broadcast by nature. In the conventional cellular framework, transmissions reusing the same time-frequency resource interfere with each other, thereby limiting the reusability of the resource in space and therefore the overall spectral efficiency. The inter-cell interference management schemes studied in Chapter 6 can be characterized as non-cooperative or passive in the sense that the inter-cell interference is treated as noise, and interference management is done via limited inter-cell coordination in terms of power and bandwidth usage. In contrast, cooperative communications take a proactive approach by coordinating coding/decoding among concurrent transmissions on the basis of global state information of channel and user data. In effect the transmitters and receivers cooperate to form a distributed network MIMO system. We will study two basic schemes, namely user cooperation assisted with relays and network cooperation with multi-cell processing.

The third idea is *cognitive radio*. The conventional cellular framework uses a static spectrum allocation policy, where licensed spectrum is assigned to and exclusively used by an operator on a long-term basis for large geographic areas. The increase in demand for mobile broadband has put great pressure on spectrum availability. On the other hand, it is found, for example, in [49], that a large portion of the assigned spectrum is not fully utilized in time or in space. The idea of cognitive radio is to share spectrum dynamically and opportunistically, thereby fundamentally overcoming the inefficiency of the static spectrum allocation policy. To realize the vision of cognitive radio, a broad interdisciplinary research effort is required, including not only traditional PHY/MAC layers (e.g., spectrum sensing, agile radio design, spectrum sharing, interference management, power control) and higher layers (e.g., networking, formal languages, architecture), but also business topics (e.g., policy reform, rule making, pricing). In this chapter, we will focus on the communications aspects in a specific framework, namely a primary/secondary system, where the primary users are incumbent licensed ones and not cognitive, while the secondary users are cognitive attempting to opportunistically utilize the licensed spectrum.

10.1 Heterogeneous topology

The initial goal of deploying a macrocellular network is mostly wide area coverage. Macrocell base stations transmit at high power and are installed on towers or atop buildings. Because of high costs of acquiring cell sites and installing and operating base stations, the operator tries to minimize the number of base stations required to cover a given geographic area. Therefore, macrocell base stations are deployed with careful cell planning and RF engineering to maximize the coverage of every cell. Macrocell network planning is usually complex and iterative because the network topology changes over time due to varying traffic demand and propagation environment, challenging indoor coverage, and difficult site acquisition.

As the network adds more users, traffic demand grows. The operator increases the network capacity to meet the demand. Wireless cellular systems have evolved to the point where the capacity of an isolated cell is closely approaching the information theoretic limits. To make further significant gains in capacity, one resorts to a different network topology ([58, 82]). With a mix of relays, microcells/picocells/femtocells, and

10.1 Heterogeneous topology

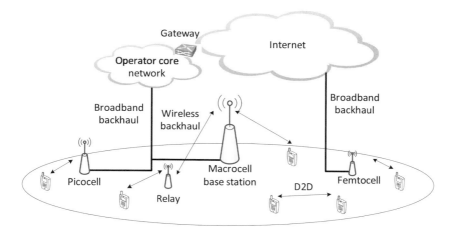

Figure 10.1 Illustration of a heterogeneous network with picocells, femtocells, relays, and D2D communications.

D2D communications, the cellular network topology evolves and becomes increasingly heterogeneous. Figure 10.1 depicts an example of a heterogeneous network. The common benefits of those approaches include improved coverage, increased capacity per area via denser spatial reuse and shorter communication distance, and lower device power consumption.

An important premise here is that the costs of deploying small cells and relays and enabling D2D communications are not high. In particular, those lower-power base stations and relays are not installed at high towers and may not be carefully tuned; rather, they are often deployed in an *ad hoc* way or even not planned or controlled by the operator. While heterogeneous networks allow a more flexible and cost effective deployment model to provide mobile broadband services, the interference patterns become more complex, necessitating more sophisticated interference management schemes, which will be the focus of this section.

10.1.1 Relays

In the conventional cellular system, the downlink and uplink channels are direct (one-hop) links between a base station and a user. A relay is an intermediate node that provides indirect (multi-hop) links that connect the base station and the user or another relay. A relay does not have a wired backhaul to the core network, but relies on the wireless channel as a backhaul to communicate with the base station. A direct link between the base station and a user is called an *access link*, a link between the base station and a relay called a *backhaul link*, and a link between a relay and a user called a *relay link*.

We focus on the deployment scenario where relays operate in-band and no additional spectrum is required to deploy relays. Specifically, the downlink channel (in either TDD or FDD systems) is used for transmissions from the base station to a relay and from a relay to a user, and the uplink channel is used from a user to a relay and from a relay to the base station. This way, a user always transmits in the uplink and receives in the downlink, whether it is communicating with a relay or with the base station, and the base station always transmits in the downlink and receives in the uplink, whether with a relay or with a user. On the other hand, unlike an ordinary user, a relay should be capable of transmitting and receiving at both the downlink and uplink channels. However, a relay cannot transmit and receive at the same time in a given channel (downlink or uplink). This is called the *half-duplex*[1] constraint and is caused by large power difference between the transmitted and received signals. Specifically, transmit and receive antennas at the relay are not sufficiently separated apart. A transmitted signal would leak into the receive path of the same device and cause severe self interference to any incoming signal, which is received at a much weaker power level. As a result, the relay cannot simultaneously receive on the backhaul link and transmit on the relay link using the downlink channel, or receive on the relay link and transmit on the backhaul link using the uplink channel.

In principle, any user in a cell can act as a relay, but in FDD, where the downlink and uplink are in two different frequencies, the RF frontend implementation costs of the user devices would have to increase because of the requirement of transmitting and receiving at the downlink and uplink. There are no additional RF costs in TDD. Thus, in this section, we focus on the scenario where relays are not ordinary users but special nodes whose only purpose is to help other users. The locations of relays are chosen so as to have good channel propagation with the base station and to be close to an area where traffic demand is expected to be high or where coverage is a concern. In general, a relay uses a lower transmit power level than the base station does. Furthermore, we focus on the *multi-hop* feature by assuming that a user communicates either directly with a base station or through a relay, *but not both*. In this case, a desired system feature is that it is transparent to users which route is taken, thereby reducing the device implementation costs and being backward compatible. We will later examine the *cooperative* feature of relays in Section 10.2.

We next elaborate on the key benefits of using relays.

Multi-hop

Figure 10.2 depicts the basic relay scenario where a single relay sits between a source node and a destination node. Clearly, multi-hop shortens the per-hop distance, thereby boosting SNR, especially when the path loss exponent is large. Is the indirect (source-to-relay-to-destination) link always superior to the direct link (source-to-destination)?

For simplicity, we assume that in the indirect link the source first sends the traffic to the relay in a fraction α of time. The relay then forwards the traffic to the destination in the remaining fraction $1 - \alpha$ of time. Denote by C_{sd}, C_{sr}, C_{rd} the capacities of the links

---

[1] More recently the authors in [26, 39] study the feasibility of a *single channel full-duplex* wireless transceiver using a combination of antenna, analog and digital cancellation, and antenna separation. However, the tolerable dynamic range of desired received signal and self interference that can be achieved so far is too limited to allow single-channel full-duplex operation over a long communication range.

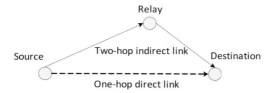

Figure 10.2 Basic relay scenario with three nodes.

from the source to the destination, from the source to the relay, and from the relay to the destination, respectively. The capacity of the direct link is equal to C_{sd}. In the indirect link, the optimal[2] time partition parameter α can be obtained by solving

$$\alpha C_{sr} = (1 - \alpha)C_{rd}. \tag{10.1}$$

Hence, the capacity of the indirect link is given by

$$C_{srd} = \left(C_{sr}^{-1} + C_{rd}^{-1}\right)^{-1} = \frac{1}{2}H(C_{sr}, C_{rd}), \tag{10.2}$$

where $H(C_{sr}, C_{rd})$ is the harmonic mean of C_{sr} and C_{rd}.

In general, the link distance associated with C_{sr} or C_{rd} is shorter than that of C_{sd}, thereby resulting in a greater channel gain and SNR. Hence, $H(C_{sr}, C_{rd}) > C_{sd}$. On the other hand, the factor of $\frac{1}{2}$ in (10.2) represents the loss in bandwidth because the total time is split between the source-to-relay and relay-to-destination transmissions.

To demonstrate the tradeoff of SNR versus bandwidth, assume a linear deployment model where the relay sits exactly on the straight line between the source and destination. Denote by d, d_1 the distances from the source to the destination and to the relay, respectively, and $d - d_1$ the distance from the relay to the destination. To emphasize the effect of shortened per-hop distance in the multi-hop link, we consider only path loss in calculating the capacity: $C_{sd} = \log(1 + \gamma_0 d^{-r})$, $C_{sr} = \log(1 + \gamma_0 d_1^{-r})$, and $C_{rd} = \log(1 + \gamma_0 (d - d_1)^{-r})$ with r the path loss exponent. Normalizing $d = 1$, γ_0 represents the baseline SNR of the direct link. It can be shown that the optimal C_{srd} is obtained at $d_1 = 0.5$ when γ_0 is small or at $d_1 = 0, 1$ when γ_0 is large. The latter case is in effect reduced to the direct link.

Figure 10.3 plots the ratio of the indirect link C_{srd} at fixed $d_1 = 0.5$ and the direct link C_{sd} as γ_0 increases. We observe that the tradeoff favors the direct link when its SNR γ_0 is already high (> 10 dB), in which case shortening the per-hop distance only marginally increases $H(C_{sr}, C_{rd})$ but cannot make up for the loss in bandwidth. However, in the low SNR regime, the benefit of SNR boost becomes more significant than the loss in bandwidth and it can be shown that C_{srd}/C_{sd} approaches 2^{r-1} when $\gamma_0 \ll 1$.

Ratio C_{srd}/C_{sd} represents a multi-hop benefit of using relays, and can translate into an increase in capacity or coverage. Relays should be deployed in hot-spots for capacity or close to cell edges for coverage. Here, cell edges are defined from the perspective

---

[2] In practice, the system has to choose a pre-determined partition parameter for *all* relay nodes and would not be able to optimize the parameter for each individual relay. Thus, this is an optimistic assumption used here to explain the *potential* benefit of a relay.

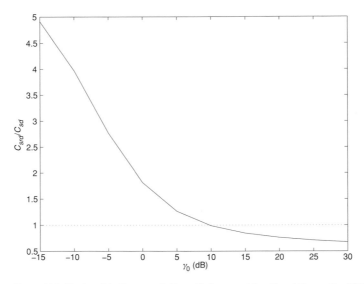

Figure 10.3 Ratio of indirect and direct link capacities C_{srd}/C_{sd} as the SNR of direct link increases. $r = 3.5$.

of wireless channel propagation and include areas far from the base station or indoors with large in-building penetration loss. Some points of the above analysis are worth mentioning:

- The source and the relay may have different transmit power levels. In the downlink the source (base station) transmits at higher power than the relay. In the uplink, the relay may be at higher power than the source (user). As a result, multi-hop will benefit the uplink more than the downlink.
- The SNR boost would be smaller if the relay does not sit exactly on the straight line between the source and destination because now $d_1 + d_2 > d$.

Spatial diversity

Another important benefit of relays is spatial diversity. There are two aspects to this benefit.

First, in a fast time scale (each data transmission), the destination can combine the signal received directly from the source and that from the relay, thereby reducing the margin required to combat against multipath fading. This is a form of cooperative diversity and will be elaborated upon in Section 10.2. Moreover, the base station or relays can employ channel-aware multiuser scheduling as a way of exploiting fading via multiuser diversity, as will be discussed later in this section.

Next we would like to focus on the second aspect, namely link selection diversity in a slow time scale. The source chooses either the direct or indirect link in a semi-static manner, thereby reducing the margin required to compensate for shadowing. Specifically, an outage is defined to be an event that the link capacity is insufficient to

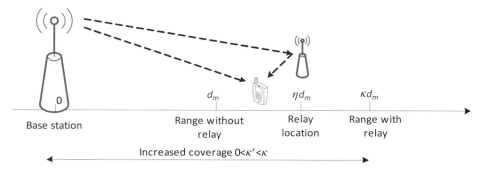

Figure 10.4 Illustration of increased range with relay link selection.

support a minimum required data rate. Here we assume that the fast fading caused by multi-path and mobility can be averaged out in the slow time scale and the link capacity is mainly determined by path loss and shadowing. Thus, the minimum required data rate can be translated into a minimum required slow time scale link budget. The outage event at distance d is equivalent to

$$\gamma_{dB}(d) + X_s < \gamma_{\min,dB}, \qquad (10.3)$$

where $\gamma_{\min,dB}$ is the SNR in dB required to achieve the minimum required data rate, X_s represents the shadowing effect of the direct link and is Gaussian random variable with zero mean and standard deviation equal to s (see Equation 4.4), and $\gamma_{dB}(d)$ is the SNR in dB at distance d given by the path loss model

$$\gamma_{dB}(d) = A - r \cdot 10 \log_{10} d, \qquad (10.4)$$

with some constant A. The coverage range d_m is the maximum value of d such that the outage probability does not exceed a certain threshold

$$\mathbb{P}\left(\gamma_{dB}(d_m) + X_s < \gamma_{\min,dB}\right) \leq p. \qquad (10.5)$$

It follows that

$$\gamma_{dB}(d_m) = \gamma_{\min,dB} + Q^{-1}(p)s, \qquad (10.6)$$

where the Q function is defined as $Q(x) = \frac{1}{\sqrt{2\pi}} \int_x^\infty e^{-\frac{x^2}{2}} dx$. The second term represents the shadowing margin. Since $\gamma_{dB}^{-1}(\gamma_{\min,dB})$ is the maximum range without taking into account shadowing, ratio $\gamma_{dB}^{-1}(\gamma_{\min,dB})/d_m$ indicates the loss of range due to shadowing.

To demonstrate how the relay helps increase the coverage range, we continue the example where the relay is deployed on the straight line connecting the source and destination. Let us start with the downlink case shown in Figure 10.4. Deploy the relay at a distance ηd_m from the base station with $\eta > 1$. Define the corresponding coverage range κd_m if for any user located at a distance $\kappa' d_m$ with $0 < \kappa' \leq \kappa$, the outage probability is less than p.

Since there are two alternative links to reach the user, outage occurs when the SNRs of both links fall below $\gamma_{\min,dB}$. The outage probability of the direct link is

$$p_0 = \mathbb{P}\left(\gamma_{dB}(\kappa' d_m) + X_s < \gamma_{\min,dB}\right)$$
$$= Q\left(Q^{-1}(p) - r \cdot 10\log_{10}\kappa'\right), \quad (10.7)$$

where random variable X_s represents the log-normal shadowing of the direct link. For the indirect link, we assume that the link between the base station and the relay is LOS subject to the same path loss model but without shadowing and that the transmit power of the relay is z_{DL} dB below that of the base station. In the low SNR regime, (10.2) is reduced to the minimum SNR requirement of the link from the relay to the user $\gamma'_{\min,dB}$

$$\frac{1}{\gamma_{\min}} = \frac{1}{\gamma(\eta d_m)} + \frac{1}{\gamma'_{\min}}, \quad (10.8)$$

where

$$\gamma(\eta d_m) = \gamma(d_m)\eta^{-r} = \gamma_{\min} B, \quad (10.9)$$

with $B = 10^{Q^{-1}(p)s/10} \cdot \eta^{-r}$. When the subscript "$dB$" is dropped, it means the same quantity in the linear scale; that is, $\gamma_{dB}(d_m) = 10\log_{10}(\gamma(d_m))$, $\gamma_{\min,dB} = 10\log_{10}(\gamma_{\min})$ and so on. It follows that

$$\gamma'_{\min,dB} = \gamma_{\min,dB} - 10\log_{10}(1 - B^{-1}), \quad (10.10)$$

The outage probability of the indirect link is

$$p_1 = \mathbb{P}\left(\gamma_{dB}(|\eta - \kappa'|d_m) - z_{DL} + X'_s < \gamma'_{\min,dB}\right),$$
$$= Q\left(Q^{-1}(p) - z_{DL} - r \cdot 10\log_{10}|\eta - \kappa'| + 10\log_{10}(1 - B^{-1})\right), \quad (10.11)$$

where random variable X'_s represents the log-normal shadowing from the relay to the user in the indirect link. Assume that X'_s is independent of X_s but has the same distribution. Hence, the overall outage probability is given by $p_0 p_1$ and should not exceed p.

Figure 10.5 plots the coverage range κ as the relay location η varies. We observe that as the relay is deployed farther away from the base station, the coverage range increases linearly. However, there is a maximum value of η beyond which a user in some area within $\kappa' \in (1, \eta)$ experiences an outage probability greater than p. What happens is that when η is large, in that area, the indirect link is sufficiently weak ($p_1 \approx 1$) and cannot help much when the direct link is in outage ($p_0 > p$). In Figure 10.5, η stops at this maximum value.

To see the benefit of link selection diversity as opposed to only multi-hop, we also plot κ versus η in Figure 10.5, assuming only the indirect link to be used. While link selection diversity achieves a slightly greater coverage range than "relay link only" when η is small, a more important advantage is that it allows the relay to be deployed farther from the base station. The maximum of η increases from 1.5 with "relay link only" to 2.5 with "link selection" in Figure 10.5, thereby resulting in a much greater coverage range where κ increases from 2 to 3.

We can apply the same analysis in the uplink, except that the transmit power of the relay is z_{UL} dB above that of the user in the indirect link. Thus, the only changes are

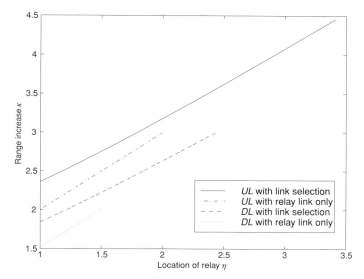

Figure 10.5 Coverage range κ versus relay location η. $r = 3.5$, $p = 0.01$, $s = 8$ dB, and $z_{DL} = 10$ dB.

that B is changed to $B = 10^{z_{UL}/10} 10^{Q^{-1}(p)s/10} \cdot \eta^{-r}$ and that the z_{DL} term in (10.11) is removed. The uplink analysis is plotted in Figure 10.5 as well. As we noted before, using the relay benefits the uplink more than the downlink. In particular, with "link selection," the relay can be deployed as far as $\eta \approx 3.4$ resulting in a coverage range increase to $\kappa \approx 4.5$.

Dense spectrum reuse

So far our focus has been on how a relay helps a link between a source and a destination. We now consider the scenario involving multiple links.

In the three-node analysis, although the indirect link boosts SNR, it suffers from a loss in bandwidth caused by the half-duplex constraint at the relay. We note, however, that such a loss is an *artifact* because of the limited view focusing only on a single source/destination pair. The loss can be mitigated by dense spectrum reuse by other pairs. For example, in the downlink, while the relay is transmitting to the user and cannot receive from the base station, the base station can *reuse* the spectrum and transmit to another relay or user rather than being idle. The word "dense" emphasizes that spectrum is reused in different areas within a macrocell. To better understand the dense spectrum reuse benefit, we next extend to a capacity analysis of a downlink four-node scenario shown in Figure 10.6. A similar analysis can be done for the uplink as well.

The base station sends the traffic to user 1 directly and to user 2 via the relay. Users 1 and 2, respectively, represent nearby and faraway users in a cell. This routing decision implies that the links from the base station to user 1 and the relay, and from the relay to user 2 are generally in the high SNR regime, and the link from the base station to user 2 in the low SNR regime.

Beyond conventional cellular frameworks

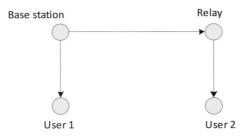

Figure 10.6 Routing topology of downlink relay scenario with four nodes.

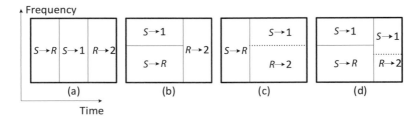

Figure 10.7 Scheduling options in the four-node downlink. In either (a), (b), (c), or (d), the x-axis represents time and the y-axis represents frequency. The dotted line in options (c) and (d) indicates that each of $S \to 1$ and $R \to 2$ can occupy either an orthogonal block of bandwidth (orthogonal multiplexing) or the entire bandwidth (reuse multiplexing).

Figure 10.7 shows four examples of scheduling options under this routing policy.

Option (a) is a simple TDM baseline scheme: time is divided into three slots, which are used, respectively, for the base station to relay transmissions (labeled as $S \to R$), the base station to user 1 transmissions ($S \to 1$), and the relay to user 2 ($R \to 2$) transmissions.

Option (b) combines the first two slots and multiplexes the $S \to R$ and $S \to 1$ transmissions in OFDMA. From Chapter 5, option (b) is in general superior to option (a) because of joint power bandwidth allocation in OFDMA user multiplexing. However, since both $S \to R$ and $S \to 1$ are in the high SNR regime, the advantage of OFDMA (b) over TDM (a) is insignificant.

Option (c) combines the last two slots and allows the $S \to 1$ and $R \to 2$ transmissions to occur simultaneously. In Figure 10.7, there are two spectrum sharing schemes to multiplex the $S \to 1$ and $R \to 2$ links. In the orthogonal multiplexing scheme, one link occupies a fraction of the bandwidth and the other link occupies the remaining bandwidth. The two links do not interfere with each other. In the reuse multiplexing scheme, either link occupies the entire bandwidth, and the two links do interfere with each other.

Option (d) is simply a combination of options (b) and (c), and its performance is close to (c) since as we argued before the difference between (b) and (a) is insignificant. So let us focus on comparing option 5(a) and (c). We next highlight the benefits of having simultaneous $S \to 1$ and $R \to 2$ transmissions.

The first benefit is *energy boost*. Suppose that the base station keeps the same transmit power in options (a) and (c), and so does the relay. Since the $S \to 1$ and $R \to 2$ links transmit in a longer time interval in (c) than in (a), a greater amount of total energy is spent in (c) for communications. It can be easily shown that with orthogonal multiplexing, option (c) achieves higher data rates of both links than (a). Of course, the cost is that higher total interference is generated to adjacent cells, which are not taken into account in the four-node analysis. Hence, in the deployment scenario where cell edge users are limited by the link budget, energy boost can improve the performance. However, if cell edge users are limited by inter-cell interference, then the gain from energy boost can be quite limited.

To make the comparison fair, we can restrict ourselves by assuming that neither the energy spent by the base station nor by the relay is allowed to change between (a) and (c). Without energy boost, it can be shown that option (c) with orthogonal multiplexing achieves the same data rates as (a). However, as shown below, option (c) with reuse multiplexing may still outperform (a) thanks to the second benefit of *bandwidth reuse*. Specifically, both the $S \to 1$ and $R \to 2$ links occupy greater bandwidth in reuse multiplexing than in orthogonal multiplexing, thereby potentially achieving higher link capacities in the high SNR regime. The cost is the cross interference between the two links. Therefore, we need to examine the SINR and study the bandwidth versus SINR tradeoff of those two parallel links.

Discussion Notes 10.1 provides a brief summary of the information theoretical results on communication schemes over parallel (interference) channels. Given the scope of this book, we next limit ourselves to practical schemes where the interference is treated as noise and cannot be exploited. In this case, we only compare the reuse and orthogonal multiplexing schemes.

Denote by P_S and P_R the maximum allowed transmit power levels at the base station and relay, respectively. Denote by G with certain subscript the channel power gain of the corresponding link, for example, G_{S1} for the link from the base station to user 1 and so on. Define SNR $\gamma_{S1} = \frac{P_S G_{S1}}{\sigma_0^2}$ and $\gamma_{R2} = \frac{P_R G_{R2}}{\sigma_0^2}$. The data rates achieved with the orthogonal multiplexing scheme are

$$R_1 = \alpha \log\left(1 + \frac{\gamma_{S1}}{\alpha}\right) \tag{10.12}$$

$$R_2 = (1-\alpha) \log\left(1 + \frac{\gamma_{R2}}{1-\alpha}\right), \tag{10.13}$$

where α is the fraction of bandwidth allocated to the $S \to 1$ link. Define interference-to-noise-ratio (INR) $\zeta_{R1} = \frac{P_R G_{R1}}{\sigma_0^2}$ and $\zeta_{S2} = \frac{P_S G_{S2}}{\sigma_0^2}$. The data rates achieved with the reuse multiplexing scheme are

$$R_1 = \log\left(1 + \frac{\beta_S \gamma_{S1}}{1 + \beta_R \zeta_{R1}}\right) \tag{10.14}$$

$$R_2 = \log\left(1 + \frac{\beta_R \gamma_{R2}}{1 + \beta_S \zeta_{S2}}\right), \tag{10.15}$$

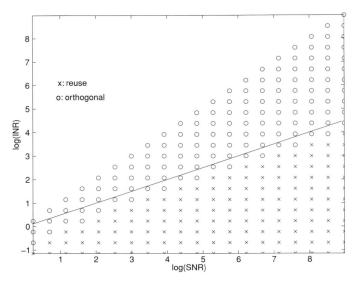

Figure 10.8 Comparison of the reuse and orthogonal multiplexing schemes in the symmetric case. For any pair of SNR and INR, "x" (or "o") indicates that the reuse (or orthogonal) scheme achieves a higher data rate. The solid line $\log \zeta = \frac{1}{2} \log \gamma$ is the borderline between the choice of the two schemes under the high SNR INR approximation.

where $\beta_S P_S$ and $\beta_R P_R$ are the actual transmit power levels used at the base station and relay, respectively, with $\beta_S, \beta_R \leq 1$. The rate regions of the two schemes can be obtained as follows. For the orthogonal scheme, vary α from 0 to 1 in (10.12) and (10.13). For the reuse scheme, first set $\beta_S = 1$ and vary β_R from 0 to 1 in (10.14) and (10.15), and then set $\beta_R = 1$ and vary β_S from 0 to 1. To gain insight, let us consider two special cases.

First, consider the symmetric case where $\gamma_{S1} = \gamma_{R2} = \gamma$ and $\zeta_{S2} = \zeta_{R1} = \zeta$. We set symmetric operating point $\alpha = \frac{1}{2}$ and $\beta_S = \beta_R = 1$. To simplify notation, we drop the subscripts. When the SNR γ and INR ζ are both high, the data rate with the orthogonal scheme now becomes

$$R = \frac{1}{2}\log(1 + 2\gamma) \approx \frac{1}{2}\log\gamma. \tag{10.16}$$

The data rate with the reuse scheme

$$R = \log\left(1 + \frac{\gamma}{1+\zeta}\right) \approx \log\left(\frac{\gamma}{\zeta}\right). \tag{10.17}$$

As long as

$$\log\zeta < \frac{1}{2}\log\gamma, \tag{10.18}$$

the reuse scheme achieves a higher data rate and therefore is superior; otherwise, the orthogonal scheme is superior.

Figure 10.8 uses "x" or "o" to indicate which scheme, reuse or orthogonal, is superior for any pair of SNR γ and INR ζ, where the data rate is calculated without the high

SNR INR approximation. As a comparison, the solid line $\log \zeta = \frac{1}{2} \log \gamma$ is plotted to represent the approximation in the high SNR INR regime. The orthogonal scheme dominates in the area above the solid line and the reuse scheme dominates in the area below. Note the slight discrepancy between the solid line and the "x," "o" points. The actual borderline between the "x" and "o" areas lies below the solid line approximation, especially when the SNR or INR is not high.

Next, we extend the above analysis to the non-symmetric case where $\gamma_{S1} \neq \gamma_{R2}$ or $\zeta_{S2} \neq \zeta_{R1}$. An interesting result from [46] is that the orthogonal scheme always dominates the reuse scheme if the the product of the cross channel gains ($G_{R1}G_{S2}$) is greater than the product of the direct channel gains ($G_{R2}G_{S1}$),

$$G_{R1}G_{S2} > G_{R2}G_{S1}, \tag{10.19}$$

or equivalently,

$$\zeta_{R1}\zeta_{S2} > \gamma_{S1}\gamma_{R2}, \tag{10.20}$$

in any SNR INR regime. Keep in mind that condition (10.20) is sufficient but not *necessary* for the orthogonal scheme to be superior. To see this, note that no matter what the relationship between the cross and direct channel gains, the orthogonalization scheme is always better if P_R and P_S go to infinity but have a constant ratio. In this case, the capacity of the orthogonal scheme is unbounded, while the capacity of the reuse scheme is bounded.

Moreover, it can be shown that in the high SNR INR regime, at the symmetric operating point of $\alpha = \frac{1}{2}$ and $\beta_S = \beta_R = 1$, the reuse scheme is superior if

$$\log \zeta_{R1} < \frac{1}{2} \log \gamma_{S1} \quad \text{and} \quad \log \zeta_{S2} < \frac{1}{2} \log \gamma_{R2}. \tag{10.21}$$

and the orthogonal scheme is superior if

$$\log \zeta_{R1} > \frac{1}{2} \log \gamma_{S1} \quad \text{and} \quad \log \zeta_{S2} > \frac{1}{2} \log \gamma_{R2}. \tag{10.22}$$

Thus, if the SNRs are much higher than the INRs, the reuse scheme is superior. Figure 10.9 shows the rate regions of two examples. In example (a), the reuse scheme is superior at the symmetric operating point; however, its rate region does not entirely dominate that of the orthogonal scheme. The INR increases slightly in example (b), and now the orthogonal scheme becomes superior.

We now apply the above analysis of reuse versus orthogonal multiplexing schemes to compare scheduling options (a) and (c) in Figure 10.7. With option (a), denote by t_1, t_2, t_3 the fractions of time allocated to $S \to R$, $S \to 1$, and $R \to 2$, respectively, with $t_1 + t_2 + t_3 = 1$. t_1, t_3 are chosen such that $t_1 \log(1 + \gamma_{SR}) = t_3 \log(1 + \gamma_{R2})$. Increasing t_1 from 0, we calculate the corresponding t_2, t_3, and obtain a rate region of (R_1, R_2) for option (a). To make a fair comparison, in option (c) we keep the same total energy spent for both the base station and the relay. Specifically, for given t_1, t_2, t_3 corresponding to an operating point (R_1, R_2) on the rate region of option (a), we let $S \to 1$ and $R \to 2$ occur simultaneously in $t_2 + t_3$ and scale back the transmit powers of the base station and relay proportionally by a factor of $\frac{t_2}{t_2+t_3}$ and $\frac{t_3}{t_2+t_3}$, respectively.

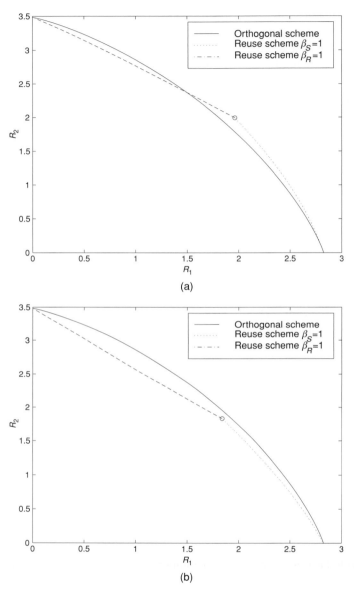

Figure 10.9 Comparison of the rate regions of the orthogonal and reuse schemes in non-symmetric SNR INR cases. In (a), $\gamma_{S1} = 12$ dB, $\gamma_{R2} = 15$ dB, $\zeta_{S2} = 6$ dB, and $\zeta_{R1} = 2$ dB. In (b), $\gamma_{S1} = 12$ dB, $\gamma_{R2} = 15$ dB, $\zeta_{S2} = 7$ dB, and $\zeta_{R1} = 3$ dB.

Given the transmit powers, we can determine the rate regions of the reuse and orthogonal schemes with (10.12) to (10.15), and then obtain the convex hull as the capacity region for option (c). We select the operating point on the convex hull where R_2 is the same as that in option (a), and thus the difference in R_1 represents the advantage of option (c)

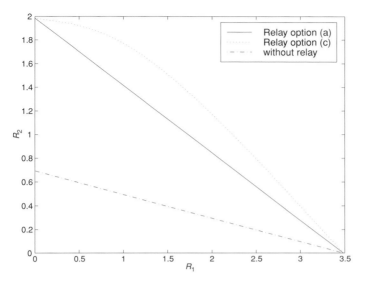

Figure 10.10 Comparison of the rate regions of scheduling options (a) and (c) and that without a relay. $\gamma_{SR} = \gamma_{S1} = 15$ dB, $\gamma_{R2} = 20$ dB, $\zeta_{S2} = 0$ dB, $\zeta_{R1} = -10$ dB.

over (a). Since the total energy spent remains the same, the difference is attributed to dense spectrum reuse rather than energy boost.

Figure 10.10 compares the rate regions of scheduling options (a) and (c), as well as the rate region of a scheme doing $S \to 1$ and $S \to 2$ in TDM without the help of a relay. The difference between the latter scheme and option (a) is attributed to multi-hop, and option (c) further enhances the data rate performance with dense spectrum reuse.

Systems picture

Finally, we shift our focus to a more complex scenario where multiple relays help multiple users in a multi-cell system, as illustrated in Figure 10.11.

Dense spectrum reuse, with a balanced tradeoff between SINR and bandwidth, is a key system benefit of deploying relays. The advantages are that long-range links are replaced with short-distance ones due to multi-hop and that multiple transmissions occur simultaneously between different relays and users (and the base station) to maximize spatial reuse. For example, in Figure 10.11, when the base station is transmitting to a relay (A) in one location of the cell, other relays (B_1, B_2, and B_3) can transmit to its users in another locations, because their intended users are nearby and they use lower transmit power and are far from receiver A. Note that when equipped with multiple antennas, the base station can also use beamforming and other multiple antenna techniques to reduce the interference to those simultaneous transmissions, thereby further encouraging dense reuse.

To fully exploit the benefits of relay enhancement, we need to address the routing and scheduling questions:

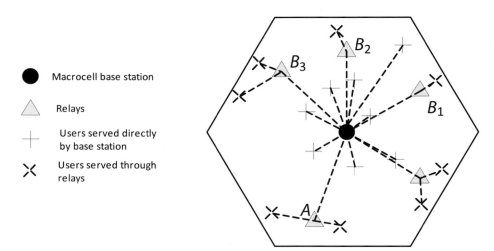

Figure 10.11 A macrocell assisted with relays. While the locations of the relays shown in the figure follow a regular pattern, they can be deployed in an *ad hoc* manner without centralized planning. Some users are served directly by the base station, while others are through relays.

- (*Routing*) Should the traffic go through a relay and if so which one?
- (*Scheduling*) Which links are to transmit at what power/bandwidth allocation and data rate?

To address the two questions jointly, an optimization problem can be formulated as an integer programming problem involving relay association and power bandwidth allocation. Consider the downlink of a single macrocell. Denote by M_R, M the numbers of relays and users, respectively. Index the base station as 0, relays $1, \ldots, M_R$, and users $M_R + 1, \ldots, M_R + M$. Suppose that the total bandwidth is equally divided into K_S subbands. At every slot, the base station can transmit to the relays and to the users, and the relays can transmit to the users, according to a routing and scheduling decision represented with three groups of variables:

- Binary routing and bandwidth allocation $\rho_{m,n,k} \in \{0, 1\}$ indicating whether or not transmitter m uses subband k to send to receiver n,
- Power allocation $\beta_{m,n,k} \in [0, 1]$ the fraction of power of transmitter m used in subband k for receiver n, and
- Traffic routing $\theta_{m,n} \in [0, 1]$ the fraction of traffic intended for user n among the total traffic sent from the base station to relay m,

with $m = 0, 1, \ldots, M_R$, $n = 1, \ldots, M_R, M_R + 1, \ldots, M_R + M$, $k = 1, \ldots, K_S$. The constraints are as follows:

- Half-duplex:

$$\rho_{0,m,k_1} + \rho_{m,n,k_2} \leq 1, \tag{10.23}$$

with $m = 1, \ldots, M_R$, $\forall n, k_1, k_2$.

- Maximum two hops from the base station to any user (no relay-to-relay forwarding):

$$\rho_{m,m',k} = 0, \tag{10.24}$$

with $m, m' = 1, \ldots, M_R, \forall k$.
- At most one transmitter for a given receiver n in a given subband k:

$$\sum_m \rho_{m,n,k} \leq 1, \forall n, k. \tag{10.25}$$

Similarly, a transmitter is constrained to send to at most one receiver in a given subband k, but can send to multiple users in a given slot:

$$\sum_n \rho_{m,n,k} \leq 1, \forall m, k. \tag{10.26}$$

- Total transmit power at transmitter m:

$$\sum_n \sum_k \beta_{m,n,k} \leq 1, \forall m. \tag{10.27}$$

- Total traffic received at relay m:

$$\sum_n \theta_{m,n} = 1, \forall m. \tag{10.28}$$

At the present slot, the amount of traffic sent from transmitter m to receiver n is given by

$$R_{m,n} = \min\left(W \sum_k \rho_{m,n,k} \log\left(1 + \frac{\beta_{m,n,k} P_m G_{m,n,k}}{\sigma_0^2 + \sum_{m' \neq m} \sum_{n' \neq n} \beta_{m',n',k} P_{m'} G_{m',n,k}}\right), Q_{m,n}^{\text{old}}\right), \tag{10.29}$$

where P_m is the total transmit power of transmitter m, σ_0^2 is the total noise power in a subband, W is the total time-frequency bandwidth of a subband over a slot, and $Q_{m,n}^{\text{old}}$ is the queue length of traffic in transmitter m intended for receiver n just before the present slot. Assuming the full buffer model at the base station, $Q_{0,n}^{\text{old}}$ is set to ∞. After transmissions are done in the present slot, the queue length at relay m is updated as follows:

$$Q_{m,n}^{\text{new}} = \max\left(Q_{m,n}^{\text{old}} + \theta_{m,n} R_{0,m} - R_{m,n}, 0\right). \tag{10.30}$$

From (10.29) and (10.30), we note that when a user n is connected through relay m, the scheduled traffic $R_{m,n}$ is limited by both the link capacity from m to n (the first term in (10.29)) and the queue length $Q_{m,n}$, which depends on the total scheduled traffic from the base station to the relay $R_{0,m}$ and the traffic routing fraction $\theta_{m,n}$. Even though the base station has a full traffic queue for user n, because of the multi-hop nature, intermediate relay m may not have a full queue to utilize the entire link capacity.

The objective function is a summation of utility functions of the traffic received at the users. The optimization problem needs to be solved at every slot. Obviously, the computational complexity is very high. In addition, the centralized solution requires

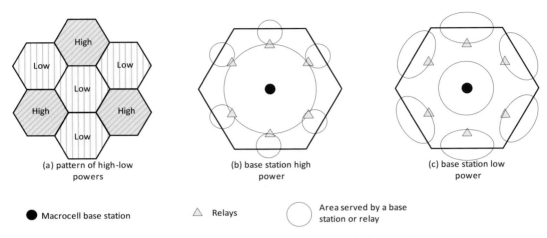

Figure 10.12 Example of semi-static power pattern among the base station and relays.

global knowledge of interference and channel state information (CSI), thereby incurring very high signaling overhead for the relays and users to report the information to the base station. If the wireless channel and traffic conditions vary rapidly, such a centralized optimization becomes less attractive.

Rather than trying to solve the joint optimization problem, we next take a more practical approach and explore suboptimal solutions.

First, *we decouple routing and scheduling with two distinct time scales*. A routing decision is made in a slow time scale (on the order of at least hundreds of milliseconds) on the basis of large-scale channel condition (path loss and shadowing) using link capacity in Equation (10.2) as well as long-term traffic loading of the relays. Given the routing topology, a scheduling decision is made in a fast time scale (on the order of a millisecond) on the basis of short-scale channel condition (multipath fading) as well as short-term traffic needs (queue length and delay).

In Figure 10.11, a number of relays are deployed in a macrocell, and users close to a relay are connected via the relay. Denote by $I_{m,n} \in \{0, 1\}$, with $m = 0, 1, \ldots, M_R$, $n = M_R + 1, \ldots, M_R + M$, the indicator function of whether user n is connected through the base station 0 or relay m. The routing decision is to determine $I_{m,n}$ such that a user is served by the base station directly or through a relay, but not both; that is, $\sum_m I_{m,n} = 1, \forall n$. Given the routing decision, if $I_{m,n} = 0$, we then set $\theta_{m,n} = 0$, $\rho_{m,n,k} = 0$, $\beta_{m,n,k} = 0$, $\forall k$, so as to reduce the variable space in the scheduling optimization.

Second, *we adopt a semi-static power pattern among the base station and relay transmitters*. The idea is similar to FFR studied in Chapter 6. Figure 10.12 shows an example of such a pattern in a given slot. A given subband is reused at neighboring macrocell base stations with different transmit power levels (high and low) according to a pattern shown in (a). At a high power level, the base station transmits to faraway users or relays as shown in (b), and at a lower power level, to nearby users as shown in (c). Meanwhile the same subband k can be reused by other relays, either at a reduced power as shown in (b) ($\sum_n \beta_{m,n,k} < 1$) or at the full power as shown in (c) ($\sum_n \beta_{m,n,k} = 1$).

Third, *we use fast scheduling at the base station and relays*. Recall from (10.29) that scheduling at different transmitters is coupled because the link capacity from transmitter m to receiver n depends on the power allocation of other transmitters $\beta_{m',n',k}$. Fixing the total transmit power of any transmitter beforehand eliminates the coupling effect and thus allows each of the base stations and relays to independently schedule the users under its control. The distributed scheduling problem is in essence the same as in a single macrocell case, and can be solved using the techniques studied in Chapters 4 and 5, such as channel-aware scheduling, adaptive coding and modulation, and user multiplexing (OFDMA power bandwidth allocation). Here, when multiple users are connected through a relay, the relay does not simply forward traffic, but functions as a multiuser scheduler, similar to the base station in this regard.

In summary, routing decision and power pattern adaptation occur in a slow time scale and therefore can be done in a centralized manner controlled by the base station. Given routing and power pattern, the base station and relays make their own scheduling decisions in a fast time scale in a distributed manner.

However, we should emphasize the power pattern needs to be adapted to interference and traffic fluctuation. Dynamic interference management with relays is in general more challenging than the one with only macrocell base stations studied in Chapter 6. The reason is twofold. First, when the relays are deployed in an *ad hoc* manner, their locations tend to be irregular, leading to more complex interference geometry. Second, a relay is typically connected with only a small number of users, and thus does not exhibit as much statistical multiplexing as a macrocell base station. Hence, interference management needs to be more agile. One example protocol will be presented in Section 10.1.3.

10.1.2 Femtocells

Dense spectrum reuse can also be realized by deploying *small* cells, as opposed to macro cells. Small cells can be considered as an extension to relays by adding wired backhaul connections. There are three types of small cells, namely *microcells*, *picocells*, and *femtocells*, which can be distinguished by their respective targeted coverage ranges. Typically the coverage range of a microcell is less than 2 km, a picocell is 200 m or less, and a femtocell is on the order of 10 m.

Microcells are usually installed in urban areas to provide capacity enhancement and in-building penetration. Picocells are mostly used in office buildings and shopping malls with very high traffic demand. They both operate in a very similar way to macro base stations. In particular, they are installed, configured, and actively managed by operators, at a significant capital expenditure (cell tower installation) and operating expenditure (electricity, site lease, backhaul).

On the other hand, femtocells represent a very different deployment model [20]. A femtocell base station is installed by the consumer at home or business to improve both coverage and capacity. The femtocell base station connects to the operator's network via the consumer's own broadband backhaul (DSL or cable). The operator may need to subsidize the femtocell base station hardware to boost market penetration. However, the overall costs can be significantly reduced as compared with the macro base stations,

Table 10.1. A comparison between macrocells, microcells, femtocells and relays.

	Macrocells	Microcells, picocells	Relays	Femtocells
Transmit power (dBm)	~43	~23 to 30		<23
Antenna gain (dBi)	~12 to 15	~0 to 5		0
Wireless channel propagation	Mostly outdoors	Mixed outdoors and indoors		Mostly indoors
Airlink access	Open			Probably closed
Backhaul	Dedicated leased lines or point-to-point microwave, part of operator's network		Over the air to macrocell base stations, probably in-band	DSL or cable broadband, shared with consumer's other traffic, connected to operator's network via Internet
Control responsibility	Operator			Consumer

microcells, and picocells. In addition to coverage and capacity, the consumers using femtocells can also benefit from prolonged battery life in their user devices (because of less transmit power required to communicate with nearby femtocells), discounted tariffs, and may even receive incentives from operators if their femtocells provides service to other passing consumers. Table 10.1 highlights the key feature differences of microcells, picocells, and femtocells, and compares them with macrocells and relays.

Among the different approaches of deploying small cells, femtocells deviate most from the traditional macrocellular framework, since femtocells are deployed and controlled by the consumer and not by the operator. As a result, the overall network coverage cannot be optimized (or at least fine-tuned) for best performance to the degree that could be achieved in a deployment where the operator owns all base stations. Furthermore, a significant disparity (> 20 dB) of transmit power between femtocell and macrocell base stations makes interference particularly challenging to deal with. The central theme of this section is to study the interference issues caused by the femtocell deployment, as seen in [21, 106]. Our goal is to develop insights using simple back-of-envelope calculations for an order-of-magnitude analysis.

Assume that macrocells and femtocells reuse the same bandwidth. With femtocells embedded in a macrocell, three new types of interference arise:

1. Macrocell-to-femtocell interference;
2. Femtocell-to-macrocell interference;
3. Femtocell-to-femtocell interference.

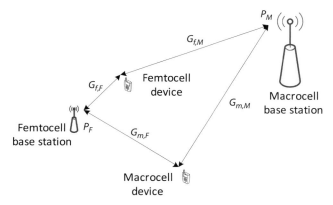

Figure 10.13 Femtocell/macrocell interference.

Note that femtocells are designed primarily for indoor use, and thus femtocell-to-femtocell interference is relatively small because femtocells are naturally isolated from each other due to penetration losses of building walls and low transmit power. We will therefore focus on the first two types of interference. However, keep in mind that when femtocells are close to each other, for example, in crowded apartments of a large building, femtocell-to-femtocell interference can be severe. Section 10.1.3 presents an example of interference management protocol to deal with such dense deployment scenarios.

A femtocell may provide access to any passing users (*open access* or *unrestricted association*) and allow them to share the broadband backhaul. However, due to privacy and security reasons, the femtocell may limit the access to a fixed set of home users (*closed access* or *restricted association*) [77, 176]. We next analyze the interference scenarios in open access and closed access.

Closed access: limits on achievable SINRs

In closed access, the macrocell user can only be served by a macrocell base station, even when it is very close to a femtocell. Figure 10.13 depicts the femtocell/macrocell interference scenario where a macrocell user gets close to a femtocell base station, which serves a femtocell user.

In the following we use subscripts f, F to represent femtocell user and base station, and m, M macrocell user and base station, respectively. For example, P_f, P_m stand for the transmit powers of the femtocell and macrocell users, respectively, P_F, P_M the total transmit powers of the femtocell and macrocell base stations, and $G_{f,M}$ the channel power gain between the femtocell user and macrocell base station. Here we assume symmetric channel gains, for example, the same channel power gain in the uplink (from the femtocell user to macrocell base station) and in the downlink (from macrocell base station to the femtocell user).

In the uplink, when both the femtocell device and the macrocell device are transmitting on the same spectrum, the SINRs are given by

$$\gamma_{f,UL} = \frac{P_f G_{f,F}}{\sigma_F^2 + P_m G_{m,F}} \tag{10.31}$$

$$\gamma_{m,UL} = \frac{P_m G_{m,M}}{\sigma_M^2 + P_f G_{f,M}}, \tag{10.32}$$

where σ_F^2, σ_M^2 are the powers of interference from other users plus noise.

Note that ignoring σ_F^2 and σ_M^2 the achieved SINRs can be upper bounded by

$$\gamma_{f,UL} \gamma_{m,UL} \leq \frac{G_{f,F} G_{m,M}}{G_{m,F} G_{f,M}}. \tag{10.33}$$

In other words, because of the macrocell/femtocell cross interference, irrespective of how P_f, P_m are controlled, $\gamma_{f,UL}$ and $\gamma_{m,UL}$ cannot *both* achieve arbitrarily high values even if the thermal noise plus interference terms (σ_F^2 and σ_M^2) are negligible. The upper bound is determined by the ratio between the product of *direct channel gains* and the product of *cross channel gains*. We have seen this ratio in the comparison of the reuse and orthogonal multiplexing schemes for relays. Specifically, in (10.19), a sufficient condition for the orthogonalization scheme to be superior to the reuse scheme is that the ratio is less than 1.

The same observation can be made in the downlink where the achieved SINRs are given by

$$\gamma_{f,DL} = \frac{P_F G_{f,F}}{\sigma_f^2 + P_M G_{f,M}} \tag{10.34}$$

$$\gamma_{m,DL} = \frac{P_M G_{m,M}}{\sigma_m^2 + P_F G_{m,F}}. \tag{10.35}$$

Thus,

$$\gamma_{f,DL} \gamma_{m,DL} \leq \frac{G_{f,F} G_{m,M}}{G_{m,F} G_{f,M}}. \tag{10.36}$$

Next we study a desired femtocell deployment where the communications over direct channels are protected by walls. Specifically, the femtocell base station and user are located indoors while the macrocell user is outdoors. Denote by $G_w \leq 1$ the wall penetration loss. When the macrocell user is close to the femtocell, we approximate $G_{f,M} \approx G_{m,M} G_w$ since the size of the femtocell is small. In the case where the femtocell and macrocell users are at an equal distance from the femtocell base station, $G_{m,F} \approx G_{f,F} G_w$. It follows that

$$\gamma_{f,DL} \gamma_{m,DL} \leq \frac{1}{G_w^2}, \tag{10.37}$$

$$\gamma_{f,UL} \gamma_{m,UL} \leq \frac{1}{G_w^2}. \tag{10.38}$$

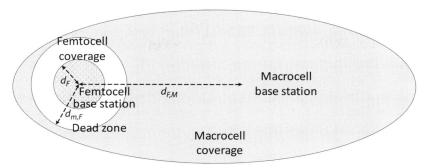

Figure 10.14 Macrocell and femtocell coverage, and the dead zone. The femtocell user stays within the femtocell coverage and the macrocell user stays outside. The dead zone is an area surrounding the femtocell coverage where femtocell/macrocell cross interference is significant. Although only one dead zone is shown, note that there are separate downlink and uplink dead zones, which are in general not identical.

Thus, G_w dictates the limit of the sum of the achieved SINRs (in the dB scale). A typical value of G_w is around -10 dB, which yields an upper bound of the sum SINR (in dB) of 20 dB. As an example, each of the femtocell and macrocell users can achieve around 10 dB, which is not bad in a cellular system. Of course, in the absence of a wall ($G_w = 0$ dB), the sum SINR is capped at as little as 0 dB.

We next examine the SINR of individual links in more detail. In the downlink, P_F and P_M are usually fixed. First consider the femtocell coverage. When σ_f^2 dominates, $\gamma_{f,DL} \approx \frac{P_F G_{f,F}}{\sigma_f^2}$ and the impact of the macrocell/femtocell interference is minor. However, when $P_M G_{f,M}$ dominates,

$$\gamma_{f,DL} \approx \frac{P_F}{P_M} \frac{G_{f,F}}{G_{f,M}} = \frac{P_F}{P_M} \frac{1}{G_w} \left(\frac{d_{f,F}}{d_{f,M}}\right)^{-r} \approx \frac{P_F}{P_M} \frac{1}{G_w} \left(\frac{d_F}{d_{F,M}}\right)^{-r}, \quad (10.39)$$

where d_F is the intended coverage size of the femtocell as depicted in Figure 10.14. Suppose that $\frac{P_F}{P_M} = -20$ dB, $G_w = -10$ dB, and the required SINR $\gamma_{f,DL} = 25$ dB at the edge of femtocell coverage. Then $d_{F,M} \approx 10 d_F$ at $r = 3.5$. In other words, to achieve a decent SINR target everywhere in the femtocell coverage, the femtocell cannot be too close to the macrocell base station. For example, if the radius of the femtocell is 20 m, then it should be at least 200 m away from the macrocell base station.

Continue on the downlink case and now consider the macrocell user. When it is close to the femtocell, its SINR degrades due to the interference from the femtocell base station. From the calculation in (10.37), the SINR of a macrocell user at the boundary of the femtocell coverage is approximated by

$$\gamma_{m,DL} \approx \frac{1}{G_w^2 \gamma_{f,DL}}. \quad (10.40)$$

Assuming wall protection ($G_w = 10$) and a moderate femtocell user SINR ($\gamma_{f,DL} = 10$ dB), the macrocell user can still achieve a decent SINR $\gamma_{m,DL} \approx 10$ dB. In this case, the femtocell does not create an excessive interference issue. However, in practice, wall protection may not exist because the femtocell attempts to extend its coverage to the outdoors. Moreover, the macrocell base station does not directly control $\gamma_{f,DL}$; the femtocell base station may set $\gamma_{f,DL}$ high enough, thereby leading to poor $\gamma_{m,DL}$.

To study the impact of the femtocell-to-macrocell interference, define a surrounding area of the femtocell as *downlink dead zone* if the SINR of the macrocell user drops below a certain SINR threshold $\gamma_{m,DL,0}$ when the macrocell user gets into that area and the interference from the femtocell dominates. It is easy to verify that the contour of the dead zone can be approximated as

$$\gamma_{m,DL} \approx \frac{P_M G_{m,M}}{P_F G_{m,F}} = \frac{P_M}{P_F} \frac{1}{G_w} \left(\frac{d_{m,M}}{d_{m,F}}\right)^{-r} = \gamma_{m,DL,0}. \qquad (10.41)$$

It follows that

$$d_{m,F} \approx d_{m,M} \left(\frac{P_F}{P_M} G_w \gamma_{m,DL,0}\right)^{\frac{1}{r}}. \qquad (10.42)$$

From Table 10.1, $\frac{P_F}{P_M} \approx \frac{1}{100}$. Assume $\gamma_{m,DL,0} = 1$. With no wall protection, $G_w = 1$. Thus, $d_{m,F} \approx \frac{1}{4} d_{m,M}$. Approximating $d_{m,M} \approx d_{F,M}$, a femtocell placed $d_{F,M}$ away from the macrocell base station creates a dead zone for macrocell users with a radius equal to $\frac{1}{4} d_{F,M}$. The dead zone radius increases with the distance between the femtocell and macrocell base stations, and can be significant when the femtocell is close to cell edges. Of course, good wall protection ($G_w \ll 1$) can reduce the size of the dead zone significantly.

Now we extend the above analysis to the uplink. When no macrocell user is nearby, P_f is usually small to achieve its SINR target. Now, suppose a macrocell user moves close and its transmit power P_m causes the macrocell-to-femtocell interference. At the femtocell base station receiver, the interference may be stronger than the desired signal because the macrocell user is far away from the macrocell base station and thus $P_m \gg P_f$. As a result, $\gamma_{f,UL}$ may drop significantly. SINR-based closed-loop power control at the femtocell will respond by increasing P_f. Apparently the increase in P_f raises interference to the macrocell base station. If such an increase is significant relative to the total interference seen at the macrocell base station, closed-loop power control at the macrocell will increase the transmit power of the macrocell user, resulting in yet more macrocell-to-femtocell interference. The vicious cycle continues.

For the sake of simplicity, we do not analyze the previous SINR-based power control iterations, but examine the drop in $\gamma_{f,UL}$ assuming power control only based on path loss. Specifically, assume an inverse power control scheme

$$P_f = \frac{A_F}{G_{f,F}}, \quad P_m = \frac{A_M}{G_{m,M}}, \qquad (10.43)$$

where A_F, A_M represent the receive power targets at the femtocell and macrocell base stations, respectively. Define a surrounding area of the femtocell as *uplink dead zone* if $\gamma_{f,UL}$ drops below a certain threshold $\gamma_{f,UL,0}$ when the macrocell user gets into that area and the macrocell-to-femtocell interference dominates. Similar to (10.42), the contour of the uplink dead zone is given as

$$d_{m,F} \approx d_{m,M} \left(\frac{A_M}{A_F} G_w \gamma_{f,UL,0}\right)^{\frac{1}{r}}. \qquad (10.44)$$

Assume $\gamma_{f,UL,0} = 1$. Without wall protection ($G_w = 1$) and setting $A_F = A_M$, it follows $d_{m,F} \approx d_{m,M}$, indicating that the uplink dead zone occupies a significant fraction of the cell. Comparison of (10.42) and (10.44) indicates that everything else being equal, to equalize the sizes of the uplink and downlink dead zones, one should set

$$\frac{A_F}{A_M} = \frac{P_M}{P_F} \qquad (10.45)$$

instead of $A_F = A_M$.

Open access: handoff between femtocell and macrocell

So far we have learned that femtocell/macrocell interference may lead to dead zones. Specifically, when a macrocell user gets very close to a femtocell, it receives significant downlink interference from the femtocell base station and meanwhile causes significant uplink interference to the femtocell base station. The problem is fundamentally attributed to the limitation of closed access; that is, the macrocell user is not allowed to access the femtocell base station even if it is nearby. A natural solution is to adopt open access instead. In open access, the macrocell user will hand off to the femtocell cell, thereby mitigating the femtocell/macrocell interference. Open access is used in a macrocell system where the user can access any macrocell base station.

Because the femtocell base station is outside the operator's network, open access in the femtocell/macrocell system faces new challenges as compared with the conventional macrocell system:

- *Handoff control.* Recall from Section 9.6 that macrocell handoff can be controlled by the user or by the network. Femtocell/macrocell handoff may be initiated and controlled by the user since the operator does not control the femtocell base station. No real-time handoff coordination between femtocell and macrocell is guaranteed.
- *Security control.* There is no trust relationship between the femtocell base station and the macrocell user established beforehand. Issues to be resolved include user authentication, registration, authorization, and traffic encryption. The macrocell user needs to authenticate the femtocell base station, too. User privacy should be protected.
- *Service control.* The femtocell base station needs to deliver QoS consistent with what the user would get from macrocells. This requires proper sharing of the airlink and backhaul resources between the macrocell user and other home users.
- *Power asymmetry.* Recall that a femtocell base station uses a much lower transmit power than a macrocell base station does. As will be shown below, this leads to very different handoff boundaries for the downlink and uplink.

In the following we will focus on the issue of power asymmetry. The question we would like to address is: *when should the macrocell user hand-off to the femtocell?*

A natural solution, as used in the macrocell system, is to connect with the femtocell or macrocell base station of a stronger downlink signal strength. The condition of 0 dB downlink SIR

$$\frac{P_M G_{m,M}}{P_F G_{m,F}} = 1 \qquad (10.46)$$

defines the boundary between the femtocell and macrocell, and handoff is triggered when the macrocell user crosses the boundary ($\gamma_{m,DL,0} = 1$). This coincides with the boundary of the downlink dead zone we discussed in the closed access case. Note that because of power asymmetry ($P_M \gg P_F$), the boundary is much closer to the femtocell base station than to the macrocell one ($G_{m,M} \ll G_{m,F}$).

The challenge is then on the uplink. When the macrocell user is close to the above boundary but still connected with the macrocell, it generates significant interference to the femtocell because of large path loss ratio $G_{m,F}/G_{m,M}$, and leads to bad femtocell uplink SINR; we have seen this before in the uplink dead zone analysis in the closed access scenario. To deal with the macrocell-to-femtocell interference, the femtocell base station has to increase the receive power target significantly. As a result, the receive power target at the femtocell A_F needs to be much higher than that at the macrocell A_M. For example, to achieve 0 dB SIR in the femtocell uplink, from (10.46) we set $A_F = \frac{P_M}{P_F} A_M$. This in turn increases femtocell-to-macrocell interference and device power consumption.

Hence, to minimize the uplink interference and power consumption, from the uplink perspective the handoff condition should be 0 dB path loss ratio

$$\frac{G_{m,M}}{G_{m,F}} = 1, \qquad (10.47)$$

instead of 0 dB downlink SIR (10.46). Note that (10.47) corresponds to the uplink dead zone boundary in the closed access case with $A_F = A_M$ and $\gamma_{f,UL,0} = 1$.

Clearly, the boundary in (10.47) moves towards the macrocell base station from the one defined in (10.46). With $P_M \gg P_F$, the discrepancy of the downlink and uplink handoff conditions is significant. The phenomenon of mismatched downlink and uplink handoff boundaries does not exist in the macrocell system, because the transmit powers of macrocell base stations are similar. One way of resolving this problem is to use different spectrum in the femtocell and macrocell, as discussed later this section.

Another possibility is to exploit the notion of make-before-break (MBB) handoff presented in Section 9.3. Specifically, as the macrocell user travels close to the femtocell, once the user crosses the uplink handoff boundary, the uplink traffic is handed off to the femtocell while the downlink traffic is kept with the macrocell. Then, after the user further crosses the downlink handoff boundary, the downlink traffic is handed off to the femtocell, as well. Figure 10.15 illustrates MBB handoff in the femtocell/macrocell scenario with separate downlink and uplink handoff boundaries.

It should be pointed out that MBB handoff does not completely solve the femtocell/macrocell interference issue. In order to send uplink traffic, the user needs to receive downlink control signals, such as assignment and acknowledgment, from the femtocell base station at a low downlink SIR – as low as P_F/P_M when the user just crosses the

10.1 Heterogeneous topology

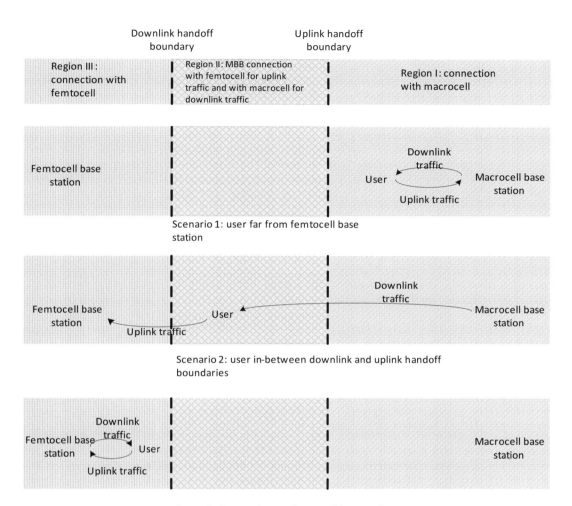

Figure 10.15 MBB handoff in a femtocell/macrocell system with separate downlink and uplink handoff boundaries. The macrocell base station is located on the right side and the femtocell base station on the left side. A user moves between the two base stations. In scenario 1, both the downlink and uplink are connected with the macrocell base station. In scenario 3, both the downlink and uplink are connected with the femtocell base station. In scenario 2, the downlink is connected with the macrocell base station and the uplink is connected with the femtocell base station.

uplink handoff boundary (10.47). Similarly, to receive downlink traffic the user needs to send uplink control signals such as SINR report and acknowledgment to the macrocell base station where the path loss ratio is high, as high as P_M/P_F when the user is about to cross the downlink handoff boundary (10.46). Therefore, specific care needs to be taken to design robust control signaling and define the handoff boundaries.

As a side benefit, we notice that only in scenario 3 of Figure 10.15 is it necessary to change the IP routing of downlink traffic packets from the macrocell to femtocell base station during which the Mobile IP home agent is updated. Therefore, as a passing macrocell user walks by different femtocells, it may perform uplink handoff multiple times without triggering the Mobile IP handoff protocol. Recall from Section 9.3.3 that the user can dynamically select the uplink among multiple base stations without informing the Mobile IP home agent.

We next study power and bandwidth allocation to further examine the tradeoffs of femtocell/macrocell handoff.

First, consider the macrocell user in the downlink shown in Figure 10.13. When it hands off to the femtocell, the tradeoff is simply resource (power and bandwidth) consumption in the femtocell and saving in the macrocell. However, to keep the same data rate for the user, the amounts of consumed and saved resource are not necessarily the same. Specifically, when the user is in the macrocell, the data rate is given by

$$R_m = \alpha_m \log\left(1 + \frac{\beta_m}{\alpha_m} \frac{P_M G_{m,M}}{\sigma_m^2 + P_F G_{m,F}}\right), \tag{10.48}$$

where α_m, β_m are the allocated fractions of bandwidth and power. After handoff, the data rate in the femtocell becomes

$$R'_m = \alpha_f \log\left(1 + \frac{\beta_f}{\alpha_f} \frac{P_F G_{m,F}}{\sigma_m^2 + P_M G_{m,M}}\right), \tag{10.49}$$

where α_f, β_f are the allocated fractions of bandwidth and power. Suppose the constraint is to keep $R_m = R'_m$. It can be shown that if $G_{m,F} P_F > G_{m,M} P_M$, then the resource saving in the macrocell is greater than the resource consumption in the femtocell. Specifically, $\beta_f < \beta_m$ if $\alpha_f = \alpha_m$, or $\alpha_f < \alpha_m$ if $\beta_f = \beta_m$. Recall that $G_{m,F} P_F > G_{m,M} P_M$ is the downlink handoff condition (10.46). Moreover, the resource in the macrocell is usually more precious than that in the femtocell, since the macrocell has larger coverage and thus higher demand. To offload the traffic from the macrocell, it is sometimes desirable to trigger macrocell-to-femtocell handoff even more aggressively than in (10.46).

When the macrocell user hands off to the femtocell, the downlink data rate of the femtocell user always decreases because a fraction of femtocell resource has to be shared with the macrocell user. However, as shown next, in the uplink there is a possibility of a win-win situation in which the the femtocell user also benefits. The reason is that the dominant uplink interference of the femtocell comes from the nearby macrocell user. Handing the user off to the femtocell significantly reduces the uplink interference. We next analyze the tradeoffs of uplink handoff.

When the macrocell user is connected with the macrocell, the uplink data rates of the macrocell and femtocell users are

$$R_m = \alpha_m \log\left(1 + \frac{1}{\alpha_m} \frac{P_m G_{m,M}}{\sigma_M^2 + P_f G_{f,M}}\right) \tag{10.50}$$

$$R_f = \log\left(1 + \frac{P_f G_{f,F}}{\sigma_F^2 + P_m G_{m,F}}\right), \tag{10.51}$$

respectively, where α_m is the fraction of bandwidth allocated to the macrocell user in the macrocell and we assume that the femtocell user is allocated all the bandwidth. After handoff, assume that in the femtocell, α'_m fraction of bandwidth is allocated to the macrocell user and $1 - \alpha'_m$ to the femtocell user. Denote by P'_m, P'_f the transmit powers after handoff. The data rates now become

$$R'_m = \alpha'_m \log\left(1 + \frac{1}{\alpha'_m} \frac{P'_m G_{m,F}}{\sigma_F^2}\right) \tag{10.52}$$

$$R'_f = (1 - \alpha'_m) \log\left(1 + \frac{1}{1 - \alpha'_m} \frac{P'_f G_{f,F}}{\sigma_F^2}\right). \tag{10.53}$$

We would like to see whether α'_m, P'_m, P'_f can be found such that rate pair (R'_m, R'_f) dominates (R_m, R_f). For simplicity, choose P'_m to keep the same SINR of the macrocell user, $\frac{P'_m G_{m,F}}{\sigma_F^2} = \frac{P_m G_{m,M}}{\sigma_M^2 + P_f G_{f,M}}$, and furthermore let $\alpha'_m = \alpha_m$. Thus $R'_m = R_m$. Note that because $G_{m,F} \gg G_{m,M}$, it follows that $P'_m \ll P_m$ as long as $\sigma_M^2 + P_f G_{f,M} \geq \sigma_F^2$. Thus, as a side benefit, the macrocell user generates far less interference to other cells after handoff.

Hence, whether macrocell-to-femtocell handoff is worthwhile depends on whether R'_f exceeds R_f. A comparison of (10.51) and (10.53) reveals the tradeoff of the femtocell user. Specifically, after macrocell-to-femtocell handoff, the femtocell user's SINR goes up thanks to the absence of the macrocell-to-femtocell interference ($P_m G_{m,F}$), but at the cost of reduced bandwidth $(1 - \alpha'_m)$. To quantitatively study the tradeoff, let $P'_f = P_f$. Define $\gamma_{f,F} = \frac{P_f G_{f,F}}{\sigma_F^2}$ and $\gamma_{m,F} = \frac{P_m G_{m,F}}{\sigma_F^2}$. Equation (10.51) and (10.53) become

$$R_f = \log\left(1 + \frac{\gamma_{f,F}}{1 + \gamma_{m,F}}\right) \tag{10.54}$$

$$R'_f = (1 - \alpha'_m) \log\left(1 + \frac{\gamma_{f,F}}{1 - \alpha'_m}\right). \tag{10.55}$$

Assuming the received power target is the same at both femtocell and macrocell $P_f G_{f,F} = P_m G_{m,M}$, it follows that $\gamma_{m,F}/\gamma_{f,F} = G_{m,F}/G_{m,M}$.

Figure 10.16 plots R'_f/R_f versus α_m. We observe that over a large range of α_m, $R'_f \gg R_f$. The rate difference increases with $G_{m,F}/G_{m,M}$. We also note that macrocell-to-femtocell handoff becomes less attractive if more bandwidth needs to be allocated to the macrocell user. Specifically, given $\gamma_{f,F}$ and $\gamma_{m,F}$, $R'_f(\alpha_m)$ decreases as α_m increases and even drops below R_f once α_m exceeds some threshold α_m^*. As shown in Figure 10.16, the threshold increases with $G_{m,F}/G_{m,M}$. We define threshold α_m^* as the critical bandwidth allocation such that $R'_f(\alpha_m^*) = R_f$:

$$\log\left(1 + \frac{\gamma_{f,F}}{1 + \gamma_{m,F}}\right) = (1 - \alpha_m^*) \log\left(1 + \frac{1}{1 - \alpha_m^*} \gamma_{f,F}\right). \tag{10.56}$$

If the actual bandwidth allocation α_m is less than α_m^*, then $R'_f(\alpha_m) > R_f$ and macrocell-to-femtocell handoff is worthwhile. The greater the critical bandwidth allocation α_m^*, the easier it is for macrocell-to-femtocell handoff to be worthwhile. Figure 10.17 plots α_m^*

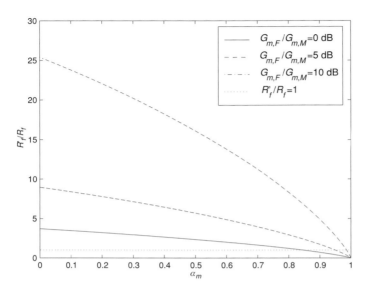

Figure 10.16 Ratio of data rates of femtocell user, before and after macrocell user hands off to femtocell. $\gamma_{m,F}/\gamma_{f,F} = G_{m,F}/G_{m,M} = 0, 5, 10$ dB. $\gamma_{f,F} = 10$ dB.

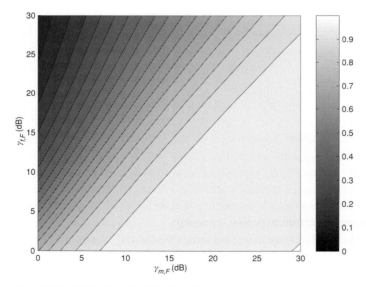

Figure 10.17 Critical bandwidth allocation α_m^*.

for a set of $\gamma_{m,F}$, $\gamma_{f,F}$. We observe that α_m^* increases as $\gamma_{m,F}$ increases and/or as $\gamma_{f,F}$ decreases, consistent with what is shown in Figure 10.16. Finally, assuming the femtocell user is in the low SINR regime before the macrocell user hands off to the femtocell and then in the high SINR regime after handoff, we can derive a simpler relationship by approximating (10.56) with

$$\frac{1}{\gamma_{m,F}} = \frac{(1 - \alpha_m^*)}{\gamma_{f,F}} \log\left(\frac{\gamma_{f,F}}{1 - \alpha_m^*}\right). \tag{10.57}$$

Mitigating femtocell/macrocell interference

Deploying femtocells inside a macrocell in the same spectrum in effect creates additional cell boundaries. The lack of real-time coordination, compounded by large power disparity, makes femtocell/macrocell interference at those cell boundaries particularly challenging to manage. In the case of closed access, femtocell/macrocell interference results in dead zones where SINR drops significantly. In the case of open access, it leads to mismatched handoff boundaries for the downlink and uplink.

Transmit power control is a well known interference management solution. Specifically, in order to maintain a similar coverage area regardless of the femtocell location within a macrocell, a femtocell base station monitors the downlink signal strength by a network listening module during initial setup or occasionally during normal operation, and adjusts its transmit power based on the path loss measurement. Thus, when the femtocell base station is closer to the macrocell base station, the femtocell base station sets a higher transmit power, and meanwhile a lower received power target as the uplink macrocell-to-femtocell interference would be smaller. However, the imbalance between the downlink and uplink could still be severe.

Femtocell/macrocell interference is the result of the same spectrum being reused in both cells. To conclude this section, we briefly present two ideas for mitigating femtocell/macrocell interference.

The first idea is *spectrum orthogonalization*. From the analysis of relays in the previous section, we have identified an important, yet challenging, problem of *reuse versus orthogonalization*. Assuming no joint encoding or decoding and treating interference as noise, two communication links can either occupy the same time-frequency spectrum (reuse) for greater bandwidth or divide the whole spectrum resource into orthogonal pieces and occupy different pieces (orthogonalize) for interference suppression. Whether to reuse or orthogonalize depends on SNR and INR. Given the challenges of femtocell/macrocell interference, it is sometimes desirable that femtocells and macrocells orthogonalize the spectrum to be used. For example, the operator can allocate a separate dedicated time-frequency spectrum to deploy femtocells.

An especially interesting scenario is when macrocells employ FFR to manage intercell interference. In particular, all the spectrum is not reused everywhere in a macrocell. In an area where an OFDMA subband is not used by the macrocell, femtocells can utilize the subband. Figure 10.18 depicts an example presented in [80]. F_1, F_2, F_3 represent three subbands F_1, F_2, F_3 to be used in the macrocell sectors and their transmit powers are set differently in sectors 1, 2, 3. $f_k, k = 1, 2, 3$, represents the same three subbands as F_k but is used in the femtocells. For example, consider sector 1. Subband F_1 is used by macrocell users everywhere and F_3 is only used by those close to the macrocell base station. Then subband f_2 is used by femtocells everywhere and f_3 is only used by those far from the macrocell base station.

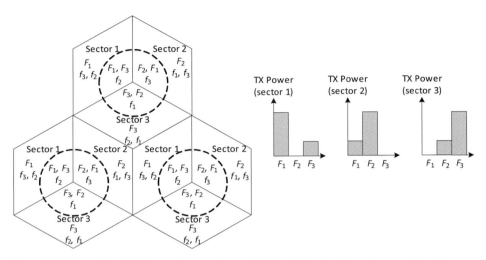

Figure 10.18 Example of static spectrum orthogonalization in femtocells and macrocells when FFR is used in macrocells.

The above example of static spectrum partition may not be efficient. In the absence of femtocells, F_2 can otherwise be used by macrocell users close to the base station. Furthermore, using f_3 at the edge of sector 1 may interfere with macrocell users using F_3 in another cell, thereby resulting in the femtocell/macrocell interference we have tried to mitigate. However, the idea of spectrum orthogonalization can be made dynamic and adaptive. For example, assume closed access. Initially, let macrocells and femtocells reuse the entire spectrum. When a macrocell user moves close to a femtocell, the macrocell user and the femtocell start to orthogonalize the occupied spectrum. As shown in Figure 10.19, after detecting a significant increase in interference in the uplink, the femtocell reduces its spectrum footprint to two subbands and leaves the third one for the macrocell user to use. The macrocell user reports that the SINR of the third subband is much superior to that in the other two, and therefore will be naturally scheduled in the third subband by a frequency selective scheduler. The two subbands used by the femtocell can be reused by other macrocell users not so close to the femtocell. Thus, the spectrum efficiency is hardly affected in the macrocell. Obviously, as the macrocell user moves away from the femtocell, the femtocell starts to utilize the third subband again. Note that the above can be achieved without real-time centralized coordination between the femtocell and macrocell base stations.

The second idea is *interference nulling*, which in effect orthogonalizes femtocells and macrocells in space using multiple antennas. Assume that the femtocell base station is equipped with multiple transmit and multiple receive antennas. In the absence of nearby macrocell users, the femtocell base station uses the multiple antennas for beamforming for power gain or spatial multiplexing for bandwidth gain. As shown in Figure 10.20, when a macrocell user gets close, the multiple antennas can be used to null the femtocell-to-macrocell interference in the downlink and the macrocell-to-femtocell interference in the uplink.

10.1 Heterogeneous topology

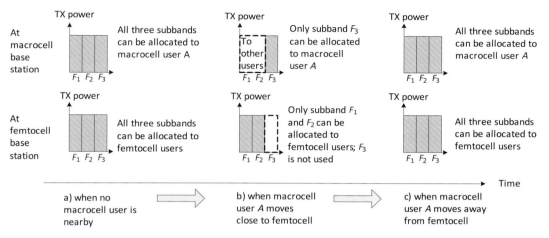

Figure 10.19 Example of dynamic spectrum orthogonalization showing that the transmit power distribution over subbands adapts over time as a macrocell user A moves close to and away from the femtocell. The example shows the downlink scenario and the same idea is applicable in the uplink.

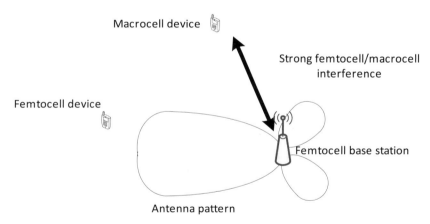

Figure 10.20 Interference nulling. Femtocell base station aligns its multiple transmit antennas to null the femtocell-to-macrocell interference in the downlink and receive antennas to null the macrocell-to-femtocell interference in the uplink.

It is interesting to contrast with the macrocell case in which interference nulling is not particularly attractive. The reason is that in a hexagonal macrocell deployment, there are usually a few dominant inter-cell interferers. However, given the typical density of macrocell users in comparison with the size of a femtocell, the femtocell/macrocell interference is often dominated by only one macrocell user, in which case the gain from interference nulling is potentially greater than the gain from beamforming or spatial multiplexing.

10.1.3 Device-to-device communications

The primary purpose of cellular systems (macrocells, picocells, or femtocells, with or without relays) is to provide network access. On the other hand, Device-to-device (D2D) communications are about direct communications between a pair of users where traffic is generated or consumed by applications that run on them.[3] This difference leads to a more complex communication topology and therefore new challenges in D2D communications, as elaborated below. In this section, we focus on single hop D2D communications where the range of D2D communications is from several meters to a few hundred meters and typically shorter than that of cellular.

Challenges

In the macrocell case, a user only communicates with one base station (except during MBB handoff), thereby naturally leading to a star topology where the base station sits at the center and connects with multiple users but does not communicate with another base station over the air. The user is allowed to connect to any base station, presumably the one with the best channel condition, a scenario referred to as *unrestricted association*. Furthermore, the base stations are deployed at well planned sites, for example, the centers of hexagons. Figure 10.21(a) depicts a regular star topology of two hexagonal macrocells where each base station is connected with a number of users close to it. As noted before, *the regular star topology is the foundation of the conventional cellular framework*.

Deploying femtocells and relays over macrocells makes the communications topology deviate from the regular star structure, as shown in Figure 10.21(a). The irregularity includes:

- In the femtocell case, a user may not be allowed to access a femtocell even though the femtocell base station has the best channel condition, a scenario referred to as *restricted association*.
- Femtocell base stations or relays are deployed in an *ad hoc* manner.
- Short links (between users and femtocell base stations or relays) are mixed with long links (between users and macrocell base stations). Those links exhibits a great level of *power asymmetry* when different transmit power levels are used.

As shown in Figure 10.21(b), the topological irregularity becomes even more significant in D2D communications, because a user may intend to communicate with any arbitrary subset of other users depending the applications. First, two users form a link because their applications intend to communicate, not just because they are nearby. A user may communicate with another faraway user in the presence of nearby interfering users. A user intending to whisper to a nearby user may be overwhelmed by yet another user using a high transmit power in a long D2D link. In D2D communications, restricted association and power asymmetry become the norm instead of an exception. Second, while the locations of femtocell base stations or relays are not optimally planned, they

---

[3] One can argue that as the ratio of base stations and users goes down, when one femtocell base station only serves one user, network access is not different from D2D communications from the interference topology perspective, especially with restricted association.

10.1 Heterogeneous topology

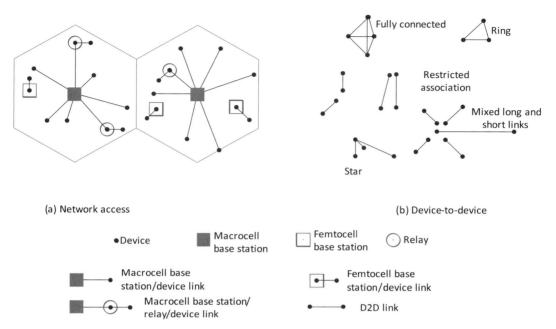

Figure 10.21 Comparison of communications topology: D2D versus network access. Network access (a) include macrocells, femtocells, and relays. D2D communications (b) exhibits an arbitrary topology, which includes stars, rings, fully connected, mixed long and short links, and restricted association.

are not completely arbitrary, either. For example, femtocell base stations are deployed in individual homes, which are naturally isolated from each other because of building walls. In D2D communications, however, users can literally be everywhere.

The topological irregularity increases the dynamic range of interference. Figure 10.22 compares the SIR distributions of three scenarios: the downlink of macrocells, the downlink of mixed macrocells and femtocells, and D2D communications. To emphasize the effect of interference, the effects of noise, fading and shadowing are ignored. The path loss channel model is the same as in Figure 6.3. In the macrocell scenario the curve is identical to the one with reuse 1 there. In the mixed macrocell and femtocell scenario, assuming closed access, the SIR of a macrocell user degrades due to the presence of femtocell base stations. The most noticeable difference lies in the longer tail in the low SIR regime caused by the event that a user is close to a femtocell base station. The tail would become worse as the density of femtocells increases. Among the three curves, the D2D one exhibits the widest spread, meaning significant disparity in SIR among the links. In particular, the low SIR tail is much heavier than that in the femtocell scenario, indicating the severe impact of restricted association and arbitrary placement of interfering users.

Large dynamic range of interference is a key challenge in D2D communications and can lead to significant ICI in OFDMA or even desensitization. Section 2.4 shows that because of real world impairments, the tone signals in OFDMA are not perfectly

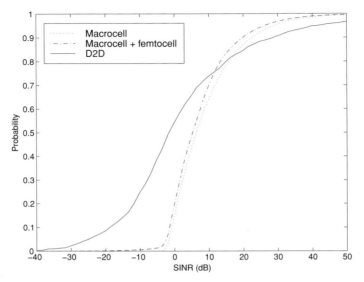

Figure 10.22 Comparison of SIR distribution of macrocell downlink, mixed macrocell and femtocell downlink, and D2D communications. In the mixed macrocell and femtocell scenario, on average two femtocell base stations are dropped uniformly in every macrocell. The transmit power of a femtocell base station is 20 dB below that of a macrocell base station. In the D2D communications scenario, assume that all D2D transmitters use the same power. The length of a D2D link is uniformly random between 0 and a quarter of the site-to-site distance of two adjacent macrocell base stations. D2D transmitters are uniformly dropped with the density of on average ten per macrocell.

orthogonal. If an interfering tone is much stronger than a desired tone, then the ICI could be significant. Sometimes, an interfering signal is strong enough to desensitize the receiver front-end and bury the desired signal in the noise. Desensitization is more likely to occur in D2D than in typical cellular communications. As an example, suppose that desensitization occurs when the interference is 60 dB stronger, which, at a path loss exponent equal to 4, translates into a factor of about 30 difference between the desired path distance and an interfering one. Such an interference scenario is certainly possible in D2D communications.

Large dynamic range of interference implies that the SIR distribution may vary substantially in space. Indeed, it often varies in time as well, because of user mobility, and more importantly, the fluctuation of traffic needs. In many applications, such as gaming and instant messaging, the traffic queue is not always full. This is very different from a cellular base station, which aggregates traffic to multiple users and usually has a full queue thanks to the statistical multiplexing effect. Figure 10.23 compares the SIR distributions under different traffic loading models and shows that in the absence of dominant interferers, the low SIR tail becomes much lighter. We thus expect large SIR fluctuation depending on whether dominant interferers are active.

Thanks to the regular star topology, interference management in the macrocell case is relatively simpler. Specifically, in OFDMA, users within a cell are scheduled by the base

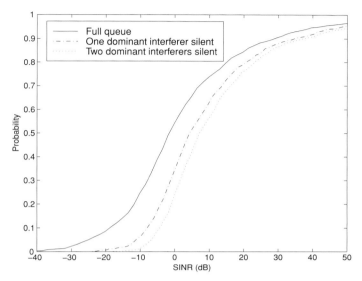

Figure 10.23 Comparison of SIR distributions of D2D communications in three scenarios: when all interferers are active (full queue), when the most dominant interferer is silent, and when the two most dominant interferers are silent.

station, in a centralized manner, to orthogonal spectrum resources so as to eliminate intra-cell interference. The dynamic range of inter-cell interference is relatively small because each user is connected to the best base station. Spectrum resources are reused among cells according to a certain reuse principle, either universal reuse or fractional frequency reuse. To a large extent, inter-cell interference management is simply reduced to interference averaging.

Here we would like to point out one implicit, yet important, tradeoff. We have assumed that the downlink in one cell never interferes with the uplink in another cell, because they do not reuse the same spectrum resource. Consider a TDD system where the time is partitioned into the downlink and uplink intervals. TDD is known to be flexible in adjusting time partition depending on traffic needs. However, such an adjustment has to be *system wide*. Specifically, at a given time, cells are either all in downlink or all in uplink. The benefit is a great reduction in the dynamic range of interference, because a base station only receives interference from users and not from other base stations, and a user only receives interference from base stations and not from other users. Otherwise, imagine a scenario where one cell is in downlink while an adjacent cell is in uplink. Then a downlink signal to an edge user in one cell may be overwhelmed by an uplink signal from another user in the other cell when the two users are sufficiently nearby. Obviously, the drawback of system wide downlink/uplink partition is that a cell cannot unilaterally changes its downlink/uplink partition without other cells all changing at the same time.

In D2D communications, because of large SIR variation in space and in time, we cannot simply apply a fixed reuse pattern across all the D2D links. Spectrum reuse

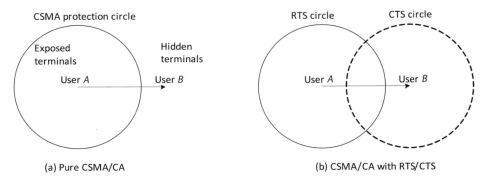

Figure 10.24 Illustration of CSMA/CA (a) with the issues of hidden terminals and exposed terminals and (b) CSMA/CA with RTS/CTS.

among links would have to be based on instantaneous local SIR geometry. The notions of downlink and uplink are not particularly helpful in D2D communications, because there are no systematic (and distributed) ways of designating one direction of any link as downlink and the other as uplink so as to greatly reduce the dynamic range of interference between links. On the other hand, any system wide resource partition potentially reduces utilization as it prevents one direction of a link from utilizing the otherwise idle resource when the other direction does not have traffic needs. Such a scenario occurs more often in D2D communications because a user, unlike a base station, does not see much statistical multiplexing of traffic and probably has an empty queue from time to time.

Interference management: CSMA/CA versus macrocell approaches

Carrier Sensing Multiple Access/Collision Avoidance (CSMA/CA) is designed to be robust against unknown interference, especially in an unlicensed spectrum environment. Designed to operate over an unpaired channel, CSMA/CA is fundamentally different from the cellular approaches studied so far. To illustrate the idea, consider a pair of users A and B, with A intending to transmit to B, as shown in Figure 10.24. User A is called transmitter and B receiver, and the link from A to B is denoted by link $A \to B$.

User A first senses whether the channel is busy. If so, presumably some other transmission is going on, and user A has to wait until the channel becomes idle. Once the channel is idle, user A waits for a random time period before transmitting. The random backoff is to reduce the probability that multiple transmitters would simultaneously start transmitting as soon as the channel is idle. For the purpose of this section, we will focus on interference management and spatial packing of links, and ignore other aspects such as collision in random access.

In pure CSMA/CA, the channel is considered idle if the received power is below some threshold γ_0. Therefore, when user A is transmitting at power P_A, any user is blocked if the channel gain from user A exceeds P_A/γ_0. This is equivalent to drawing around user A a "CSMA protection circle" (the left one in Figure 10.24) whose radius is a function of P_A/γ_0. The notion of distance here corresponds to a wireless channel propagation distance that depends on both physical distance and channel fading. Users in the CSMA protection circle are blocked and will not interfere with link $A \to B$.

However, the CSMA protection circle is neither sufficient nor necessary to protect link $A \to B$. Specifically, the CSMA approach suffers from two problems.

- A *hidden terminal*, which is a user close to user B but still outside of the CSMA protection circle around user A, will sense that the channel is idle and start to transmit, thereby interfering link $A \to B$.
- An *exposed terminal*, which is a user within the protection circle of user A but far from user B, will cause no harm to link $A \to B$, but be blocked from transmitting.

Collision avoidance with RTS (request to send) and CTS (clear to send) is used to address the drawbacks of pure CSMA/CA. After sensing that the channel is idle and waiting for a random time period, user A first transmits an RTS instead of traffic itself. Upon receiving the RTS, user B then responds by transmitting a CTS. Finally, user A transmits traffic. The RTS draws a protection circle around user A as before. The CTS draws a second protection circle around user B, as shown in Figure 10.24. While RTS/CTS resolves the hidden terminal problem, the exposed terminal problem still exists.

RTS and CTS are transmitted at the maximum power even when A and B are nearby each other. Therefore, the sizes of the two protection circles are big and do not vary with the length of link $A \to B$. The medium access principle of CSMA/CA works as follows. When the channel is idle, users A and B use RTS/CTS to exclude other users inside the protection circles from reusing the spectrum. When the link length is short, the SINR, and thus the burst data rate, are high. This way, user A can quickly finish its transmission and relinquish the channel to other users. However, in a dense deployment every user takes a turn to transmit, and because the protection circles are big, user A only occupies the channel for a small fraction of time. Hence, from the perspective of the SINR and bandwidth tradeoff, CSMA/CA achieves high SINR at the cost of low bandwidth for a given link.

It should be pointed out that operating in the high SINR regime is not power efficient. Specifically, when the link length is short, one may be better off making the protection circles smaller (sacrificing SINR) and allowing more users to transmit simultaneously (encouraging reuse). Indeed, a more beneficial way of managing interference is to control SINR instead of drawing fixed-size protection circles. From communications theory, the CSMA/CA protection is, in general, neither necessary nor sufficient. These problems with CSMA/CA and RTS/CTS have been noted in literature. In particular, the insufficient protection provided by RTS/CTS has been observed in [177]. Also, the unnecessary excess protection provided by RTS/CTS has been studied in [182]. In essence, successful decoding occurs at user B as long as the SINR is sufficient. Assuming the noise is negligible compared to the interference, this implies that the protection circle drawn by user B should be of a variable radius proportional to the RF-distance between users A and B. This is illustrated in Figure 10.25. As we will show next, this condition ensures a protection of fixed $A \to B$ SIR. With this SIR-based

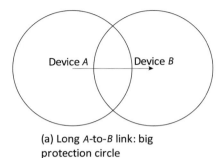 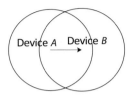

(a) Long A-to-B link: big protection circle

(a) Short A-to-B link: small protection circle

Figure 10.25 Illustration of protection circles in SIR-based mechanism. The radius is proportional to the RF-distance between A and B. As a comparison, in CSMA/CA, the protection circles are of a fixed maximum radius.

mechanism, spatial packing is channel/interference-aware and much more efficient than CSMA/CA, leading to significant spatial throughput gains.

Synchronous SIR-based distributed scheduling

In the following we present an SIR-based scheduling scheme proposed in [172]. Similar to CSMA/CA, we consider an unpaired channel for D2D communications. A user is subject to the half-duplex constraint such that it cannot simultaneously transmit and receive. Unlike the traditional cellular TDD that operates over an unpaired channel, there is no system wide notion of downlink and uplink.

To illustrate the main elements of the scheme, first look at a simple two-link example shown in Figure 10.26(a). A intends to transmit to B, and C to D. The two links are denoted by $A \to B$ and $C \to D$. Users A, C are called transmitters and B, D are receivers. Here, two links have direct channel gains G_{AB}, G_{CD}, and cross channel gains G_{AD}, G_{BC}. Because of TDD, we assume channel reciprocity (e.g., $G_{AB} = G_{BA}$). In this setting, it is clear from Discussion Notes 10.1 that if the cross channel gains are small, the links $A \to B$ and $C \to D$ will not significantly interfere with each other, and thus should be simultaneously scheduled (reuse). On the other hand, if the cross channel gains are large, only one of the links should be scheduled at any time (orthogonalization).

Recall that we have already encountered the reuse versus orthogonalization problem in the study of relay scheduling options of Figure 10.7. In D2D, without a centralized scheduler, a simple way of determining reuse versus orthogonalization in a distributed way is to preassign priorities to the two links. The understanding is that the high priority link always gets scheduled. However, for the low priority link to also be scheduled, it has to check whether its transmission is going to cause excessive SIR damage to the high priority link. This is done by comparing with an SIR threshold the *predicted* SIR of the high priority link if the low priority link did proceed with its transmission. This mechanism ensures that the high priority link is protected, and both links get scheduled only if the cross channel gains are "weak enough." By randomizing the preassigned priorities of links over time, fairness across links can be maintained.

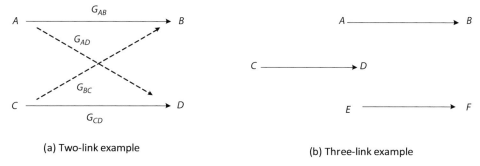

(a) Two-link example (b) Three-link example

Figure 10.26 Scheduling of two links and three links.

Specifically, in Figure 10.26(a), assuming link $A \to B$ currently has high priority, link $C \to D$ can potentially be scheduled simultaneously if C *does not cause too much interference to* B; that is, the SIR of link $A \to B$ due to C's transmission should be at least γ_{TX}:

$$\frac{P_A G_{AB}}{P_C G_{BC}} > \gamma_{TX}, \qquad (10.58)$$

where P_A represents the power used by A and P_C the power by node C. Since (10.58) is a constraint for transmitter C, it is referred to as the *TX yielding* condition.

The TX yielding condition suffices in the simple two-link example above. In a more general multiple-link network, we need another condition to determine whether to schedule the $C \to D$ link. Consider Figure 10.26(b), where the priority order from the highest to lowest is $A \to B$, $C \to D$, and $E \to F$. Link $C \to D$ does not significantly interfere with link $A \to B$, because C is far from B. However, since D is close to A, the SIR seen by D would be low. In such a scenario, it may be better off for link $C \to D$ not to transmit from the overall network perspective. This will potentially allow for much better packing where another link $E \to F$ can be scheduled with a much higher SIR. Thus, for link $C \to D$ to be scheduled, D *should see a reasonable SIR if scheduled*; that is, the SIR of link $C \to D$ due to A's transmission should be at least γ_{RX}:

$$\frac{P_C G_{CD}}{P_A G_{AD}} > \gamma_{RX}. \qquad (10.59)$$

Since (10.59) is a constraint for receiver D, it is referred to as the *RX yielding* condition.

In summary, the SIR-based scheduling scheme consists of the following three main elements:

1. a fair mechanism for assigning priorities to links and randomizing the link priorities over time,
2. a TX yielding criterion to protect high priority links, and
3. an RX yielding criterion to further improve network spatial packing in a multi-link scenario.

It is desirable that these criteria can be checked in a distributed way. At first, it may appear that C needs to know P_A, G_{BC} and G_{AB}, and D needs to know P_A, G_{CD} and G_{AD}, thereby involving significant signaling overhead. Fortunately, as shown next, C and D can estimate the SIRs by measuring the power of two signals from A and B, namely the *direct power signal* and the *inverse power echo*.

A direct power signal is sent by a transmitter at a power directly proportional to the power at which the transmitter intends to transmit the traffic. Thus, A sends a direct power signal at power P_A, and C sends one at power P_C. D receives the signal from A at power $P_A G_{AD}$ and that from C at power $P_C G_{CD}$. Taking the ratio of the measured powers of these two received signals, D can estimate the SIR $\frac{P_C G_{CD}}{P_A G_{AD}}$ and check the RX yielding condition (10.59).

After receiving the direct power signal from its own transmitter, if the receiver decides that it does not have to yield, it sends an inverse power echo at a power inversely proportional to the received power of the direct power signal. Thus, B receives the direct power signal from K at power $P_A G_{AB}$, and sends an inverse power echo at power $\frac{K}{P_A G_{AB}}$ for a fixed system constant K. This signal is received by C at power $r_p = \frac{K G_{BC}}{P_A G_{AB}}$, from which C can estimate the SIR at B to be $\frac{K}{r_p P_C}$ and check the TX yielding condition (10.58).

In summary, the direct power signal sent by a high priority transmitter allows a low priority receiver to predict its own SIR so as to check the RX yielding condition. The inverse power echo sent by a high priority receiver allows a low priority transmitter to predict the SIR at the high priority receiver so as to check the TX yielding condition.

The idea of using those two signals to check the TX and RX yielding conditions can be extended to a more general multi-link network. Because of the broadcast nature of the wireless channel, a transmitter only sends one direct power signal rather than multiple ones to individual receivers, and a receiver only sends one inverse power echo rather than multiple ones to individual transmitters. We consider a cascaded scheduling algorithm where in the present slot, links are assigned with a strict priority order, and scheduled in a sequential manner starting from the one with the highest priority. The priority order varies in a pseudo-random manner over time to ensure fairness. The flow chart of the algorithm is shown in Figure 10.27. A link at priority level L is scheduled if and only if the transmitter of link L satisfies the TX yielding condition and the receiver satisfies the RX yielding condition:

- *TX yielding*: link L's transmitter does not cause too much interference to any higher priority link; that is, the SIR of *any* higher priority link is at least γ_{TX} dB assuming the interference comes *solely* from the transmitter of link L.
- *RX yielding*: link L's receiver will see a reasonable SIR; that is, the SIR of link L is at least γ_{RX} dB, where the interference comes from *all* the higher priority links.

Here we implicitly assume that a user belongs to only one link. In a complex topology such as a fully connected graph, ring, or star shown in Figure 10.21(b), a user may simultaneously have links with multiple other users. In that case, the user may RX yield in one link because it intends to transmit in another link.

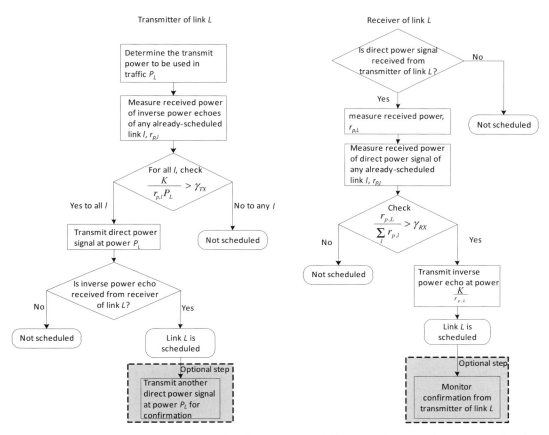

Figure 10.27 Flow chart of SIR-based cascaded scheduling algorithm. Steps in gray are optional.

Because of the distributed nature, the algorithm suffers from two problems. The first problem is *cascaded yielding*. Consider the three links in Figure 10.26(b), but now move user F close to C. C does not TX yield to B and thus sends a direct power signal. D RX yields to A and link $C \to D$ is not scheduled, which is however not known to F. Since the received power at F of the direct power signal from C is high, F unnecessarily RX yields to C. The cascaded yielding problem can be resolved if we add one step where C confirms whether the link is scheduled depending on whether it receives the inverse power echo from D, as shown in the gray box of Figure 10.27. The second problem is that *no minimum SIR is guaranteed for any link*. Multiple low priority transmitters may interfere with a high priority receiver simultaneously. Since each of the transmitters independently checks the TX yielding condition and is not aware of others, the actual SIR of the receiver is not guaranteed to be at least γ_{TX}. When their interference adds together, the SIR may fall below γ_{TX}. Moreover, since the RX yielding condition does not take into account interference from low priority transmitters, γ_{RX} is not guaranteed, either. As a result, the SIR is not precisely controlled in link scheduling. However,

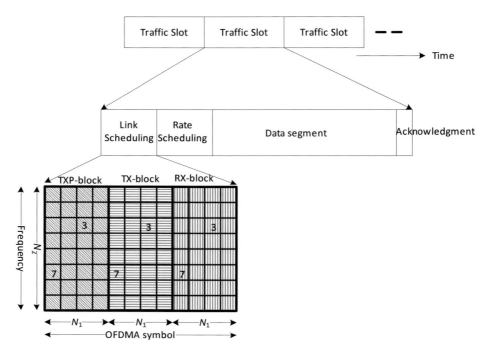

Figure 10.28 Traffic channel structure.

because of subsequent rate scheduling, the SIR can be accurately estimated for the purpose of rate selection.

Next we outline a practical channel signaling structure that implements the above SIR-based scheduling scheme. We assume that the users derive fine-grained timing and frequency synchronization from external sources such as cellular base stations or GPS, which enables synchronous operation of traffic scheduling. As shown in Figure 10.28, the channel is divided into a sequence of traffic slots. The duration of a traffic slot is a few milliseconds. A traffic slot is further divided in a TDM manner into four subchannels – link scheduling, rate scheduling, data segment, and acknowledgment – each of which occupies the entire bandwidth:

- *Link scheduling.* All the links that intend to transmit in the present traffic slot compete with each other in order to get scheduled. The reuse versus orthogonalization decision is made in a distributed manner by individual users. Not all the links get scheduled, as some of them decide to TX yield and/or RX yield to others. Only those that have passed the TX yielding and RX yielding checks are scheduled in the present slot and proceed to the next step of rate scheduling.
- *Rate scheduling.* The scheduled links determine the coding and modulation scheme and burst data rate to be used in the present data segment. The transmitters send pilots, based on which the receivers estimate and report the SINR. The SINR estimation in rate estimation is based on the total interference that will actually show up in the data

segment and is therefore more accurate than the SIR estimation and prediction in link scheduling.
- *Data segment.* All the scheduled links transmit actual traffic in the data segment using the coding and modulation scheme determined in rate scheduling. In this simple protocol, the data segments of all the links reuse the entire bandwidth of the present slot.
- *Acknowledgment.* The receivers acknowledge whether the data segments have been received correctly or not. The acknowledgments allow fast retransmissions if needed.

In principle, link scheduling is done independently in every traffic slot. The memoryless nature makes it suitable to deal with rapidly changing interference and traffic pattern.

We next focus on link scheduling. As shown in Figure 10.28, the link scheduling subchannel consists of three blocks, each of which is N_1 OFDMA symbols. An OFDMA symbol consists of N_2 tones. Thus, each block consists of $N_1 N_2$ tone-symbols. Every link has a link-ID, which corresponds to a tone-symbol in each of the three blocks. For example, Figure 10.28 shows the corresponding tone-symbols of link-IDs 3 and 7. There are $N_1 N_2$ link-IDs in total. A link management protocol is used to allocate *locally unique* link-IDs among links in a distributed manner and reuse link-IDs spatially among links that are far away from each other. Because the duration over which a link communicates is usually at least a few seconds, the link management protocol operates at a much slower time scale than the traffic scheduling, which occurs every slot of a few milliseconds. The link-ID reuse is based on an SIR criterion analogous to that used in link scheduling. The details are omitted due to limited space.

Within each block, define a priority order where the highest priority is the top-left tone-symbol and the lowest is the bottom-right one, and with lexicographical ordering and the x-axis having higher priority than the y-axis. Thus, in the present slot, link-ID 7 is of higher priority than 3. The mapping from a link-ID to a tone-symbol in the link scheduling blocks varies every slot in a pseudo-random manner to ensure fairness across links.

In the TX-block, a transmitter sends a direct power signal in the corresponding tone-symbol if it intends to compete for scheduling in the present slot. All the receivers listen to the TX-block and determine if they receive the direct power signals from the corresponding transmitters, and if so, whether to RX yield. If the receiver decides "not scheduled" according to the flow chart in Figure 10.27, it does not respond; that is, it sends no power in the corresponding tone-symbol in the RX-block. Otherwise, it "lights up" the tone-symbol to send an inverse power echo. The transmitters that have sent the direct power signals listen to the RX-block to determine if they receive the inverse power echoes from the corresponding receivers, and if so, whether to TX yield.

Note that the link scheduling subchannel in Figure 10.28 does not strictly follow the cascaded scheduling logic of Figure 10.27 in that the entire TX-block occurs prior to the RX-block. Thus, a low priority link transmitter sends its direct power signal in the TX-block *before* it measures the RX-block to check the TX-yielding condition. The rationale of the design of Figure 10.28 is to tightly pack the signaling resources of the direct power signals and inverse power echoes.

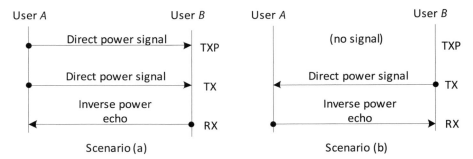

Figure 10.29 Use of the TXP-block. In the present slot, user A has the primary right of being the transmitter. In scenario (a), user A intends to be the transmitter. In scenario (b), user A does not intend to be the transmitter, but user B intends to be the transmitter. In either scenario, the present slot is not wasted whether user A has traffic to send and there is no collision between users A and B.

The TX-block and RX-block are used for interference management among competing D2D links. Here we assume that between the two users of a given link, which user is the transmitter and which one is the receiver have been known to both. How do the two users reach this consensus? One way is to designate one user to be the transmitter and the other the receiver for a given slot beforehand. However, the user designated as the transmitter in the present slot may have no traffic to send, thereby wasting the slot. Such a hard-coded designation results in channel under-utilization. The TXP-block in Figure 10.28 is used to resolve this problem. In a slot, one of the two users is designated to have the *primary right* of being the transmitter beforehand. If that user (for example, user A) does intend to be the transmitter, it sends a direct power signal in the corresponding tone-symbol in the TXP-block to announce its intention, and then another direct power signal in the TX-block; otherwise, it simply keeps silent in the TXP-block. The other user (B) is required to monitor the TXP-block to see whether user A intends to be the transmitter, and if not, it can become the transmitter and send a direct power signal in the TX-block. This logical flow is shown in Figure 10.29. This way, no traffic slot is wasted when the primary user has no traffic to send.

Figure 10.30 presents a simulation comparison of the SIR-based distributed scheduling scheme with IEEE 802.11g representing a real-world CSMA/CA protocol. The simulation is based on a detailed software implementation of both schemes, and all signaling overheads are fully accounted. Both outdoor and indoor settings are simulated. Link lengths are chosen to be 20 m. For the outdoor deployment, links are dropped randomly in a 1000 m × 1000 m square with wrap-around. For the indoor deployment, links are dropped randomly in a 50 m × 100 m × 20 m building with five floors, each of height 4 m. The detailed simulation parameters can be found in [172]. By varying the number of links from 1 to 256, Figure 10.30 shows that the sum rate saturates much earlier and at a much lower value with 802.11g, and that the SIR-based scheme results in a factor of 4 to 5 increase in sum throughput with 256 links.

10.1 Heterogeneous topology

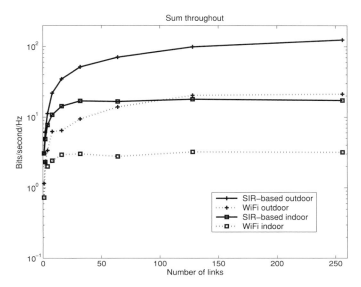

Figure 10.30 Sum rate comparison of SIR-based scheme and IEEE 802.11g protocol.

Finally, it is worthwhile to pay attention to the signaling scheme used in link scheduling. While the position of a direct power signal or inverse power echo in the time-frequency resource grid represents a corresponding link-ID, the information for interference management is exchanged not via explicit bits but via power measurement in an *analog* fashion. In the simplest form shown in Figure 10.28, a direct power signal or inverse power echo is a single tone-symbol. This of course minimizes the resource overhead. Furthermore, the fact that a user transmits a single tone signal at a given OFDMA symbol leads to two salient features that are important in practice [121]:

- The PAPR of a single tone signal is ideal: the baseband power does not fluctuate except for the transition between successive OFDMA symbols. This allows the user to boost its transmit power without a large power backoff required for linearity protection at the power amplifier. Depending on how stringent the out-of-band emission requirement is, the user can operate beyond the conventional linear region of the power amplifier to maximize power efficiency.
- During link scheduling, the signal dynamic range is potentially very large, since competing links can show up with arbitrary geometry before interference management takes effect. When one signal is strong enough to desensitize another weak signal, the only way of telling them apart is to let them transmit at different symbol times, for example, via pseudo-random time hopping. Even when desensitization does not happen, the in-band ICI can still be significant enough to cause errors. With a single tone signal, the nonlinearity of the power amplifier does not generate in-band ICI but higher-order harmonics at the multiples of the carrier frequency, which are far from the carrier frequency (on the order of GHz) and can therefore be filtered out easily.

Discussion notes 10.1 Gaussian interference channel capacity

Several of the interference scenarios encountered in this chapter, including relay-to-relay interference, femtocell-to-femtocell interference, and D2D interference, exhibit unique characteristics:

- Interference dominates noise, because the communication distances in those scenarios are usually shorter than the counterparts in typical macrocell communications.
- Interference mostly comes from a small number of dominant interfering transmitters; as a result, the assumption that interference is averaged from many sources and can be treated as Gaussian noise is no longer valid.
- Sometimes interference can be stronger than the desired signal because of restricted association.

The capacity of the interference channel is not known in general. For the special two-user case, a recent work [48] obtained the capacity within one bit by calculating a new capacity upper bound and demonstrating that a special communication scheme achieves a rate within one bit/s/Hz of the upper bound. In the following, we provide a brief summary of that work and connect it with the discussion of reuse versus orthogonalization in this chapter.

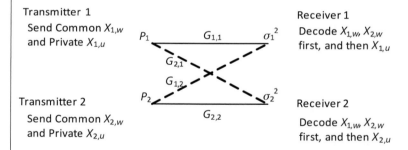

Figure 10.31 Two-user Gaussian interference channel and Han-Kobayashi scheme.

Figure 10.31 shows the model of the two-user Gaussian interference channel where transmitter i intends to communicate to receiver i, $i = 1, 2$. Denote by R_i the achieved data rate from transmitter i to receiver i given a communication scheme. The channel from transmitter j to receiver i is an interference channel for $i \neq j$. Denote by $G_{i,j}$ the channel power gain from transmitter j to receiver i, P_j the average power constraint of transmitter j, and σ_i^2 the Gaussian noise power at receiver i.

We focus on the case of a symmetric channel and a symmetric data rate where $G_{1,1} = G_{2,2}$, $G_{1,2} = G_{2,1}$, $P_1 = P_2$, $\sigma_1^2 = \sigma_2^2$, $R_1 = R_2$. The nonsymmetric cases are examined in [48]. Define signal-to-noise-ratio (SNR) $\gamma = G_{i,i} P_i / \sigma_i^2$ and interference-to-noise-ratio (INR) $\zeta = G_{i,j} P_j / \sigma_i^2$ for $i \neq j$. We are interested in the high SNR INR regime ($\gamma \gg 1$, $\zeta \gg 1$) to capture the interference-limited behavior.

We study both the weak interference regime where $\zeta < \gamma$ and the strong interference regime where $\zeta \geq \gamma$.

For the two-user interference channel, there are two suboptimal communication schemes. In the first scheme (reuse), the two transmitters completely reuse the spectrum and each receiver treats interference as noise. In the second scheme (orthogonalization), each transmitter only uses the non-overlapping half of the bandwidth to completely avoid interference. The performance of the two suboptimal schemes is easy to characterize. We are interested in the performance of an optimal communication scheme. Denote by C the capacity of the symmetric channel where $R \leq C$ for any communication scheme. An upper bound of C given in [48] is written as follows:

$$C \approx \begin{cases} \log(\gamma/\zeta), & \text{if } \log \zeta < \frac{1}{2} \log \gamma, \\ \log \zeta, & \text{if } \frac{1}{2} \log \gamma \leq \log \zeta < \frac{2}{3} \log \gamma, \\ \log(\gamma/\sqrt{\zeta}), & \text{if } \frac{2}{3} \log \gamma \leq \log \zeta < \log \gamma, \\ \log \sqrt{\zeta}, & \text{if } \log \gamma \leq \log \zeta < 2 \log \gamma, \\ \log \gamma, & \text{if } \log \zeta \geq 2 \log \gamma. \end{cases} \quad (10.60)$$

Furthermore, it is shown that except for $\frac{2}{3} < a < 1$, the difference between C and the approximation in (10.60) asymptotically goes to zero when SNR and INR go to infinity and the ratio of INR in dB and SNR in dB is fixed.

To formalize the above approximation in the high SNR INR regime, fix the ratio of INR in dB and SNR in dB

$$a = \frac{\log \zeta}{\log \gamma} \geq 0 \quad (10.61)$$

as γ, ζ grow large. For a given a, define the generalized degrees of freedom

$$d(a) = \lim_{\gamma,\zeta \to \infty} \frac{C}{C_0}, \quad (10.62)$$

where $C_0 = \log(1 + \gamma)$ is the capacity of the AWGN channel without interference. The insight is that $d(0) = 1$ without interference and $d(a) \leq 1$ for $a > 0$ because interference does not increase the capacity. Therefore, $d(a)$ reflects the loss in degrees of freedom in the high SNR INR regime. From (10.60), it follows that

$$d(a) = \begin{cases} 1 - a, & \text{if } 0 \leq a < \frac{1}{2}, \\ a, & \text{if } \frac{1}{2} \leq a < \frac{2}{3}, \\ 1 - \frac{a}{2}, & \text{if } \frac{2}{3} \leq a < 1, \\ \frac{a}{2}, & \text{if } 1 \leq a < 2, \\ 1, & \text{if } a \geq 2. \end{cases} \quad (10.63)$$

Figure 10.32 plots the generalized degrees of freedom $d(a)$ as a function of a, together with the performance of the two suboptimal schemes. A few points are worth highlighting:

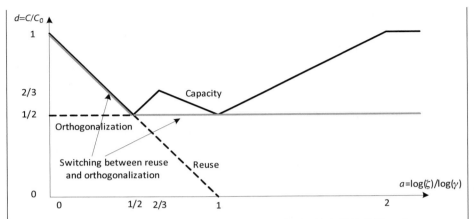

Figure 10.32 Generalized degrees of freedom versus ζ/γ in the high SNR INR regime.

- When interference is very weak ($a \leq \frac{1}{2}$), the capacity decreases as interference increases. The reuse scheme, which treats interference as noise, turns out to be optimal.
- When interference is weak ($\frac{1}{2} \leq a < 1$), the capacity no longer decreases monotonically. The reason will be explained below.
- When interference is strong ($a > 1$), the capacity increases as interference increases and eventually ($a > 2$) reaches its maximum, which is equal to the AWGN capacity as if interference does not exist. The reason is that strong interference – at least part of it – can be eliminated with interference cancellation.
- The orthogonalization scheme is never the optimal solution, except at two points of $a = \frac{1}{2}, 1$.

It is shown in [48] that a simple Han-Kobayashi scheme can achieve the capacity with one bit per second per Hz. Specifically, the signal of transmitter i consists of two streams

$$x_i = x_{i,u} + x_{i,w} \tag{10.64}$$

for $i = 1, 2$. $x_{i,w}$ comes from a common codeword to be decoded by both receivers. $x_{i,u}$ comes from a private codeword to be decoded only by receiver i, but appears as noise in the other receiver j. Clearly, the total transmit power is split between the two streams $P = P_u + P_w$. Furthermore, set the private power so that the private codeword signal $x_{i,u}$ is received at the noise power level at the other receiver ($\zeta_p = G_{i,j} P_u / \sigma^2 = 1$). Each receiver decodes all the common codewords first, cancels them, and then decodes its own private codeword, as shown in Figure 10.31.

The observations of Figure 10.32 can be explained from the use of the simple Han-Kobayashi scheme. When interference is very weak ($a \leq \frac{1}{2}$), the common codeword is negligible; the private rate, and thus the capacity, decrease as ζ increases. When $\frac{1}{2} \leq a < 1$, the common codewords are used to partially cancel interference. Increasing ζ shows two opposing effects. Initially ($\frac{1}{2} \leq a < \frac{2}{3}$), as ζ increases, more common information can be decoded and canceled; thus the capacity increases. Then, as ζ

further increases ($\frac{2}{3} \leq a < 1$), less private information can be sent because of the constraint $\zeta_p = 1$; thus the capacity decreases. Finally, when the strong interference regime is reached ($1 \leq a < 2$), all the information is common; the capacity increases again with ζ. When the interference is very strong ($a = 2$) all the interference can be canceled before decoding its own information and thus interference no longer has any effect on the capacity.

In most of this chapter, we limit ourselves to not use interference cancellation. We examine the tradeoff of the two suboptimal schemes of reuse and orthogonalization, and aim at switching between the reuse and orthogonalization schemes at an appropriate switching point around $a = \frac{1}{2}$, as shown in Figure 10.32. The practical challenge is to apply the idea to the real world where the channel is not as simple as the symmetric two-user interference channel model and furthermore interference management/scheduling is done in a distributed manner. One such example is the TX yielding and RX yielding protocols that choose between reuse and orthogonalization in the synchronous SIR-based distributed scheduling presented in Section 10.1.3.

10.2 Cooperative communication

Signals sharing the same bandwidth interfere with each other, and interference is a key challenge in wireless communications. So far we have studied three basic approaches of dealing with interference in a conventional cellular system:

1. Spatial spectral reuse ($K > 1$) where neighboring cells do not use the same spectral resource but instead use non-overlapping orthogonal resources.
2. Interference averaging, which allows full reuse ($K = 1$) of the same resource in each cell using spread spectrum techniques such as direct sequence CDMA or frequency hopping OFDMA.
3. Interference coordination, such as FFR, which further improves the performance by coordinating the power and bandwidth allocation among neighboring cells.

The above approaches take a non-cooperative view, where we assume that interference can only be treated as noise to be *mitigated* rather than *exploited*. However, the view is evolving. More recently, researchers have proposed to exploit the broadcast nature, instead (see [54, 85, 86, 103]). Specifically, when a signal transmitted to a receiver is overheard by a third communication node, that node can help the receiver to recover the signal. This new cooperative view forms the basis of cooperative communication. Recall that a central theme in Chapter 4 is that the wireless fading channel is considered something to be mitigated in a conventional view, but now something to be exploited in mobile broadband communications. Cooperative communications represents the same evolution of the thought process, this time applied to interference instead of channel.

Compared with conventional cellular systems, cooperative communication relies on more sophisticated transmission and coding techniques, and is still in an active research

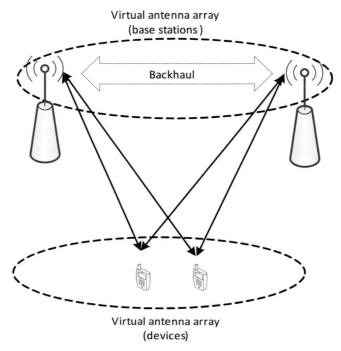

Figure 10.33 Illustration of network cooperation (network MIMO).

stage. There are two basic forms of cooperative communication: *user cooperation* and *network cooperation*.

User cooperation is very similar to the relay idea shown in Figure 10.2. In Section 10.1.1, we studied the multi-hop feature of relays and assumed that a user communicates either directly with a base station or indirectly through a relay, but *not both*. User cooperation takes a further step and allows the receiver to use the signals from *both* the direct and indirect paths for decoding.

Network cooperation exploits the pre-existing backhaul infrastructure that has linked the base stations together such that the base stations cooperate to encode the downlink signals and to decode the uplink signals. This form of cooperation is more powerful than interference coordination, because in the latter case the base stations share only channel power gain and transmit power information, but not the user signals themselves. In network cooperation, rather than just a single base station, a network of base stations as a whole is serving a user. As shown in Figure 10.33, the antennas of the base stations in different cells form a *virtual antenna array*. So do the antennas of the users. Transmitting/receiving data streams belonging to multiple users is very similar to multiuser MIMO studied in Chapter 7. Therefore, network cooperation can also be viewed as network or multi-cell MIMO. Compared with multiuser MIMO, network cooperation has the advantage that the base station antennas are usually far apart, thus achieving better spatial diversity and multiplexing. On the other hand, to achieve network cooperation, the base stations share information through the backhaul. In practice, the

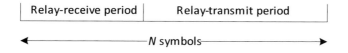
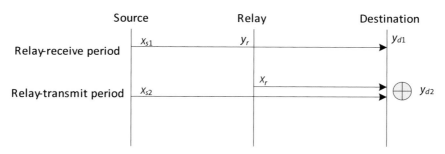

Figure 10.34 Time-division half-duplex relay operation.

backhaul is not ideal and subject to limited capacity, latency, and errors. Those factors have to be taken into account when network cooperation is applied in the real world. [54] presents a survey of the theory and techniques of network cooperation.

Cooperative communication is closely related to multiple antenna techniques with one important advantage that a user is not required to be equipped with multiple antennas (similar to MU-MIMO), therefore no additional hardware costs or power consumptions. However, it requires extensive exchange of user data and CSI and relies on sophisticated encoding and decoding schemes.

10.2.1 User cooperation

User cooperation exploits ([31, 96, 133]) a form of space diversity, referred to as *cooperative diversity*, in which users share their antennas, transmit power, and signal processing ability to create a distributed virtual array. In the following we study cooperative diversity in the relay channel in Figure 10.2.

In the literature, it is generally assumed that the relay can operate in either the full-duplex or half-duplex mode. In the full-duplex mode, the relay is able to transmit and receive simultaneously on the same frequency channel. However, as pointed out before, implementing full-duplex in practice is very challenging due to RF front-end design constraints. Therefore, we will focus on the half-duplex mode. In particular, we consider time-division half-duplex relaying shown in Figure 10.34, in which a block of N symbols is divided into two parts: a relay-receive period of length αN followed by a relay-transmit period of length $(1 - \alpha)N$, with $0 \leq \alpha \leq 1$. In the relay-receive period, the source transmits a codeword x_{s1} to the destination. After overhearing this transmission and processing the received signal y_r, the relay then constructs and transmits a codeword x_r on the basis of y_r in the relay-transmit period. In the meantime, the source may continue to transmit another codeword x_{s2}. The destination uses the signals received in both the relay-receiver and relay-transmit periods for decoding.

Denote by H_{sr}, H_{sd}, H_{rd} the complex channel coefficients between the source and the relay, between the source and the destination, and between the relay and the destination, respectively. It follows that at symbol n in the relay-receive period with $1 \leq n \leq \alpha N$,

$$y_r[n] = H_{sr} x_{s1}[n] + w_r[n] \tag{10.65}$$

$$y_{d1}[n] = H_{sd} x_{s1}[n] + w_d[n], \tag{10.66}$$

and in the relay-transmit period with $\alpha N < n \leq N$,

$$y_{d2}[n] = H_{sd} x_{s2}[n] + H_{rd} x_r[n] + w_d[n], \tag{10.67}$$

where w_r, w_d are independent additive complex white Gaussian noise with zero mean and variance σ_0^2. Both the source and relay transmitters are subject to an average power constraint $\frac{1}{N}(\sum_{n=1}^{\alpha N} |x_{s1}[n]|^2 + \sum_{n=\alpha N+1}^{N} |x_{s2}[n]|^2) \leq P_s$ and $\frac{1}{N} \sum_{n=\alpha N+1}^{N} |x_r[n]|^2 \leq P_r$. We assume that the channel coefficients are constant in the duration of a block of N symbols, and that all the nodes are symbol synchronized.

To illustrate the main concepts, we next compare three schemes, namely amplify-and-forward, decode-and-forward, and compress-and-forward.

First consider a simple amplify-and-forward cooperative diversity scheme analyzed in [87]. Let $\alpha = \frac{1}{2}$. The source only transmits in the relay-receive period. In the relay-transmit period, the relay amplifies the received signal

$$x_r[n] = A_r y_r[n - N/2], \tag{10.68}$$

with the amplifier gain set to meet the average power constraint

$$A_r = \sqrt{\frac{2 P_r}{2|H_{sr}|^2 P_s + \sigma_0^2}}. \tag{10.69}$$

Here the factor 2 in both numerator and denominator is because either the source or relay only transmits in half of the N-symbol block and therefore can double its power from the average power constraint P_s or P_r, respectively, when it transmits.

Define SNR $\gamma_{sd} = |H_{sd}|^2 P_s / \sigma_0^2$, $\gamma_{sr} = |H_{sr}|^2 P_s / \sigma_0^2$, and $\gamma_{rd} = |H_{rd}|^2 P_r / \sigma_0^2$. [87] shows that the channel capacity of the amplify-and-forward scheme is given by

$$C_{AF} = \frac{1}{2} \log(1 + 2\gamma_{sd} + f(2\gamma_{sr}, 2\gamma_{rd})), \tag{10.70}$$

where $f(x, y) = \frac{xy}{x+y+1}$. As a comparison, the channel capacity of the direct transmission is

$$C_D = \log(1 + \gamma_{sd}), \tag{10.71}$$

To study cooperative diversity, we are often interested in diversity gain and data rate gain in the high SNR regime. The *diversity gain* is used to show how fast the probability of decoding error $P_e(\gamma)$ decays with SNR γ:

$$d = -\lim_{\gamma \to \infty} \frac{\log P_e(\gamma)}{\log \gamma}. \tag{10.72}$$

The *data rate gain* has two components: a multiplexing gain and an additive gain. The *multiplexing gain* shows how fast the data rate $R(\gamma)$ increases with γ:

$$r = \lim_{\gamma \to \infty} \frac{R(\gamma)}{\log \gamma}. \tag{10.73}$$

The *additive gain* is

$$a = \lim_{\gamma \to \infty} R(\gamma) - r \log \gamma. \tag{10.74}$$

If all the limits exist, we can approximate the data rate as follows:

$$R(\gamma) \approx r \log \gamma + a, \text{ as } \gamma \to \infty. \tag{10.75}$$

Let us compare the diversity gains of the amplify-and-forward and direct transmission schemes. Suppose the channel coefficients are subject to fading. A decoding error occurs when the actual channel capacity drops below a predefined data rate R. Consider Rayleigh fading where $|H_{sd}|^2$, $|H_{sr}|^2$, $|H_{rd}|^2$ are all exponentially distributed. Denote expected SNR $\bar{\gamma}_{sd} = \mathbb{E}(|H_{sd}|^2)P_s/\sigma_0^2$, $\bar{\gamma}_{sr} = \mathbb{E}(|H_{sr}|^2)P_s/\sigma_0^2$, and $\bar{\gamma}_{rd} = \mathbb{E}(|H_{rd}|^2)P_r/\sigma_0^2$.

For the direct transmission scheme,

$$P_{e,D} = \mathbb{P}(C_D < R) = \mathbb{P}\left(\frac{P_s|H_{sd}|^2}{\sigma_0^2} < e^R - 1\right) \tag{10.76}$$

$$\approx \frac{e^R - 1}{\bar{\gamma}_{sd}}. \tag{10.77}$$

Similarly, for the amplify-and-forward scheme,

$$P_{e,AF} = \mathbb{P}(C_{AF} < R)$$

$$= \mathbb{P}\left(\frac{2P_s|H_{sd}|^2}{\sigma_0^2} + f\left(\frac{2P_s|H_{sr}|^2}{\sigma_0^2}, \frac{2P_r|H_{rd}|^2}{\sigma_0^2}\right) < e^{2R} - 1\right) \tag{10.78}$$

$$\approx \frac{(e^{2R} - 1)^2}{2^2 \bar{\gamma}_{sd} \bar{\gamma}_{srd}}, \tag{10.79}$$

where $\bar{\gamma}_{srd}$ is the harmonic mean of $\bar{\gamma}_{sr}$ and $\bar{\gamma}_{rd}$

$$\bar{\gamma}_{srd} = 2\left(\frac{1}{\bar{\gamma}_{sr}} + \frac{1}{\bar{\gamma}_{rd}}\right)^{-1}. \tag{10.80}$$

Hence, the diversity gain is $d = 2$ for the amplify-and-forward scheme, greater than $d = 1$ for the direct transmission scheme.

It should be pointed out that having a relay channel does not automatically ensure $d = 2$. For example, consider a decode-and-forward scheme. Similarly to the amplify-and-forward scheme, let $\alpha = \frac{1}{2}$, and the source only transmits in the relay-receive period. The relay attempts to decode the received signal, and transmits the decoded signal in the relay-transmit period:

$$x_r[n] = \hat{x}_{s1}[n - N/2]. \tag{10.81}$$

Assuming the relay is required to fully decode the source message, [87] shows that the channel capacity is given by

$$C_{DF} = \frac{1}{2} \min(\log(1 + 2\gamma_{sr}), \log(1 + 2\gamma_{sd} + 2\gamma_{rd})). \tag{10.82}$$

Thus,

$$P_{e,DF} = \mathbb{P}(C_{DF} < R) \approx \frac{e^{2R} - 1}{\bar{\gamma}_{sr}}. \tag{10.83}$$

Unlike $P_{e,AF}$, $P_{e,DF}$ is limited by the link between the source and relay, thereby only achieving $d = 1$.

A major cost of achieving $d = 2$ diversity gain in the above amplify-and-forward scheme is that the multiplexing gain is only $r = 1/2$. Note the factor $1/2$ in (10.70), which arises because the relay spends half of the time repeating what it overhears in the other half. Is this inefficiency mostly attributable to the use of repetition coding or more fundamentally to the half-duplex constraint? The answer turns out to be the former, as shown in [148]. With proper coding schemes, the highest multiplexing gain is $r = 1$. The argument is based on the following capacity bounds of the relay channel calculated in [69].

The capacity of the time-division half-duplex Gaussian relay channel is not precisely known in general. However, it is possible to obtain upper and lower bounds (see [30]). When the difference between an upper and a lower bounds is small, the capacity is practically known. In the following, we present the information theoretical results from [69]. We consider a synchronous model, where the transmitters also know all the coefficients, and an asynchronous model, where the transmitters only know the amplitudes but not the phases. Note that the results presented below assume that the source and the relay are subject to a *total* power constraint, which may not be the case in practice. However, the same approach as in [69] can be applied to study the model with separate power constraints for the source and the relay.

For the synchronous model, the capacity of the relay channel is upper-bounded by

$$C_{ub} = \max_{\rho,\alpha,k \in [0,1]} \min(C_{ub1}, C_{ub2}), \tag{10.84}$$

where

$$C_{ub1} = \alpha \log\left(1 + (\gamma_{sr} + \gamma_{sd})\frac{k}{\alpha}\right)$$
$$+ (1-\alpha) \log\left(1 + (1-\rho^2)\gamma_{sd}\frac{1-k}{1-\alpha}\right), \tag{10.85}$$

$$C_{ub2} = \alpha \log\left(1 + \gamma_{sd}\frac{k}{\alpha}\right) + (1-\alpha) \log\left(1 + \gamma_{sd}\frac{1-k}{1-\alpha} + \gamma_{rd}\frac{1}{1-\alpha}\right.$$
$$\left. + \frac{2}{1-\alpha}\sqrt{\rho^2 \gamma_{sd}\gamma_{rd}(1-k)}\right), \tag{10.86}$$

where the source spends kP_s transmit power in the relay-receive period and $(1-k)P_s$ in the relay-transmit period, and ρ reflects the correlation between x_{s2} and x_r:

$\rho = \mathbb{E}(x_{s2}x_r^*)/\sqrt{(1-k)P_sP_r}$. The optimum value of ρ can be found in closed form (see [70]). An upper bound for the asynchronous model is obtained by setting $\rho = 0$.

We next consider two cooperative diversity schemes and calculate the corresponding achievable rates, which can be used to determine a lower bound of the relay channel capacity.

The first scheme is *decode-and-forward*. It has a few enhancements over the one we studied previously. We are not constrained to $\alpha = \frac{1}{2}$. After the relay fully decodes x_{s1}, it does not have to use repetition coding (10.81), but can use a different code book to generate x_r. More importantly, since the source can completely predict x_r, the signal to be sent by the relay in the relay-transmit period, it can apply the beamforming principle and generate x_{s2} as a phase-shifted version of x_r such that x_{s2} and x_r add up coherently at the destination. Then, (10.67) becomes

$$y_{d2} = (H_{sd}A + H_{rd})x_r[n] + w_d[n], \qquad (10.87)$$

where A is a complex number to phase-align H_{rd} and H_{sd} and magnitude is subject to the average power constraint $P_s(1-k)/(1-\alpha)$. The destination combines y_{d1} and y_{d2} to decode the original message from the source.

The achievable rate of the decode-and-forward scheme of the synchronous model is given by

$$C_{df} = \max_{\rho, \alpha, k \in [0,1]} \min(C_{df1}, C_{df2}), \qquad (10.88)$$

where

$$C_{df1} = \alpha \log\left(1 + \gamma_{sr}\frac{k}{\alpha}\right) + (1-\alpha)\log\left(1 + (1-\rho^2)\gamma_{sd}\frac{1-k}{1-\alpha}\right), \qquad (10.89)$$

$$C_{df2} = C_{ub2}. \qquad (10.90)$$

The achievable rate of the asynchronous model is obtained by setting $\rho = 0$.

Since the relay has to perfectly decode the source message, the bottleneck is the link from the source to the relay. If it is inferior to the direct link from the source to the destination, then the decode-and-forward scheme performs worse than the direct transmission scheme. To overcome this bottleneck, consider an alternative strategy in which the relay is not required to fully decode the message in the relay-receive period. The amplify-and-forward scheme studied before is one example. A more sophisticated scheme is *compress-and-forward*. The idea was originally suggested in [30] and then generalized in [69]. The scheme works as follows.

The source transmits codewords x_{s1}, x_{s2} in the relay-receive and relay-transmit periods as before. The relay uses a source coding scheme to compress the received signal y_r, and then encodes the compressed signal to become x_r. Note that y_r (received at the relay) and y_{d1} (received at the destination) both come from x_{s1}, thereby being correlated. Thus, the destination can decode y_r from y_{d2} while treating y_{d1} as side information. [174] studied the source coding problem where side information is only available at the decoder (the destination in this case) but not at the encoder (the relay) and developed the Wyner-Ziv lossy source coding scheme. Here, the relay can apply that scheme to compress y_r.

The destination recovers x_{s1}, x_{s2} with successive cancellation decoding as follows. First, decode $\hat{x}_r$ from y_{d_2} treating x_{s2} as noise. Next, subtract $\hat{x}_r$ from y_{d_2} and decode $\hat{x}_{s2}$. Then, estimate $\hat{y}_r$ from $\hat{x}_r$ with Wyner-Ziv decoding, treating y_{d1} as side information. Finally, decode $\hat{x}_{s1}$ from combined $\hat{y}_r$ and y_{d1}.

The achievable rate of the compress-and-forward scheme for both the synchronous and asynchronous models is given by

$$C_{cf} = \max_{\alpha, k \in [0,1]} \left(C_{cf1}(\alpha, k) + C_{cf2}(\alpha, k) \right), \tag{10.91}$$

where

$$C_{cf1} = \alpha \log \left(1 + \gamma_{sd} \frac{k}{\alpha} + \gamma_{sr} \frac{k}{\alpha(1 + \sigma_w^2)} \right) \tag{10.92}$$

$$C_{cf2} = (1 - \alpha) \log \left(1 + \gamma_{sd} \frac{1 - k}{1 - \alpha} \right), \tag{10.93}$$

with "compression noise" σ_w^2 given by

$$\sigma_w^2 = \frac{\alpha + (\gamma_{sd} + \gamma_{sr}) k}{\left(\left(1 + \frac{\gamma_{rd}}{1 - \alpha + \gamma_{sd}(1-k)} \right)^{\frac{1-\alpha}{\alpha}} - 1 \right) (\alpha + \gamma_{sd} k)}. \tag{10.94}$$

Unlike the decode-and-forward scheme, the compress-and-forward scheme always outperforms the direct transmission scheme. However, neither of the two schemes dominates the other in all conditions. As will be shown next, the decode-and-forward scheme is superior when the link between the source and the relay is better than that between the relay and the destination, and the compress-and-forward scheme is superior when the link between the relay and the destination is better. In general, a lower bound of the relay channel capacity can be obtained by taking the maximum of the two achievable rates,

$$C_{lb} = \max(C_{df}, C_{cf}). \tag{10.95}$$

To get an idea of the gap between C_{ub} and C_{lb} and a comparison between the decode-and-forward and compress-and-forward schemes, we next examine a simple scenario where the source, relay, and destination nodes are all located in a straight line. We fix the source and destination, and vary the position of the relay. Let the source and destination be at coordinate 0 and 1, respectively, and denote by p the coordinate of the relay. Furthermore, assume $P_r = P_s$ and H_{sd}, H_{sr}, H_{rd} only include the path loss effect without fading or shadowing. It follows that

$$\gamma_{sr} = |p|^{-r} \gamma_{sd}, \tag{10.96}$$

$$\gamma_{rd} = |1 - p|^{-r} \gamma_{sd}, \tag{10.97}$$

where r is the path loss exponent.

We consider the direct transmission scheme as the baseline, and plot the rate gains of C_{ub}, C_{df}, C_{cf} relative to the baseline capacity

$$C_d = \log(1 + \gamma_{sd}) \tag{10.98}$$

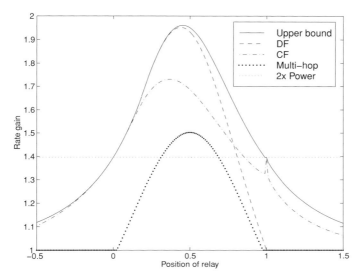

Figure 10.35 Capacity bounds and achievable rates of cooperative communications and multi-hop schemes as compared with the baseline capacity.

in Figure 10.35. We also plot the rate gain of the conventional *non-cooperative* multi-hop scheme as studied in Section 10.1.1. The achievable rate is

$$C_{mh} = \max\left(C_d, \alpha \log\left(1 + \frac{\gamma_{sr}}{\alpha}\right)\right), \qquad (10.99)$$

where α is such that

$$\alpha \log\left(1 + \frac{\gamma_{sr}}{\alpha}\right) = (1 - \alpha) \log\left(1 + \frac{\gamma_{sr}}{1 - \alpha}\right). \qquad (10.100)$$

Since the cooperative and multi-hop schemes utilize twice as much power (total power equal to $P_s + P_r$) in total, we consider the hypothetical case where the source transmits directly to the destination with $2P_s$ power for which the capacity is

$$C_{2x} = \log(1 + 2\gamma_{sd}). \qquad (10.101)$$

Figure 10.35 shows a few interesting points.

First, when the relay co-locates with the source, $C_{ub} = C_{df} = C_{cf} = C_{2x}$, but C_{mh} has no gain over C_d. When the relay co-locates with the destination, $C_{ub} = C_{cf} = C_{2x}$, but C_{df}, C_{mh} have no gain over C_d. The optimum position of the relay is exactly the middle between the source and destination for C_{mh} but somewhat biased towards the source for C_{ub}, C_{df}, C_{cf}.

Second, the decode-and-forward scheme outperforms the compress-and-forward scheme when the relay is close to the source; however, the reverse is true when the relay is close to the destination. In this linear example, as the relay moves from the

Table 10.2. A comparison of cooperative and multi-hop communications.

	Cooperative communications	Multi-hop communications
Combining gain	Destination combines signals from both source and relay	Destination uses one of the two but not both
Soft decoding gain	Relay does not require to fully decode; instead it can forward (compressed) "soft" information to help decoding at destination	Relay has to fully decode before it can help
Multiplexing gain	Source can transmit new information simultaneously with relay's transmission to obtain multiplexing gain $r = 1$	TDM between source's and relay's transmissions resulting in loss in bandwidth, thus $r < 1$ in general
Power gain	Source's and relay's signals can coherently add up at destination to achieve beamforming gain	No beamforming gain

source to destination, after the relay crosses over the middle point the performance of the decode-and-forward scheme drops quickly. Once $G_{sr} \leq G_{sd}$ (when $p > 1$), the decode-and-forward scheme has no gain; however, the relay continues to help in the compress-and-forward scheme.

Third, the gap between C_{ub} and C_{lb} is quite small in most of the regime. In particular, the gap is negligible when the relay is close to the source.

Finally, the two cooperative schemes are universally superior to the multi-hop scheme. When the relay is not between the source and destination ($p < 0$ or $p > 1$), the multi-hop scheme has no gain and yet the cooperative schemes achieve remarkable gains. Table 10.2 summarizes the general advantages of cooperative communication over conventional non-cooperative multi-hop transmissions.

The model we have studied for user cooperation assumes that the source and the relay transmit on the same channel. Other models are also possible. For example, suppose there are two non-overlapping (orthogonal) channels. The source transmits to the relay in one orthogonal channel, and the source and the relay both transmit to the destination in the other orthogonal channel. As another example, the source transmits to the relay and destination in one orthogonal channel, and the relay transmits to the destination in the other orthogonal channel. Those two models are studied in [42] and [95], respectively, and the capacity can be established for some Gaussian cases.

We have concentrated on physical layer cooperative diversity. While cooperative diversity takes advantage of the broadcast nature of the wireless channel, conventional MAC and network layer protocols often treat it as the source of interference and thus a drawback to be overcome. To apply the ideas to a practical mobile broadband system, new MAC (resource allocation, scheduling) and network (routing) layer protocols will be needed to exploit cooperative diversity, and a cross-layer approach of design and optimization is expected to be a key ([104]).

10.2.2 Network cooperation

Network cooperation mostly benefits users in the interference limited regime, in particular, at the boundary areas of adjacent cells. The potential gain is expected to increase as the cell size shrinks.

The notion of network cooperation can be traced back to soft handoff in CDMA, in which a cell boundary user communicates with multiple base stations simultaneously. The key difference is that while soft handoff aims at providing macro-diversity to combat channel fading, network cooperation attempts to eliminate the cost of inter-cell interference. The tradeoff of bandwidth and SINR is fundamental to conventional cellular frequency reuse. A universal frequency reuse gains more bandwidth per cell but suffers from severe inter-cell interference at cell edges. Inter-cell interference can be controlled with spatial reuse partitioning, but at the cost of the frequency reuse factor being greater than one. Fractional frequency reuse (FFR) can significantly improve the performance of conventional cellular systems by coordinating spectrum reuse and power patterns among neighboring cells. Extending the example of FFR leads to a general idea of *interference coordination*, in which neighboring base stations share channel power gain and transmit power information and coordinate their scheduling strategies; such as user selection, power and bandwidth (frequency/time) allocation, and beamforming directions.

Network cooperation takes a further step by additionally sharing channel state information and jointly encoding/decoding data (*multi-cell processing*). With network cooperation, the per cell capacity under universal frequency reuse is similar to or even greater than that of an isolated cell with no inter-cell interference. However, the real-world performance gain is limited by practical constraints. One predominant constraint is backhaul capacity, because sharing user encoding/decoding data among base stations puts a significant burden on backhaul. In a conventional cellular system, the backhaul capacity is usually budgeted for user traffic, which is less than for jointly encoding/decoding data. In the following, we will study the per cell capacity assuming finite backhaul capacity.

Let us consider the "soft handoff" system model shown in Figure 10.36. The cellular system consists of single-antenna base stations arranged in a line (from $-\infty$ to ∞). One single-antenna user is served in each cell at a time. A user is close to the edge between two adjacent cells and we assume that the channel gains between the user and any other cells are 0. For simplicity, we focus on the non-fading Gaussian channel, and assume that the channel coefficients are all constant and known. For the user closest to the mth base station, referred to as user m, the channel coefficient with that base station is normalized to 1, and with the other, $(m-1)$th, base station is equal to a, with $0 \leq a \leq 1$. When $a = 1$, the user is exactly in the middle of the two base stations, and $a = 0$ corresponds to the case of isolated cells. The base stations are all connected via a backhaul to a central server for joint multi-cell processing (encoding for the downlink or decoding for the uplink) and the backhaul capacity is limited to C_b. The soft handoff model is studied in [147] as a variant of the Wyner model originally proposed in [175]. The Wyner model has the same linear layout of the locations of the base stations as in Figure 10.36; however, each user has an additional nonzero channel coefficient with

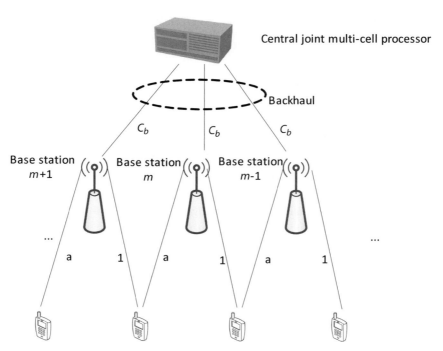

Figure 10.36 "Soft handoff" model.

a third base station. In particular, for user m, the channel coefficients with both the $(m-1)$th and $(m+1)$th base stations are of the same statistics.

First, consider the downlink. The signal received at user m is given by

$$y_m = x_m + a x_{m-1} + w_m, \qquad (10.102)$$

where x_m, x_{m-1} are the symbols transmitted by mth and $(m-1)$th base stations, respectively, subject to a per-cell power constraint $\mathbb{E}(|x_m|^2) = P$, and w_m a white Gaussian noise with unit power. In [146], it is shown that an upper bound on the per-cell capacity is given by

$$C_{ub} = \min(C_b, C_\infty(P)), \qquad (10.103)$$

where

$$C_\infty(P) = \log_2 \left(\frac{1 + (1+a^2)P + \sqrt{1 + 2(1+a^2)P + (1-a^2)^2 P^2}}{2} \right) \qquad (10.104)$$

is the per-cell capacity when the backhaul capacity is unlimited. We observe that in the high SNR regime, in order not to be the bottleneck of maximizing the per-cell capacity, the backhaul capacity C_b should grow as $C_\infty(P) \approx \log_2 P$.

To derive an achievable rate, we next consider a centralized scheme of oblivious base stations. Specifically, the base stations are not aware of any codebook used in the system and encoding takes place at the central server, which performs joint dirty paper coding ([19]) and produces one codeword block for every base station. Since the backhaul capacity is finite, the central server compresses the codeword block before sending it to the corresponding base station. Upon the reception of the compressed codeword block, the base station simply forwards it over the air.

Compression introduces quantization noise, whose effect is twofold. First, the effective power budget for the desired signal is reduced to $P/(1 + 1/(2^{C_b} - 1))$ because part of the total power budget P is used to transmit the quantization noise. Second, the quantization noise power $P/2^{C_b}$ is added to the channel noise w_m. As a result, [146] shows that the equivalent SNR is reduced from P to

$$\tilde{P} = \frac{P}{(1 + (1 + a^2)P)/(2^{C_b} - 1) + 1}, \quad (10.105)$$

and the achievable per-cell capacity is given by

$$C_{DL}(P) = C_\infty(\tilde{P}). \quad (10.106)$$

In other words, replace P with $\tilde{P}$ in (10.104).

As before, we are interested in the high SNR regime and would like to see how the quantization noise due to finite backhaul capacity affects the achievable per-cell capacity. Assuming C_b is fixed, it follows that

$$\lim_{P \to \infty} C_{DL}(P) = C_b - 1 + \log_2\left(1 + \sqrt{1 - \frac{4a^2}{(1 + a^2)^2}(1 - 2^{-C_b})^2}\right), \quad (10.107)$$

which is at most 1 bit away from the upper bound $C_{ub} (= C_b$ as $P \to \infty)$. Another interesting asymptotic behavior is to let C_b grow with $\log_2 P$:

$$C_b = \eta \log_2 P. \quad (10.108)$$

In this case,

$$\lim_{P \to \infty} C_{DL}(P) \sim \min(\eta, 1) \log_2 P. \quad (10.109)$$

Hence, the optimal multiplexing gain of 1 can be achieved by letting $\eta = 1$, as we have concluded previously from (10.103) and (10.104).

The uplink case can be studied similarly. The signal received at the mth base station is given in exactly the same way as (10.102), with x_m, x_{m-1} the symbols transmitted by users m and $m - 1$, respectively, subject to a per-user power constraint $\mathbb{E}(|x_m|^2) = P$. [135] derives the same upper bound as (10.103) and (10.104). Moreover, we consider a similar centralized scheme of oblivious base stations, which are not aware of any codebook and cannot perform decoding locally. Each base station forwards a compressed version of the received signal y_m to the central server, which then performs joint optimal multiuser detection. Clearly the power of quantization noise introduced in compression depends on the backhaul capacity. In [135], it is shown that the achievable per-cell capacity is given by

$$C_{UL}(P) = \log_2\left(\frac{1+(1+a^2)P+2a^22^{-C_b}P^2+\sqrt{1+2(1+a^2)P+((1-a^2)^2+4a^22^{-C_b})P^2}}{2(1+2^{-C_b}P)(1+a^22^{-C_b}P)}\right). \quad (10.110)$$

Now consider the high SNR regime. First, fix C_b and it follows that

$$\lim_{P\to\infty} C_{UL}(P) = C_b, \quad (10.111)$$

which is equal to the upper bound C_{ub}. Second, let C_b grow as in (10.108). Then, same as (10.109), we have

$$\lim_{P\to\infty} C_{UL}(P) \sim \min(\eta, 1)\log_2 P. \quad (10.112)$$

For the sake of comparison, we also consider the following two schemes

- the non-cooperative universal reuse scheme

$$C_1 = \min\left(\log_2\left(1+\frac{P}{1+a^2P}\right), C_b\right), \quad (10.113)$$

- the co-phased reuse factor 1/2 scheme (that is, at a given time, a user in each even or odd cell is served by two adjacent reachable base stations, with the downlink or uplink signals being coherently combined before decoding)

$$C_2 = \min\left(\frac{1}{2}\log_2\left(1+(1+a^2)P\right), C_b\right). \quad (10.114)$$

As a baseline, we use the capacity C_0 of an ideal isolated cell with unlimited backhaul capacity $C_0 = \log_2(1+P)$.

In Figure 10.37, we plot $C_{ub}, C_{DL}, C_{UL}, C_1, C_2$, relative to C_0, versus a^2, for $P = 10$ dB, $C_b = 4$ bits per second per Hz. Note that $C_0 = 3.5$ bits per second per Hz, which is smaller than C_b. $C_{ub} = C_\infty$. Interestingly, $C_{ub} \geq C_0$ for all a^2 and indeed increases slightly with a^2, indicating potentially significant performance gain in the interference-limited regime. C_{DL} and C_{UL} are very close, and both below C_0. In contrast, C_1 of the conventional universal reuse scheme drops quickly as the interference interference power a^2 increases. The co-phased reuse factor 1/2 scheme sacrifices bandwidth upfront; however, C_2 increases with a^2 and is not too far from C_{DL} or C_{UL} at cell edges $a^2 = 1$.

To demonstrate the impact of limited backhaul capacity C_b, we consider four operating regimes: high/low SNR ($P = 10, 0$ dB) and high/low inter-cell interference power ($a^2 = 1, 0.1$), and plot relative data rates in Figure 10.38. To scale C_b properly, we let $C_b = \kappa C_0$, and vary κ. The observation is that the cooperative scheme works well in the high SNR regime and closely approaches the capacity upper bound with $\kappa = 2$. Moreover, the performance is not sensitive to a^2. However, in the low SNR regime, a much simpler scheme (C_1 or C_2 depending on a^2) is preferred when κ is limited, although the potential

10.2 Cooperative communication

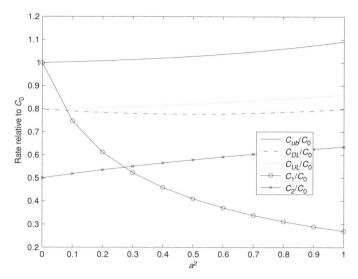

Figure 10.37 Relative achievable rates of network cooperative and other schemes compared to the baseline capacity as inter-cell interference varies. $P = 10$ dB, $C_b = 4$ bits per second per Hz.

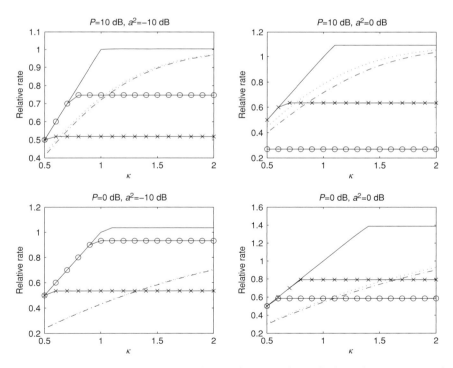

Figure 10.38 Relative achievable rates of network cooperative and other schemes compared to the baseline capacity as backhaul capacity varies. The same legends are used as in Figure 10.37.

gain of C_{ub} over C_0 is the greatest in the low SNR and high inter-cell interference regime if κ is allowed to grow large. Hence, the scenario in which network cooperative can deliver the most capacity gain in practice is likely in the interference-limited regime, which is the case as the cell size shrinks and interference instead of coverage becomes the major capacity limiting factor. In this case, network cooperation can still be much superior even when a user is not at cell edges ($a^2 = 0.1$) provided that the backhaul capacity is not too restrictive ($\kappa \geq 1$).

In addition to finite backhaul capacity studied so far, other practical constraints also limit the performance. Understanding their impact is a key to advancing cooperative communications from theory to practice. Below we briefly touch on two main system issues.

First, *robustness*. Network cooperation requires acquiring and sharing detailed CSI among base stations. For a fading channel, it means not only channel statistics but also instantaneous channel knowledge, thereby imposing greater burden than interference coordination does. This leads to pilot overhead for channel estimation and signaling overhead for CSI feedback. In practice, because of constraints on the overhead cost and channel uncertainty inherent in the mobile environment, the CSI used in cooperative encoding and decoding is subject to inaccuracy and delay. In particular, delay potentially presents a much more significant challenge to network cooperation than finite backhaul capacity does; even with unlimited backhaul capacity, the use of stale CSI may make cooperative communications perform worse than simple interference coordination even at modest Doppler. Moreover, network cooperation also requires tight synchronization. Carrier frequency synchronization can be achieved if base stations are all synchronized to a common external timing source such as GPS. However, in general the symbol times of signals from multiple transmitters are not completely aligned at every receiver due to different propagation delays, although the time difference can be covered with a sufficiently long cyclic prefix. Hence, it is important to understand the sensitivity of the performance with respect to various model mismatch conditions.

Second, *scalability*. The capacity-achieving schemes for the downlink (broadcast) channel and uplink (multi-access) channel require nonlinear computationally intensive operations and scale poorly with the size of the network, as the algorithm complexity often grows exponentially. Direct implementation of those optimal schemes would be practically challenging even for a modest number of base stations, users, and antennas. Linear suboptimal schemes, such as MMSE or zero-forcing, are thus attractive alternatives ([74]). In addition, an architecture relying on a centralized server to process encoding and decoding for all base stations is not scalable and is subject to a single point of failure. On the other hand, as captured in the Wyner or soft handoff model, inter-cell interference is very much confined within neighboring cells. The localized nature of interference motivates the use of distributed algorithms, such as turbo decoding ([110, 162]) and clustering ([81, 93]). Another interesting idea is to combine the ideas of network cooperation and interference coordination such that intra-cluster interference is eliminated with cooperative schemes and inter-cluster interference is mitigated via coordination.

One way to model the above effect of robustness and scalability is to introduce the notion of cluster as in [108]. Specifically, the signal seen at receiver n is given by

$$Y_n = \sum_{k=1}^{M_t} H_{n,k} \sqrt{P} X_k + \sum_{k=M_t+1}^{\tilde{M}_t} H_{n,k} \sqrt{P} X_k + W_n, \qquad (10.115)$$

where $\tilde{M}_t$ is the total number of transmitters, M_t the number of transmitters within a cluster, $H_{n,k}$ the channel from transmitter k to receiver n, X_k the transmitted symbol, P the transmit power, and W_n the noise. We assume that the transmitters and receivers within the cluster cooperate with each other perfectly and the channel $H_{n,k}$ for $k = 1, \ldots, M_t$ is known. However, there is no cooperation outside the cluster. The second term of (10.115), $\sum_{k=M_t+1}^{\tilde{M}_t} H_{n,k} \sqrt{P} X_k$, represents the out-of-cluster interference that cannot be eliminated. In contrast, the study so far has not taken into account the out-of-cluster interference. Note that the out-of-cluster interference grows with P. We have seen a similar effect when studying the self-noise model. Indeed, a comparison of (10.115) and (2.76), shows the similarity of the out-of-cluster interference and self-noise. With both the out-of-cluster interference and noise, it is shown in [108] that the capacity scales as $\log_2 P$ initially when the noise dominates the out-of-cluster interference, and then saturates when eventually the out-of-cluster interference dominates as P grows large.

So far our discussion has been focused on the physical layer. To apply the ideas of cooperative communications to a real mobile broadband system, one needs to think through cross-layer design issues, in particular of user and resource scheduling. For example, in the problem formulation so far, we assume that the set of users to be scheduled in a given time-frequency resource is given beforehand. From the study of MU-MIMO we have learned that the performance gain heavily depends on the geometric characteristics of the selected users to be multiplexed. Similar scheduling principles are expected to extend from MU-MIMO to network MIMO.

10.3 Cognitive radio

In a conventional licensed spectrum cellular communication usage scenario, spectrum bands are licensed to operators under a *static* spectrum allocation policy. Specifically, governmental agencies assign a spectrum band exclusively to a license holder (an operator in this case) on a long-term basis (multiple years/decades) for large geographic areas (hundreds of miles across). Often, only one single system is deployed to legally transmit in that band in a given location, the concern being that simultaneous transmissions of multiple systems may interfere with each other and degrade the performance. Similar spectrum allocation policies have also been used to regulate licensed spectrum for other wireless applications, such as television broadcasting, public safety, and satellite communications. The business rationale of such a static approach is that because the license holders usually make large capital investments to build wireless systems and services, static spectrum allocation is needed to ensure a sufficiently long life span of the systems and services so as to justify the investments.

Although this approach is effective in preventing interference, it also results in underutilization of spectrum. Indeed it is found that *"most bands in most places are underused most of the time"* ([79, 153], and the references therein). There are system and device level reasons for such observations. Many systems, such as those for public safety, have a large peak-to-average usage ratio. Specifically, the spectrum is for a large part reserved for emergency use and the duty cycle of such usage is inherently low. Similarly, the spatial reuse of the spectrum allocated to TV broadcasting is very sparse. Even for cellular systems with high traffic demand on average, the spectrum use is still very non-uniform in time and in space. In addition, some spectrum bands are not used due to practical device implementation constraints such as limited rejection at adjacent/image frequency channels.

A large body of studies have shown that spectrum scarcity is largely because of underutilization rather than physical shortage ([57]). As the demand for wireless communications services increases drastically, finding and utilizing underused spectrum bands will certainly help address the challenge of spectrum scarcity. The idea of *cognitive radio* is to make spectrum allocation dynamic and smart. Specifically, the radio uses real-time interaction with its environment to determine its transmitter parameters such as frequency, time and power ([65, 112]). Similar ideas have been applied in the past in the unlicensed spectrum domain, for example, for cordless telephones to sense the environment and select to use frequency channels experiencing the least amount of interference. The goal here, however, is to address licensed spectrum bands. As a starting point, the Federal Communications Commission (FCC) has expressed its intent of permitting cognitive access to white spaces in the TV bands (at 700 MHz), part of the reason being the excellent propagation characteristics due to the low carrier frequency of the TV bands and the relatively predictable usage pattern in time and space. If the cognitive use of TV white spaces becomes successful, similar principles can potentially be applied to other bands as well.

In the following, we explore a specific framework of cognitive radio, namely primary/secondary spectrum management, in which two classes of users – non-cognitive primary users (incumbent license holders) and cognitive secondary users – share the same spectrum. In this framework, primary users operate as if secondary users did not exist, and the burden is on secondary users to minimize any adverse impact on primary users and to maximize their own (secondary) use of the spectrum. In general, cognitive secondary users take the following four major steps ([3]):

1. *Spectrum analyzing*. Finding underused spectrum bands in the local neighborhood at a given time by analyzing the radio activity of primary users. Spectrum sensing is a main approach.
2. *Spectrum decision*. Deciding whether the spectrum bands can be used for cognitive secondary communications based on spectrum availability and policies.
3. *Spectrum sharing*. Maximizing the use of the spectrum without affecting the performance of the primary users as much as possible.

4. *Spectrum mobility*. Continuing to monitor whether the spectrum usage of primary users has changed, and if so, adjusting the transmitter parameters accordingly or even suspending from transmitting.

Our focus in this section is on spectrum sensing and sharing.

10.3.1 Spectrum sensing

There are three main approaches that allow secondary users to determine whether a given licensed spectrum is available for cognitive use.

First, primary users explicitly broadcast the current spectrum availability information in some known control channel, which can be either an existing channel with new control information defined or a new dedicated channel. Secondary users decode the control channel and get the information. This approach helps simplify the implementation of secondary users at the costs of modifying incumbent primary users and being incompatible with legacy users.

Second, primary users register at a database server the information about primary transmission such as location, power, and duration. Secondary users need to determine their own location and compare with the information retrieved from the database server to determine the spectrum availability. Mobile users have to check the location frequently. Retrieving the information from the server requires secondary users to have Internet access. Moreover, if the primary transmission information changes in real time, then secondary users need to access the server frequently even if they are stationary.

We are primarily interested in the third approach, namely spectrum sensing, in which secondary users determine the spectrum availability by direct over-the-air sensing of the spectrum bands of interest. Spectrum sensing does not require new infrastructure, modifications to primary users, or ubiquitous Internet access. In addition, it potentially provides more granular spatial and temporal spectrum reuse. For example, when a secondary user is located in an isolated area, with spectrum sensing, it can correctly realize that a spectrum band is available. On the other hand, the database registry approach is probably too coarse to provide different spectrum availability information when the user is in a basement versus in a living room, and as a result, tends to be conservative and blocks the entire neighborhood area.

However, there are technical challenges to making spectrum sensing useful and efficient in practice. Because of the half-duplex constraint discussed earlier in this chapter, secondary users cannot transmit and sense (receive) in the same band simultaneously. While utilizing a white space, they have to dedicate *quiet periods* during which they suspend secondary transmission and sense the spectrum in case primary transmission changes or secondary users move. Suppose that quiet periods occur periodically every T_p and each quiet period lasts for T_q. T_p is the maximum time for which secondary users are unaware of a reappearing primary transmission and harmfully interfere with it. T_p depends on the characteristics of service degradation that primary transmission can tolerate as well as the spatial/temporal pattern of primary transmission. T_q is the minimum time required to achieve satisfactory performance of primary signal detection.

Overall, $1 - T_q/T_p$ defines the efficiency of spectrum sensing. Hence, it is desirable to minimize T_q in primary signal detection.

Primary signal detection can be formulated as the following two hypotheses:

$$y(t) = \begin{cases} w(t), & \mathcal{H}_0 \\ hs(t) + w(t), & \mathcal{H}_1 \end{cases}, \qquad (10.116)$$

with $y(t)$ the received signal, $s(t)$ the primary signal, h the channel response representing the path loss, shadowing, and fading effects, and $w(t)$ the background noise assumed to be additive white Gaussian with power σ_0^2.

The simplest sensing scheme is energy detection. Assuming σ_0^2 is known, a normalized output Y is calculated from the received signal $y(t)$

$$Y = \frac{2}{\sigma_0^2} \int_0^{T_q} |y(t)|^2 \, dt \qquad (10.117)$$

and then compared with a decision threshold Θ to determine whether $s(t)$ is present. The SNR is defined as $\gamma = \frac{P}{\sigma_0^2}$ with P the power of $hs(t)$. The probabilities of false alarm P_F and miss P_M are two important performance metrics:

$$P_M = \mathbb{P}\{Y \leq \Theta | \mathcal{H}_1\} \qquad (10.118)$$

$$P_F = \mathbb{P}\{Y > \Theta | \mathcal{H}_0\}, \qquad (10.119)$$

P_F, P_M depends on T_q, γ and Θ. Given T_q and γ, Θ is a control parameter; as Θ increases, P_F decreases while P_M increases. A plot of P_M versus P_F, called a complementary receiver operating characteristics (ROC) curve, describes the tradeoff between P_F and P_M. P_F, P_M can both decrease when γ and T_q increase.

Though simple, the energy detection scheme is not efficient. If the primary signal $s(t)$ is known, then the optimal detector is to calculate Y with a matched filter

$$Y = \frac{2}{\sigma_0^2} \int_0^{T_q} |y(t)s^*(t)| \, dt. \qquad (10.120)$$

Given SNR γ, to achieve a given pair of P_F, P_M, a smaller T_q is required in the matched filter scheme than is required in the energy detection one. Even if the knowledge of the primary signal is not entirely available, as long as some features of it are known, they can be exploited to reduce T_q from what is achieved in the energy detection scheme. Examples of such features include cyclostationarity, covariance, and modulation type ([57, 109]).

We next examine the SNR γ encountered in spectrum sensing. Consider a typical scenario of sensing TV white spaces depicted in Figure 10.39, where a single primary transmitter in an area transmits high power TV signals and provides coverage up to a distance of R (see also [109]). Denote by Γ the lowest SNR required for successful primary signal reception. Assume that without secondary transmissions, a primary receiver at the coverage boundary has an SNR margin of δ:

$$\delta = \frac{P_p G(R)}{\sigma_p^2 \Gamma} > 1, \qquad (10.121)$$

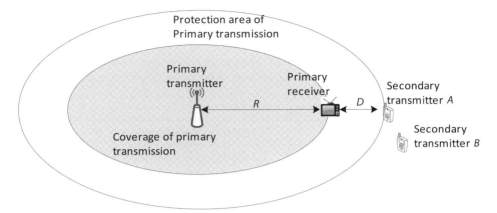

Figure 10.39 Spectrum sensing scenario.

with P_p the primary transmit power, σ_p^2 the noise power at the primary receiver, and $G(R)$ the channel gain at distance R including path loss, shadowing and fading effects.

To protect primary signal reception, no secondary transmitters are allowed to be too close to the coverage area. Denote by D the nearest distance that any secondary transmitter is allowed to be from the primary coverage boundary, as depicted in Figure 10.39. Now with a secondary transmission at distance D, the SNR of the primary receiver degrades to its lowest acceptable value

$$\frac{P_p G(R)}{\sigma_p^2 + P_s G(D)} = \Gamma, \tag{10.122}$$

with P_s the secondary transmit power. It follows that D has to satisfy

$$\frac{G(D)}{G(R)} = \left(1 - \frac{1}{\delta}\right) \frac{P_p}{P_s \Gamma}. \tag{10.123}$$

In effect $R + D$ defines a protection area shown in Figure 10.39 such that secondary transmitters have to be at least $R + D$ distance away from the primary transmitter while primary receivers reside within the coverage area, which is smaller than the protection area. Given the coverage area, (10.123) indicates that the protection area increases with secondary transmit power P_s or minimum SNR Γ, and decreases with SNR margin δ. Obviously, the larger the protection area, the smaller the area in which the secondary use of the spectrum is allowed. Moreover, secondary transmitters are located at a greater distance (larger than $R + D$) from the primary transmitter, and therefore have to detect the primary signal at a much lower power level than primary receivers do, although unlike primary receivers, secondary users only need to detect the presence of the signal, but are not required to decode the signal. Hence, the worst case SNR to detect the signal is

$$\gamma_{\min} = \frac{P_p G(R + D)}{\sigma_s^2}, \tag{10.124}$$

with σ_s^2 the noise power at secondary users.

In the above spectrum sensing scenario, primary receivers (for example, TVs) are completely silent. Secondary users cannot directly measure their potential interference impact on primary receivers; they can only indirectly assess the interference situations from signals received from the primary transmitter. This leads to the hidden terminal problem. In order to make sensing robust, secondary users make worst case assumptions regarding various sources of uncertainty, some of which are listed next:

Uncertainty	Worst case assumption
Wireless channel	Deep fading or heavy shadowing from primary transmitter to secondary user, but favorable condition from secondary user to primary receiver
Primary receiver location	Primary receivers exist anywhere within the coverage
SNR margin	Minimum value of δ equal to the worst case implementation

As a result, secondary users have to be very conservative in declaring that the spectrum is a white space. To meet the detection performance requirement of P_F and P_M, they need to implement more sophisticated detection schemes or spend a greater T_q in sensing the spectrum, which means higher device costs or lower efficiency for secondary use of the spectrum. Moreover, as demonstrated in [154], when the noise power σ_s^2 is uncertain and the SNR is below a certain threshold, called the *SNR wall*, robust detection is impossible even with infinite sensing times for the energy detection, matched filter, and other feature based schemes. Cooperative sensing aims to mitigate some of those adverse impacts by exploiting spatial and multiuser diversity among multiple secondary users.

So far we have only considered a single secondary transmitter. Multiple secondary transmitters may operate in a nearby neighborhood, as illustrated in Figure 10.39, which results in "*multiuser uncertainty*" due to the distributive nature of their operations:

1. *Sum interference.* In (10.122), the assumption is that the interference only comes from a single secondary transmitter. The sum interference will certainly go up in the presence of multiple secondary transmitters. However, the challenge is that the number of secondary transmitters and their geographic distribution are unknown to any individual one. For example, the two secondary users in Figure 10.39 may not see each other although they both interfere with a given primary receiver.
2. *Unsynchronized quiet periods.* In primary signal detection (10.116), we assume no interference from secondary transmissions. In other words, the quiet periods of individual secondary users have to be synchronized. The challenge is that without synchronization, a secondary user may be desensitized by nearby secondary transmitters and unable to detect the primary signal, which is weak. For example, unless their quiet periods overlap, two nearby secondary transmitters may desensitize each other and both reach a wrong conclusion that no primary signal is present.

Multiuser uncertainty is worsened if secondary users belong to different secondary systems all of which attempt to utilize the spectrum in the same geographic area. Therefore, *coordinated* spectrum sensing seems desirable among those secondary systems to

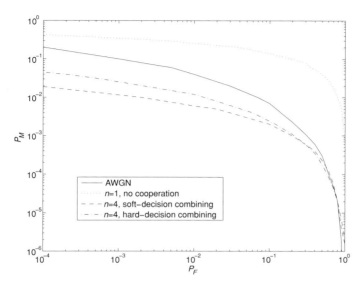

Figure 10.40 Complementary ROC curves under i.i.d. 6 dB log-normal shadowing. Average SNR $\gamma = 5$ dB. Normalized sensing period $T_q \mathcal{W} = 5$ with $\mathcal{W}$ being the bandwidth.

resolve multiuser uncertainty, for example, to synchronize quiet periods and to exchange information about the total number of active secondary transmitters. Taking this step further, we can exploit spatial diversity in multiuser scenarios and develop *cooperative spectrum sensing* to overcome wireless channel uncertainty. The basic idea is to allow secondary users to share their sensing results and cooperatively decide on the spectrum availability.

To show the diversity gain of cooperative spectrum sensing, we continue to use the simplest energy detection scheme, keeping in mind that similar gain will be achieved with more sophisticated detection schemes. We assume that cooperation is done in a centralized way where each secondary user sends its measurement output Y to a central controller, which then makes a final decision after combining all the received Ys and broadcasts the decision to the secondary users. We consider two combining schemes:

1. *Soft-decision combining.* Define equal gain combining $Y_0 = \sum_{i=1}^{n} Y_i$, where n is the number of secondary users, i the user index and Y_i the measurement output of user i. Y_0 is compared with a threshold to decide $\mathcal{H}_1$ or $\mathcal{H}_0$.
2. *Hard-decision combining.* Each secondary user first makes a one-bit hard decision on the basis of its own measurement Y_i with the same decision threshold, and then sends it to the central controller. The controller decides $\mathcal{H}_1$ if at least one of the decisions are $\mathcal{H}_1$.

Assume that the transmission between secondary users and the controller is error free. Assuming that multipath fading is averaged out, we focus on shadowing and assume that all secondary users experience i.i.d. log-normal shadowing with the same average SNR. The complementary ROC performance is studied in [56]. Some of the numerical results are plotted in Figure 10.40 for four cases: AWGN (no shadowing, single user), $n = 1$

(shadowing, no cooperation), and $n = 4$ (shadowing, cooperation) with soft-decision or hard-decision combining.

As expected, in the single user case spectrum sensing is more difficult with shadowing than with AWGN. In the multiuser case, even at $n = 4$, cooperative sensing exploits the shadowing effect and outperforms single user AWGN over a wide range of operation. We have seen a similar benefit in opportunistic multiuser scheduling studied in Chapter 4. However, note that the performance difference diminishes when the required P_M is very small, an operating regime that is particularly important for primary/secondary spectrum management. Moreover, the performance advantage of soft-decision over hard-decision combining is only marginal.

Finally, it should be pointed out that the diversity gain achieved in cooperative spectrum sensing comes at the cost of additional communications overhead to share sensing results among secondary users. Secondary users can report their sensing results to a centralized server if they all have access to the Internet, which makes the solution look similar to the database approach of spectrum analyzing mentioned earlier. Alternatively, a separate small piece of spectrum can be dedicated as a common control channel that secondary users are granted the right to access without sensing. Secondary users can share spectrum sensing information (presumably at a low data rate) over the air on the common control channel in the absence of other network infrastructure.

10.3.2 Spectrum sharing

As stated before, in primary/secondary spectrum sharing, the burden is on secondary users to not violate a tolerance constraint of the interference to primary users. If the primary system uses one-way communications, such as in TV transmission, then secondary users make a binary decision of whether to transmit on the basis of spectrum sensing ($\mathcal{H}_0$ or $\mathcal{H}_1$). Assuming ideal sensing, primary and secondary transmissions are *interweaved* in that secondary transmissions only occur when primary ones stop, thereby being inherently opportunistic.

However, if the primary system use two-way communications, primary receivers also transmit some signals, such as acknowledgments. Measuring those signals allows secondary users to more accurately assess the potential interference caused to primary receivers and to accordingly adjust transmitter parameters such as power, antenna direction, and channel coding. As a result, the spectrum sharing decision is not limited to be binary. In the following, we examine two basic approaches, namely overlay and underlay, that both allow *concurrent* primary and secondary transmissions in the same spectrum band.

A key assumption of the overlay approach ([36, 37, 78, 173]) is that a secondary transmitter has knowledge of the primary signal. One such scenario is that the secondary transmitter is much closer to the primary transmitter than a primary receiver is. Thus, the secondary transmitter can successfully decode the primary message within a small fraction of a primary transmission duration and perfectly predict the primary signal in the remaining time. Knowledge of the primary message allows the secondary transmitter

10.3 Cognitive radio

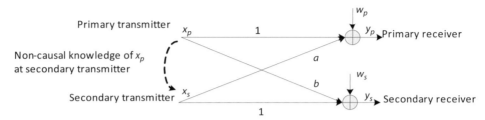

Figure 10.41 Basic overlay network.

to selflessly assist the primary receiver, just as in the relay case, or to split the transmit power so as to achieve its own data rate to the secondary receiver while assisting the primary receiver. In the latter case, the secondary signal interferes with the primary receiver, and the power can be split such that the assistance and interference from the secondary transmitter completely offset each other, and therefore do not affect the rate of the primary transmission. Then, the achieved rate of the secondary transmission defines the gain of the cognitive radio scheme.

Let us consider the basic overlay network depicted in Figure 10.41. With the direct channel gains normalized to be 1, the received signals are given by

$$y_p = x_p + ax_s + w_p \qquad (10.125)$$

$$y_s = bx_p + x_s + w_s, \qquad (10.126)$$

with x_p, x_s being power constrained by P_p, P_s, respectively, and w_p, w_s being AWGN with unit power. For simplicity, consider a real channel where a, b are real and positive, and x_p, x_s, w_p, w_s are all real numbers. The extension to a complex channel can be found in [78].

In the absence of the secondary signal x_s, the rate of the primary user is limited by its SNR and is denoted by $R_p = \frac{1}{2}\log(1 + P_p)$, where the coefficient $\frac{1}{2}$ is because the channel is real. Assume that the secondary transmitter and receiver know all the channel gains precisely. The capacity of the cognitive radio channel R_c is defined to be the largest rate achievable from the secondary transmitter to the receiver while keeping the primary channel rate at R_p. The capacity is obtained in [78, 173] in a *low interference gain* regime, where $a \leq 1$.

The secondary signal x_s consists of two parts:

$$x_s = \hat{x}_s + \sqrt{\frac{\beta P_s}{P_p}} x_p. \qquad (10.127)$$

The second part represents the amount of power (βP_s) to assist the primary receiver. The first part, with power $(1 - \beta)P_s$, encodes the message intended to the secondary receiver and is generated by performing dirty-paper precoding [29] and treating $(b + \sqrt{\frac{\beta P_s}{P_p}})x_p$ as known interference to the secondary receiver. From the information theoretic result of dirty-paper precoding, the achieved rate at the secondary receiver is $\frac{1}{2}\log(1 + (1 - \beta)P_s)$. Power allocation parameter β is set such that the SNR is

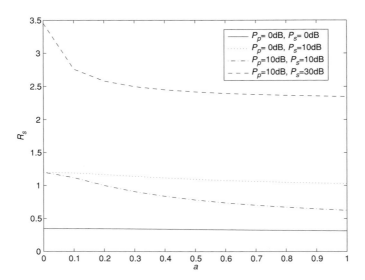

Figure 10.42 R_s versus a.

unaffected at the primary receiver

$$\frac{(\sqrt{P_p} + a\sqrt{\beta P_s})^2}{1 + a^2(1-\beta)P_s} = P_p. \quad (10.128)$$

After some algebra, we solve β as follows:

$$\beta^* = \left(\frac{\sqrt{P_p}\left(\sqrt{1 + a^2 P_s(1 + P_p)} - 1\right)}{a\sqrt{P_s}(1 + P_p)} \right)^2. \quad (10.129)$$

Hence, the capacity is given by

$$R_s = \frac{1}{2}\log\left(1 + (1-\beta^*)P_s\right). \quad (10.130)$$

Figure 10.42 plots R_s versus a for a range of P_p, P_s. We observe that with the overlay approach the degradation in R_s from $a = 0$ to $a = 1$ is not significant, especially when P_p is low.

A somewhat surprising observation is that in order to achieve the above capacity, the primary users only need to use regular single-user encoding and decoding, as if x_s did not exist. Furthermore, the capacity is independent of b. As shown in [78], neither holds true in a high interference gain regime, where $a \gg 1$. In that regime, the primary decoder depends on the secondary encoder in order to achieve the capacity and the capacity is a function of b.

Because it requires access to channel/message side information and sophisticated coding, the overlay approach may be difficult to apply in practice, at least in an early deployment of cognitive radio systems. However, its importance lies in the way that it attempts to characterize the upper limits of the cognitive radio channel. On the other

10.3 Cognitive radio

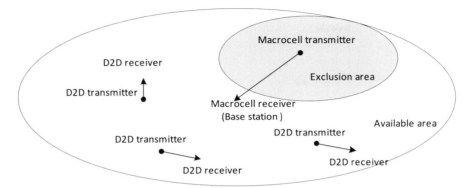

Figure 10.43 Underlay network sharing a cellular uplink. For simplicity, only one cell is shown. The primary receiver is a base station and the primary transmitters are macrocell users. The secondary system is for short range D2D communications.

hand, the underlay approach has much less stringent requirements and is thus easier to implement. In particular, the secondary transmitter is unaware of the primary message and thus does not assist the primary receiver. The secondary signal is treated as the interference at the primary receiver. To allow concurrent primary and secondary transmissions, the underlay approach limits the interference power at the primary receiver that is caused by the secondary transmitter by controlling its transmit power, assuming that the channel gain from the secondary transmitter to the primary receiver is known. This constraint in turn limits the communication range and rate of the secondary transmissions.

Let us consider a specific example illustrated in Figure 10.43 where the primary system is a cellular uplink and the secondary system is for short range D2D communications, similar to what is studied in [38].

Recall from the previous discussion on spectrum sensing that the hidden terminal problem is a major challenge when the primary receivers is completely silent. In that case, the secondary users have to meet stringent detection sensitivity requirements, leading to severe underestimation of the spectrum availability. In the cellular uplink case, however, the primary receivers (base stations) are easily detectable from the downlink signals and there are no hidden terminals. In particular, the downlink pilot signal is usually transmitted at a constant power. A secondary user estimates the downlink channel gain from the received power of the pilot signal. Furthermore, a base station can broadcast its interference budget, a single number indicative of the acceptable interference level. Then, the secondary user can determine its maximum transmit power to meet the requirement assuming the uplink and downlink channel gains are equal.

To make our study concrete, we consider the following protocol. The interference budget broadcast by a base station is represented by a power backoff value Ψ_B, where $\Psi_B \geq 1$. A secondary user measures the received power of the downlink pilot, P_r, and would set the transmit power to be K/P_r (inverse power control) if it were to communicate to the base station. For short range D2D communications, the secondary user uses a fixed transmit power equal to $(K/P_r)/\Psi_B$. The secondary user uses a distributed interference management scheme, such as CSMA/CA or SIR-based approach studied in Section 10.1.3, to share the uplink channel with other short range D2D users. Here, the interference comes from not only other secondary users but also macrocell transmitters. Therefore, if the secondary user is too close to a macrocell transmitter, it yields because the SINR is too low. On the other hand, the macrocell transmitters do not yield to any short range D2D transmitters.

Denote by P_t the transmit power of the downlink pilot signal. The received interference power from a single D2D transmitter is equal to $K/(P_t \Psi_B)$. Denote by L the number of active D2D transmitters at a given time. The total D2D interference power at the base station is

$$\sigma_{D2D}^2 = \frac{KL}{P_t \Psi_B}. \tag{10.131}$$

The power backoff Ψ_B is for a single short range D2D transmitter. Given a total interference budget σ_{D2D}^2, the base station adjusts Ψ_B depending on L. Alternatively, if D2D scheduling is distributed and L is not known to the base station, the base station measures the total D2D interference power, and adjusts Ψ_B in a closed-loop manner to meet the target σ_{D2D}^2.

We next estimate the sum rate achieved in D2D communications. Qualitatively, the uplink spectrum is used sparsely by macrocell transmitters in a cell. For example, in OFDMA without multiuser spatial multiplexing, a degree of freedom (tone-symbol) is used by one macrocell transmitter, which transmits at a relatively higher power than any short range D2D link. The D2D links close to the macrocell transmitter are blocked because of its strong interference. That area is labeled as *exclusion area* in Figure 10.43. Meanwhile, the D2D links in the remaining area, referred to as *available area*, can reuse the spectrum. Define available area to be a collection of locations of a D2D receiver at which the receiver does not see too much interference from the macrocell transmitter, say, 0 dB SIR. The D2D sum rate depends on the spectrum reuse in the available area. Denote by d the D2D link length and r the macrocell radius. As an order of magnitude approximation, we argue below that if d/r is sufficiently small and the total number of D2D links is sufficiently large, the sum D2D rate R scales with $(r/d)^2$,

$$R \propto \left(\frac{r}{d}\right)^2. \tag{10.132}$$

For simplicity, let us consider an isolated cell and assume that all the D2D links are of the same length d. Our argument consists of two parts. First, the size of the available area is at least $c_1 r^2$ for some constant c_1, irrespective of the location of the macrocell transmitter. This is true because it can be shown that a point belongs to the available area if it is at least $2r/3$ from the base station and at least $r/3$ from the macrocell transmitter. Second, consider sphere packing. Within the available area, randomly select a first D2D link for transmission. To ensure an SIR target, a circle area around the receiver is wiped out for other D2D links to be selected. The size of that circle area is proportional to d^2. We continue to select another D2D link outside the first circle area, and the process goes

10.3 Cognitive radio

Table 10.3. Numerical results of D2D interference budget versus macrocell rate.

I_0 (dB)	Macrocell rate (bits/second/Hz)
0	2.29
−10	3.00
−20	3.12
−30	3.13
−∞ (no D2D transmissions allowed)	3.13

on until the available area is all covered by circle areas. Thus, the number of circles is equal to the available area divided by a circle area. The SIR target of each scheduled D2D link ensures a minimum achieved rate. Hence, the total rate $R = c_2 c_1 r^2 / d^2$ with c_2 another constant.

We simulate a single 120° sector cell with 16 macrocell transmitters uniformly distributed in the cell and scheduled in a round-robin TDM manner. A large number of D2D links are dropped in the cell with a fixed length d. After a macrocell transmitter is scheduled, a D2D scheduler selects D2D links for transmission such that a scheduled D2D link does not get too much interference from the macrocell transmitter (0 dB threshold) or cause too much interference to another already-scheduled D2D link (10 dB threshold). The macrocell transmitters use inverse power control to reach an SNR target of 10 dB with the thermal noise power σ_0^2. Denote relative D2D interference budget $I_0 = \sigma_{D2D}^2 / \sigma_0^2$. The power backoff Ψ_B is set for a variety of I_0. Obviously, the smaller the I_0, the less the degradation of the macrocell rate (Table 10.3). However, we observe that the effect is diminishing when I_0 is already small (say, −10 to −20 dB).

Figure 10.44 plots R versus d and illustrates the scaling relationship of (10.132). Clearly, the larger the I_0, the greater the sum D2D rate to be achieved. Overall, when d is small, at a minor cost of degrading the macrocell rate, the underlay approach achieves a significant sum D2D rate. We observe that a modest I_0, say, −10 dB, well balances the cost of the macrocell rate and the benefit of the D2D rate.

Our focus has so far been on spectrum sharing between primary and secondary users. To conclude this section, we would like to briefly comment on the multiuser environment. Similar to multiuser uncertainty in spectrum sensing, we need to consider the impact of multiple secondary users on spectrum sharing. For example, in the underlay approach, we have seen that the sum secondary-to-primary interference needs to be taken into account in order to determine the power budget of secondary users. Moreover, the coexistence issue needs to be addressed when multiple secondary systems intend to operate in the same area over the same spectrum once it is deemed available for secondary use. The unlicensed spectrum world has studied the coexistence among Wi-Fi and Bluetooth ([25, 75]) and learned that multi-radio coexistence in a dense spectrum environment could be very challenging. An effective solution is to coordinate them via a common control channel ([76]) such that they either orthogonalize with each other in time-frequency or reuse the spectrum with SINR-based power control.

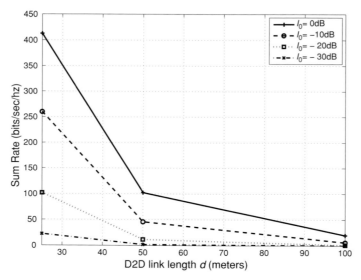

Figure 10.44 Sum rate R versus D2D link length d for a variety of total D2D interference budget I_0. Cell radius $r = 1000$ m. The maximum transmit power is 23 dBm. The antenna gain is 10 dB for macrocell and -2 dB for D2D transmissions. The path loss model is $28.6 + 35 \log_{10} d$ for macrocell and $20 + 40 \log_{10} d$ for D2D transmissions. No shadowing or fading is taken into account. The rate is calculated using the AWGN channel capacity formula with a cap at 6 bits/second/Hz.

Practical example 10.1 LTE-Advanced

LTE is an evolving technology [7, 16]. LTE Release 8 (December 2008) is the first release for this OFDMA-based mobile broadband technology. Subsequently, Release 9 (December 2009) adds relatively minor features such as time-difference-of-arrival-based location techniques and dual-layer beamforming. Release 10 (March 2011), known as LTE-Advanced (LTE-A), represents a major technical advancement to satisfy the IMT-Advanced system requirements defined in [71]. The new features include carrier aggregation, heterogenous networks including relays and femtocells, enhanced downlink and uplink MIMO, and network MIMO. The examples on LTE so far are based on Release 8. We will provide a brief overview of LTE-A in this example.

Enhanced DL/UL MIMO spatial multiplexing

Example 7.1 summarizes the multiple antenna techniques specified in LTE Release 8. We next describe the enhancements of MIMO spatial multiplexing in LTE-A.

Downlink MIMO. LTE-A increases the maximum number of layers of spatial multiplexing from 4 to 8, thereby doubling the peak data rate. Recall that spatial multiplexing in LTE Release 8 is based on predefined codebook precoding using CS-RS. When the number of layers increases to 8, using CS-RS causes excessive overhead. In addition, coarse quantization of any predefined codebook limits the benefit of MU-MIMO. To address those issues, LTE-A takes a very different approach of non-codebook-based precoding. The key idea is to not rely on CS-RS. Note that downlink reference signals in general serve two purposes: channel state information measurement for scheduling and channel estimation for demodulation. In LTE Release 8, both purposes are served by the same signal (CS-RS). LTE-A introduces two new signals to address the two needs separately [114].

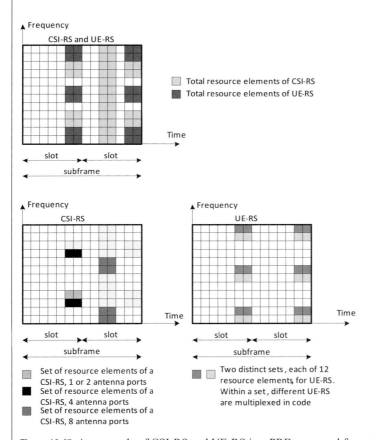

Figure 10.45 An example of CSI-RS and UE-RS in a PRB over a subframe. Normal CP.

First, low duty-cycle wideband signals, called channel-state-information reference signals (CSI-RS), are used for UEs to measure CSI. Figure 10.45 shows the physical resource elements that can be potentially used for CSI-RS in a PRB over a subframe. An eNB can be configured with 1, 2, 4, or 8 antenna ports. With 2, 4, 8 antenna ports, there are 2, 4, 8 corresponding CSI-RS, which are multiplexed to a set of 2, 4, 8 resource elements and separated in the code space with orthogonal cover codes (e.g., $[+, +], [+, -]$ in the case of 2 antenna ports). With 1 antenna port, CSI-RS occupies a set of 2 resource elements, consecutive in time, using only one of the orthogonal cover codes. Figure 10.45 shows that multiple CSI-RS resource sets are available to use among different cells and the total number depends on the number of antenna ports; for example, the number is 20 in the case of 2 antenna ports. Neighboring eNBs should use different resource sets to avoid interference between their CSI-RS.

CSI-RS is not sent in every subframe. An eNB can configure the time periodicity of its CSI-RS from 5 ms to 80 ms. In a given cell, CSI-RS only occupies one of the resource patterns shown in Figure 10.45 in a subframe where CSI-RS is transmitted. While Figure 10.45 shows CSI-RS in one PRB, it is wideband and the pattern repeats over every PRB in frequency. Clearly, the presence of CSI-RS takes away some time-frequency resource space from the PDSCH that shares the same PRB. If the PDSCH is assigned to a Release 8 legacy UE for the sake of backward compatibility, CSI-RS is sent on top of PDSCH modulation symbols when they share the same resource elements without modifying Release 8 PDSCH mapping. However, if the PDSCH is assigned to an LTE-A UE, which understands the CSI-RS design, then PDSCH mapping avoids the resource elements occupied by CSI-RS, which is different from Release 8.

Second, UE specific reference signals (UE-RS) provide precoded pilots to enable data demodulation. UE-RS is sent only on the PDSCH PRBs assigned to a UE's data and precoded with the same precoding matrix used for the data as shown in Figure 10.46. As a result, the UE receiver can perform channel estimation based on precoded UE-RS and coherently demodulate different layers without explicit knowledge of the precoding used at the eNB transmitter, as long as the UE knows the transmission rank. Clearly, the eNB is no longer constrained to a predefined codebook and can apply any appropriate precoding, thus non-codebook-based precoding. Note the difference from codebook-based precoding of LTE Release 8 in Figure 7.22, which relies on CS-RS. Unlike UE-RS, CS-RS are from different antenna ports and do not go through the same precoding matrix as the PDSCH data.

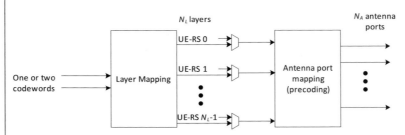

Figure 10.46 Basic transmit signal processing structure of spatial multiplexing in LTE-A downlink using non-codebook-based precoding. $1 \leq N_L \leq N_A \leq 8$.

The notion of UE-RS already exists in Release 8 and is used for non-codebook-based beamforming. Although the basic idea of UE-RS is found to be useful, the particular waveform design in Release 8 (Figure 7.21) is considered inadequate for LTE-A [32]. Instead, a new UE-RS waveform design shown in Figure 10.45 is adopted. A UE-RS occupies 12 resource elements in a PRB over a subframe. UE-RS up to 8 different antenna ports are separated either in frequency or code. For example, in the case of 8 antenna ports, four UE-RS share the same set of 12 resource elements and are multiplexed using four orthogonal cover codes ($[+, +, +, +], [+, -, +, -], [+, +, -, -], [+, -, -, +]$), and the other four occupy a different set, again multiplexed in code. The actual number of UE-RS varies with the transmission rank used in the present PDSCH. If the transmission rank is 1 or 2, then only the first set of 12 resource elements is used for UE-RS; otherwise, UE-RS occupies both sets. Similar to CSI-RS, UE-RS takes away resource from the PDSCH. However, since the new UE-RS design is only used where an LTE-A UE is assigned, PDSCH mapping can avoid the resource elements occupied by UE-RS without the concern of backward compatibility.

The performance difference between non-codebook-based and codebook-based precoding is relatively small for SU-MIMO; however, it turns out that non-codebook-based precoding becomes a key enabling feature for MU-MIMO [101]. To enable MU-MIMO, different UE-RS are assigned to the simultaneous transmissions to different UEs. In LTE-A, up to two orthogonal UE-RS can be used for orthogonal user multiplexing, and they can be further scrambled by a pseudo-random sequence to support nonorthogonal user multiplexing. Hence, a maximum of four users can be multiplexed in MU-MIMO.

Uplink MIMO. LTE Release 8 only supports MU-MIMO and not SU-MIMO. The use of multiple receive antennas at the eNB increases the cell capacity but not the peak data rate of a given UE. LTE-A extends spatial multiplexing to SU-MIMO of up to 4 layers, thereby boosting the peak data rate. To achieve SU-MIMO, a UE in LTE-A uses multiple transmit antennas simultaneously, a device capability not required in LTE Release 8. Obviously, SU-MIMO and MU-MIMO can be combined in the sense that a given PUSCH can be used by one UE with four-layer SU-MIMO or by two UEs each with two-layer SU-MIMO.

Figure 10.47 shows the basic transmit signal processing of uplink SU-MIMO in LTE-A. Note its similarity to downlink SU-MIMO shown in Figure 10.46. Layer mapping is the same as Figure 7.23. An importance difference lies in DFT precoding needed to generate SC-FDMA waveforms as explained in Figure 2.24.

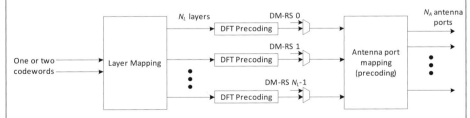

Figure 10.47 Basic transmit signal processing structure of spatial multiplexing in LTE-A uplink. $1 \leq N_L \leq N_A \leq 4$.

Just like UE-RS in the downlink, DM-RS are applied with the same precoding matrix used for the PUSCH data. The eNB receiver can estimate the MIMO channel based on precoded DM-RS without explicit knowledge of the precoding used at the UE transmitter. Uplink MIMO could adopt non-codebook-based precoding. However, LTE-A chooses to take the codebook-based precoding approach where the eNB selects the uplink precoding matrix and informs the UE of the index of the selected one among the predefined precoding matrices as part of the scheduling grant. Next are some examples of precoding matrices in the case of transmission rank equal to 2:

$$\begin{bmatrix} +1 & 0 \\ +1 & 0 \\ 0 & +1 \\ 0 & -j \end{bmatrix}, \begin{bmatrix} +1 & 0 \\ +1 & 0 \\ 0 & +1 \\ 0 & +j \end{bmatrix}, \begin{bmatrix} +1 & 0 \\ -j & 0 \\ 0 & +1 \\ 0 & +1 \end{bmatrix}, \begin{bmatrix} +1 & 0 \\ -j & 0 \\ 0 & +1 \\ 0 & -1 \end{bmatrix},$$

and in the case of transmission rank equal to 3:

$$\begin{bmatrix} +1 & 0 & 0 \\ +1 & 0 & 0 \\ 0 & +1 & 0 \\ 0 & 0 & +1 \end{bmatrix}, \begin{bmatrix} +1 & 0 & 0 \\ -1 & 0 & 0 \\ 0 & +1 & 0 \\ 0 & 0 & +1 \end{bmatrix}, \begin{bmatrix} +1 & 0 & 0 \\ 0 & +1 & 0 \\ +1 & 0 & 0 \\ 0 & 0 & +1 \end{bmatrix}, \begin{bmatrix} +1 & 0 & 0 \\ 0 & +1 & 0 \\ -1 & 0 & 0 \\ 0 & 0 & +1 \end{bmatrix}.$$

Unlike downlink MIMO, the uplink MIMO precoding is designed to preserve the single-carrier property in all the antenna ports. Specifically in Figure 10.47, DFT precoding is done separately in each layer. When the single-carrier waveforms of N_L layers are combined in antenna port mapping, to prevent the PAPR from increasing, all the above MIMO precoding matrices contain one and only one non-zero element in each row. This way, the signal to be sent on any antenna port only comes from one layer. No single-carrier signals from multiple layers are added together on any antenna port.

In spatial multiplexing, each layer needs a separate DM-RS. To support four MIMO layers, four distinct DM-RS are needed in LTE-A. Multiple orthogonal DM-RS can be generated by using different cyclic shifts of the same base sequence or applying orthogonal cover codes ([+, +], [+, −]) to the two OFDM symbols in a subframe.

Uplink SRS is used for the eNB to derive the channel state information and serves a similar purpose to CSI-RS in the downlink. Figure 2.30 depicts SRS. LTE Release 8 specifies periodic SRS transmission configured in a semi-static manner, which is not flexible enough to respond to fast traffic and channel fluctuations. LTE-A further defines aperiodic SRS transmission, which is configured beforehand and can be dynamically triggered as needed without much delay.

Coordinated Multipoint Processing (CoMP). The basic idea of CoMP is that multiple points (e.g., eNBs) coordinate or cooperate with each other such that the transmissions from/to them do not result in significant interference and can even be useful signals. Since coordination and cooperation rely on the backhaul connections among those points, the applicability and performance of CoMP greatly depends on the capacity, latency, and reliability of the backhaul technology. Recall from Example 6.2 that LTE Release 8 supports ICIC via message exchanges between eNBs through an interface called X2. Since the latency of X2 interface is not guaranteed to be low, ICIC are designed for semi-static inter-cell coordination. CoMP intends to overcome that limitation and achieve dynamic coordination and even cooperation by utilizing much lower latency and higher capacity backhaul. 3GPP has recently completed a study on CoMP in terms of deployment scenarios, techniques, standard impact, and performance [91]. CoMP is not included in LTE Release 10 but the study has shown its great promise for the future evolution of LTE, especially in irregular, nonuniform deployment scenarios with complex cell boundaries and user distribution.

In addition to connecting macrocell eNBs with high speed backhaul, a new deployment architecture seems interesting. In this architecture the eNB functionalities are split into a baseband unit (BBU) that performs baseband signal processing and carries out MAC and upper layer protocols and one or multiple remote radio heads (RRHs) for RF processing, including power amplification, filtering, and up/down frequency conversion. An RRH is next to the transmit/receive antennas, thereby reducing cable loss. RRHs can be deployed close to traffic hot spot areas to increase capacity while being controlled by a BBU performing centralized scheduling or a few BBUs rapidly exchanging coordination messages. Often, RRHs are low-power nodes, and together with macrocell eNBs, form a heterogenous network. The backhaul between an RRH and a BBU is via optical fiber to guarantee adequate capacity and latency performance. Such a distributed deployment of RRHs facilitates the use of CoMP.

The 3GPP study has focused on four deployment scenarios: between sectors of an eNB requiring no signaling exchanges via backhaul, between eNBs of a macrocell network, and between a macrocell eNB and multiple low-power RRHs where the RRHs either function with their own cell identities or are associated with the macrocell eNB with one macrocell identity. The CoMP techniques studied are categorized into three types depending on backhaul requirements and scheduling complexity .

- *Coordinated scheduling and coordinated beamforming*. Multiple points do not share data packets. Encoding/decoding take place only at one point. Coordinated scheduling deals with which points to transmit to which UE and coordinated beamforming determines transmit power and beamforming coefficients. Joint coordinated scheduling and coordinated beamforming further reduce multiuser and intercell interference.
- *Joint transmission and reception*. Similar to network cooperation (Section 10.2.2), multiple points cooperate in encoding/decoding and simultaneously transmit to or receive from a UE. Non-coherent schemes such as single-frequency network (SFN) or cyclic delay diversity (CDD) achieve diversity gain and combine transmit power from multiple points. Coherent schemes further achieve network MIMO spatial multiplexing gain. In addition to requiring low latency high capacity backhaul, coherent schemes need tight synchronization among the points and accurate channel estimation.
- *Transmission point selection*. The idea is similar to downlink macro-diversity in the framework of MBB handoff (Section 9.3.4). While the data to be sent to a UE is available at multiple transmission points of a CoMP set, the signal to the UE is transmitted from only one transmission point. The UE dynamically selects the

preferred transmission point – for example, with the best SINR – in a time scale as fast as a subframe. The remaining transmission points in the CoMP set keep silent in the same time-frequency resource to improve the SINR. Transmission point selection is simpler than joint transmission since only antenna selection is employed, as opposed to network MIMO.

Heterogeneous networks
The idea of heterogeneous networks is to overlay low-power nodes within a macrocell network for coverage enhancement, and more importantly, for capacity improvement. 3GPP has considered a variety of low-power nodes including pico eNBs, home eNBs (HeNB, femtocell), relays, and distributed antenna systems (RRH used in CoMP). In an out-of-band deployment, the low-power nodes use different frequency bandwidth from the macrocell network. The drawback is that additional bandwidth is required. In this section, we focus on an in-band deployment where the low-power nodes share the same spectrum as the macrocell network.

Interference management is critical in an in-band deployment, as studied in Sections 10.1.1 and 10.1.2. In addition, because of the coexistence of LTE Release 8 and LTE-A UEs, two important design considerations are backward compatibility and efficiency in multiplexing legacy Release 8 and new LTE-A UEs. In the following, we use relays as an example to illustrate how LTE-A addresses those design considerations. A more detailed description can be found in [67].

Relays. LTE-A has studied two types of relays. Type 1 relays are non-transparent to the UEs and behave like regular eNBs in the sense that they have their own physical layer cell identity and transmit all the necessary channels. Type 2 relays have no cell identity themselves. A type 2 relay attempts to decode as early as possible a signal from an eNB and then transmits the signal, together with the eNB's transmission or retransmission, so as to boost the received signal power at the UE in a transparent way. Type 2 relays exploit the user cooperation idea of Section 10.2.1. Type 1 relays work similar to those studied in Section 10.1.1. The 3GPP specifications support type 1 relays.

Figure 10.48 illustrates an in-band deployment example of a relay, which is denoted by RN. The eNB that serves the RN is called the donor eNB (DeNB). The backhaul link is between the DeNB and the RN, the access link is between the RN and UE 1, and the direct link is between the DeNB and another UE 2, which is not served by any RN. In the downlink, the DeNB-to-RN transmission takes place in the same frequency as the RN-to-UE transmission. In the uplink, the UE-to-RN and RN-to-DeNB transmissions share the same frequency. To avoid desensitization, at a given time a half-duplex RN either receives on the backhaul link from the DeNB or transmits on the access link to the UE, but not both. A full-duplex RN can perform both simultaneously, for example, thanks to spatial separation between transmit and receive antennas. To minimize the change in the DeNB and UE, the RN exhibits a dual personality. It behaves like an eNB to the UE, which can be a Release 8 UE, and like a UE to the eNB. The RN schedules its own UEs.

10.3 Cognitive radio

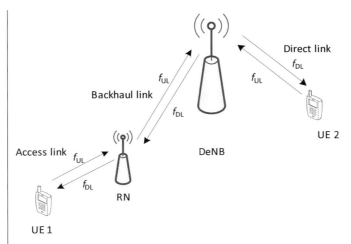

Figure 10.48 Illustration of LTE relays.

To illustrate the issue of backward compatibility, consider the downlink for example. In a regular subframe, the DeNB transmits to UE 2 on the direct link and the RN transmits to UE 1 on the access link. When should the backhaul link occur? In any subframe, the Release 8 UE 1 is expecting the downlink transmission from the RN, which cannot receive on the backhaul link if it is half-duplex. To solve this problem, the RN configures a subset of subframes as multicast broadcast single frequency network (MBSFN) subframes. In an MBSFN frame, UE 1 processes the first few OFDM symbols that carry control information and then ignores the remaining OFDM symbols in the subframe if it is not configured to receive multicast broadcast services. Now, the RN uses those symbols to quickly switch from the transmit mode to the receive mode and then receives on the backhaul link from the DeNB. Meanwhile, the DeNB treats the same subframe as a normal (not MBSFN) one, and sends its control information in the first few OFDM symbols. In the remaining symbols, the DeNB possibly multiplexes the direct link PDSCH to the Release 8 UE 2 and the backhaul transmission to the RN.

Figure 10.49 shows an example of how those remaining symbols are used. R-PDCCH is a new relay physical downlink control channel to transport the grant information of the backhaul link PDSCH. The R-PDCCH is only defined in LTE-A; the Release 8 UEs do not need to process the R-PDCCH. A few interesting points are worth noting.

First, the R-PDCCH is frequency multiplexed with the direct link PDSCH, which is granted with the Release 8 PDCCH in the first few OFDM symbols of the subframe. The FDM nature of resource partition minimizes scheduling complexity and makes it transparent to support the Release 8 UEs.

Second, the backhaul link PDSCH is also frequency multiplexed with the R-PDCCH. Since the RN does not know any grant information beforehand, the RN

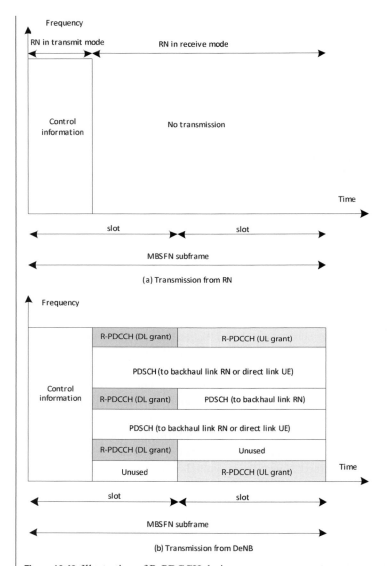

Figure 10.49 Illustration of R-PDCCH design.

needs to buffer the PDSCH while receiving the R-PDCCH. Note that the R-PDCCH for DL grants ends in the first slot of the subframe thereby allowing early decoding of the R-PDCCH. This way, the RN knows how to further process the corresponding PDSCH without having to wait until the end of the subframe before R-PDCCH decoding. In contrast, the R-PDCCH for UL grants occupies the second slot, because it is not time critical to decode UL grants.

Overall the design example shown in Figure 10.49 helps to efficiently utilize the time-frequency resource of the whole subframe. Note, however, that if there is no DL grant in the first slot but there is UL grant in the second slot, then the first slot is wasted.

10.3 Cognitive radio

Enhanced ICIC/Interference cancellation (IC). eICIC is the enhancement in LTE-A of the ICIC techniques defined in Release 8 for interference management in a heterogeneous network. As studied in Section 10.1.2, an in-band deployment with mixed low power nodes and high power macrocell eNBs generally leads to challenging inter-cell interference issues. Consider femtocell/macrocell interference, for example. With closed access, a macrocell UE near an HeNB experiences excessive downlink interference. With open access, the downlink and uplink handoff points are severely imbalanced.

We next take the downlink as an example. The same idea is applicable to the uplink. Orthogonal resource partitioning is an effective approach for dealing with severe inter-cell interference. Following the idea of time domain FFR, ideally the eNB completely mutes for certain subframes during which the low-power node sees no interference from the macrocell. In the case of open access, the low-power node can, in effect, expand its coverage, as shown in Figure 10.50, and serve those UEs that otherwise would be severely affected by inter-cell interference. For the sake of backward compatibility, LTE-A introduces a subframe called Almost Blank Subframes (ABS) in which the eNB transmits only Release 8 CS-RS, primary/secondary synchronization signal and physical broadcast channel for the sake of legacy UEs. The time-frequency resource for other control and data channels are left idle. The same ABS idea can be used in the low-power node so as to expand the coverage of the macrocell.

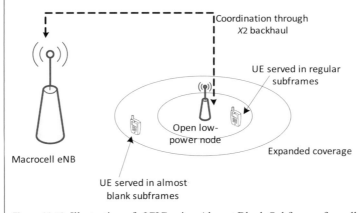

Figure 10.50 Illustration of eICIC using Almost Blank Subframes for cell coverage expansion.

Moreover, to partition the resources adaptively, LTE-A defines a bitmap exchange between a low-power node and a macrocell eNB through X2 backhaul similar to the mechanism used in ICIC. A bitmap consists of 40 bits with each bit representing a subframe. The pattern repeats every 40 ms. Depending on traffic demand, the pattern can be updated as often as every 40 ms. Obviously, with the use of ABS, the inter-cell interference may vary rapidly from one subframe to another. Thus, any estimation algorithm based on measurements has to take the rapidly time-varying pattern into account.

The macrocell signals that remain in the ABS (e.g., CS-RS, PSS/SSS, PBCH) can still cause significant residual interference, because the power from the macrocell eNB is dominant and strong. The interference will appear impulsive since those signals only occupy a small fraction of the time-frequency resource. The receiver should consider non-Gaussian decoding metrics such as the one shown in Figure 2.10. In addition, an advanced UE receiver may further perform interference cancellation (IC) to decode and remove the strong interference before processing the desired signal.

Carrier aggregation

LTE Release 8 supports scalable contiguous bandwidths of up to 20 MHz. LTE-A introduces the feature of carrier aggregation such that a UE or eNB can simultaneously utilize multiple (up to 5) component carriers [144]. Two main motivations drive the study of carrier aggregation. First, carrier aggregation increases the bandwidth and thus the peak data rates required by IMT-Advanced. Second, carrier aggregation allows the operators to pool together fragmented spectrum to provide mobile broadband services, and to dynamically allocate traffic across the entire spectrum to improve utilization. Three spectrum scenarios are targeted: contiguous component carriers in a single band, non-contiguous component carriers in a single band, and non-contiguous component carriers in multiple bands.

The use of carrier aggregation is very flexible. In an eNB, the number of downlink component carriers may be the same as or different from the number of uplink component carriers. UEs may be assigned different subsets of downlink and uplink component carriers. For example, suppose there are three downlink and three uplink component carriers in an eNB. A UE may be assigned with two downlink and two uplink component carriers, while another UE assigned with two downlink and one uplink component carriers. Yet another UE – a Release 8 UE that does not support carrier aggregation – uses only one downlink and one uplink component carriers.

An LTE-A UE has one downlink primary component carrier and an associated uplink primary component carrier. The association between downlink and uplink primary component carriers is cell-specific and signaled in the system information. Different UEs in a cell may use different pairs of carriers as their primary component carriers. In addition, the UE may have one or multiple secondary component carriers in either downlink or uplink. For a given UE, the difference between the primary and secondary component carriers is that the primary ones provide system and control information such as PUCCH configuration and connection maintenance. Using the upper layer signaling protocols, secondary component carriers can be added, removed, and reconfigured, and furthermore be activated/deactivated in order to reduce UE battery consumption.

To accomplish the above flexibility, each component carrier has to be backward compatible with Release 8/9, although other possibilities are envisioned for future evolution including non–backward-compatible carriers (which can be accessed only by LTE-A UEs) and extension carriers (which operate as an extension of some other carriers). In addition, for a given UE, a component carrier is just a data transmission pipe independently running its own H-ARQ protocol. Multiple component carriers are managed by a single scheduler, which multiplexes/demultiplexes data traffic to/from the upper layer protocols, such as Radio Link Control (RLC) and Packet

Data Convergence Protocol (PDCP). The use of carrier aggregation is transparent to those protocols.

Finally we would like to highlight a few interesting design aspects.

Inter-band carrier aggregation. One example is to aggregate a higher frequency band such as 3.5 GHz and a lower frequency band such as 700 MHz. The component carrier in the higher frequency band may be much wider than that in the lower frequency band. The difference in carrier frequency furthermore results in large difference in path loss. Therefore, the two component carriers provide very different coverage. One deployment scenario is to use the lower frequency component carrier for ubiquitous coverage and high mobility support and the higher frequency component carrier for higher throughput at hot spots. Moreover, the higher frequency component carrier can be deployed along the cell boundaries of the lower frequency component carrier to improve the cell edge performance.

Cross-carrier scheduling. Where should the PDCCH of DL/UL grants be sent for the PDSCH/PUSCH? One obvious choice is to configure each component carrier to transmit the PDCCH and the scheduled PDSCH on the same carrier, and transmit the PDCCH on the downlink carrier that is linked to the uplink carrier of the scheduled PUSCH according to the system information. However, the FFR-based ICIC techniques used to coordinate the PDSCH/PUSCH are not applicable to the PDCCH, because the PDCCH is not confined to any frequency subband but is instead transmitted across the whole component carrier. The PDCCH inter-cell interference issue becomes particularly severe in a heterogenous network. Cross-carrier scheduling addresses this issue by transmitting the PDCCH on a component carrier to schedule the PDSCH/PUSCH on a different different component carrier. Thus the PDCCH inter-cell interference can be mitigated in a manner similar to FFR, except that the coordination is based on component carriers instead of subbands. Figure 10.51 illustrates an example of a heterogenous network with two component carriers F_1, F_2. A macrocell eNB transmits its PDCCH on F_1 to schedule the PDSCH/PUSCH on both carriers, while a low-power node (picocell/femtocell eNB) transmits its PDCCH on F_2 to schedule the PDSCH/PUSCH on both carriers. The macrocell eNB reduces its transmit power on F_2 to mitigate interference to the PDCCH of the picocell/femtocell eNB.

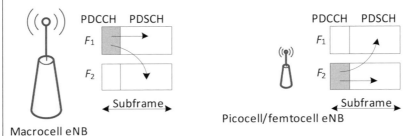

Figure 10.51 Illustration of cross-carrier scheduling achieving FFR inter-cell interference coordination on PDCCH.

Uplink transmit power management. Carrier aggregation requires more transmission power in the uplink. First, when a UE uses multiple UL component carriers, the single carrier property is no longer maintained, thereby increasing the PAPR. Second, in the case of asymmetric carrier aggregation where there are more DL component carriers than UL ones, the UE needs to send additional control signaling of H-ARQ-ACK and CQI/PMI/RI. For example, CSI is fed back for each component carrier because of different channel and interference conditions among component carriers. Hence, it is important to manage uplink transmit power carefully. As in Release 8, the uplink power control in LTE-A consists of open-loop and closed-loop portions. The process is done independently on individual component carriers. When the sum of the required transmit powers of all the activated component carriers exceeds the total maximum output power of the UE, the UE has to distribute its power budget among the PUCCH/PUSCH on different component carriers. For example, suppose the priority is descending from the PUCCH to the PUSCH with uplink control information and then to the PUSCH without uplink control information. Then the UE attempts to satisfy any high priority transmission before giving any power to a low priority transmission. Once all the high priority transmissions are satisfied, the UE distributes the remaining power budget to the PUSCH transmissions without uplink control information by scaling back by the same factor.

Practical example 10.2 Cognitive radio RAN in TV white spaces (IEEE 802.22)

The TV VHF/UHF broadcast bands cover 54 to 862 MHz, depending on the regulation in various countries and regions. Because of low industrial noise and good propagation, those bands are considered ideal to provide broadband coverage in sparsely populated rural areas. IEEE 802.22 is the first cognitive radio standard for wireless regional area networks [149]. The goal is to apply cognitive radio techniques and share unused TV bands (TV white spaces) on a non-interfering basis; the interweaved approach. In particular, the standard allows the base stations and users to operate in the TV broadcast bands without causing harmful interference to incumbent TV broadcasting and low-power licensed devices such as wireless microphones.

To achieve the above goal, the 802.22 system has three main design features:

1. Cognizance of incumbent operations in the vicinity
2. Self coexistence among multiple 802.22 systems operating in the same area
3. Support of large cells with coverage up to 100 km, with stationary users

For the sake of studying cognitive radio, we are interested in the first two features.

The standard specifies two tools to achieve the cognitive functions.
First, knowledge of geolocation and access to a channel availability database.

The location of every device needs to be determined. The location of a base station is required to be known within 15 m and that of a user to be known within 100 m. Every device is installed in a fixed location and equipped with satellite-based geolocation technology, for example, GPS. In the initialization stage, after it determines its geolocation, a user attempts to connect to a base station and to send the location information. The base station sends to the database its own location information and that of the connected users. The database is built to provide accurate and up-to-date information about the incumbent operations in the area to be protected. On the basis of interference analysis, the database informs the base station of the available channels as well as the corresponding maximum transmit power. The base station can also provide other information, such as antenna height and pattern, to improve interference analysis.

Second, spectrum sensing. Sensing is performed in both the base station and all the users. The users report the sensing results to the base station, which makes the final decision on whether a channel is available. The quiet periods for sensing are synchronized between the base station and all the connected users, and among neighboring base stations. There are three incumbent transmissions to be protected: analog TV, digital TV, and low-power auxiliary devices such as wireless microphones. The standard specifies a number of sensing requirements. For example, a sensing antenna must have a gain of at least 0 dBi in all directions, and must be outdoors, clear of obstructions as much as possible, and at least 10 m above ground level. For 0 dBi antenna gain, the sensing receiver sensitivity is -116 dBm for digital TV, -94 dBm for analog TV, and -107 dBm for wireless microphones. The channel detection time is 2 s. The probability of detection is 0.9 and that of false alarm is 0.1.

Taking into account the results from both database access and spectrum sensing, the base station informs the connected users of what channels to be used for broadband access.

Similarly, the database access and spectrum sensing approaches can be employed to enable self coexistence among multiple 802.22 systems operating in the same vicinity. Although inter-system interference may appear similar to inter-cell interference in a conventional cellular system, the deployment scenario is quite different since multiple 802.22 operators may independently deploy their own systems.

Inter-system interference could be severe. For example, 802.22 is a TDD system since it is not always possible to find unused paired TV channels. Therefore, those 802.22 systems have to synchronize the downlink and uplink time partition, not only for co-channel operations but also for adjacent channels because of out-of-band leakage. In addition, assuming closed access between those systems, restricted association is a significant problem, as we have seen in D2D communications.

Clearly, those 802.22 systems can coordinate among each other via a common database. In addition, the 802.22 standard specifies a coexistence beacon protocol (CBP) to allow over-to-air spectrum sharing schemes to address this issue. Specifically, in a dedicated self-coexistence window at the end of some frames, a base station or user either transmits a CBP packet in the operating channel or monitors to receive CBP packets on any channel. The self-coexistence window is synchronized across the TV channels for efficient inter-system coordination. A CBP packet consists of all necessary information for system discovery and sharing.

Finally, since the TV channel bandwidths vary in different parts of the world, to maintain a common signal frame/symbol structure, the 802.22 standard defines a single mode (2048) FFT and scales the sampling frequency and thus tone spacing and symbol length depending on the actual channel bandwidth (6, 7, or 8 MHz).

10.4 Summary of key ideas

- The conventional cellular framework is optimized to maximize coverage given the number of base stations to be deployed in an area. To drastically improve system capacity, this chapter explores ideas beyond the conventional cellular framework.
- The first idea is to adopt heterogeneous network topology by shortening the communication distance. This includes the use of relays, small cells, and D2D communications.
- We consider in-band relays subject to the half-duplex constraint. A relay divides a single hop (source to destination) into multi-hop (source to relay to destination), thereby shortening the per hop communication distance. For a given source/destination link, the tradeoff is boosted SNR versus reduced bandwidth due to half-duplex. However, in a system with multiple links, such bandwidth loss is offset by dense spectrum reuse. A link can select between the direct one-hop path and the indirect multi-hop path via a relay for spatial diversity. In general, routing and scheduling should be addressed in two different time scales to reduce complexity.
- We have not explored out-of-band relays where backhaul links and relay links operate at two disjoint bandwidths. Two frequency configurations are possible. In the first, backhaul links use the same bandwidth as access links and relay links are allocated a separate frequency. In the second, relay links use the same bandwidth as access links and backhaul links are allocated a separate frequency. Assuming more bandwidths become available for future mobile broadband communications, out-of-band relays overcome many technical and performance limitations of in-band relays. For example, in the second configuration above, the users operate seamlessly between relay links and access links in a conventional uplink bandwidth while backhaul links utilize TV white spaces. The fact that relays and base stations are fixed makes it easier to apply cognitive radio techniques.
- Femtocells are small cells deployed by end users instead of operators for cost reduction. We consider femtocells sharing spectrum with macrocells and investigate interference between macrocells and femtocells. There are two basic deployment models of femtocells. In the closed access model, a passing macrocell user is not allowed to access the femtocell. This causes dead zones for the macrocell user in the downlink and for the femtocell base station in the uplink. In the open access model, a passing macrocell user can hand off to the femtocell. However, the downlink and uplink require very different optimal handoff points. The interference management techniques studied

10.4 Summary of key ideas

in the previous chapters, such as FFR and beamforming, can be applied to mitigate macrocell/femtocell interference.

- We have not explored out-of-band femtocells where femtocells and macrocells use two disjoint bandwidths. Obviously, many of the femtocell/macrocell interference issues disappear naturally. In this case, one needs to design efficient cell search and handoff protocols to facilitate inter-frequency handoff between femtocells and macrocells.
- D2D communications allow nearby users to communicate directly without going through an access point. The greatest challenge comes from very irregular interference topology, which results in large dynamic signal range. Conventional cellular techniques do not seem adequate. On the other hand, CSMA/CA with RTS/CTS is designed to deal with unknown/uncontrolled interference in the unlicensed spectrum and seems robust. We take a step further to study a synchronous SIR-based distributed scheduling that uses a single-tone analog power measurement for interference control.
- The second idea we have explored is to cooperate among concurrent transmissions such that interference is not treated as noise. The notion of cell boundary disappears. Instead, transmitters and receivers of the links form a distributed network MIMO system. One use case is user cooperation with relays where the destination combines the signals from both the source and the relays. Network cooperation is another use case where one exploits the fact that base stations are connected via backhaul and can cooperate for encoding in the downlink and decoding in the uplink. An important practical constraint is limited backhaul capacity, because cooperation in encoding/decoding requires an excessive amount of data exchange among base stations.
- The third idea is to make spectrum allocation dynamic and smart such that a cognitive radio is able to determine the spectrum availability in a real time manner. We consider a special but practical framework of primary/secondary spectrum management, for which spectrum sensing and sharing are two important steps. While other approaches are possible to determine the spectrum availability, spectrum sensing does not require modification of primary users or ubiquitous Internet access. When a secondary user senses by itself, it has to be conservative and yet suffers from multiuser uncertainty. Those issues can be resolved with coordinated or cooperative spectrum sensing where multiple secondary users help each other. There are three basic schemes of spectrum sharing: interweaved, overlay, and underlay. The interweaved scheme orthogonalizes primary and secondary users. In the overlay scheme, the secondary user splits its power and transmits its own signal with a fraction of its total power, which interferes with the primary user, and assists the primary user with the remaining power. The assistance and interference completely offset each other so that the primary user is not degraded. The overlay scheme is difficult to implement in practice but can serve as a performance benchmark. The underlay scheme is more practical. It simply controls the transmit power of secondary users so as to bound the rise-of-thermal at the primary user. However, the drawback is that secondary users may be constrained to short communication range.

- The above ideas can be combined to offer greater gains. One example mentioned above is the use of out-of-band relays where backhaul links utilize TV white spaces cognitively. Another example is to combine relays and D2D communications such that relays do not have to be specially installed nodes in fixed locations. Instead, every user can potentially be a relay for another user using a D2D link.

A   Overview of system operations

Strictly speaking, OFDMA is merely a multiple access principle of sharing spectrum. To build a full-fledged mobile broadband communication system, one needs to design a whole set of system operations. We highlight a few important elements of the required system operations in Figure A.1 and provide a high level overview in this section.

A.1   Cell search, synchronization, and identification

After power-up, the user first detects the existence of a base station in the area by searching for some signal signature in the downlink. Below are some examples of commonly used signal signature:

- Correlation between cyclic prefix and the last portion of the OFDM symbol (see Section B.2 for cyclic prefix correlation).
- Synchronization channel, which consists of pre-defined signal waveforms with special properties, such as autocorrelation function close to a delta function (see Section 9.5 for an example of synchronization channel).
- Common pilot channel, which consists of a subset of high power tone-symbols distributed according to pre-defined patterns (see Section B.3 for an example of pilot channel).
- Beacon channel, which consists of a special OFDM symbol in which most of signal power is concentrated on one tone-symbol (see Section 9.5 for the use of beacon channel).

Once the user has discovered a base station, it synchronizes its carrier frequency, clock, and symbol time with the received downlink signal and then retrieves from the broadcast channel certain system information, such as system time, base station identifier, operator identifier. When multiple base stations have been discovered, the user may select one of them to connect.

The above operations are illustrated in Figure A.2 and are carried out solely in the user requiring no interaction with a base station.

Overview of system operations

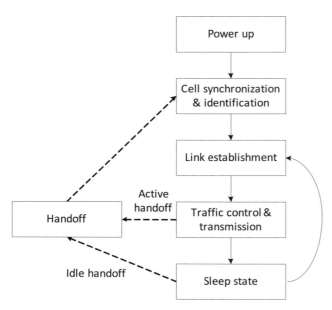

Figure A.1 Sequence of system operations.

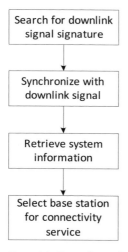

Figure A.2 Operations of "cell synchronization and identification."

A.2 Link establishment

To establish a link, the user first goes through the physical layer access by sending an access probe. In general, access probes can be random access or contention-free access. For initial access, the access probe is random access. Recall from Section 2.2 that the user is not time synchronized with the base station yet. To avoid causing intra-sector interference to other uplink channels, the access probe channel can use separate channel

A.3 Traffic control and transmission

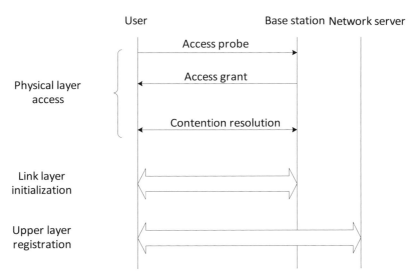

Figure A.3 Operations of "link establishment."

resource. The base station keeps on monitoring the access probe channel to detect potential access probes. If an access probe is detected, the base station responds with an access grant. One function of the access grant is to correct the transmit symbol time of the user such that the uplink signal from the user is synchronized with the base station receiver. When multiple users send access probes simultaneously, collision may occur in the access probe channel. Therefore, the user and the base station need additional handshaking to resolve potential collision.

Once the physical layer access is over, the user and the base station now have access to the traffic channel to initialize the link layer protocols and parameters, such as fragmentation and reassembly, and ARQ. Furthermore, the user communicates via the base station with some network server to establish security between the user and the network, such as user/network authentication, service authorization, and encryption key exchange. The base station fetches the user's service profile from the network server so as to provide appropriate QoS control. The user also registers with a mobility management server in order to receive incoming traffic as it moves.

The above operations are illustrated in Figure A.3.

A.3 Traffic control and transmission

Once the link has been established, the user is ready to utilize the downlink and uplink traffic channels. The operations are illustrated in Figure A.4.

Packets pending for downlink transmission are queued at the base station. The user periodically sends channel quality indicator (CQI) reports to inform the base station of the downlink channel condition. CQI reports may cover the entire bandwidth or a portion of it, and furthermore include MIMO related information when applicable. If the base

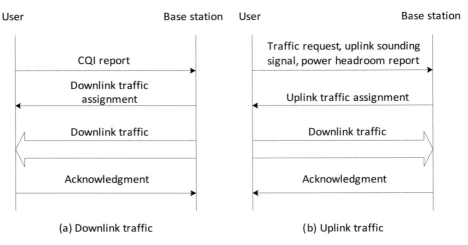

Figure A.4 Operations of "traffic control and transmission."

station decides to schedule the user, it sends an assignment message so that the user knows to receive the corresponding traffic channel. The user then sends acknowledgment about whether the traffic channel has been received successfully.

Packets pending for uplink transmission are queued at the user. The user sends traffic requests to inform the base station of its queue status. A traffic request may simply be an ON-OFF bit or include additional information about the queue length or traffic type. To assist uplink scheduling, the user may further send uplink sounding signal to allow the base station to measure the uplink channel condition, and information about its available power headroom such that the scheduled uplink traffic transmission is within the transmit power capability of the user. If the base station decides to schedule the user, it sends an assignment message such that the user knows to send the corresponding traffic channel. The base station then sends acknowledgment about whether the traffic channel has been received successfully.

Below we list a few important concepts that enhance the performance of traffic control and transmission.

- *Scheduling*. The base station scheduler decides what users are to be allocated in both the downlink and the uplink. Power and bandwidth allocation can be viewed as part of the scheduling function. Scheduling is done in a packet-switched manner (slot-by-slot) such that traffic from different users can be multiplexed efficiently.
- *Adaptive coding and modulation*. The choice of coding and modulation used in the traffic channel transmission is part of the scheduling decision, and depends on the time-varying channel condition of the user as well as the power/bandwidth allocated to the user.
- *Hybrid ARQ*. When a first traffic channel transmission fails, the sender makes a retransmission subsequently. The receiver combines the two transmissions to improve the chance of successfully receiving the traffic.

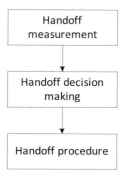

Figure A.5 Operations of "handoff."

A.4 Sleep state

When the user is actively communicating with the base station, it spends a fair amount of power in sending/receiving control signaling, such as CQI reports and traffic channel assignment, in addition to actual traffic transmission. Moreover, closed-loop control mechanisms, required to correct the power, symbol time, and/or frequency of the user transmitter, also consume power. Meanwhile, in a data session, the user's traffic often exhibits an ON-OFF duty cycle pattern. The user can reduce its power consumption and occupy less control resource by going to a sleep state when there is no traffic. During sleep, the user turns off the closed-loop control mechanisms and most of control signaling, except for receiving paging from the base station with relatively long periodicity on the order of seconds.

The base station pages the user when new packets arrive to be sent in the downlink. When paged, or when new packets are generated to send in the uplink, the user will need to return from the sleep state and establish a new link as shown in Figure A.1. As compared with initial access, certain steps in link establishment can be simplified or skipped here. For example, the base station may assign a contention-free access probe in the paging message so as to avoid collision in physical layer access. In addition, the security association with the base station or network user established before the user goes to sleep may still be valid.

A.5 Handoff

A unique phenomenon in a mobile communication system, handoff allows the user to communicate with the base station of the "best" channel condition when the user moves from one cell to another. As shown in Figure A.1, there are two types of handoff, namely *idle handoff* (e.g., for location update), when the user is in the sleep state, and *active handoff*, when the user has active traffic session with the serving base station.

Handoff consists of three major steps, as illustrated in Figure A.5.

In order to detect potential handoff candidates, the user keeps on searching for new base stations and monitoring the change in signal strength from neighboring base stations as a background process while communicating with the serving base station. The downlink signal to be monitored can be the synchronization, common pilot, or beacon channel mentioned earlier.

Based on the handoff measurement, either the user or the base station (network) makes a decision to trigger handoff. The former case is called mobile controlled handoff, and the latter case, network controlled. Handoff decision is often based on a hysteresis test to avoid the ping-pong effect between base stations.

The procedure of handoff to the new base station is similar to that of initial access, except that certain steps in link establishment can be simplified or skipped here. For example, the new base station may assign a contention-free access probe in the handoff message so as to avoid collision in physical layer access. In addition, security association with the old base station may be transferred to the new base station via backhaul. During active handoff, the packets queued in the old base station may be transferred to the new base station and new packets are routed directly to the new base station.

In addition to the above handoff scenario between base stations (intra-technology handoff), inter-technology handoff is also possible. For example, the user may hand off from a cellular base station to a Wi-Fi access point, for example, to offload traffic.

B OFDM point-to-point communications

Point-to-point communication techniques are the basic building blocks to carry out the system operations outlined in Section A. They have been well covered in many textbooks on digital or wireless communications, such as [61, 125, 126, 159]. This section is not intended to be a comprehensive survey of those techniques, but rather to cover a few areas directly related to the system level topics studied in this book. We provide a brief summary of commonly used schemes with an emphasis on the applications in an OFDM system.

B.1 Signal-presence detection

Signal detection is used by a receiver to decide the current "state" among a finite number of possible ones from a received signal corrupted by the wireless channel and noise.

Hypothesis testing

Consider the AWGN channel (C.1). A commonly cited example is detection of an uncoded binary antipodal signal where $x[t] = \pm 1$. In a practical wireless system, most signals are coded, and as a result, detection (or demodulation) is done jointly with decoding, as will be studied in Section B.4. In the following we will focus on a special case, namely signal-presence detection. There are two possible states or hypotheses:

$$\mathcal{H}_0 : x[t] = 0,$$
$$\text{versus} \quad (B.1)$$
$$\mathcal{H}_1 : x[t] = s[t],$$

where $s[t]$ is a known waveform. Signal-presence detection is used by users for cell search in initial system acquisition (Section A.1) and recurrent monitoring for handoff (Section A.5), and by base stations in detecting access probes (Section A.2).

In the above hypothesis testing, there are two types of errors. A type I error (false alarm) occurs when $\mathcal{H}_0$ is falsely rejected: the receiver erroneously declares a valid $s[t]$ be detected although there is just noise. A type II error (miss) occurs when $\mathcal{H}_1$ is falsely rejected: the receiver fails to detect the presence of $s[t]$ though it does exist. Note that the cost of a type I or II error on the system operation is usually different. Take the detection of access probes as an example. A missed detection means that a user sends an access probe to a base station but gets unnoticed. The consequence is that the user needs to send another access probe subsequently, thereby increasing access latency. On the other hand, a false alarm means the base station is going to respond to a non-existent access probe and allocate system resource to it. While the base station will eventually recover from the error, the allocated system resource goes wasted for some period of time.

Denote by P_F, P_M the false alarm and miss probabilities, respectively. $P_D = 1 - P_M$ is the detection probability. A meaningful design objective involves a trade-off between P_F and P_M, because it is undesirable to make one arbitrarily small at the expense of the other. A widely used design objective is the Neyman-Pearson criterion,

$$\max P_D, \text{ subject to } P_F \leq \hat{P}_F, \tag{B.2}$$

where $\hat{P}_F$ is a bound on the false alarm probability. Suppose the observed variable $y[t]$ follows distribution function $p_0(\cdot)$ under hypothesis $\mathcal{H}_0$ or $p_1(\cdot)$ under $\mathcal{H}_1$. Define *likelihood ratio*

$$L(y) = \frac{p_1(y)}{p_0(y)}. \tag{B.3}$$

The solution to the Neyman-Pearson problem is that

$$\text{Accept } \begin{cases} \mathcal{H}_1, & \text{if } L(y[t]) > \Theta, \\ \mathcal{H}_0, & \text{if } L(y[t]) < \Theta, \end{cases} \tag{B.4}$$

where Θ is the decision threshold. If $L(y[t]) = \Theta$, then accept $\mathcal{H}_1$ with some appropriate tie-breaking probability. However, for continuous signal $y[t]$, we assume that the probability of $L(y[t]) = \Theta$ is zero. This form of solution is not limited to signal-presence detection in the AWGN channel but applicable to much more general problems.

Define region $\Gamma(\Theta) = \{y | L(y) > \Theta\}$. Threshold Θ is chosen to satisfy the constraint of the false alarm probability

$$P_F = \int_{\Gamma(\Theta)} p_0(y) \, dy = \hat{P}_F. \tag{B.5}$$

The achieved detection probability is

$$P_D = \int_{\Gamma(\Theta)} p_1(y) \, dy. \tag{B.6}$$

Matched filter detector

In the AWGN channel (C.1), assuming L samples of $y[t]$ are observed, $t = t_1, \ldots, t_L$, the optimal solution (B.4) becomes

$$\text{Accept } \begin{cases} \mathcal{H}_1, & \text{if } \Lambda > \Theta', \\ \mathcal{H}_0, & \text{if } \Lambda < \Theta', \end{cases} \tag{B.7}$$

with Θ' a different threshold from Θ, and

$$\Lambda = \left| \sum_{t=t_1}^{t_L} s^*[t] y[t] \right|^2, \tag{B.8}$$

the test statistics of the matched filter detector. Threshold Θ' depends on not only the false alarm probability target $\hat{P}_F$ but also the noise power σ_0^2, which is not known beforehand and varies in time in a cellular system, as the noise term in (C.1) also includes interference from various sources. Therefore, the receiver needs to estimate real time σ_0^2 in order to implement the optimal detector (B.7).

The above matched filter detector can be extended to the fading channel. In a flat fading channel (C.6), assuming fading coefficient $h[t]$ remains unchanged from t_1 to t_L, the matched filter detector is still optimal, except that Θ' further depends on the statistics of fading coefficient $h[t]$. If the interval $[t_1, t_L]$ spans over multiple channel coherence times, then the detector cannot coherently correlate $s^*[t]$ and $y[t]$ over the entire interval. Instead, coherent correlation should be done in each coherence time and then the outputs are added non-coherently for all the coherence times. The test statistics in (B.8) is changed to

$$\Lambda = \sum_{k=1}^{K} \left| \sum_{t=t_{k,1}}^{t_{k,M}} s^*[t]y[t] \right|^2, \tag{B.9}$$

where the interval $[t_1, t_L]$ is divided into K coherence times, each consisting of M samples $t_{k,1}, \ldots, t_{k,M}$.

Signal-presence detection in a frequency selective channel (4.10) is apparently more complex because it involves multiple taps. Without the knowledge of the channel power delay profile, we take a heuristic approach. The RAKE receiver in a CDMA system is such an example. Suppose that more than L samples $y[t]$ are observed, $t = t_1, \ldots$. Calculate a time-shifted correlation

$$\Lambda(l) = \left| \sum_{t=t_1}^{t_L} s^*[t]y[t+l] \right|^2, \quad l = 0, 1, \ldots, \tag{B.10}$$

where l indicates the time shift measured in samples. Some $\Lambda(l)$s are significantly stronger than others and thus considered to represent potential channel taps containing signal energy. To collect the signal energy from multiple channel taps, those $\Lambda(l)$s are added to form the test statistics and be compared against the detection threshold (B.7).

Note that the signal-presence detector relies on the model assumptions of the wireless channel. In practice, we can further introduce closed-loop mechanisms to ensure the false alarm probability target $\hat{P}_F$ (B.5) is met. For example, adjust the threshold Θ' depending on how frequently false alarm events occur. However, the challenge is that because $\hat{P}_F$ is usually set low, false alarm events are too rare to be an effective feedback.

Detection of access probes

Finally let us apply the signal-presence detector to the problem of detecting access probes at a base station. A similar idea can be applied to the problem of cell search at a user.

As described in Section 2.2, in an OFDMA uplink, the base station uses closed-loop timing control to synchronize all the arriving signals from the users that have already been connected with the base station. When a new user sends an access probe, it is not yet connected with the base station and closed-loop timing control has not yet taken place. The user employs open-loop timing control illustrated in Figure 2.17. Figure B.1

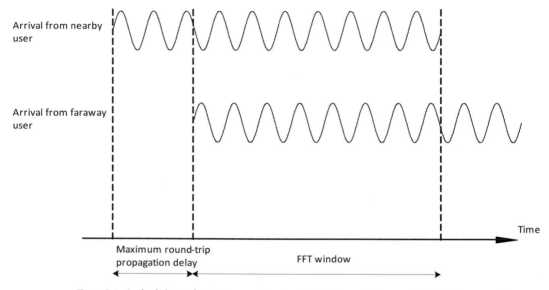

Figure B.1 Arrival time of an access probe signal at the base station receiver. The user employs open-loop timing control to determine its transmit symbol time. As a result, the difference between the arrival time of a nearby user and that of a faraway user can be as large as the maximum round-trip propagation delay between the user and the base station.

shows a possible arrival time of the access probe signal, which depends on the distance between the user and the base station.

The difference between the arrival time from a nearby user and that from a faraway one is bounded by the maximum round-trip delay of the cell. To simplify the receiver implementation, the access probe signal uses a much longer cyclic prefix than a normal OFDMA signal does to cover the maximum round-trip delay. Recall that the cyclic prefix of a normal OFDMA signal is used to cover the maximum delay spread of the channel. This way, the base station receiver can apply a single FFT window, as shown in Figure B.1, irrespective of the arrival time of the access probe signal, which is unknown to the base station.

Let $y[t], t = 1, \ldots, L$, to be the L samples received in the FFT window in Figure B.1, and $s[t]$ the known access probe signal. Because of the cyclic nature of the OFDM signal, the base station calculates a time-shifted correlation similar to (B.10)

$$\Lambda_c(l) = \left| \sum_{t=1}^{L} s^*[t+l \mod L] y[t] \right|^2, \quad l = 0, 1, \ldots, L-1, \tag{B.11}$$

Direct calculation of $\Lambda_c(l)$ as in (B.11) for all l involves $O(L^2)$ operations. Recall that the cyclic convolution of two sequences a, b is defined as

$$[a[t] * b[t]](l) = \sum_{t=1}^{L} a[l-t \mod L] b[t]. \tag{B.12}$$

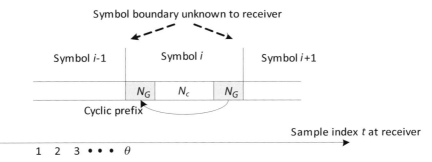

Figure B.2 OFDM symbols with unknown arrival time θ.

It follows that

$$\Lambda_c(l) = |[y[t] * s^*[-t]](l)|^2. \tag{B.13}$$

Taking advantage of efficient FFT, $\Lambda_c(l)$ can be calculated in the frequency domain with only $O(L \log(L))$ operations

$$[\Lambda_c(0), \ldots, \Lambda_c(L-1)] = |\text{IFFT}\{Y[k]S[k]^*\}|^2, \tag{B.14}$$

where $Y[k]$ and $S[k]$ are the discrete Fourier transform of $y[t]$ and $s[t]$, respectively.

For signal-presence detection, the base station compares $\max_l \Lambda_c(l)$ with a threshold as in (B.7). If $\mathcal{H}_1$ is declared, the time index l^* corresponding to $\max_l \Lambda_c(l)$ indicates the arrival time of the detected access probe signal, based on which the base station sends a closed-loop timing control command to adjust the transmit symbol time of the user. Estimation of l^* is an example of the symbol timing synchronization problem, which will be addressed next.

B.2 Synchronization

Synchronization is an important part of receiver design. Consider a sample sequence $x[t]$ of an OFDM signal with t the sample index. An OFDM symbol consists of N_c samples of the FFT body and N_G samples of the cyclic prefix, as shown in Figure B.2. For the sake of simplicity, consider the following AWGN channel with unknown arrival delay θ and unknown carrier frequency offset $f_0 + \epsilon$,

$$y[t] = x[t - \theta]e^{j2\pi(f_0+\epsilon)t/N_c} + w[t], \tag{B.15}$$

with integer f_0 and $-\frac{1}{2} < \epsilon \leq \frac{1}{2}$. ϵ represents a fine frequency offset equal to a fractional of the normalized tone spacing ($= 1/N_c$), and f_0 a coarse frequency offset representing an integer number of tone spacing. Note that the fractional frequency offset ϵ causes the ICI and f_0 causes the uncertainty of tone numbering. For a tone numbered as k at the transmitter, the receiver may treat it as another tone $k' \neq k$.

We consider two major synchronization problems in an OFDM system:

1. *Symbol timing synchronization*: to determine the correct symbol start position θ, based on which the receive symbol window is placed. In addition, several consecutive OFDM symbols often form a frame, in which case *framing synchronization* is also needed to determine the starting OFDM symbol of a frame.
2. *Carrier frequency synchronization*: to eliminate any carrier frequency offset $f_0 + \epsilon$ due to the discrepancy of local oscillators between the transmitter and receiver or the wireless channel such as the Doppler shift.

A third synchronization problem is sampling clock synchronization, which is used to eliminate mismatched sampling clocks between the transmitter and receiver. See [83, 128].

Synchronization generally involves acquisition followed by tracking. Acquisition obtains initial coarse estimates of θ, f_0, ϵ, while tracking is an ongoing process to refine the estimates.

Acquisition using cyclic prefix correlation

First, consider acquisition. Note that the first and last N_G samples in an OFDM symbol are identical. Therefore, it follows that

$$\mathbb{E}\left(y[t]y^*[t+N_c]\right) = \begin{cases} \sigma_x^2 e^{-j2\pi\epsilon}, & \text{if } t \text{ is a sample of the cyclic prefix} \\ 0, & \text{otherwise,} \end{cases} \quad (B.16)$$

with $\sigma_x^2 = \mathbb{E}(|x[t]|^2)$ the average signal power. By exploiting redundant information contained within the cyclic prefix, [35] derives an maximum likelihood (ML) estimation of θ and ϵ as follows.

Suppose that $2N_c + N_G$ consecutive samples $y[1], \ldots, y[2N_c + N_G]$ are observed. Assume that these samples contain one complete OFDM symbol of $N_c + N_G$ samples. As shown in Figure B.2, the position of that symbol is unknown. Define

$$\Lambda(l) = \sum_{t=l}^{l+N_G-1} y[t]y^*[t+N_c], \quad (B.17)$$

$$\Phi(l) = \frac{1}{2}\sum_{t=l}^{l+N_G-1} |y[t]|^2 + |y[t+N_c]|^2, \quad (B.18)$$

$$\rho = \left|\frac{\mathbb{E}(y[t]y^*[t+N_c])}{\sqrt{\mathbb{E}(|y[t]|^2)\mathbb{E}(|y[t+N_c]|^2)}}\right| = \frac{\sigma_x^2}{\sigma_x^2 + \sigma_w^2} = \frac{\gamma}{\gamma+1}, \quad (B.19)$$

with $\sigma_w^2 = \mathbb{E}(|w[t]|^2)$ the average noise power, and SNR $\gamma = \sigma_x^2/\sigma_w^2$. The joint ML estimation is given by

$$\hat{\theta} = \arg\max_l \left(|\Lambda(l)| - \rho\Phi(l)\right), \quad (B.20)$$

$$\hat{\epsilon} = -\frac{1}{2\pi}\angle\Lambda(\hat{\theta}), \quad (B.21)$$

where $\angle$ represents the angle of a complex number in the interval of $[-\pi, \pi)$.

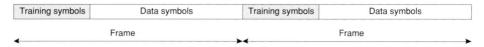

Figure B.3 Illustration of using training symbols in a frame to facilitate synchronization.

Furthermore, if θ, ϵ are constant over a long period, then we can accumulate the observation of multiple OFDM symbols to improve the performance. The estimation becomes

$$\hat{\theta} = \arg\max_{l} \left(\sum_{l} |\Lambda\left(l + l(N_c + N_G)\right)| - \rho\Phi\left(l + l(N_c + N_G)\right) \right), \quad (B.22)$$

$$\hat{\epsilon} = -\frac{1}{2\pi} \angle \left(\sum_{l} \Lambda\left(\hat{\theta} + l(N_c + N_G)\right) \right), \quad (B.23)$$

with l representing the consecutive OFDM symbols.

The above synchronization scheme utilizes the inherent redundancy structure of an OFDM signal and uses a correlation with the cyclic prefix to find the symbol timing without additional overhead in the signal. However, a major drawback is that in a dispersive channel with frequency selective multipath fading, the cyclic prefix portion is contaminated by ISI. In particular, suppose that the channel impulse response lasts for L samples. Then (B.16) holds only for $N_G - L$ samples. As a result, the estimate of θ tends to fluctuate significantly. In addition, an obvious limitation is that only the fractional frequency offset ϵ is estimated. The integer frequency offset f_0 cannot be estimated, thereby leaving us with the uncertainty of tone numbering. Similarly, only symbol timing synchronization, and not framing synchronization, is achieved.

Acquisition using matched filter or half-symbol correlation

To overcome the above problems, we insert training symbols in the signal. Suppose that the training symbols are placed in a fixed location within a frame, as shown in Figure B.3. Then frame synchronization can be obtained once the location of the training symbols is determined. As training symbols are the overhead, it is desirable to keep the ratio of the training and data symbols low.

A training symbol is a known waveform. For a general waveform, we can use the matched filter scheme to correlate the received samples with the known waveform, and set

$$\hat{\theta} = \arg\max_{l}(|\Lambda(l)|), \quad (B.24)$$

where $\Lambda(l)$ is given in (B.10). Obviously, it is desirable that the waveform have good autocorrelation properties in the time domain. To make the correlation effective, the frequency offset has to be kept small so that the phase rotation due to the frequency offset is negligible within a training symbol.

The drawback of the matched filter approach is high computational complexity, as the time-shifted correlation $\Lambda(l)$ needs to be re-computed each time when a new sample arrives. To reduce the complexity, let us specially design the training symbol waveform.

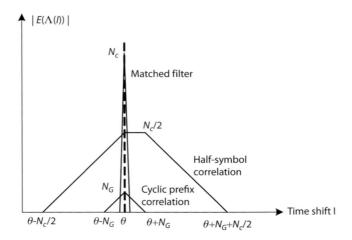

Figure B.4 Comparison of $|\mathbb{E}(\Lambda(l))|$ in the AWGN channel (B.15) with σ_x^2 normalized to 1.

One such example proposed in [136] is an OFDM symbol with two identical time domain halves in the body, which is generated by transmitting pseudo-random QPSK symbols on the even tones and zeros on the odd tones. To exploit redundant information in the training symbol, check the correlation

$$\mathbb{E}\left(y[t]y^*\left[t+\frac{N_c}{2}\right]\right) = \begin{cases} \sigma_x^2 e^{-j\pi(f_0+\epsilon)}, & \text{if } t \text{ is one of the first } N_G + \frac{N_c}{2} \text{ samples} \\ 0, & \text{otherwise.} \end{cases} \quad (B.25)$$

Therefore, modify (B.17) and (B.18) to

$$\Lambda(l) = \sum_{t=l}^{l+\frac{N_c}{2}-1} y[t]y^*\left[t+\frac{N_c}{2}\right], \quad (B.26)$$

$$\Phi(l) = \frac{1}{2}\sum_{t=l}^{l+\frac{N_c}{2}-1} |y[t]|^2 + \left|y\left[t+\frac{N_c}{2}\right]\right|^2. \quad (B.27)$$

Note that $\Lambda(l)$ in (B.26) can be recursively computed when a new sample arrives:

$$\Lambda(l+1) = \Lambda(l) + y\left[l+\frac{N_c}{2}\right]y^*[l+N_c] - y[l]y^*\left[l+\frac{N_c}{2}\right] \quad (B.28)$$

We can apply them in (B.20) to determine $\hat{\theta}$. Here $\Lambda(l)$ provides the estimates of θ and ϵ. Its magnitude, compensated by an energy term, determines θ, and its phase determines ϵ.

In general $\Lambda(l)$ consists of a "signal" component and the remaining "noise" component, where the signal component is equal to $\mathbb{E}(\Lambda(l))$ and reflects the redundancy signal structure to be exploited. Figure B.4 compares $|\mathbb{E}(\Lambda(l))|$ of half-symbol correlation versus cyclic prefix correlation, as well as the matched filter scheme assuming a training symbol consisting of N_c samples of a pseudo-noise sequence. We observe that

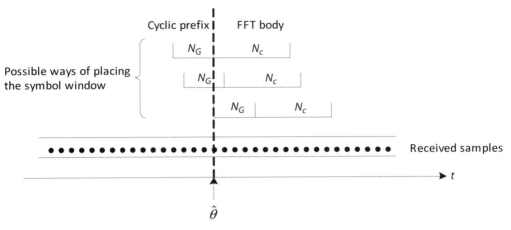

Figure B.5 Ways of placing symbol window in the AWGN channel.

at $l = \theta$, $|\mathbb{E}(\Lambda(l))|$ accumulates $N_c/2$ pieces of sample energy in (B.26), as opposed to N_G pieces in (B.17). Since typically $N_c/2 \gg N_G$, half-symbol correlation facilitates timing acquisition as compared with cyclic prefix correlation because of the greater signal portion. As l moves away from θ, $|\mathbb{E}(\Lambda(l))|$ decays; however, the rate of delay (slope) in both schemes is not as steep as that in the matched filter scheme. In fact, $|\mathbb{E}(\Lambda(l))|$ exhibits a plateau with half-symbol correlation. Because $|\Lambda(l)|$ is corrupted with noise, such a gradual decay makes it hard to pinpoint θ. This naturally leads to a two-step timing acquisition strategy. First, coarse timing synchronization using half-symbol correlation to determine an initial estimate $\hat{\theta}$ with low computational complexity. Second, fine timing synchronization by searching around the initial estimate $\hat{\theta}$ using a matched filter.

Determination of receiver symbol windows

Once $\hat{\theta}$ is determined, how should the receiver place the symbol window? In the AWGN channel (B.15), the receiver can directly take the next N_c samples as the FFT body, or skip the next N_G samples and take the subsequent N_c samples as the FFT body of the OFDM symbol. There are other possible ways of placing the symbol window, as shown in Figure B.5. As long as the symbol start $\hat{\theta}$ resides in the cyclic prefix, there is no ISI in the FFT body.

In a dispersive channel there are multiple arrival paths, each with a distinct symbol start. If the receiver estimates all the $\hat{\theta}$s as well as the corresponding signal energy, then it should place the FFT body such that the cyclic prefix interval contains the $\hat{\theta}$s of the maximum total signal power, as shown in Figure B.6(a). Equation [115] shows that this solution maximizes the SIR, where the signal power is what is captured in the FFT body and the interference power is ICI and ISI caused by paths whose symbol starts fall outside the cyclic prefix interval. Multiple paths interfere with each other, making it hard in practice for the receiver to precisely estimate all the $\hat{\theta}$s. A suboptimal solution shown in Figure B.6(b) is to estimate the dominant or weighted average $\hat{\theta}_0$, and place it

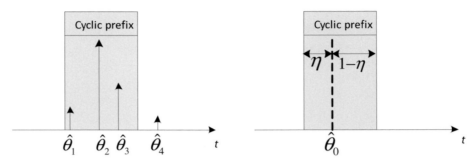

Figure B.6 Placing symbol window in the dispersive channel. In (a), each arrow is an estimated arrival path with the height representing the corresponding signal power.

in the cyclic prefix interval according to a fixed proportion $\eta : 1 - \eta$, where $0 < \eta < 1$ is determined heuristically from empirical power delay profiles.

Estimation of integer frequency offset

Now let us turn to frequency offset estimation using half-symbol correlation (B.26). Similar to (B.21), it follows from (B.25) that

$$f_0 + \epsilon = -\frac{1}{\pi}\angle\Lambda(\hat{\theta}) + 2z, \qquad (B.29)$$

where $\hat{\theta}$ is determined in the above timing synchronization. If we know a priori that $-1 < f_0 + \epsilon \leq 1$, then $z = 0$. However, if the frequency offset is potentially large, then it is not necessary that $z = 0$, in which case $\phi = -\frac{1}{\pi}\angle\Lambda(\hat{\theta})$ represents a partial frequency offset. In general, z is an integer representing the uncertainty of tone numbering. For a tone numbered as k at the transmitter, the receiver may treat it as tone $k + 2z$, with $z = 0, \pm 1, \ldots$.

To determine z, immediately subsequent to the first training OFDM symbol, we transmit a second one, which is generated by transmitting a different set of pseudo-random QPSK symbols on the even tones. Denote by $X[1, k]$, $X[2, k]$ the QPSK symbols transmitted on even tone k in the first and second training symbols, respectively. What is important is differential encoding $(X[1, k]X^*[2, k])$. Symbols transmitted on the odd tones can be arbitrary.

The receiver algorithm works as follows. First, correct the partial frequency offset ϵ by multiplying each received sample $y[t]$ with $e^{-j2\pi\epsilon t}$. This eliminates the ICI due to the frequency offset. Then, perform the FFT in the first and second symbol. Denote by $Y[1, k]$ and $Y[2, k]$ the FFT outputs on even tone k. We estimate z by phase-comparing differential $(Y[1, k + 2z]Y^*[2, k + 2z])$ with $(X[1, k]X^*[2, k])$ for any shift in tone numbering z

$$\hat{z} = \arg\max_{z=0,\pm 1,\ldots} \left(\left| \sum_k (Y[1, k + 2z]Y^*[2, k + 2z])(X[1, k]X^*[2, k])^* \right| \right). \qquad (B.30)$$

Tracking using channel estimation

Finally, let us consider tracking. Recall from Figure B.3 that while the training symbols facilitate acquisition, they are a relatively small fraction of a frame since they are the overhead and should be kept to a minimum. While acquisition uses the training symbols and gets the receiver roughly timing and frequency synchronized, the tracking operation continues to utilize the data symbols and fine-tune synchronization.

As will be elaborated in Section B.3, the data symbols often consist of pilots to be used for channel estimation. Channel estimation can be further improved in a decision-directed manner once the data symbols are successfully decoded. In addition to assisting decoding and demodulation, channel estimates can also be used for synchronization tracking as follows.

Denote by $\hat{H}[s,k]$ the channel estimate at tone k of OFDM symbol s. The estimate of the residual fractional frequency offset $\hat{\epsilon}$ can be adjusted by phase-comparing the channel estimates on the same tone k between two adjacent OFDM symbols $s, s+1$

$$\hat{\epsilon}[s+1] = (1-\delta)\hat{\epsilon}[s] + \delta\Delta_\epsilon, \tag{B.31}$$

$$\Delta_\epsilon = \angle\left(\sum_k \hat{H}[s+1,k]\hat{H}^*[s,k]\right), \tag{B.32}$$

with $0 < \delta < 1$ a step size.

It may appear that we should similarly adjust $\hat{\theta}$ by phase-comparing the channel estimates at the same OFDM symbol between two adjacent tones $\left(\sum_k \hat{H}[s,k+1]\hat{H}^*[s,k]\right)$. However, we have to be careful here, because this adjustment would directionally move the start of the FFT body towards the dominant or weighted average symbol start, but not necessarily minimize the ICI and ISI. Instead, we can transform the frequency domain channel estimate $\hat{H}[s,k]$ to the time domain and estimate the channel impulse response. From the discrete path model (4.10), it follows that the channel estimate at symbol s is related to the channel impulse response

$$\hat{H}[s,k] = \sum_{l=1}^{L} \bar{\alpha}_l e^{-j\bar{\phi}_l} e^{-j2\pi \tau_l k/T_s}, \tag{B.33}$$

where T_s is the length of the FFT body and the time variable t is omitted to simplify the notations. Timing tracking is used to estimate $\bar{\alpha}_l$ and τ_l, and to adjust the symbol window to maximize the total signal energy captured in the cyclic prefix interval, as shown in Figure B.6(a).

B.3 Channel estimation

In OFDM, the received symbol $Y[s,k]$ on tone k in OFDM symbol s is given by (2.9)

$$Y[s,k] = H[s,k]X[s,k] + W[s,k], \tag{B.34}$$

where $X[s,k]$ and $H[s,k]$ are both complex numbers representing the transmitted symbol and the frequency response coefficient of the wireless channel, respectively, and $W[s, k]$ the additive noise. In general, information is transported with the amplitude/phase of $X[s, k]$. However, the wireless channel modifies $X[s, k]$ by introducing amplitude and phase distortion $H[s, k]$. The goal of channel estimation is to estimate $H[s, k]$ in order for coherent demodulation of $X[s, k]$. Channel estimation for wireless OFDM is well studied in the literature (see [119] and the references therein).

The wireless channel is often modeled using a discrete path impulse response model (4.10)

$$h(\tau, t) = \sum_{l=1}^{L(t)} \bar{\alpha}_l(t) e^{-j\bar{\phi}_l(t)} \delta(\tau - \tau_l(t)). \tag{B.35}$$

The frequency response is given by (4.28)

$$H(f, t) = \sum_{l=1}^{L(t)} \bar{\alpha}_l(t) e^{-j\bar{\phi}_l(t) - j2\pi f \tau_l(t)}. \tag{B.36}$$

$H[s, k]$ is equal to $H(f, t)$ with t equal to OFDM symbol s and f tone frequency f_k.

The number of discrete paths $L(t)$ is usually much smaller than the number of tones N_c, and the channel impulse response does not change rapidly from one OFDM symbol to another. Therefore, the frequency response coefficients are correlated in time and in frequency. The idea of pilot-aided channel estimation is to transmit known symbols (pilots) in a subset $[s_p, k_p]$, referred to as pilot tone-symbols. The receiver estimates the coefficients in those tone-symbols and then the channel in other tone-symbols (data tone-symbols). In addition to pilot-aided channel estimation, decision-directed channel estimation treats data symbols that have been previously successfully decoded as pilots, thereby increasing pilot density and improving channel estimation quality. We can take a step further and perform channel estimation and data decoding in an iterative (turbo) fashion such that the quality of estimation and decoding improves as the iterations go on. However, for simplicity, our focus here is the basics of pilot-aided channel estimation.

The least square (LS) estimate at a pilot tone-symbol is

$$\hat{H}_{LS}[s_p, k_p] = \frac{Y[s_p, k_p]}{X[s_p, k_p]}. \tag{B.37}$$

Then, the channel estimates at the remaining data tone-symbols, represented by vector $\hat{\mathbf{H}}$, are obtained by interpolation/extrapolation exploiting the time-frequency correlation. A linear estimator is given by

$$\hat{\mathbf{H}} = \mathbf{Q}\hat{\mathbf{H}}_{LS}, \tag{B.38}$$

with $\hat{\mathbf{H}}_{LS}$ the vector consisting of the LS estimates at the pilot tone-symbols (B.37) and $\mathbf{Q}$ the interpolation/extrapolation matrix to be described next.

Pilot pattern

Before we study a linear estimator $\mathbf{Q}$, let us first consider the pilot pattern in the time-frequency two-dimensional domain. The density of the pilot tone-symbols must satisfy

B.3 Channel estimation

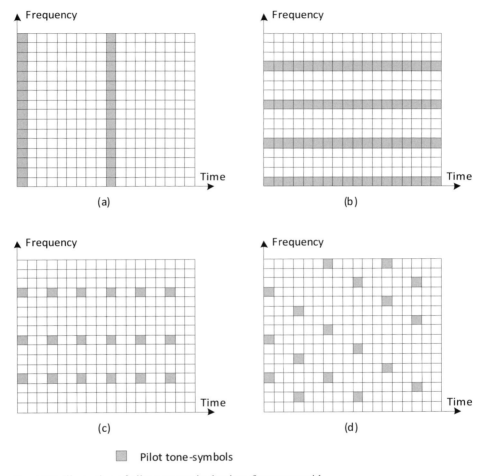

Figure B.7 Illustration of pilot patterns in the time-frequency grid.

the two-dimensional Nyquist sampling theorem in order to recover channel parameters:

$$D_f \leq \frac{1}{\tau_{\max}\Delta_f} \qquad (B.39)$$

$$D_t \leq \frac{1}{2f_{Dm}(T_s + T_{cp})}, \qquad (B.40)$$

where D_f (or D_t) represents the number of tones (or OFDM symbols) between adjacent pilot tone-symbols in the frequency (or time) domain. For robustness, the pilot density in practice is higher than the minimum required, and over-sampling by a factor of 2 is sometimes recommended.

Figure B.7 shows a few pilot patterns in the time-frequency grid. In (a), the pilots occupy all the tones but only occur once every few OFDM symbols. The pilots in this pattern are also referred to as training symbols in the literature. In (b), the pilots occupy a subset of tones in every OFDM symbol. A particular example is that the pilot tones are equally spaced, often referred to as comb-type. The pilots in (c) and (d) are more uniformly distributed. In (d), the pilot tone-symbols follow a modular slope hopping pattern, where the pilot tone-symbols in one OFDM symbol are the cyclic shift of those in another OFDM symbol with the shift equal to a slope. As a special case of (d) with slope equal to 0, the pattern in (c) is called rectangular grid.

Roughly speaking, the channel estimation error of data tone-symbol $[s, k]$ increases with the weighted distance

$$\sqrt{(s - s_p)^2 f_{Dm}(T_s + T_{cp}) + (k - k_p)^2 \tau_{\max} \Delta_f}$$

from a nearby pilot $[s_p, k_p]$. The pilot pattern should be designed to minimize the worst case distance of all the data tone-symbols. In general, for channel estimation, the pattern of (c) or (d) performs better than that of (a) or (b), because it is intuitively desirable to make the pilot spacing uniform in time or frequency. In [117], it is found that indeed the best performance is achieved with pilots uniformly spaced in frequency. However, the channel model so far has not included the transmitter/receiver filters. The filters make the frequency response of the overall channel decay much faster at the edge of the bandwidth than what would manifest in the wireless channel itself. As a result, one should either pre-compensate for the effect of the filters if they are known beforehand, or add more pilots at the edges of the bandwidth. In [113], it is reported that non-uniformly spaced pilots minimize the channel estimation error.

In addition to the 2-D density requirement, the pilots are often transmitted at a higher average power than data symbols to suppress noise. In a synchronized cellular deployment, if adjacent cells use the same pilot pattern, then boosting the pilot transmit power itself does not increases the SINR of the pilot tone-symbols much when inter-cell interference dominates, because the pilots of adjacent cells interfere with each other. In this case, we need to further randomize the inter-cell pilot interference. One way is to use the slope hopping pilot pattern shown in Figure B.7(d) and assign different slope values in adjacent cells. Another way is to use the rectangular grid shown in Figure B.7(c) and shift the grid in time or frequency with different offsets in adjacent cells to completely prevent the pilots from interfering with each other.

Non-parametric channel estimation

Non-parametric channel estimation does not explicitly assume the discrete path impulse response model (4.10) in the time domain. Instead, it works entirely in the frequency domain as follows.

Direct interpolation is perhaps the simplest way to derive the channel estimates. Assume the pilot pattern to be a rectangular grid; that is, the pilot tone-symbols are $[D_t l_t, D_f l_f]$, with l_t, l_f integers. From the 2-D sampling theorem, the channel estimate is given by

$$\hat{H}[s, k] = \sum_{l_t, l_f} \hat{H}_{LS}[D_t l_t, D_f l_f] \text{sinc}\left(2\pi \left(\frac{s}{D_t} - l_t\right)\right) \text{sinc}\left(2\pi \left(\frac{k}{D_f} - l_f\right)\right). \tag{B.41}$$

The above 2-D interpolation can be implemented with two separable 1-D interpolations, as shown in Figure B.8.

B.3 Channel estimation

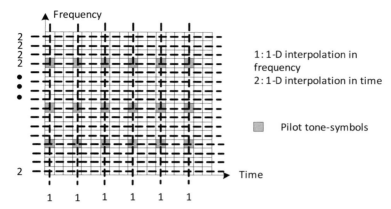

Figure B.8 Illustration of two separable 1-D interpolations to obtain 2-D interpolation.

First, at OFDM symbol $D_t l_t$ where pilot tone-symbols are present, estimate the channel at any tone k from $\hat{H}_{LS}[D_t l_t, D_f l_f]$ for all l_f

$$\hat{H}[D_t l_t, k] = \sum_{l_f} \hat{H}_{LS}[D_t l_t, D_f l_f]\operatorname{sinc}\left(2\pi\left(\frac{k}{D_f} - l_f\right)\right). \tag{B.42}$$

Second, on tone k, estimate the channel at any OFDM symbol s from $\hat{H}[D_t l_t, k]$ for all l_t

$$\hat{H}[s, k] = \sum_{l_f} \hat{H}[D_t l_t, k]\operatorname{sinc}\left(2\pi\left(\frac{s}{D_t} - l_t\right)\right). \tag{B.43}$$

Breaking the 2-D interpolation into two 1-D ones reduces the complexity. In practice, we can replace the sinc function with polynomials and filters, and the scheme is not limited to the pilot pattern being a rectangular grid. For example, when OFDM symbol s carrying pilot tone-symbols $[s, k_{p,1}], \ldots, [s, k_{p,M}]$ is received, a temporary estimate is first obtained by linearly interpolating $\hat{H}_{LS}[s, k_{p,i}]$ in frequency to any tone k

$$\tilde{H}[s, k] = \sum_{i=1}^{M} \hat{H}_{LS}[s, k_{p,i}] \max\left(1 - a|k - k_{p,i}|, 0\right), \tag{B.44}$$

where $a > 0$ is the slope of the linear interpolation. Sometimes, it is better to interpolate amplitude and phase separately

$$|\tilde{H}[s, k]| = \sum_{i=1}^{M} |\hat{H}_{LS}[s, k_{p,i}]| \max\left(1 - a|k - k_{p,i}|, 0\right), \tag{B.45}$$

$$\angle \tilde{H}[s, k] = \sum_{i=1}^{M} \angle \hat{H}_{LS}[s, k_{p,i}] \max\left(1 - a|k - k_{p,i}|, 0\right). \tag{B.46}$$

Then a first-order filter is used to update the channel estimate in time

$$\hat{H}[s, k] = (1 - \delta)\hat{H}[s - 1, k] + \delta \tilde{H}[s, k], \tag{B.47}$$

with $0 < \delta < 1$ a step size.

In addition to the above first order approach, higher-order polynomials can approximate the channel more accurately as the channel itself is smooth in time and frequency. Moveover, knowing some characteristics of the wireless channel helps determine the parameters heuristically. For example, increase δ as the channel variation in time increases, and decrease a as the channel variation in frequency increases. To more explicitly exploit the channel characteristics, we next consider linear MMSE estimation [34].

Suppose we are interested in estimate $H[s, k]$ where $0 \le k \le K-1$ and $0 \le s \le S-1$. Index those tone-symbols linearly as $[s_1, k_1] \ldots, [s_{KS}, k_{KS}]$. Denote by M the number of the pilot tone-symbols and index them $[s_{p,1}, k_{p,1}], \ldots, [s_{p,M}, k_{p,M}]$. The linear MMSE estimator in the matrix form is given in (B.38), where $\hat{\mathbf{H}}$, $\hat{\mathbf{H}}_{LS}$ are $KS \times 1$ and $M \times 1$ vectors, respectively, and $\mathbf{Q}$ a $KS \times M$ matrix. The coefficient matrix $\mathbf{Q}$ is chosen to minimize the mean square error $\mathbb{E}(|\mathbf{H} - \hat{\mathbf{H}}|^2)$. From the orthogonality principle, it follows that

$$\mathbb{E}\left((\mathbf{H} - \hat{\mathbf{H}})\hat{\mathbf{H}}_{LS}^H\right) = \mathbb{E}\left(\mathbf{H}\hat{\mathbf{H}}_{LS}^H\right) - \mathbf{Q}\mathbb{E}\left(\hat{\mathbf{H}}_{LS}\hat{\mathbf{H}}_{LS}^H\right) = 0. \quad (B.48)$$

Recall from (B.37),

$$\hat{H}_{LS}[s_p, k_p] = H[s_p, k_p] + \frac{W_{s_p,k_p}}{X_{s_p,k_p}}. \quad (B.49)$$

Assuming X_{s_p,k_p} are of constant amplitude and W_{s_p,k_p} AWGN, it follows that

$$\mathbb{E}\left(\hat{\mathbf{H}}_{LS}\hat{\mathbf{H}}_{LS}^H\right) = \mathbf{R}_{H_p H_p} + \frac{1}{\gamma}\mathbf{I}_{M \times M}, \quad (B.50)$$

$$\mathbb{E}\left(\mathbf{H}\hat{\mathbf{H}}_{LS}^H\right) = \mathbf{R}_{H H_p}, \quad (B.51)$$

with the SNR defined as $\gamma = \mathbb{E}(|X_{s_p,k_p}|^2)/\mathbb{E}(|W_{s_p,k_p}|^2)$. $\mathbf{R}_{H_p H_p}$ is the auto-covariance where the (i, j) element is $\mathbb{E}(H[s_{p,i}, k_{p,i}]H^*[s_{p,j}, k_{p,j}])$. $\mathbf{R}_{HH_p}$ is the cross-covariance where the (i, j) element is $\mathbb{E}(H[s_i, k_i]H^*[s_{p,j}, k_{p,j}])$. Hence,

$$\mathbf{Q} = \mathbf{R}_{HH_p}\left(\mathbf{R}_{H_p H_p} + \frac{1}{\gamma}\mathbf{I}_{M \times M}\right)^{-1}. \quad (B.52)$$

An interesting observation is that unless at high SNR ($\gamma \gg 1$), $\hat{H} \ne \hat{H}_{LS}$ at any pilot tone-symbol $[s_p, k_p]$. Different from direct interpolation (B.41), the MMSE estimator tries to improve upon the initial LS result by exploiting the channel correlation in time and frequency.

Similar to Figure B.8, we can use two cascading 1-D linear MMSE estimators to replace the 2-D one (B.52), thanks to the separability of the channel correlation in frequency and in time [66, 134]. To further reduce the complexity, we can use a linear MMSE estimator in one domain (say, frequency) and another simple linear interpolator in the other domain (time) [145].

Subspace methods are an alternative approach to reduce computational complexity. In the meantime, they can further suppress noise. For simplicity, consider the 1-D linear MMSE estimator in frequency ($S = 1$). We assume that the tone-symbols are all pilots as shown in Figure B.7(a). We can drop the subscript p in (B.52). Extension to the case where the pilot tones are equally spaced (comb-type), as shown in Figure B.7(b), can be found in [120].

Perform singular value decomposition (SVD) over Hermitian $\mathbf{R}_{HH}$

$$\mathbf{R}_{HH} = \mathbf{U}\Lambda\mathbf{U}^H, \tag{B.53}$$

where $\mathbf{U}$ is a unitary matrix and Λ is a diagonal matrix with the singular values $\lambda_1, \lambda_2, \ldots, \lambda_K$ in a descending order. Then (B.52) becomes

$$\mathbf{Q} = \mathbf{U}\Delta\mathbf{U}^H, \tag{B.54}$$

with Δ a diagonal matrix with entries

$$\delta_i = \frac{\lambda_i}{\lambda_i + \frac{1}{\gamma}}, \quad i = 1, \ldots, K. \tag{B.55}$$

Suppose r represents the number of the significant singular values where $\lambda_r \gg \lambda_{r+1}$. The subspace method in [40] is to replace Δ with a low-rank diagonal matrix Δ' with entries

$$\delta_i' = \begin{cases} \delta_i, & i = 1, \ldots, r \\ 0, & i = r+1, \ldots, K. \end{cases} \tag{B.56}$$

The subspace estimator can be viewed as projecting $\hat{\mathbf{H}}_{LS}$ onto a subspace and performing the estimation there. In spite of a smaller dimension (r instead of K), the subspace describes the channel well so that there is little loss of channel information. Meanwhile, since the noise is uniformly distributed with equal energy in all the dimensions, performing the estimation in the subspace eliminates the noise in the other dimensions, thereby achieving noise reduction. On the other hand, the subspace estimator has an irreducible error floor because part of channel that does not belong to the subspace gets lost in the subspace representation (B.56). Therefore, the choice of r has to balance the needs of noise reduction and channel representation.

Parametric channel estimation

The basic assumption of parametric channel estimation is that the wireless channel can be fully characterized with a small number of possibly independent parameters. One commonly used channel model is the discrete path impulse response model (4.10). From (4.28), the frequency domain channel response coefficient of the time domain model is

$$H[s, k] = \sum_{l=1}^{L} h_l[s] e^{-j2\pi k \frac{\tau_l[s]}{T_s}}, \tag{B.57}$$

with complex number $h_l[s] = \bar{\alpha}_l(t) e^{-j\bar{\phi}_l(t)}$. Here we assume that the channel parameters $L(t), \bar{\alpha}_l(t), \bar{\phi}_l(t), \tau_l(t)$ are constant in one OFDM symbol duration and all the taps are within the cyclic prefix. For the sake of comparison with non-parametric channel estimation, we consider the 1-D estimation in frequency and therefore drop time index s, and assume that all the tone-symbols are pilots. Extension to the case with the comb-type pilots can be found in [28].

First assume that the tap delays τ_l's are equally spaced with the spacing equal to the sampling time at the receiver. Thus, $\tau_l = l$, and $L \leq K_{cp}$, with K_{cp} the number of samples in the cyclic prefix. In the matrix form, (B.57) becomes

$$\mathbf{H} = \mathbf{F}\mathbf{h}, \tag{B.58}$$

with $K \times 1$ vectors $\mathbf{H} = [H[1], \ldots, H[K]]^T$ and $\mathbf{h} = [h_1, \ldots, h_{K_{cp}}, 0, \ldots, 0]^T$. For now, we assume $L = K_{cp}$. $\mathbf{F}$ is a $K \times K$ Fourier matrix with the (k, i) element equal to $e^{-j2\pi \frac{ki}{K}}$. The above assumption leads to the so-called Fourier Transform domain estimator:

$$\tilde{\mathbf{h}} = \frac{1}{K}\mathbf{F}^H \hat{\mathbf{H}}_{LS} \tag{B.59}$$

$$\hat{\mathbf{h}} = \mathbf{D}\tilde{\mathbf{h}} \tag{B.60}$$

$$\hat{\mathbf{H}} = \mathbf{F}\hat{\mathbf{h}}, \tag{B.61}$$

where $\mathbf{D}$ is

$$\mathbf{D} = \begin{bmatrix} \mathbf{D}_{K_{cp}} & \mathbf{0}_{K_{cp} \times (K-K_{cp})} \\ \mathbf{0}_{(K-K_{cp}) \times K_{cp}} & \mathbf{0}_{(K-K_{cp}) \times (K-K_{cp})} \end{bmatrix}. \tag{B.62}$$

$\mathbf{D}_{K_{cp}}$ can be an identity matrix or take a different form incorporating the correlation between the taps to reduce the channel estimation error [41].

With the above subspace Fourier Transform estimator, $\hat{\mathbf{H}}_{LS}$ is first transformed to the time domain (B.59) to obtain a temporary estimate. The last $K - K_{cp}$ time domain estimate samples are then truncated (B.60). Finally, the remaining K_{cp} taps of the channel impulse response are transformed back to the frequency domain (B.61). Figure B.9(a) illustrates the strengths of typical time domain temporary estimate samples ($\tilde{\mathbf{h}}$) where the first K_{cp} taps carry significant channel energy and the last $K - K_{cp}$ ones are caused by noise. Zeroing out those taps and only keeping the significant taps reduces noise, similar to what we have seen with the subspace method (B.56) in non-parametric channel estimation.

We can also derive a linear MMSE estimator using the time domain model

$$\hat{\mathbf{h}} = \mathbf{R}_{hh}\mathbf{F}^H \left(\mathbf{F}\mathbf{R}_{hh}\mathbf{F}^H + \frac{1}{\gamma}\mathbf{I}_{K \times K}\right)^{-1} \hat{\mathbf{H}}_{LS} \tag{B.63}$$

$$\hat{\mathbf{H}} = \mathbf{F}\hat{\mathbf{h}} \tag{B.64}$$

Note the equivalence between (B.52) and (B.63), (B.64), as $\mathbf{R}_{hh}$ and $\mathbf{R}_{HH}$ are related as follows:

$$\mathbf{R}_{HH} = \mathbf{F}\mathbf{R}_{hh}\mathbf{F}^H. \tag{B.65}$$

Similar to (B.53), we can perform SVD over $\mathbf{R}_{hh}$ and obtain a subspace estimator. From (B.65), that subspace estimator should use the same rank r as in (B.56). In [40], it is shown that r is approximately equal to K_{cp}. Hence, the noise reduction factor is the same for all the subspace methods we have discussed so far and roughly equal to K/K_{cp}.

So far we have assumed that the channel taps are exactly sample spaced. If this is not the case, then the energy of the non-sample spaced taps will leak to the other taps, as

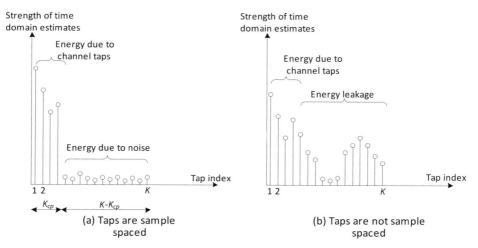

Figure B.9 Comparison of typical time domain estimates when the channel taps are sample spaced or not sample spaced.

illustrated in Figure B.9(b), in which case simply zeroing out some taps will lose the channel information and cause an error floor.

This leads us to another two important steps in parametric channel estimation:

1. Estimation of the number of significant paths L. In [179], the authors use the minimum description length criterion to detect the number of paths in the channel.
2. Estimation of the tap delays τ_ls. In [179], the authors use the ESPRIT method to acquire the initial values of τ_ls and then a delay locked loop to track them by utilizing the slow time-varying nature of tap delays.

Denote by $\hat{L}$ the estimate of L, and $\hat{\tau}_l$ the estimate of τ_l. With the tap delay information, we can use a linear MMSE estimator to obtain $\hat{\mathbf{H}}$. Now (B.58) becomes

$$\mathbf{H} = \mathbf{F}_\tau \mathbf{h}_\tau, \tag{B.66}$$

with $\mathbf{h}_\tau = [h_1, \ldots, h_{\hat{L}}]^T$ and $\mathbf{F}_\tau$ a $K \times \hat{L}$ Fourier matrix with the (k, l) element equal to $e^{-j2\pi \frac{k\hat{\tau}_l}{T_s}}$. The linear MMSE estimator of $\mathbf{h}_\tau$ is the same as in (B.63), except that $\mathbf{F}$ is replaced with $\mathbf{F}_\tau$. Furthermore, we can similarly apply the SVD technique to reduce the channel estimation dimension from K to $\hat{L}$. The noise reduction factor is roughly $K/\hat{L}$. Note that $K_{cp} > \hat{L}$ for a sparse multipath channel; thus, it follows that $K/\hat{L} > K/K_{cp}$. This explains the observation in [179] that the parametric scheme achieves better noise reduction than a non-parametric scheme as long as the tap delay estimation is reasonably accurate.

MIMO OFDM

The channel estimation techniques of SISO-OFDM can be extended to MIMO-OFDM where the goal is to estimate a $K_t \times K_r$ channel matrix, with K_t, K_r the numbers of transmit and receive antennas, respectively. Each matrix element is the coefficient $\mathbf{H}$ of the channel response from a transmit antenna to a receive antenna.

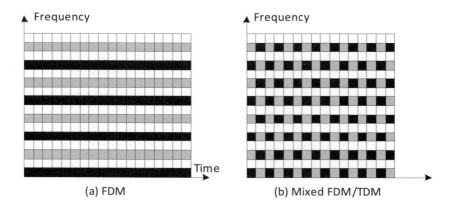

■ Pilot tone-symbol of transmit antenna 1

▦ Pilot tone-symbol of transmit antenna 2

Figure B.10 Pilot allocation for MIMO-OFDM channel estimation: FDM and TDM.

A simple idea for MIMO-OFDM channel estimation is to repeat SISO-OFDM channel estimation $K_t K_r$ times. Specifically, different transmit antennas are allocated with disjoint sets of pilot tone-symbols. At a pilot tone-symbol allocated to one transmit antenna, all other transmit antennas keep silent (no pilot or data) so as to avoid interference and K_r receive antennas independently perform SISO-OFDM channel estimation. Disjoint sets of pilot tone-symbols can be constructed in a TDM or FDM manner as shown in Figure B.10.

The drawback of the above scheme is the increase in pilot overhead by a factor of K_t. To reduce the overhead, an alternative pilot allocation scheme is to let transmit antennas share the pilot tone-symbols and separate them in a CDM (code division multiplexing) manner. For the sake of description, we continue to consider the 1-D estimation in frequency and assume that all the tone-symbols are pilots. Assume that the tap delays τ_ls are equally spaced with the spacing equal to the sampling time and denote $L_{\max}$ the maximal length of all the channels. Denote by $X_1[k], k = 1, \ldots, K$ the pilot symbol to be transmitted on tone k by transmit antenna 1. Then from [10], an optimal pilot symbol on tone k by transmit antenna m, $m = 2, \ldots, K_t$ that minimizes the mean square error of the LS channel estimator is

$$X_m[k] = X_1[k] e^{-j2\pi \frac{k(m-1)L_{\max}}{K}}. \tag{B.67}$$

Note that the number of transmit antennas is limited by

$$K_t \leq \frac{K}{L_{\max}}. \tag{B.68}$$

Recall that a phase rotation in the frequency domain corresponds to a time shift in the time domain. The rationale of (B.67) is that the pilot phases are rotated such that the time domain channel response from transmit antenna m is shifted by $(m-1)L_{\max}$ samples in time. Since the channel response from each antenna is limited to $L_{\max}$ samples, the shifted channel responses from different antennas do not overlap.

The above FDM, TDM, and CDM schemes can be flexibly combined. Below is an example presented in [111] of the symbols to be sent in a set of pilot tone-symbols from different antennas.

	Tone index	Antenna 1	Antenna 2	Antenna 3	Antenna 4
CDM	0	α_1	α_1	α_1	α_1
	1	α_1	$\alpha_1 e^{-j\pi/2}$	$\alpha_1 e^{-j2\pi/2}$	$\alpha_1 e^{-j3\pi/2}$
	2	α_1	$\alpha_1 e^{-j2\pi/2}$	$\alpha_1 e^{-j4\pi/2}$	$\alpha_1 e^{-j6\pi/2}$
	3	α_1	$\alpha_1 e^{-j3\pi/2}$	$\alpha_1 e^{-j6\pi/2}$	$\alpha_1 e^{-j9\pi/2}$
	4	α_1	$\alpha_1 e^{-j4\pi/2}$	$\alpha_1 e^{-j8\pi/2}$	$\alpha_1 e^{-j12\pi/2}$
	5	α_1	$\alpha_1 e^{-j5\pi/2}$	$\alpha_1 e^{-j10\pi/2}$	$\alpha_1 e^{-j15\pi/2}$
	6	α_1	$\alpha_1 e^{-j6\pi/2}$	$\alpha_1 e^{-j12\pi/2}$	$\alpha_1 e^{-j18\pi/2}$
	7	α_1	$\alpha_1 e^{-j7\pi/2}$	$\alpha_1 e^{-j14\pi/2}$	$\alpha_1 e^{-j21\pi/2}$
FDM	0	$2\alpha_1$	0	0	0
	1	0	$2\alpha_3$	0	0
	2	0	0	$2\alpha_5$	0
	3	0	0	0	$2\alpha_3$
	4	$2\alpha_2$	0	0	0
	5	0	$2\alpha_4$	0	0
	6	0	0	$2\alpha_6$	0
	7	0	0	0	$2\alpha_8$
FDM + CDM	0	$\sqrt{2}\alpha_1$	$\sqrt{2}\alpha_1$	0	0
	1	0	0	$\sqrt{2}\alpha_5$	$\sqrt{2}\alpha_5$
	2	0	0	$\sqrt{2}\alpha_6$	$-\sqrt{2}\alpha_6$
	3	$\sqrt{2}\alpha_2$	$-\sqrt{2}\alpha_2$	0	0
	4	$\sqrt{2}\alpha_3$	$\sqrt{2}\alpha_3$	0	0
	5	0	0	$\sqrt{2}\alpha_7$	$\sqrt{2}\alpha_7$
	6	0	0	$\sqrt{2}\alpha_8$	$-\sqrt{2}\alpha_8$
	4	$\sqrt{2}\alpha_4$	$-\sqrt{2}\alpha_4$	0	0

Here α_i can be any constant-modulus symbol.

The above CDM scheme works well in a sample spaced channel. However, it should be pointed out that when the channel is not sample spaced, the taps from different antennas interfere with each other, thereby degrading the performance of the CDM scheme. In this case, the techniques designed to address non-sample spaced taps in the SISO case can be applied [178].

Finally, we note that in the MIMO case, the receiver can exploit spatial correlation in the channel estimation in the same way as with the frequency and time correlation in the linear MMSE channel estimation.

B.4 Error correction

Fading, interference and noise potentially cause errors to communications over the wireless channel. Two basic error correction approaches are forward (channel coding) and feedback (ARQ).

Forward channel coding

Consider a point-to-point communications scenario where a source transmits source information to a destination via a channel. In [141], Shannon showed that the problem can be decomposed into two separate problems, namely *source coding*, where redundancy in the source information is removed to increase communications efficiency while preserving accuracy, and *channel coding*, where carefully chosen redundancy is added to combat noise and fading due to the channel. Channel coding is mostly used for error correction and detection.

The input of channel coding is referred to as information bits, and the output as coded bits. Furthermore, modulation maps coded bits to constellation points (complex numbers). Typical constellations include BPSK, QPSK, 16-QAM, 64-QAM, and 256-QAM. The output of modulation is modulation symbols.

In an OFDMA system, coding and modulation operates in a block-by-block manner. At a transmitter, a coding and modulation module transforms a block of information bits into modulation symbols, which are mapped to a block of tone-symbols to form OFDM symbols. A reverse sequence of operations take place at a receiver. Figure 2.5 shows the coding and modulation module in the transmitter chain. The counterpart module for the reverse operation at the receiver is decoding and demodulation.

Denote by k, n the numbers of information and coded bits, respectively, in a coding block. There are 2^k distinct codewords and the size of the code space is 2^n. The *coding rate* is defined to be k/n, which is less than 1. A lower coding rate means more redundancy being added and greater distances between codewords, and therefore the code is better protected against noise and fading. Denote by M the constellation size. Each modulation symbol carries $\log_2 M$ coded bits. Hence, the spectral efficiency is equal to $k/n \log_2 M$ information bits per tone-symbol.

The benefit of channel coding can be measured by a reduction in the error probability. An error occurs when the decoded codeword is different from the transmitted one. the farther apart two codewords are in the code space, the less likely it is that an error will occur. The *coding gain* measures the difference between the SNRs or E_b/N_0 of the uncoded and coded schemes required to achieve a given error probability (bit or block). Figure B.11 depicts typical curves of error probability versus E_b/N_0 of uncoded and coded schemes.

The error probability drops as E_b/N_0 increases. For a given code, the coding gain varies depending on the required error probability. At a given error probability, different codes (turbo/LDPC or convolutional codes) achieve different coding gains. In Figure B.11, a turbo/LDPC code typically exhibits a two-phase behavior: a very sharp waterfall region and a relatively flat error-floor region. In contrast, the curve of a convolutional code is more gradual, and far less steep than the waterfall region of a turbo/LDPC code for large block lengths.

The benefit of error correction comes at a cost of data rate reduction (or bandwidth expansion) by a factor of k/n, although by jointly optimizing channel coding and modulation, the coded modulation scheme introduced in [161] can achieve significant coding gain without bandwidth expansion. Given a coding rate, the channel capacity defines the minimum E_b/N_0 required to drive the error probability to zero. The vertical

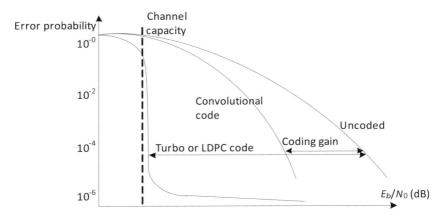

Figure B.11 Illustration of error probability versus required E_b/N_0 in the AWGN channel.

line labeled "channel capacity" in Figure B.11 represents that asymptotic limit. For a practical code, a higher E_b/N_0 is required to drive the error probability close to zero. The extra amount of E_b/N_0 is indicative of the efficiency of the code. In the example of Figure B.11, the convolutional code is less efficient than the turbo or LDPC code, which operates within a fraction of a dB from the capacity curve in the waterfall region [15, 53].

Note that the capacity-achieving performance of turbo/LDPC codes usually requires a large block length n. A large n generally boosts the coding gain. However, the transmitter has to accumulate enough information bits before a codeword is formed, thereby causing delay. Furthermore, when the transmitter only intends to send a small control message, the block length has to be kept small due to coding granularity.

Computational complexity is another important concern in practice. A brute-force maximum likelihood decoding algorithm compares a received signal with all 2^k possible codewords, resulting in the complexity exponential with k. Such an algorithm can only work for short codes. When the block length is large, the complexity has to be linear to be practical. Good codes not only need to achieve good coding gains but also be well suited to hardware or software implementation.

In addition to error correction, channel coding can also be used for error detection. In data communications, even a single bit error causes an error of the entire coding block (block error). Error detection is primarily based on parity checks.[1] Figure B.12 shows a typical flow chart that combines the functions of error correction and detection.

In-depth description of channel coding can be found in many excellent books such as [97, 127]. While in general there is no need to design codes specially for OFDM systems, per se, we should be aware of the assumptions of noise and fading characteristics behind the code design. For example, codes designed for the AWGN channel may not perform well in the fading channel due to bursts of errors, and decoding metrics may be different

---

[1] Decoding algorithms for LDPC codes can determine when a correct codeword has been decoded. Therefore, LDPC codes may not need separate parity check bits for error detection.

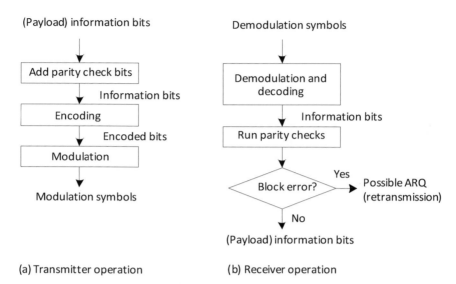

Figure B.12 Typical flow chart of using channel coding for error correction and detection.

for Gaussian noise versus peaky interference. One such example of receiver design is studied in Figure 2.10.

Adaptive modulation and coding

Adaptive modulation and coding (AMC) means that the modulation and coding scheme changes over time in response to the variations of channel, interference, and transmit power. In practice, for a coding block, AMC selects a modulation and coding scheme from a finite set of choices according to SINR and an error probability target. The SINR at a tone-symbol k is given by

$$\gamma_k = \frac{G_k P_k}{\sigma_k^2}, \tag{B.69}$$

where G_k is the channel power gain, σ_k^2 the interference plus noise power and P_k the transmit power in the tone-symbol k. Because of channel/interference selectivity, γ_k may differ for different k. The so-called *effective SINR* γ measures an "average" SINR for the entire coding block.

Figure B.13 illustrates the performance curves of a set of modulation and coding schemes, labeled as "QPSK, rate 1/4" to "64-QAM, rate 3/4." Parameters $\gamma_{1,2}, \ldots, \gamma_{7,8}$ represent the SINR switching points. For example, as γ increases, if it exceeds $\gamma_{1,2}$, the "QPSK, rate 1/4" scheme is selected, if it further exceeds $\gamma_{2,3}$, the "QPSK, rate 1/2" scheme is selected, and so on.

We next list a few design considerations of the set of modulation and coding schemes for AMC.

B.4 Error correction

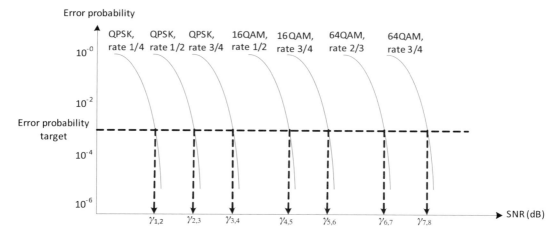

Figure B.13 Illustration of error probability versus SINR of a variety of modulation and coding schemes.

- The schemes should span a sufficiently large dynamic range of SINR to cover possible operating conditions. In a typical macrocellular system, the lowest SINR is about −10 to −5 dB and the highest is about 25 to 30 dB.
- The SINR gaps between adjacent SINR switching points should be somewhat uniformly sized among the schemes. The same modulation and coding scheme is selected when SINR falls within a gap, say, between $\gamma_{1,2}$ and $\gamma_{2,3}$. Thus a large gap results in a loss in data rate due to wasted SINR. On the other hand, small gaps require a large size of the set to cover the same total SINR dynamic range, and therefore large signaling overhead of AMC assignment. If γ cannot be predicted accurately, there is no point of making the gaps too small. In a typical mobile cellular system, the SINR gaps are about 2–4 dB.
- As discussed before, the performance curves depend on the channel model (AWGN/fading, fading channel parameters, etc.) and code design (e.g., convolutional/turbo/LDPC, block length). For example, QPSK rate 2/3 coding and 16-QAM rate 1/3 coding have the same data rate, but depending on the fading channel model, one outperforms the other. Therefore, a device in a different fading environment can be configured to use a different set of modulation and coding schemes.

SINR prediction is a key to AMC. In the downlink, the base station transmits the pilot at a fixed power P_p. The device measures the SINR of the pilot, and reports it to the base station. To predict the SINR γ of a coding block, the base station usually linearly scales the pilot SINR by P/P_p, where P is the transmit power allocated to the coding block. Nonlinear scaling is needed for SINR prediction in the presence of self noise, as shown in Section 2.5. In the uplink, the device transmits a sounding signal at a power P_s controlled by the base station. The base station measures the SINR of the sounding signal, and predicts the SINR γ of a coding block to be the sounding signal SINR linearly scaled by P/P_s, where P is the transmit power of the coding block controlled by the base station.

Clearly, the accuracy of SINR prediction depends on the quality of SINR measurement, and more importantly on the channel variation in the delay interval from the measurement to the transmission of the coding block. In the downlink case, the quantization errors of the SINR report and the errors in the feedback channel also affect the accuracy. Moreover, the prediction of error probability given SINR is subject to modeling errors, because the performance curve of error probability versus SINR is usually obtained by offline simulation with certain channel and noise models, which may not precisely match the reality.

Link adaptation is a closed-loop technique to deal with the above real world uncertainty. In a very simple form, link adaptation introduces an SINR correction term Δ_γ to improve SINR prediction

$$\gamma' = \gamma + \Delta_\gamma, \text{ in dB}. \tag{B.70}$$

γ' is the corrected predicted SINR and is used to select a modulation and coding scheme. Correction term Δ_γ (in dB) can be positive or negative, and is increased if the error target fails to meet or decreased otherwise. Similarly, we can introduce one adaptive correction term for each modulation and coding scheme to provide fine adjustment.

ARQ

ARQ is an error control mechanism that takes advantage of a feedback channel. In Figure B.12, after running parity checks the receiver sends a one-bit ACK or NAK to inform the transmitter of whether the coding block is in error or not. If NAK, the transmitter can retransmit the coding block. In a fading channel, ARQ achieves time diversity. As noted previously, a key challenge of the forward approach of modulation and coding is the real world uncertainty of SINR prediction and channel/noise modeling. With the feedback approach of ARQ, we do not have to be overly conservative in selecting the AMC scheme.

Suppose the coding block is used for traffic. In a packet-switched OFDMA system, the ACK/NAK channel resource can be slaved to the resource of the coding block, referred to as traffic channel resource, with a fixed one-to-one relationship [88], as shown in Figure B.14. The traffic channel resource can be allocated to any users, and the assignment information is contained in a control message. The ACK/NACK channel resource is only used for the assigned user. The delay from traffic to ACK/NAK is kept to be a minimum, constrained mostly by the processing delay of the decoder: typically on the order of a few milliseconds. For this reason, the ACK/NAK-based ARQ is called *fast loop ARQ*.

What happens if the feedback channel is in error? In the case of ACK being interpreted as NAK, the transmitter may retransmit the coding block, thereby wasting the traffic channel resource. But a more sophisticated protocol is needed to recover from a NAK-to-ACK error, because when that error happens, the transmitter stops retransmission while the receiver does not yet successfully get the information bits. Suppose that the payload information bits are indexed with some link layer sequence numbers. When

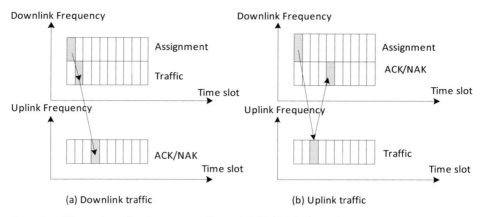

Figure B.14 Illustration of assignment, traffic, and ACK/NAK channel resources.

a NAK-to-ACK error occurs, the corresponding sequence numbers are missing at the receiver for some period of time. Upon a timer expiration, the receiver then sends a control message to explicitly request the missing pieces. Because of relatively long delay involved, that retransmission procedure is called *slow loop ARQ*.

Note that ARQ comes at the cost of extra delays due to retransmissions. To avoid excessive delays, the ARQ target is usually set to succeed in m transmissions for every coding block. If it takes m' transmissions for a given coding block, then correction term Δ_γ in link adaptation (B.70) is increased if $m' > m$, or decreased otherwise. In addition, to reduce ARQ delay, after receiving a NAK the transmitter targets a lower error probability in AMC selection or boosts the transmit power so as to increase the success rate of the subsequent retransmissions.

Hybrid-ARQ

The native ARQ protocol is not spectral efficient, because a single bit error causes the entire coding block to be retransmitted. Combining the advantages of both ARQ and channel coding, hybrid-ARQ (H-ARQ) lets the receiver combine the received coding blocks.

There are two popular combining schemes. In *chase combining*, the transmitter sends the same coding block in a retransmission, and the receiver soft-combines the received symbols with previous transmissions. Because the noise and interference in the transmissions are independent, soft combining increases the SINR. In *incremental redundancy*, the transmitter sends new additional redundancy bits corresponding to the same information bits in each retransmission, thereby in effect reducing the coding rate. Chase combining can be viewed as a special case of incremental redundancy where repetition coding is used between transmissions. Chase combining achieves power gain in addition to time diversity, and incremental redundancy has coding gain in addition. In either case, conditional on unsuccessful decoding from the past m transmissions, the error probability drops when the $(m+1)$th transmission takes place. As a comparison, in native ARQ, the conditional error probability remains the same in an i.i.d. channel.

In H-ARQ, the transmitter can purposely select a high coding rate or modulation order in the first transmission. If the SINR condition happens to be favorable, then the first transmission will luckily succeed. Otherwise, the subsequent transmissions will eventually bring down the error probability. In a sense, at the cost of delays, H-ARQ is a way of gradually probing the channel, and opportunistically exploiting favorable SINR conditions. As a final note, a multi-level NAK that informs the transmitter of "how much additional redundancy is needed" can better prepare the transmitter for retransmission and further reduce H-ARQ delays.

C Brief review of channel capacity

The *capacity* of a channel is defined to be the maximum rate at which data is sent over the channel where the error probability of communication can be driven to arbitrarily small. The framework for studying channel capacity is information theory established by Claude Shannon ([140, 141]). In this section, we will review some important information theoretical results on channel capacity.

C.1 AWGN channel

The AWGN channel is the best studied point-to-point channel model. Consider a discrete-time complex baseband channel:

$$y(t) = x(t) + w(t), \qquad (C.1)$$

where t is the time index, $x(t)$ and $y(t)$ are the complex input and output signals, respectively, and $w(t)$ the additive noise assumed to be i.i.d. complex white Gaussian with power spectral density $N_0/2$. Denote by W the channel bandwidth (in Hz) and P the average power constraint of $x(t)$ (in watts). The channel capacity of the AWGN channel (in nats[1] per second) is given by

$$C = W \log\left(1 + \frac{P}{N_0 W}\right). \qquad (C.2)$$

Define γ the signal-to-noise-ratio (SNR)

$$\gamma = \frac{P}{N_0 W}, \qquad (C.3)$$

where $N_0 W$ is the total noise power. The channel capacity depends on two system resources, namely bandwidth and power. When the SNR is low,

$$C \approx \frac{P}{N_0}, \text{ if } \gamma \approx 0. \qquad (C.4)$$

In this low SNR regime, the capacity is limited by the power in the sense that C increases *linearly* with P; however, the effect of increasing W is negligible. When the SNR is

---

[1] The basic unit of information in computing and communication is a bit (binary digit, 0 or 1). A nat is another unit equal to $\log_2 e \approx 1.443$ bits. If $\log_2$ is used in (C.2) instead of the natural logarithm, then the unit of the capacity is bits per second.

high,

$$C \approx \mathcal{W} \log\left(\frac{P}{N_0 \mathcal{W}}\right), \text{ if } \gamma \gg 1. \tag{C.5}$$

In this high SNR regime, the capacity is limited by the bandwidth in the sense that C increases *logarithmically* with P, but *linearly* with $\mathcal{W}$.

C.2 Flat fading channel

In the flat fading model,

$$y(t) = h(t)x(t) + w(t), \tag{C.6}$$

where $h(t)$ represents the multiplicative complex gain due to the wireless channel, following some channel distribution. For simplicity, assume a block fading model where $h(t)$ is constant in each channel coherence time and changes to a new independent value from one coherence time to another.

$h(t)$ is often referred to as the *channel side information* or *channel state information* (CSI). The channel capacity depends on whether the CSI is available at the receiver only or at both the receiver and transmitter. We assume that the CSI is always available at the receiver via channel estimation. In a time-invariant channel, we assume that the CSI is also available at the transmitter because the cost for the receiver to measure and send the CSI to the transmitter is negligible. In a time-varying channel, however, the cost is not negligible. Moreover, the transmitter may be unable to keep track of the CSI if it varies rapidly. Therefore, we next consider two cases depending on whether the transmitter has the CSI or not.

C.2.1 Channel side information only at receiver

In information theory, the capacity is achieved with a sufficiently large coding block. In practice, the coding block size is finite and limited by implementation complexity and decoding delay. In general, a coding block experiences one or multiple instantaneous realizations of the channel distribution depending on the coding block size and channel variation rate. Consider two extreme cases of fast fading and slow fading.

In fast fading, the coding block spans a large number of channel coherence times. The channel capacity is given by

$$C = \mathcal{W} \mathbb{E}\left(\log(1 + |h|^2 \gamma)\right), \tag{C.7}$$

where the expectation is taken over the channel distribution. Here the capacity-achieving code is sufficiently long to experience all possible independent fades. The capacity is thus referred to as ergodic or Shannon capacity.

In slow fading, the coding block is within a coherence time and h is constant. Since the transmitter does not know about the CSI beforehand, it has to fix a transmission rate R_0 without the knowledge of the instantaneous h. Outage occurs when the transmission rate exceeds the capacity achievable by the instantaneous h. The outage probability is

$$P_{\text{out}} = \mathbb{P}\left(|h|^2 < |h_0|^2\right), \tag{C.8}$$

where $|h_0|^2$ is the channel gain corresponding to the capacity equal to R_0

$$R_0 = \mathcal{W} \log\left(1 + |h_0|^2 \gamma\right). \tag{C.9}$$

The capacity with outage is given by

$$C = R_0 \left(1 - P_{\text{out}}\right). \tag{C.10}$$

Unlike the ergodic capacity, the capacity with outage involves a design parameter $|h_0|^2$. One can find the optimal $|h_0|^2$ to maximize C subject to an acceptable outage probability.

C.2.2 Channel side information at both receiver and transmitter

The knowledge of the CSI allows the transmitter to adapt the power and rate in response to the instantaneous channel realization. Denote by L the number of coherence times experienced by a coding block. The channel can be modeled as L parallel subchannels that fade independently. Denote by $P_1, \ldots, P_L$ the transmit powers allocated to the subchannels. Since the transmitter knows about the CSI of each subchannel, it will transmit at the rate exactly equal to the capacity of that subchannel. Hence, the overall channel capacity is obtained by optimizing the power allocation:

$$C = \max_{P_1,\ldots,P_L} \frac{1}{L} \sum_{l=1}^{L} \mathcal{W} \log\left(1 + |h_l|^2 \frac{P_l}{N_0 \mathcal{W}}\right), \tag{C.11}$$

subject to

$$\frac{1}{L} \sum_{l=1}^{L} P_l = P. \tag{C.12}$$

The optimal power allocation is water-filling:

$$P_l = \left(\frac{1}{\lambda} - \frac{N_0 \mathcal{W}}{|h_l|^2}\right)^+, \tag{C.13}$$

where function $(x)^+$ is equal to x if $x \geq 0$ or 0 otherwise. Parameter λ is solved by plugging (C.13) into (C.12).

In fast fading, $L \to \infty$, and the summation in (C.12) becomes the expectation. λ can be solved without knowing specific channel gain realizations $|h_l|^2$

$$\mathbb{E}\left(\left(\frac{1}{\lambda} - \frac{N_0 \mathcal{W}}{|h|^2}\right)^+\right) = P. \tag{C.14}$$

λ is calculated beforehand from the channel distribution. At a given coherence time l, the calculation of the optimal power allocation (C.13) only depends on instantaneous $|h_l|^2$. Since a coding block spans over many coherence times, the capacity remains the same in every coding block.

Now consider slow fading. If the average power constraint of every coding block is fixed ($=P$), then the capacity is simply equal to

$$C = \mathcal{W} \log\left(1 + |h|^2 \gamma\right), \tag{C.15}$$

with $|h|^2$ the instantaneous channel gain of the present coherence time. Obviously C varies with different channel gain realizations. However, there is no outage, as the transmitter will not transmit at a rate exceeding the capacity.

Alternatively, suppose that P is the long-term average power constraint over many coding blocks and the power budget can be shifted from one coding block to another. There are two basic scenarios.

If a constant rate is required, then we apply the idea of channel inversion. Specifically, set a target SNR γ_0 for every coding block. The transmit power is given by $\frac{\gamma_0 N_0 \mathcal{W}}{|h|^2}$. From the average power constraint

$$\mathbb{E}\left(\frac{\gamma_0 N_0 \mathcal{W}}{|h|^2}\right) = P, \tag{C.16}$$

it follows that

$$\gamma_0 = \gamma^{-1}\left(\mathbb{E}\left(\frac{1}{|h|^2}\right)\right). \tag{C.17}$$

Note that $\mathbb{E}\left(\frac{1}{|h|^2}\right)$ can be very large in a fading channel. For example, it goes to infinite in Rayleigh fading. To avoid the excessive cost of inverting the channel at a deep fade, we modify the idea to truncated channel inversion. Set a target SNR γ_0 for a coding block when the channel gain is at least $|h_0|^2$. When the instantaneous channel power gain $|h|^2$ is lower than $|h_0|^2$, the transmitter does not transmit any bits and outage occurs. The threshold $|h_0|^2$ is a control parameter to trade off the capacity versus outage probability.

On the other hand, if a variable rate is allowed, then we can apply the same water-filling idea as in the fast fading case to increase the capacity, except that now the water-filling is over many coding blocks rather than many coherence times within a coding block. Here the transmit power and rate are constant for a given coding block and may vary from one coding block to another. Outage occurs when the instantaneous channel gain is low enough that

$$\frac{1}{\lambda} - \frac{N_0 \mathcal{W}}{|h|^2} < 0, \tag{C.18}$$

and thus the transmit power is equal to 0. In that case, the transmitter simply gives up the present coherent time.

Figure C.1 compares the water-filling and truncated channel inversion power allocation schemes. Outage may occur when the channel gain drops below a certain threshold. The cutoff threshold may be different in either scheme. Outside the outage region, the strategy is very different. As the channel gain gets smaller, less power is allocated in the water-filling scheme ($P_2 < P_1$), but more power is allocated in the channel inversion scheme ($P_2' > P_1'$). At the boundary of the outage region, the allocated power drops suddenly (discontinuously) in the channel inversion scheme but continuously in the water-filling

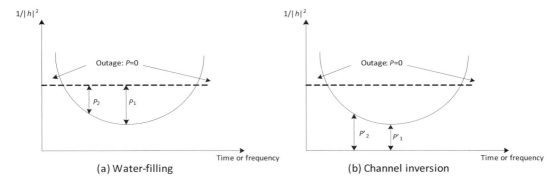

Figure C.1 Comparison of the water-filling and truncated channel inversion power allocation schemes.

scheme. In general, the channel inversion scheme achieves a much lower rate than the water-filling scheme, which is the cost of maintaining the rate constant.

C.3 Frequency selective fading channel

First, consider a time-invariant frequency selective channel. We assume that the CSI is available at both the transmitter and receiver. Suppose the bandwidth W is divided into L parallel subchannels, where the channel gain in each subchannel is constant. This is similar to the flat fading case. The optimal power allocation is water-filling over frequency instead of over time as in (C.13). Moreover, in flat fading, solving λ from (C.12), (C.13) requires non-causal knowledge of future channel gain realizations $|h_l|^2$. Such a requirement disappears when $L \to \infty$, and the solution of λ only depends on the channel distribution. For the time-invariant frequency selective channel, all the channel gains are causally known and we do not require $L \to \infty$ or the knowledge of the channel distribution.

In general, a time-varying frequency selective fading channel can be divided in frequency and time into two-dimensional parallel subchannels: a set of frequency subchannels in each coherence time, and multiple sets of frequency channels over different coherence times. The capacity results of the one-dimensional flat fading or time-invariant frequency selective channel can be extended to the two-dimensional frequency selective fading channel. For example, in fast fading, when the CSI is available at the transmitter, the capacity is achieved by the water-filling power allocation over time and frequency.

C.4 Multiuser capacity

In the single user case, the capacity is a single number representing the maximum rate achievable over the channel. In the multiuser case, the capacity is not just a number, but a region. For simplicity, consider only two users. Denote by R_1, R_2 a pair of rates

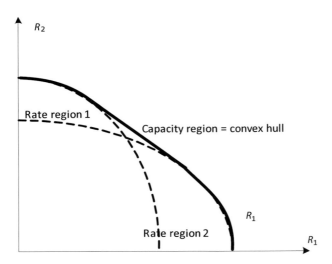

Figure C.2 Illustration of rate regions and capacity region. In the figure, the two-user capacity region is the convex hull of two rate regions.

simultaneously achieved by users 1, 2, respectively. $\{R_1, R_2\}$ is called a feasible rate pair. Define *rate region* the set of all feasible rate pairs. *Capacity region* is defined to be the *Pareto-boundary* of the rate region, as shown in Figure C.2. A rate pair $\{R_1, R_2\}$ belongs to the capacity region if and only if it is not strictly dominated by another rate pair of the rate region. For any other rate pair $\{R_1', R_2'\}$,

$$R_1 \geq R_1' \text{ or } R_2 \geq R_2' \tag{C.19}$$

Note that the capacity region is the convex hull or convex envelope of the rate region. The reason is that if two rate pairs $\{R_1, R_2\}, \{R_1', R_2'\}$ are feasible, then any rate pair on the segment connecting those two rate pairs,

$$\{tR_1 + (1-t)R_1', tR_2 + (1-t)R_2'\} \tag{C.20}$$

for $0 \leq t \leq 1$, is also feasible and can be achieved by time multiplexing $\{R_1, R_2\}$ and $\{R_1', R_2'\}$.

We next consider the downlink and uplink AWGN channels and refer the study of the fading channels to [61, 159].

In the downlink (broadcast) AWGN channel, a single transmitter (base station) sends separate information to the two users. Similar to (C.1), the received baseband signal at user k $(k = 1, 2)$ is

$$y_k(t) = h_k x(t) + w_k(t), \tag{C.21}$$

with $x(t)$ the signal and h_k the fixed channel complex gain. Suppose $|h_1| < |h_2|$. Split the power budget P between the two users

$$P = P_1 + P_2, \tag{C.22}$$

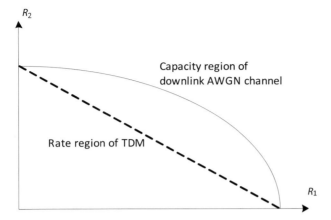

Figure C.3 Capacity region of the downlink AWGN channel.

with P_k the average transmit power constraint of user k. The following rate pair can be achieved:

$$R_1 = W \log\left(1 + \frac{P_1|h_1|^2}{P_2|h_1|^2 + N_0 W}\right), \quad \text{(C.23)}$$

$$R_2 = W \log\left(1 + \frac{P_2|h_2|^2}{N_0 W}\right). \quad \text{(C.24)}$$

To achieve the rate pair, use superposition coding

$$x(t) = x_1(t) + x_2(t), \quad \text{(C.25)}$$

with $x_k(t)$ the signal intended to user k. User 1 treats $x_2(t)$ as noise and decodes $x_1(t)$ from $y_1(t)$, thereby achieving rate R_1 (C.23). User 2 applies the idea of successive interference cancellation (SIC). Since user 2 has a better channel than user 1, it first decodes $x_1(t)$ reliably, then subtracts $h_2 x_1(t)$ from $y_2(t)$, and decodes $x_2(t)$ only dealing with noise $w_2(t)$, thereby achieving rate R_2 (C.24). The capacity region is the convex hull of $\{R_1, R_2\}$ for all possible power splits (C.22), as illustrated in Figure C.3. There are two end-points $\{R_{1,\max}, 0\}$ and $\{0, R_{2,\max}\}$, and the capacity region is in general a convex curve connecting them. As a comparison, the rate region of time division multiplexing the two end-points $\{t R_{1,\max}, (1-t) R_{2,\max}\}$ is shown to be dominated by the capacity region.

We now turn to the uplink (multiple access) channel, where the two users send separate information to a single receiver (base station). Similar to (C.1), the received baseband signal at the base station is

$$y(t) = h_1 x_1(t) + h_2 x_2(t) + w(t). \quad \text{(C.26)}$$

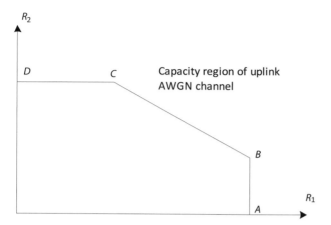

Figure C.4 Capacity region of the uplink AWGN channel.

Assume that user k has an average power constraint P_k. The capacity region is defined by the following constraints:

$$R_1 \leq W \log\left(1 + \frac{|h_1|^2 P_1}{N_0 W}\right) \tag{C.27}$$

$$R_2 \leq W \log\left(1 + \frac{|h_2|^2 P_2}{N_0 W}\right) \tag{C.28}$$

$$R_1 + R_2 \leq W \log\left(1 + \frac{|h_1|^2 P_1 + |h_2|^2 P_2}{N_0 W}\right). \tag{C.29}$$

As shown in Figure C.4, the capacity region consists of three segments, AB, BC, CD. Segment AB is dominated by corner point B and segment CD by C. Segment BC is obtained by time multiplexing B and C. Thus, we are particularly interested in B and C, whose coordinates are given by

$$B : \left\{W \log\left(1 + \frac{|h_1|^2 P_1}{N_0 W}\right), W \log\left(1 + \frac{|h_2|^2 P_2}{|h_1|^2 P_1 + N_0 W}\right)\right\} \tag{C.30}$$

$$C : \left\{W \log\left(1 + \frac{|h_1|^2 P_1}{|h_2|^2 P_2 + N_0 W}\right), W \log\left(1 + \frac{|h_2|^2 P_2}{N_0 W}\right)\right\}. \tag{C.31}$$

Similar to the downlink case, either B or C is obtained by letting $x_1(t)$, $x_2(t)$ occupy the whole spectrum W and mix over the wireless channel ("superposition coding" over the air). The base station receiver uses SIC. To obtain B, it first decodes $x_2(t)$ treating $x_1(t)$ as noise, then subtracts $h_2 x_2(t)$ from $y(t)$, and decodes $x_1(t)$ only dealing with noise $w(t)$. C is obtained by changing the order of SIC.

Note that the difference in the capacity region between the downlink and uplink channels is attributed to the different form of power constraints. In the downlink, the power constraint is the total budget to be shared among two users. In the uplink, each user has its own power budget, which is not shared with the other one.

References

[1] F. Adachi, M. Sawahashi, and K. Okawa, "Tree-structured generation of orthogonal spreading codes with different lengths for forward link of DS-CDMA mobile radio," *Electronics Letters*, vol. 33, no. 1, 1997.

[2] R. Agrawal and V. Subramanian, "Optimality of certain channel aware scheduling policies," in *Proceedings of Allerton Conference*, Oct., 2002.

[3] I. F. Akyildiz, W. Lee, M. C. Vurand, and S. Mohanty, "A survey on spectrum management in cognitive radio networks," *IEEE Communications Magazine*, vol. 46, no. 4, pp. 40–8, Apr., 2008.

[4] S. Alamouti, "A simple transmit diversity technique for wireless communications," *IEEE Journal of Selected Areas in Communications*, vol. 16, no. 8, pp. 1451–8, Oct., 1998.

[5] J. Andrews, F. Baccelli, and R. Ganti, "A tractable approach to coverage and rate in cellular networks," *Communications, IEEE Transactions on*, vol. 59, no. 11, pp. 3122–34, Nov., 2011.

[6] F. Baccelli and B. Blaszczyszyn, *Stochastic Geometry and Wireless Networks Volume I: Theory*. Foundations and Trends in Networking, NOW Publishers, 2009.

[7] M. Baker, "From LTE-Advanced to the future," *IEEE Communications Magazine*, vol. 50, no. 2, pp. 116–20, Feb., 2012.

[8] H. Balakrishnan, S. Seshan, E. Amir, and R. H. Katz, "Improving TCP/IP performance over wireless networks," in *Proceedings of MobiCom'95*, Nov., 1995.

[9] H. Balakrishnan, V. N. Padmanabhan, S. Seshan, and R. H. Katz, "A comparison of mechanisms for improving TC performance over wireless links," *IEEE/ACM Transactions on Networking*, vol. 5, no. 6, pp. 756–69, Dec., 1997.

[10] I. Barhumi, G. Leus, and M. Moonen, "Optimal training design for MIMO OFDM systems in mobile wireless channels," *IEEE Transactions on Signal Processing*, vol. 51, no. 6, pp. 1615–24, June, 2003.

[11] D. Baum, D. Gore, R. Nabar, S. Panchanathan, K. Hari, V. Erceg, and A. Paulraj, "Measurement and characterization of broadband MIMO fixed wireless channels at 2.5 GHz," in *IEEE International Conference on Personal Wireless Communications*, 2000, pp. 203–6.

[12] P. Bender, P. Black, M. Grob, R. Padovano, N. Sindhushayana, and A. Viterbi, "CDMA/HDR: a bandwidth-efficient high-speed wireless data service for nomadic users," *IEEE Communications Magazine*, vol. 38, no. 7, pp. 70–7, 2000.

[13] D. Bertsekas and R. Gallager, *Data Networks*. Prentice Hall, 1992.

[14] S. Beyme and C. Leung, "Efficient computation of dft of Zadoff–Chu sequences," *Electronics Letters*, vol. 45, no. 9, pp. 461–3, 2009.

[15] C. Berrou, A. Glavieux, and P. Thitimajshima, "Near Shannon limit error-correcting coding and decoding," in *Proceedings of ICC*, May, 1993.

[16] P. Bhat, S. Nagata, L. Campoy, I. Berberana, T. Derham, G. Liu, X. Shen, P. Zong, and J. Yang, "LTE-Advanced: an operator perspective," *IEEE Communications Magazine*, vol. 50, no. 2, pp. 104–14, Feb., 2012.

[17] S. Bodas, S. Shakkottai, L. Ying, and R. Srikant, "Scheduling in multi-channel wireless networks: Rate function optimality in the small-buffer regime," in *ACM Proceedings of the Eleventh International Joint Conference on Measurement and Modeling of Computer Systems (SIGMETRICS)*, 2009, pp. 121–32.

[18] S. Bodas, S. Shakkottai, L. Ying, and R. Srikant, "Low-complexity scheduling algorithms for multi-channel downlink wireless networks," in *Proceedings of IEEE INFOCOM, 2010*, pp. 1–9, 2010.

[19] G. Caire and S. Shamai (Shitz), "On the achievable throughput of a multi-antenna Gaussian broadcast channel," *IEEE Transactions on Information Theory*, vol. 49, no. 7, pp. 1691–706, July, 2003.

[20] V. Chandrasekhar, J. G. Andrews, and A. Gatherer, "Femtocell networks: a survey," *IEEE Communications Magazine*, vol. 46, no. 9, pp. 59–67, Sept., 2008.

[21] V. Chandrasekhar and J. G. Andrews, "Uplink capacity and interference avoidance for two-tier femtocell networks," *IEEE Transactions on Wireless Communications*, vol. 8, no. 7, pp. 3498–509, July, 2009.

[22] M. C. Chan and R. Ramjee, "Improving TCP/IP performance over third-generation wireless networks," *IEEE Transactions on Mobile Computing*, vol. 7, no. 4, pp. 430–43, Apr., 2008.

[23] E. F. Chaponniere, P. Black, J. M. Holtzman, and D. Tse, "Transmitter directed, multiple receiver system using path diversity to equitably maximize throughput," *U.S. Patent 6 449 490*, 2002.

[24] M. Chiang, P. Hande, T. Lan, and C. Tan, "Power control in wireless cellular networks," *Foundation and Trends in Networking*, vol. 2, no. 4, pp. 381–533, July, 2008.

[25] C. F. Chiasserini and R. Rao, "Performance of IEEE 802.11 WLANs in a Bluetooth environment," in *Proceedings of IEEE Wireless Communications and Networking Conference*, 2000.

[26] J. Choi, M. Jain, K. Srinivasan, P. Levis, and S. Katti, "Achieving single channel, full duplex wireless communication," in *Proceedings of MobiCom'10*, pp. 1–12, Sept., 2010.

[27] D. Chu, "Polyphase codes with good periodic correlation properties," *IEEE Transactions on Information Theory*, vol. 18, no. 4, pp. 531–32, 1972.

[28] S. Coleri, M. Ergen, A. Puri, and A. Bahai, "Channel estimation techniques based on pilot arrangement in OFDM systems," *IEEE Transactions on Broadcasting*, vol. 48, no. 3, pp. 223–9, Sept., 2002.

[29] M. Costa, "Writing on dirty-paper," *IEEE Transactions on Information Theory*, vol. 29, p. 439–41, 1983.

[30] T. Cover and A. El Gamal, "Capacity theorems for the relay channel," *IEEE Transactions on Information Theory*, vol. 25, no. 5, pp. 572–84, 1979.

[31] R. Dabora and S. Servetto, "Broadcast channels with cooperating decoders," *IEEE Transactions on Information Theory*, vol. 52, no. 12, pp. 5438–54, Dec., 2006.

[32] E. Dahlman, S. Parkvall, and J. Sköld, *4G LTE/LTE-Advanced for Mobile Broadband*. Academic Press, 2011.

[33] H. Dahrouj and W. Yu, "Coordinated beamforming for the multicell multi-antenna wireless system," *IEEE Transactions on Wireless Communications*, vol. 9, no. 5, pp. 1748–59, May, 2010.

[34] J.-J. van de Beek, O. Edfors, M. Sandell, S. Wilson, and P. O. Börjesson, "On channel estimation in OFDM systems," in *Proceedings of 45th IEEE Vehicular Technology Conference*, pp. 815–19, July, 1995.

[35] J.-J. van de Beek, M. Sandell, and P. O. Börjesson, "ML estimation of time and frequency offset in OFDM systems," *IEEE Transactions on Signal Processing*, vol. 45, no. 7, pp. 1800–5, July, 1997.

[36] N. Devroye, P. Mitran, and V. Tarokh, "Achievable rates in cognitive radio channels," *IEEE Transactions on Information Theory*, vol. 52, no. 5, pp. 1813–27, May, 2006.

[37] N. Devroye, P. Mitran, and V. Tarokh, "Limits on communications in a cognitive radio channel," *IEEE Communications Magazine*, vol. 44, no. 6, pp. 44–9, June, 2006.

[38] K. Doppler, M. Rinne, C. Wijting, C. B. Ribeiro, and K. Hugl, "Device-to-device communication as an underlay to LTE-Advanced networks," *IEEE Communications Magazine*, vol. 47, no. 12, pp. 42–9, 2009.

[39] M. Duarte and A. Sabharwal, "Full-duplex wireless communications using off-the-shelf radios: Feasibility and first results," in *Proceedings of the Forty Fourth ASILOMAR Conference on Signals, Systems and Computers*, pp. 1558–62, Nov., 2010.

[40] O. Edfors, M. Sandell, J.-J. van de Beek, S. Wilson, and P. O. Börjesson, "OFDM channel estimation by singular value decomposition," *IEEE Transactions on Communications*, vol. 46, no. 7, pp. 931–9, July, 1998.

[41] O. Edfors, M. Sandell, J.-J. van de Beek, S. Wilson, and P. O. Börjesson, "Analysis of DFT-based channel estimators for OFDM," *Wireless Personal Communications*, vol. 12, no. 1, pp. 55–70, Jan., 2000.

[42] A. El Gamal and S. Zahedi, "Capacity of a class of relay channels with orthogonal components," *IEEE Transactions on Information Theory*, vol. 51, no. 5, pp. 1815–17, May, 2005.

[43] A. Eryilmaz and R. Srikant, "Fair resource allocation in wireless networks using queue-lenghtbased scheduling and congestion control," in *Proceedings of IEEE INFOCOM*, pp. 1794–803, 2005.

[44] A. Eryilmaz and R. Srikant, "Fair resource allocation in wireless networks using queue-lenghtbased scheduling and congestion control," *IEEE/ACM Transactions on Networking*, vol. 15, no. 6, pp. 1333–44, Dec., 2007.

[45] A. Eryilmaz, R. Srikant, and J. Perkins, "Stable scheduling policies for fading wireless channels," *Networking, IEEE/ACM Transactions on*, vol. 13, no. 2, pp. 411–24, Apr., 2005.

[46] R. Etkin, A. Parekh, and D. Tse, "Spectrum sharing for unlicensed bands," *Selected Areas in Communications, IEEE Journal on Selected Areas in Communications*, vol. 25, no. 3, pp. 517–28, Apr., 2007.

[47] R. Etkin and D. Tse, "Degrees of freedom in some underspread MIMO fading channels," *IEEE Transactions on Information Theory*, vol. 52, no. 4, pp. 1576–608, 2006.

[48] R. Etkin, D. Tse, and H. Wang, "Gaussian interference channel capacity to within one bit," *IEEE Transactions on Information Theory*, vol. 54, no. 12, pp. 5534–62, 2008.

[49] FCC, "ET docket no. 03-322 notice of proposed rule marking and order," Dec., 1972.

[50] G. J. Foschini, "Layered space-time architecture for wireless communication in a fading environment when using multi-element antennas," *Bell Labs. Tech. J.*, vol. 1, no. 2, pp. 41–59, 1996.

[51] G. Foshini and M. J. Gans, "On limits of wireless communication in a fading environment when using multiple antennas," *Wireless Personal Communications*, vol. 6, no. 3, pp. 311–35, 1998.

[52] G. Foschini and Z. Miljanic, "A simple distributed autonomous power control algorithm and its convergence," *IEEE Transactions on Vehicular Technology*, vol. 42, no. 4, pp. 641–6, Nov., 1993.

[53] R. G. Gallager, "Low-density parity-check codes," *IRE Transactions on Information Theory*, vol. 8, pp. 21–8, 1962.

[54] D. Gesbert, S. Hanly, H. Huang, S. Shamai, O. Simeone, and W. Yu, "Multi-cell MIMO cooperative networks: a new look at interference," *IEEE Journal on Selected Areas in Communications*, vol. 28, no. 9, pp. 1380–408, 2010.

[55] D. Gesbert, M. Kountouris, R. Heath, C.-B. Chae, and T. Sälzer, "Shifting the MIMO paradigm," *IEEE Signal Processing Magazine*, vol. 24, no. 5, Sept., 2007.

[56] A. Ghasemi and E. Sousa, "Opportunistic spectrum access in fading channels through collaborative sensing," *Journal of Communications*, vol. 2, no. 2, pp. 71–82, Mar., 2007.

[57] A. Ghasemi and E. Sousa, "Spectrum sensing in cognitive radio networks: requirements, challenges and design trade-offs," *IEEE Communications Magazine*, vol. 46, no. 4, pp. 32–9, Apr., 2008.

[58] A. Ghosh, J. Zhang, J. Andrews, and R. Muhamed, *Fundamentals of LTE*. Prentice Hall, 2010.

[59] T. Goff, J. Moronski, D. S. Phatak, and V. Gupta, "Freeze-TCP: a true end-to-end enhancement mechanism for mobile environments," in *Proceedings of IEEE INFOCOM*, 1537–45, 2000.

[60] G. Golden, G. J. Foschini, R. Valenzuela, and P. Wolniansky, "Detection algorithm and initial laboratory results using V-BLAST space-time communication architecture," *Electronic Letter*, vol. 35, pp. 14–16, Jan., 1999.

[61] A. Goldsmith, *Wireless Communications*. Cambridge University Press, 2005.

[62] H. Jin, R. Laroia, and T. Richardson, "Superposition by position," in *Proceedings of IEEE Information Theory Workshop (ITW'06)*, Mar., 2006.

[63] P. Hande and M. Chiang, "Distributed uplink power control for optimal SIR assignment in cellular data networks," in *Proceedings of IEEE INFOCOM*, Apr., 2006.

[64] P. Hande, S. Rangan, M. Chiang, and X. Wu, "Distributed uplink power control for optimal SIR assignment in cellular data networks," *IEEE/ACM Transactions on Networking*, vol. 16, no. 6, pp. 1430–43, Nov., 2008.

[65] S. Haykin, "Cognitive radio: brain-empowered wireless communications," *Journal on Selected Areas in Communications*, vol. 23, no. 2, pp. 201–20, 2005.

[66] P. Hoeher, S. Kaiser, and P. Robertson, "Two-dimensional pilot-symbol-aided channel estimation by Wiener filtering," in *Proceedings of IEEE International Conference on Acoust., Speech, and Signal Processing*, vol. 2, pp. 1845–8, 1997.

[67] C. Hoymann, W. Chen, J. Montojo, A. Golitschek, C. Koutsimanis, and X. Shen, "Relaying operation in 3GPP LTE: challenges and solutions," *IEEE Communications Magazine*, vol. 50, no. 2, pp. 156–62, Feb., 2012.

[68] J. Huang, V. Subramanian, R. Agrawal, and R. Berry, "Downlink scheduling and resource allocation for OFDM systems," *Wireless Communications, IEEE Transactions on Wireless Communications*, vol. 8, no. 1, pp. 288–96, Jan., 2009.

[69] A. Høst-Madsen and J. Zhang, "Capacity bounds and power allocation for wireless relay channels," *IEEE Transactions on Information Theory*, vol. 51, no. 6, pp. 2020–40, 2005.

[70] A. Høst-Madsen, "On the capacity of wireless relaying," in *Proceedings of IEEE Vehicular Technology Conference (VTC'02 Fall)*, pp. 1333–7, 2002.

[71] ITU-R Rep. M. 2134, "Requirements related to technical performance for IMT-Advanced radio interface(s)," *Tech. Rep.*, 2008.

[72] N. Jindal, S. Vishwanath, and A. J. Goldsmith, "On the duality of Gaussian multiple-access and broadcast channels," *IEEE Transactions on Information Theory*, vol. 50, no. 5, 768–83, 2004.

[73] N. Jindal, "MIMO broadcast channels with finite-rate feedback," *IEEE Transactions on Information Theory*, vol. 52, no. 11, pp. 5045–60, Nov., 2006.

[74] S. Jing, D. Tse, J. Soriaga, J. Hou, J. Smee, and R. Padovani, "Downlink macro-diversity in cellular networks," in *Proceedings of IEEE International Symposium on Information Theory (ISIT'07)*, June, 2007.

[75] X. Jing and D. Raychaudhuri, "Spectrum co-existence of IEEE 802.11b and 802.16a networks using reactive and proactive etiquette policies," *ACM Journal Mobile Network Applications*, vol. 11, no. 4, pp. 539–54, 2006.

[76] X. Jing, S. Anandaraman, M. A. Ergin, I. Seskar, and D. Raychaudhuri, "Distributed coordination schemes for multi-radio co-existence in dense spectrum environments: an experimental study on the ORBIT testbed," in *Proceedings of IEEE DySPAN 2008*, pp. 1–10, 2008.

[77] H. S. Jo, P. Xia, and J. G. Andrews, "Open vs. closed access femtocells in downlink," in *Proceedings of IEEE ICC*, pp. 1–5, June, 2011.

[78] A. Jovičić and P. Viswanath, "Cognitive radio: an information-theoretic perspective," *IEEE Transactions on Information Theory*, vol. 55, no. 9, pp. 3945–58, Sept., 2009.

[79] F. Jondral, "Cognitive radio: a communications engineering view," *IEEE Wireless Communications*, vol. 14, no. 4, pp. 28–33, Aug., 2007.

[80] R. Juang, P. Ting, H. Lin, and D. Lin, "Interference management of femtocell in macro-cellular networks," in *Proceedings of Wireless Telecommunications Symposium (WTS)*, Apr., 2010.

[81] E. Katranaras, M. Imran, and R. Hoshyar, "Sum-rate of linear cellular systems with clustered joint processing," in *Proceedings of the IEEE Vehicular Technology Conference (VTC'09 Spring)*, vol. 8, 2009, pp. 1910–21.

[82] A. Khandekar, N. Bhushan, T. Ji, and V. Vanghi, "LTE-Advanced: heterogeneous networks," in *European Wireless Conference*, pp. 978–82, 2010.

[83] D. K. Kim, S. Do, H. Cho, H. Chol, and K. Kim, "A new joint algorithm of symbol timing recovery and sampling clock adjustment for OFDM systems," *IEEE Transactions on Consumer Electronics*, vol. 44, no. 3, pp. 1142–9, Aug., 1998.

[84] S. Kittipiyakul and T. Javidi, "Delay-optimal server allocation in multiqueue multiserver systems with time-varying connectivities," *IEEE Transactions on Information Theory*, vol. 55, no. 5, p. 2319–33, May, 2009.

[85] G. Kramer, M. Gastpar, and P. Gupta, "Cooperative strategies and capacity theorems for relay networks," *IEEE Transactions on Information Theory*, vol. 51, no. 9, pp. 3037–63, Sept., 2005.

[86] G. Kramer, I. Maric, and R. Yates, "Cooperative communications," *Foundations and Trends in Networking*, vol. 1, no. 3, pp. 271–425, August 2006.

[87] J. N. Laneman, D. Tse, and G. Wornell, "Cooperative diversity in wireless networks: efficient protocols and outage behavior," *IEEE Transactions on Information Theory*, vol. 50, no. 12, pp. 3062–80, 2004.

[88] R. Laroia, S. Uppala, and J. Li, "Designing a mobile broadband wireless access network," *IEEE Signal Processing Magazine*, vol. 21, no. 5, pp. 20–8, 2004.

[89] R. Laroia, J. Li, S. Rangan, and M. Srinivasan, "Enhanced opportunistic beamforming," in *Proceedings of IEEE Vehicle Technology Conference*, Oct., 2003.

[90] J. W. Lee, R. Mazumdar, and N. B. Shroff, "Opportunistic power scheduling for multi-server wireless systems with minimum performance constraints," in *Proceedings of IEEE INFOCOM*, Mar., 2004.

[91] D. Lee, H. Seo, B. Clerckx, E. Hardouin, D. Mazzarese, S. Nagata, and K. Sayana, "Coordinated multipoint transmission and reception in LTE-Advanced: deployment scenarios and operational challenges," *IEEE Communications Magazine*, vol. 50, no. 2, pp. 148–55, Feb., 2012.

[92] H. Lei, M. Yu, A. Zhao, Y. Chang, and D. Yang, "Adaptive connection admission control algorithm for LTE systems," in *Proceedings of the IEEE Vehicular Technology Conference (VTC'08 Spring)*, pp. 2336–40, May, 2008.

[93] N. Levy and S. Shamai (Shitz), "Clustered local decoding for Wyner-type cellular models," *IEEE Transactions on Information Theory*, vol. 55, no. 11, pp. 4967–85, 2009.

[94] J. Li, N. Shroff, and E. Chong, "A reduced-power channel reuse scheme for wireless packet cellular networks," *IEEE/ACM Transactions on Networking*, vol. 7, no. 6, pp. 818–32, Dec., 1999.

[95] Y. Liang and V. V. Veeravalli, "Gaussian orthogonal relay channel: optimal resource allocation and capacity," *IEEE Transactions on Information Theory*, vol. 51, no. 9, pp. 3284–9, Sept., 2005.

[96] Y. Liang and V. V. Veeravalli, "Cooperative relay broadcast channels," *IEEE Transactions on Information Theory*, vol. 53, no. 3, pp. 900–28, Mar., 2007.

[97] S. Lin and J. D. J. Costello, *Error Control Coding*. Prentice Hall, 2004.

[98] X. Lin, N. B. Shroff, and R. Srikant, "On the connection-level stability of congestion-controlled communication networks," *IEEE Transactions on Information Theory*, vol. 54, no. 5, pp. 2317–38, May, 2008.

[99] X. Lin and N. Shroff, "Joint rate control and scheduling in multihop wireless networks," in *Proceedings of 43rd IEEE Decision and Control Conference*, pp. 1484–9, 2004.

[100] X. Lin and N. Shroff, "The impact of imperfect scheduling on cross-layer rate control in multihop wireless networks," in *Proceedings of IEEE INFOCOM*, pp. 1804–14, 2005.

[101] L. Liu, R. Chen, S. Geirhofer, K. Sayana, Z. Shi, and Y. Zhou, "Downlink MIMO in LTE-Advanced: SU-MIMO vs. MU-MIMO," *IEEE Communications Magazine*, vol. 50, no. 2, pp. 140–7, Feb., 2012.

[102] X. Liu, E. K. P. Chong, and N. B. Shroff, "Opportunistic transmission scheduling with resource sharing constraints in wireless networks," *IEEE Journal on Selected Areas in Communications*, vol. 19, no. 10, pp. 2053–65, 2001.

[103] K. Liu, A. Sadek, W. Su, and A. Kwasinski, *Cooperative Communications and Networking*. Cambridge University Press, 2009.

[104] P. Liu, Z. Tao, Z. Lin, E. Erkip, and S. Panwar, "Cooperative wireless communications: a crosslayer approach," *IEEE Wireless Communications*, vol. 13, no. 4, pp. 84–92, Aug., 2006.

[105] S. Liu, L. Ying, and R. Srikant, "Throughput-optimal opportunistic scheduling in the presence of flow-level dynamics," *IEEE/ACM Transactions on Networking*, vol. 19, no. 4, pp. 1057–70, Aug., 2011.

[106] D. López-Pérez, A. Valcarce, G. D. L. Roche, and J. Zhang, "OFDMA femtocells: A roadmap on interference avoidance," *IEEE Communications Magazine*, pp. 41–8, Sept., 2009.

[107] D. J. Love, R. Heath, W. Santipach, and M. L. Honig, "What is the value of limited feedback for MIMO channels," *IEEE Communications Magazine*, vol. 42, no. 10, pp. 54–9, Oct., 2004.

[108] A. Lozano, R. H. Jr., and J. Andrews, "Fundamental limits of cooperation," *submitted to IEEE Transactions on Information Theory*, Mar., 2012.

[109] J. Ma, G. Y. Li, and B. H. Juang, "Signal processing in cognitive radio," *Proceedings of IEEE*, vol. 97, no. 5, pp. 805–23, 2009.

[110] T. Mayer, H. Jenkac, and J. Hagenauer, "Turbo base-station cooperation for intercell inteference cancellation," in *Proceedings of the IEEE International Conference on Communications*, vol. 11, June, 2006, pp. 4977–82.

[111] H. Minn and N. Al-Dhahir, "Optimal training signals for MIMO OFDM channel estimation," *IEEE Transactions on Wireless Communications*, vol. 5, no. 5, pp. 1158–68, May, 2006.

[112] J. Mitola III and G. Maguire, "Cognitive radio: making software radios more personal," *IEEE Personal Communications*, vol. 6, no. 4, pp. 13–18, Aug., 1999.

[113] M. Morelli and U. Mengali, "A comparison of pilot-aided channel estimation methods for OFDM systems," *IEEE Transactions on Signal Processing*, vol. 49, no. 12, pp. 3065–73, Dec., 2001.

[114] Y. Nam, Y. Akimoto, Y. Kim, M. Lee, K. Bhattad, and A. Ekpenyong, "Evolution of reference signals for LTE-Advanced systems," *IEEE Communications Magazine*, vol. 50, no. 2, pp. 132–8, Feb., 2012.

[115] R. van Nee and R. Prasad, *OFDM for Wireless Multimedia Communications*. Artech House Publishers, 2000.

[116] M. J. Neely, E. Modiano, and C. Li, "Fairness and optimal stochastic control for heterogeneous networks," in *Proceedings of IEEE INFOCOM*, 1723–34, 2005.

[117] R. Negi and J. Cioffi, "Pilot tone selection for channel estimation in a mobile OFDM system," *IEEE Transactions on Consumer Electronics*, vol. 44, pp. 1122–8, Aug., 1998.

[118] D. Niyato and E. Hossain, "Call admission control for QoS provisioning in 4G wireless networks: issues and approaches," *IEEE Networks*, pp. 5–11, Sept./Oct., 2005.

[119] M. K. Özdemir and H. Arslan, "Channel estimation for wireless OFDM systems," *IEEE Communications Surveys & Tutorials*, vol. 9, no. 2, pp. 18–48, 2007.

[120] M. K. Özdemir, H. Arslan, and E. Arvas, "Toward real-time adaptive low-rank LMMSE channel estimation of MIMO-OFDM systems," *IEEE Transactions on Wireless Communications*, vol. 5, no. 10, pp. 2675–78, 2006.

[121] S. Patil, M. Anand, X. Wu, and J. Li, "Effective OFDMA based signaling in ad hoc wireless networks," in *Proceedings of IEEE Globecom*, pp. 1–6, Dec., 2011.

[122] A. Paulraj, D. Gore, and R. Nabar, *Introduction to Space-Time Wireless Communications*. Cambridge University Press, 2003.

[123] C. E. Perkins, *Mobile IP: Design Principles and Practices*. Addison-Wesley Press, 1997.

[124] B. M. Popović, "Generalized chirp-like polyphase sequences with optimum correlation properties," *IEEE Transactions on Information Theory*, vol. 38, no. 4, pp. 1406–9, 1972.

[125] J. Proakis and M. Salehi, *Digital Communications*. McGraw-Hill Science/Engineering/Math, 2007.

[126] T. Rappaport, *Wireless Communications: Principles and Practice*. Prentice Hall, 2002.

[127] T. Richardson and R. Urbanke, *Modern Coding Theory*. Cambridge University Press, 2008.

[128] M. Rice, "Loop control architecture for symbol timing synchronization in sampled data receivers," in *Proc. of MILCOM*, vol. 2, pp. 987–91, Oct., 2002.

[129] B. Sadiq, R. Madan, and A. Sampath, "Downlink scheduling for multiclass traffic in LTE," *EURASIP Journal on Wireless Communications and Networking*, vol. 2009, Article ID 510617, 2009.

[130] B. Sadiq, S. J. Baek, and G. de Veciana, "Delay-optimal opportunistic scheduling and approximations: the Log rule," in *Proceedings of IEEE INFOCOM*, pp. 1–9, 2009.

[131] B. Sadiq and G. de Veciana, "Throughput optimality of delay-driven maxweight scheduler for a wireless system with flow dynamics," in *47th Annual Allerton Conference on Communication, Control, and Computing, 2009. Allerton*, 2009, pp. 1097–102, Oct. 2, 2009.

[132] B. Sadiq and G. de Veciana, "Large deviations sum-queue optimality of a radial sum-rate monotone oppportunistic scheduler," *IEEE Transactions on Information Theory*, vol. 56, no. 7, pp. 3395–412, 2010.

[133] A. Sendonaris, E. Erkip, and B. Aazhang, "User cooperation diversity-part i and ii," *IEEE Transactions on Communications*, vol. 51, no. 11, pp. 1927–48, Nov., 2003.

[134] M. Sandell and O. Edfors, "A comparative study of pilot-based channel estimators for wireless OFDM," Luleå University of Technology, Luleå, Sweden, *Tech. Rep.*, Sept., 1996.

[135] A. Sanderovich, O. Somekh, and S. Shamai (Shitz), "Uplink macro diversity with limited backhaul capacity," in *Proceedings of IEEE International Symposium on Information Theory (ISIT'07)*, pp. 11–15, June, 2007.

[136] T. M. Schmidl and D. C. Cox, "Robust frequency and timing synchronization for OFDM," *IEEE Transactions on Communications*, vol. 45, no. 12, pp. 1613–21, Dec., 1997.

[137] M. Schwartz, *Broadband Integrated Networks*. Prentice Hall, 1996.

[138] S. Sesia, I. Toufik, and M. Baker, *LTE – The UMTS Long-term Evolution, From Theory to Practice*. John Wiley, 2009.

[139] M. Shafi, M. Zhang, A. Moustakas, P. Smith, A. Molisch, F. Tufvesson, and S. Simon, "Polarized MIMO channels in 3-D: models, measurements and mutual information," *IEEE Journal of Selected Areas in Communications*, vol. 24, no. 3, pp. 514–27, Mar., 2006.

[140] C. Shannon, "Communication in the presence of noise," *Proceedings of the IRE*, vol. 37, pp. 10–21, 1949.

[141] C. Shannon, "A mathematical theory of communication," *Bell System Technical Journal*, vol. 27, pp. 379–423, 623–56, 1948.

[142] M. Sharif and B. Hassibi, "On the capacity of MIMO broadcast channels with partial side information," *IEEE Transactions on Information Theory*, vol. 51, no. 2, pp. 506–22, Feb., 2005.

[143] M. Sharma and X. Lin, "OFDM downlink scheduling for delay-optimality: many-channel manysource asymptotics with general arrival processes," in *Information Theory and Applications Workshop (ITA)*, 2011.

[144] Z. Shen, A. Papasakellariou, J. Montojo, D. Gerstenberger, and F. Xu, "Overview of 3GPP LTE Advanced carrier aggregation for 4G wireless communications," *IEEE Communications Magazine*, vol. 50, no. 2, pp. 122–30, Feb., 2012.

[145] F. Shu, J. Lee, L. Wu, and G. Zhao, "Time-frequency channel estimation for digital amplitude modulation broadcasting systems based on OFDM," *IEE Proceedings Communications*, vol. 150, no. 4, pp. 259–64, Aug., 2003.

[146] O. Simeone, O. Somekh, H. Poor, and S. Shamai (Shitz), "Downlink multicell processing with limited-backhaul capacity," *EURASIP Journal on Advances in Signal Processing*, vol. 2009, no. Article ID 840814, 2009.

[147] O. Somekh, B. Zaidel, and S. Shamai (Shitz), "Sum rate characterization of joint multiple cell-site processing," *IEEE Transactions on Information Theory*, vol. 53, no. 12, pp. 4473–97, 2007.

[148] V. Stanković, A. Høst-Madsen, and Z. Xiong, "Cooperative diversity for wireless ad hoc networking," *IEEE Signal Processing Magazine*, vol. 23, no. 5, pp. 37–49, 2006.

[149] C. Stevenson, G. Chouinard, Z. Lei, W. Hu, S. Shellhammer, and W. Caldwell, "IEEE 802.22: the first cognitive radio wireless regional area network standard," *IEEE Communications Magazine*, vol. 47, no. 1, pp. 130–8, Jan., 2009.

[150] A. L. Stolyar, "Maximizing queueing network utility subject to stability: Greedy primal-dual algorithm," *Queueing Systems*, vol. 50, no. 4, pp. 401–57, 2005.

[151] A. L. Stolyar, "Large deviations of queues sharing a randomly time-varying server," *Queueing Systems*, vol. 59, no. 1, pp. 1–35, 2008.

[152] A. Stolyar and H. Viswanathan, "Self-organizing dynamic fractional frequency reuse in OFDMA systems," in *INFOCOM 2008. The 27th Conference on Computer Communications. IEEE*, Apr., 2008, pp. 691–9.

[153] R. Tandra, S. M. Mishra, and A. Sahai, "What is a spectrum hole and what does it take to recognize one?" *Proceedings of IEEE*, vol. 97, no. 5, pp. 824–48, 2009.

[154] R. Tandra and A. Sahai, "SNR walls for signal detection," *IEEE Journal of Selected Topics in Signal Processing*, vol. 2, no. 1, pp. 4–17, Feb., 2008.

[155] V. Tarokh, N. Seshadri, and A.R. Calderbank, "Spacetime codes for high data rate wireless communication: Performance analysis and code construction," *IEEE Transactions on Information Theory*, vol. 44, no. 2, pp. 744–65, Mar., 1998.

[156] L. Tassiulas and A. Ephremides, "Stability properties of constrained queuing systems and scheduling policies for maximum throughput in multihop radio networks," *IEEE Transactions on Automatic Control*, vol. 37, no. 12, pp. 1936–48, Dec., 1992.

[157] I. Telatar, "Capacity of multi-antenna gaussian channels," *European Transactions on Telecommunications*, vol. 10, pp. 585–95, Nov./Dec., 1999.

[158] E. Tragos, G. Tsiropoulos, G. Karetsos, and S. A. Kyriazakos, "Admission control for QoS support in heterogeneous 4G wireless networks," *IEEE Networks*, pp. 30–6, May/June, 2008.

[159] D. Tse and P. Viswanath, *Fundamentals of Wireless Communication*. Cambridge University Press, 2005.

[160] A. M. Tulino, A. Lozano, and S. Verdú, "Capacity-achieving input covariance for single-user multi-antenna channels," *IEEE Transactions on Wireless Communications*, vol. 5, no. 3, pp. 662–71, Mar., 2006.

[161] G. Ungerboeck, "Channel coding with multi-level/phase signals," *IEEE Transactions on Information Theory*, vol. 28, no. 1, pp. 55–67, 1982.

[162] M. Valenti and B. Woerner, "Iterative multiuser detection, macro-diversity combining, and decoding for the TDMA cellular uplink," *IEEE Journal on Selected Areas in Communications*, vol. 19, pp. 1570–83, Aug., 2001.

[163] P. van de Ven, S. Borst, and S. Shneer, "Instability of MaxWeight scheduling algorithms," in *INFOCOM 2009, IEEE*, Apr., 2009, pp. 1701–9.

[164] V. Veeravalli, Y. Liang, and A. Sayeed, "MIMO wireless channels with uniform linear arrays: capacity, optimal signaling, and asymptotics," *IEEE Transactions on Information Theory*, vol. 51, no. 6, pp. 2058–72, June, 2005.

[165] S. Verdú, *Multiuser Detection*. Cambridge University Press, 1998.

[166] S. Verdú, "On channel capacity per unit cost," *IEEE Transactions on Information Theory*, vol. 36, no. 9, pp. 1019–30, September, 1990.

[167] S. Verdú, "Spectral efficiency in the wideband regime," *IEEE Transactions on Information Theory*, vol. 48, no. 6, pp. 1319–43, 2002.

[168] A. Viterbi, *CDMA: Principles of Spread-Spectrum Communication*. Addison-Wesley Publishing Company, 1995.

[169] P. Viswanath, D. Tse, and R. Laroia, "Opportunistic beamforming using transmit antenna arrays," *IEEE Transactions on Information Theory*, vol. 48, no. 6, pp. 1277–94, 2002.

[170] P. W. Wolniansky, G. Foschini, G. Golden, and R. Valenzuela, "V-BLAST: an architecture for realizing very high data rates over the rich-scattering wireless channels," in *Proceedings of the URSI International Symposium on Signals, Systems, and Electronics Conference*, 1998, pp. 295–300.

[171] S. Wu, K. Wong, and B. Li, "A dynamic call admission policy with precision QoS guarantee using stochastic control for mobile wireless networks," *IEEE/ACM Transactions on Networking*, vol. 10, no. 2, pp. 257–71, Apr., 2002.

[172] X. Wu, S. Tavildar, S. Shakkottai, T. Richardson, J. Li, R. Laroia, and A. Jovicic, "Flashlinq: a synchronous distributed scheduler for peer-to-peer ad hoc networks," in *Proceedings of Allerton Conference on Communication, Control, and Computing*, pp. 514–21, 2010.

[173] W. Wu, S. Vishwanath, and A. Arapostathis, "Capacity of a class of cognitive radio channels: interference channels with degraded message sets," *IEEE Transactions on Information Theory*, vol. 53, no. 11, pp. 4391–9, Nov., 2007.

[174] A. Wyner and J. Ziv, "The rate-distortion function for source coding with side information at the decoder," *IEEE Transactions on Information Theory*, vol. 22, no. 1, pp. 1–10, 1976.

[175] A. Wyner, "Shannon-theoretic approach to a Gaussian cellular multiple-access channel," *IEEE Transactions on Information Theory*, vol. 40, no. 6, pp. 1713–27, 1994.

[176] P. Xia, V. Chandrasekhar, and J. G. Andrews, "Open vs. closed access femtocells in the uplink," *IEEE Transactions on Wireless Communications*, vol. 9, no. 10, pp. 3798–809, Dec., 2010.

[177] K. Xu, M. Gerla, and S. Bae, "How effective is the IEEE 802.11 RTS/CTS handshake in ad hoc networks?" in *Proceedings of GLOBECOM*, pp. 72–6, 2002.

[178] B. Yang, Z. Cao, and K. B. Letaief, "Analysis of low-complexity windowed DFT-based MMSE channel estimator for OFDM systems," *IEEE Transactions on Communications*, vol. 49, no. 11, pp. 1977–87, Nov., 2001.

[179] B. Yang, K. B. Letaief, R. Cheng, and Z. Cao, "Channel estimation for OFDM transmission in multipath fading channels based on parametric channel modeling," *IEEE Transactions on Communications*, vol. 49, no. 3, pp. 467–79, Mar., 2001.

[180] R. Yates, "A framework for uplink power control in cellular radio systems," *IEEE Journal of Selected Areas in Communications*, vol. 13, no. 7, pp. 1341–7, Sept., 1995.

[181] M. Yavuz and F. Khafizov, "TCP over wireless links with variable bandwidth," in *Proceedings of IEEE Vehicular Technology Conference (VTC'02 Fall)*, pp. 1322–7, 2002.

[182] F. Ye, S. Yi, and B. Sikdar, "Improving spatial reuse of IEEE 802.11 based ad hoc networks," in *Proceedings of GLOBECOM*, pp. 1013–17, 2003.

[183] L. Ying, R. Srikant, A. Eryilmaz, and G. Dullerud, "Distributed fair resource allocation in cellular networks in the presence of heterogeneous delays," *IEEE Transactions on Automatic Control*, vol. 52, no. 1, pp. 129–34, Jan., 2007.

[184] W. Yu, G. Ginis, and J. Cioffi, "Distributed multiuser power control for digital subscriber lines," *Selected Areas in Communications, IEEE Journal on Selected Areas in Communications*, vol. 20, no. 5, pp. 1105–15, June, 2002.
[185] S. Zadoff, "Phase coded communication system," *U.S. Patent 3 099 796*, 1963.
[186] L. Zheng and D. Tse, "Diversity and multiplexing: A fundamental tradeoff in multiple antenna channels," *IEEE Transactions on Information Theory*, vol. 49, pp. 1073–96, May, 2003.

Index

access link 367, 458
additive white Gaussian noise (AWGN) model 495
   channel capacity 495
   multiuser capacity region 499
      downlink 500
      uplink 501
   power and bandwidth 495–6
admission control 88, 283, 299–300, 314
amplify-and-forward scheme 418–19
analog power signaling 411, 459
antenna diversity 84, 115–20, 122, 123, 239
   antenna selection diversity 240
ARQ 5, 75, 82, 293, 323, 492–3
   fast loop ARQ 492
   hybrid ARQ 75, 293, 310–11, 464, 493–4
   slow loop ARQ 493

backhaul 1, 235, 317, 321, 367, 383, 384, 385, 416, 459
   overhead 269, 323
   capacity constraint 425–30
   delay constraint 430
backhaul link 367, 368, 458
beacon signal 345–50
   beacon insertion 356
beamforming 52, 84, 118, 239, 255, 421, 424
   coordinated beamforming 268–70
   discontinous beamforming 142
   downlink beamforming 261–7
   frequency selective beamforming 87
   inter-sector beamforming 271–2
   receive beamforming 85, 239, 259, 264
   transmit beamforming 84, 239, 262, 264
binding update 328, 338, 339

carrier diversity 225
CDMA 4, 4, 21, 21–3, 70, 77, 80, 91, 92, 193, 197, 203–5, 206–8, 254
   RAKE receiver 91, 160, 322, 469
   spreading code 21, 130, 160, 203, 254, 1, 1
cellular systems and services 1, 18
   3G 4, 52
   4G 2–3

broadband data 79, 138 ,92, 81, 19
cells 1
cellular telephony 1, 3–4
circuit-switched voice 24, 70, 77, 79, 81, 88, 92, 92, 96, 197, 282, 319, 79
conventional cellular framework 365
sectors 19
sectorization 19, 189–90
synchronized sectors 190–2, 271
technology evolution 1
channel coding 488–90
   adaptive modulation and coding 490–92
   coding gain 488, 493, 121
   convolutional codes 352, 488
   Turbo or LDPC codes 488
      error-floor region 488
      waterfall region 488, 107
channel estimation 25, 122, 189, 274, 358, 430, 477–87
   imperfect channel estimation 64–5, 259
   least square estimation 478
   MIMO OFDM estimation 485–487
   MMSE estimation 482
   non-parametric channel estimation 480–3
   parametric channel estimation 483–5
   pilot pattern 478–80
   subspace method 482
channel fading 77, 87, 93, 94
   large-scale fading 94
   path loss 95
   shadowing 95
   small-scale fading 94
channel gain ratio 135, 213
channel prediction 124, 146
   margin 107
   uncertainty 146, 114, 125, 146–8
closed access 385–9, 396, 399
cognitive radio 366, 432, 456
   spectrum sensing 432, 433–8
   spectrum sharing 432, 438–43
      interweaved 438
      overlay 438–40
      underlay 441–3

Index

compress-and-forward scheme 421
    Wyner-Ziv lossy source coding scheme 421
congestion control 4, 290, 283
    joint congestion control and scheduling 290–2
constellation 488
    Gaussian 184
    Gaussian+ 184
    QPSK 184
    QPSK+ 184
context transfer 328, 339, 340
cooperative communication 415–31
    network cooperation 417–24
    user cooperation 425–31
cooperative diversity 370, 417, 418
cooperative spectrum sensing 437, 438
    hard-decision combining 437
    soft-decision combining 437
core network 1, 317, 317
coverage 19, 365, 365, 367
    femtocell coverage 387
    macrocell coverage 387
    primary/secondary coverage 434
    relay coverage 371–3
cross channel gains 377, 386, 404
cross interference model 63, 167–72, 180–3
CSMA/CA 402–4

dead zone 388, 388
decode-and-forward scheme 419, 421
dense spectrum reuse 373–9, 383
detection 343, 349
    cyclic prefix correlation 472–3
    energy detection 434
    half-symbol correlation 475
    matched filter detection 473, 434, 468–9
    sample-level search 347, 349
    signal-presence detection 467–71
    symbol-level search 347
device-to-device communications 398–411
    communication topology 398
direct channel gains 377, 386, 404
dirty-paper precoding 264, 439
duality 39, 263

effective SINR 192, 490
eigenfunction 10, 70
elastic traffic 283
equalization 17, 65, 160
    frequency domain equalization 40, 42
EV-DO 130
    downlink multiuser diversity 130–2
    downlink user multiplexing 160
exposed terminal problem 403

fading channel capacity 496–9
fading exploitation 75

fading mitigation 75
fairness and utility function 283–6
FDMA/TDMA 196
filtering 16–17
Flash-OFDM 133
    beacon signal 352–3
    flash signaling 352
    handoff in a railway network 353
    HOLD MAC state 313
    superposition-by-position coding 352
    uplink multiuser diversity 134
flash signaling 351
flow level traffic 283
    long-lived flows 302
    mixed long-lived and short-lived flows 304
    short-lived flows 302
frequency diversity 20, 46, 120, 136, 146, 203
frequency reuse 196
    fractional frequency reuse 7, 74, 83, 219–36, 354
        adaptive FFR 233–6
        breathing cells 230–3
        power scaling 228, 355
        static FFR 226–30
    frequency reuse factor 21
    universal frequency reuse 21, 80, 82, 197, 342
frequency synchronization and control 32, 63, 324

Non-Gaussian interference 22, 23
    decoding metrics 23–4
Gaussian interference model 412
    channel capacity 413

half-duplex constraint 368, 373, 404, 433
handoff 1, 91–2, 196
    break-before-make handoff 337–8
    expedited break-before-make handoff 339–42
    handoff condition 389
        downlink and uplink boundary mismatch 390
        downlink condition 390
        uplink condition 390
    handoff initiation 342–56
        handoff hysteresis 321, 334, 340
    macrocell-femtocell handoff 389–95
    make-before-break handoff 323–36
    soft handoff 4, 91, 319–23, 425
    softer handoff 193, 321
heterogeneous network 365, 366
hidden terminal problem 403
high SINR regime 84, 85, 153, 160, 240, 394
home agent 327–8, 334, 338
HSDPA 160

Index

HSUPA 160
hypothesis testing 467

IEEE 802.22 456–8
   coexistence beacon protocol 457
   spectrum sensing 457
   TV channels 456
inelastic traffic 283
inter-cell interference management 1, 80–4
   inter-cell interference coordination 72
   interference averaging 4, 20, 21, 21–5, 81, 197
interference 1, 5, 21
   inter-cell interference 20, 19
   inter-sector interference 19
   intra-cell interference 19, 19
interference budget 213, 215, 220, 330, 441, 163
interference limited regime 164, 425, 430
interference nulling 396
IP access router 89, 317, 331
IP QoS 88
IS-95 130

joint power and rate control 111

layered network architecture models 315
   open system interconnection model 315
   TCP/IP model 315
linear antenna array 241–6
   inter-element spacing 241
   normalized antenna size 241
   receive directional cosine 242
   transmit directional cosine 242
link adaptation 82, 492
link layer registration 328
link reliability 4, 239, 315, 323
link scheduling 329, 408–10
link selection 329, 330, 331, 334
   link selection diversity 372, 370
load 163, 213
log-likelihood ratio 18, 24
low SINR regime 84, 153, 160, 239, 394
LTE 2
   ARQ/H-ARQ 30, 46–50, 308, 309, 454
   Backhaul 236
   cell search 357–9
   CQI 46
   CS-RS 47, 274, 278
   DM-RS 47, 280
   DRX mode 312
   downlink user multiplexing 161
   frame structure 27
   frequency diversity 24–8, 46, 49, 162
   general concepts 26
   H-ARQ-ACK 46
   handoff procedure 362
   inter-cell interference coordination 236
      HII 236
      OI 236
      RNTP 236
   MIMO 273–80
      antenna port 274
      antenna port mapping 277
      beamforming 279
      closed-loop spatial multiplexing 278
      codebook-based transmission 278
      cyclic delay diversity 278
      layer mapping 277
      MU-MIMO 279, 280
      non-codebook-based transmission 279
      open-loop spatial multiplexing 278
      transmit antenna selection 280
      transmit diversity 276
   PRB 27
   PBCH 359
   PCFICH 162
   PDCCH 162
   PDSCH 27
   PHICH 162
   PRACH 359
   PSS/SSS 357
   PUCCH 47
   PUSCH 46
   PMI 46
   random access 359
   RI 46
   semi-persistent scheduling 311
   signaling for scheduling 307–9
   SR 46
   SRS 50
   synchronization 358
   UE-RS 276
   uplink multiuser diversity 135
   uplink power control 217–19
      fractional power control 218
   X2 interface 236
LTE-Advanced 444–56
   Beamforming 449
   carrier aggregation 454
      component carrier 454
         primary component carrier 454
         secondary component carrier 454
   codebook-based transmission 448
   coodinated multipoint processing
      backhaul 448–9
      BBU 449
      RRH 449
   downlink MIMO 444
      CSI-RS 445
      UE-RS 446
   enhanced ICIC/IC 453
      ABS 453
      cell coverage expansion 377

Index

heterogeneous networks 450
   relays 450: access link 450–1; backhaul link 450, 451; donor eNB 450; MBSFN 451; R-PDCCH 451
   inter-cell interference coordination 4
   interference cancelation 454
   non-codebook-based transmission 445
   uplink MIMO 447
      DM-RS 447
macrocell 336, 365, 366, 384
macro-diversity 134, 320, 324
   downlink macro-diversity 333–5
   uplink macro-diversity 328–33
maximal ratio combining 116, 118, 120, 239
MIMO 74, 115, 240
   MIMO modeling 240–51
   Multicell MIMO 267–73
   multiuser MIMO 85, 254–67
   single user MIMO 85, 251–4
MIMO channel matrix 239, 243, 249
   angular domain representation 244
   angular spread 193, 245, 255
   frequency selectivity 245
   multipath scenario 243
   single path scenario 242
   spatial resolvability 244
   transmit/receive bins 244, 245
Minkowski sum of rate regions 221
mobile IP 91, 327
multi-cell processing 425
   upper bound on capacity 426
      downlink 427
      uplink 427
multi-carrier 336, 338
multi-hop 368–70, 424
multipath fading 94, 97–106
   channel tap 98
   Clarke's model 100
   coherence bandwidth 104
   coherence time 101
   delay spread 104
   discrete path model 98
   Doppler spread 101
   Gaussian Markov model 105
   impulse response 97
   multipath resolvability 98
   narrowband fading 99
   power delay profile 103
   Rayleigh fading 99
   Ricean fading 100
   uncorrelated scattering assumption 100
   wideband fading 99
   wide-sense stationary assumption 99
multiple antennas 84–7, 115, 239
   diversity gain 116
   power gain 116

   spatial multiplexing gain 240
multiuser diversity 76, 126–44, 286, 330
   multiuser diversity with multiple transmit antennas 137–44
multiuser uncertainty 436

narrowband FDMA 21, 80, 203, 205, 320
   hopping 34
near-far effect 79
network architecture 89, 315
   IP-based network architecture 89, 317–19
   proprietary network architecture 89
network MIMO 239
nominal SINR 152, 164, 164

OFDM 9
   cyclic prefix 10
   DFT and IDFT 14, 15
   FFT and IFFT 15
   modulation symbol 9
   OFDM symbol 9
   sinusoids 9
   tone 9
   tone-symbol 13
OFDMA 18
   time-frequency resource 18, 13
      resource granularity 74
      time-frequency grid 13
   tone hopping 20, 24–5
   user demultiplexing module 33
   user multiplexing module 33
open access 389–95
opportunistic beamforming 137
   enhanced opportunistic beamforming 142
opportunistic scheduling 76, 145, 286
   downlink opportunistic scheduling 126–30
   multi-cell opportunistic scheduling 129, 132
   proportional fair scheduling 127, 228, 235, 284, 285, 288, 297
   reliable opportunistic communications 79
   single cell opportunistic scheduling 126, 132
   uplink opportunistic scheduling 132–3
opportunistic spatial multiplexing 266
orthogonality 5, 10, 13, 31
   level of orthogonality 92
outage 95, 107, 370, 496
   deep fade 107

parallel independent links 324–6
Pareto-optimality 155, 163
   Pareto-boundary 500
   Pareto-optimal power and bandwidth allocation 155
peak-to-average power ratio 34, 35
   PAPR reduction 35
penetration loss 370, 386

pilot and null pilot 25, 64, 68
polarized antennas 247–51, 254, 273
   cross-polarized antenna 248
   cross polarization discrimination 250
   dipole moment 247
   dual-polarized antenna 248
   electrical dipoles 247
   free space propagation model 249
   polarization moment 249
   reflection 249
power asymmetry 389, 398
power and bandwidth allocation 83, 83, 151–9, 234, 392, 415, 464
power budget 78, 82, 110, 151, 160, 163, 263, 443
power control 5, 75, 79, 109, 134
   channel inversion 109
   distributed power control 209–11
   opportunistic truncation 110, 164
   truncated channel inversion 109
power limited regime 110, 164
power spillage 252
precoding matrix 252
primary/secondary spectrum management 366, 432
primary transmission coverage 435

quiet periods 433, 436

radio access network 317
radio network controller 320
random access probes 339, 462
rate control 4, 75, 110
rate region 150, 221, 283, 500
rate scheduling 408
real-world impairment 52–61
   carrier frequency offset 52
   Doppler spread 53
   imperfect channel estimation 64
   imperfect sectorization 189
   imperfect time synchronization 55
   I/Q imbalance 60
   phase noise 58
   power amplifier nonlinear distortion 61
   sampling rate mismatch 56
   receive diversity 80, 115, 122
reduced-power channel reuse 226
registration request 328
relay 365, 367–83, 416
   in-band relay 367–417, 458
   out-of-band relay 458, 460
   relay link 367, 368, 458
   cooperative relay channel capacity 420, 422
restricted association 385, 398, 412, 457
reuse versus orthogonalization 395, 404, 412
routing and scheduling 379
RTS/CTS 403

SC-FDMA 35–46
   distributed SC-FDMA 39
   localized SC-FDMA 36
   SINR degradation in SC-FDMA 42
scheduling 4, 74, 87, 126, 195, 282, 318, 323, 380
   channel-aware scheduling 132, 135, 282
   control signaling for scheduling 304–6
   coordinated scheduling 193, 269, 449
   dynamic packet scheduling 304–6
   exponential scheduler 296
   frequency diversity scheduling 136
   frequency selective scheduling 136
   gradient-based search scheme 286–9
   log scheduler 296
   max rate scheduling 303
   MAC state scheduling 89, 311–12
   max weight scheduler 296
   multi-class scheduling 300–1
     strict priority scheduling 300
   queue-aware scheduling 295
   queue length 290, 291, 296, 297, 304
   scheduling latency 141, 267
   semi-static scheduling 310
   throughput optimal scheduling 294–6
SDMA 255, 256–61
   frequency selective SDMA 255
   scheduled SDMA 259
   SDMA with multiuser diversity 144, 256
self-noise model 63, 64
   noise characteristic line 66
   signal dependent noise 66
   signal independent noise 66
signal dynamic range 5, 74, 77, 78, 79, 195, 411
   dynamic range of interference 399
single tone-symbol signal 346, 411
single user diversity 75
singular value decomposition 251
SINR distribution 198–209
SINR feasibility region 210–11
SIR-based distributed scheduling 404–11
   direct power signal 406
   inverse power echo 406
   TX yielding 405
   RX yielding 405
Siri application 302
small cell 336, 365, 383
   femtocell 383–97, 383
   microcell 383
   picocell 383
spatial signature sequence 254, 256
spectral radius 211
spillage 213
spread spectrum 24, 70
statistical multiplexing 87, 300, 383, 400, 402
   scheduled statistical multiplexing 87
   unscheduled statistical multiplexing 88

stochastic geometry 198
subband 135
successive interference cancellation 150, 175, 176, 254, 501
superposition coding 71, 174, 175
　superposition-by-position coding 183–9
synchronization 30, 30, 471–7
　acquisition 472–5
　frequency synchronization 472
　　fractional offset 473
　　integer offset 476
　symbol time synchronization 472
　　receive symbol window 475–6
　tracking 477
systems approach 6

TCP 4, 5, 91
　congestion control 283
　TCP over wireless 292–3
temporary tunnel 328, 337, 346
thermal limited regime 220
time diversity 121, 122, 122, 123, 145, 492
time scales 94, 138, 146, 216, 231, 234, 312, 301
timing synchronization control 30, 471–7
　closed-loop timing control 30, 31, 471
　open-loop timing control 469, 30
traffic 3, 74, 81, 87, 282
　delay requirement 122, 123, 5, 88, 141, 232, 334

elastic traffic 289
flow level traffic 301
inelastic traffic 293
infinitely backlogged traffic 283
transition regime 164
transmit diversity 118, 122, 239, 322
　Alamouti scheme 118, 120
　space-time coding 84, 118

unplanned deployment 90, 202, 71
unrestricted association 385, 398, 196,365
user multiplexing 71, 77–80, 150
　non-orthogonal multiplexing 174
　orthogonal multiplexing 151
utility maximization 216, 234, 283, 286, 300
　a-utility function 284
　local marginal utility 235

V-BLAST architecture 253, 256
virtual antenna array 416
voice activity 4, 88

water-filling scheme 17, 73, 111, 252, 497, 499
　iterative water-filling 234
white space 432, 436
　TV white space 434, 456
wideband broadcast signal 343, 344

Zadoff-Chu sequences 52, 344